About the Authors

A science librarian for more than thirty years, **James Bobick** is Department Head in the Science and Technology Department at the Carnegie Library of Pittsburgh. He has master's degrees in biology and library science.

Naomi Balaban, a science reference librarian for fifteen years, also works at the Carnegie Library. She coedited, with James Bobick, *The Handy Science Answer Book* and has a background in linguistics and a master's degree in library science.

A college biology teacher for fifteen years, **Sandra Bobick** currently teaches general biology and genetics at the Community College of Allegheny County, Pennsylvania, and coordinates a summer "Biotechnology Boot Camp" for middle school girls. She holds master's degrees in biology and education and a doctoral degree in science education.

Laurel Bridges Roberts has been a college biology instructor since 1992 and currently teaches introductory biology and human physiology at the University of Pittsburgh. She holds both master's and doctoral degrees in behavioral ecology and has research experience in experimental pathology and molecular biology.

The Handy Answer Book® Series

The Handy Answer Book for Kids
(and Parents)

The Handy Biology Answer Book

The Handy Bug Answer Book

The Handy Dinosaur Answer Book

The Handy Geography Answer Book

The Handy Geology Answer Book

The Handy History Answer Book

The Handy Ocean Answer Book

The Handy Physics Answer Book

The Handy Politics Answer Book

The Handy Presidents Answer Book

The Handy Religion Answer Book

The Handy Science Answer Book

The Handy Space Answer Book

The Handy Sports Answer Book

The Handy Weather Answer Book

Please visit us at visibleink.com.

THE
HANDY
BIOLOGY
ANSWER
BOOK

THE HANDY BIOLOGY ANSWER BOOK

James Bobick, Naomi Balaban,
Sandra Bobick, and Laurel Bridges Roberts

VISIBLE INK PRESS

Detroit

THE HANDY BIOLOGY ANSWER BOOK

Visible Ink Press®
43311 Joy Road #414
Canton, MI 48187–2075

Visible Ink Press is a registered trademark of Visible Ink Press LLC.

Most Visible Ink Press books are available at special quantity discounts when purchased in bulk by corporations, organizations, or groups. Customized printings, special imprints, messages, and excerpts can be produced to meet your needs. For more information, contact Special Markets Director, Visible Ink Press, at www.visibleink.com.

Front cover: Image of whale courtesy Digital Source; DNA strand courtesy Electronic Illustrators Group; image of cells courtesy AP/Wide World Photos.

Back cover: DNA strand courtesy Electronic Illustrators Group; bee/flower and octopus courtesy AP/Wide World Photos.

Art Director: Mary Claire Krzewinski
Typesetting: The Graphix Group
ISBN 1–57859–150–3

Library of Congress Cataloging-in-Publication Data

The handy biology answer book / James Bobick ... [et al.].— 1st ed.
    p. cm.
  Includes bibliographical references (p.   ).
  ISBN 1–57859–150–3
  1. Biology—Miscellanea. I. Bobick, James E.
  QH349.H36 2004
  570—dc22

                                                2004017808

Contents

INTRODUCTION *ix*
ACKNOWLEDGMENTS *xi*
PHOTO AND ILLUSTRATION CREDITS *xiii*
MILESTONES OF MAJOR EVENTS IN BIOLOGY *xv*

CHEMISTRY FOR BIOLOGY ... 1

Basics ... Molecules ... Metabolic Pathways ... Enzymes ... Applications

CELL STRUCTURE ... 31

Introduction and Historical Background ... Prokaryotic and Eukaryotic Cells ... Organization within Cells ... Nucleus ... Cytoplasm ... Cell Wall and Plasma Membrane ... Cilia and Flagella ... Miscellaneous

CELL FUNCTION ... 55

Basic Functions ... Cell Jobs ... Cellular Specialization ... Applications

BACTERIA, VIRUSES, AND PROTISTS ... 85

Introduction and Historical Background ... Bacteria ... Viruses ... Single-celled Organisms ... Applications

FUNGI ... 125

Introduction and Historical Background ... Structure ... Function ... Mushrooms and Edible Fungi ... Lichens ... Yeasts ... Applications

PLANT DIVERSITY ... 159

Introduction and Historical Background ... Bryophytes ... Ferns and Related Plants ... Gymnosperms ... Angiosperms (Flowering Plants) ... Applications

PLANT STRUCTURE AND FUNCTION ... 197

Plant Cells and Tissues ... Seeds ... Shoots and Leaves ... Roots ... Soils ... Applications

ANIMAL DIVERSITY ... 233

Introduction and Historical Background ... Sponges and Coelenterates ... Worms ... Mollusks ... Arthropods ... Echinoderms ... Chordates

ANIMAL STRUCTURE AND FUNCTION ... 277

Introduction and Historical Background ... Tissue ... Organs and Organ Systems ... Digestion ... Respiration ... Circulation ... Excretory System ... Endocrine System ... Nervous System ... Immune System ... Reproduction ... Skeletal System ... Locomotion

ANIMAL BEHAVIOR ... 323

Introduction and Historical Background ... Learning ... Behavioral Ecology ... Applications

DNA, RNA, AND CHROMOSOMES ... 359

Introduction and Historical Background ... DNA ... RNA ... Genes and Chromosomes ... Transcription and Translation

GENETICS ... 385

Mendelian Genetics ... New Genetics ... Human Genome Project ... Applications

BIOTECHNOLOGY AND GENETIC ENGINEERING ... 413

Introduction and Historical Background ... Methods ... Applications

EVOLUTION ... 445

Introduction and Historical Background ... Darwin-Wallace Theory ... Post-Darwinian Evolution ... Evolutionary History of Life on Earth ... Applications

ENVIRONMENT ... 477

Environmental Cycles and Related Concepts ... Biomes and Related Concepts ... Extinct and Endangered Plants and Animals ... Conservation, Recycling, and Other Applications

LABORATORY TOOLS AND TECHNIQUES ... 513

Scientific Method ... Basic Laboratory Chemistry ... Measurement ... Microscopy ... Microtechnique ... Centrifugation ... Chromatography ... Electrophoresis ... Spectroscopy ... X-Ray Diffraction ... Nuclear Magnetic Resonance and Ultrasound ... Methods in Microbial Studies ... Methods in Ecological Studies

FURTHER READING   539

INDEX   553

Introduction

From the molecular and subcellular level to ecosystems and global conditions, biology holds and demands our attention. The spectacular discoveries and achievements in molecular biology in the last fifty years have created a gene-based medical revolution, with implications from crime-scene testing to stem cell research. The discovery of the structure of DNA (or deoxyribonucleic acid) in 1953 by Dr. James D. Watson and Dr. Francis Crick was one of the great advances in science, providing a universal key to understanding all forms of life. Crick and Watson discovered that DNA consisted of two complementary strands, a structure that solved the mystery of how the genetic material could be copied when a cell divides. Their original research has lead to our deciphering of the human genome, which consists of some three billion units of DNA encoded with all the biological information needed to generate and maintain a living person.

The Handy Biology Answer Book explores the quantum leap in our understanding of biology, answering more than 1,600 questions in plain English on all aspects of human, animal, plant, and microbial biology. In the new millennium, biology continues to generate hot-button medical and political issues, including cloning, stem cell therapy, and genetic manipulation.

In this informative book, you'll find answers to such intriguing questions as: What is cell cloning? What are DNA and RNA? When and how did cells first evolve? Can two blue-eyed people have a brown-eyed child? How serious a disease is influenza? Do overweight children have a greater risk of being overweight adults? What is an atom? Why are some fats "good" and others "bad"? Why do humans need cholesterol? How do northern birds know to fly south in winter? Can the environment determine the sex of an animal? When was the first microscope developed?

Easy to use with special appeal to the general science reader and student, *Handy Biology* is organized in sixteen chapters with 100 illustrations and 150 tables and charts. *Handy Biology* covers such topics as cell structure and function, bacteria and viruses, plant and animal diversity, DNA and chromosomes, genetics, biotechnology and genetic engineering, evolution and adaptation, the environment, laboratory tools and techniques, and related topics. A chronology lists major accomplishments and milestones in biological research, and a resources section lists books, periodicals, and Web sites where readers can explore each subject in greater depth.

The selection of chapter titles and the content within each chapter provide a solid overview of the entire subject of biology. The information in *Handy Biology* will appeal to those with a background in biology as well as those seeking an introduction to biology. We've included questions that are interesting, unusual, frequently asked at the reference desk or in the classroom, or difficult to answer. The questions range from the history and development of biology to current topics and controversies. Each chapter was a collaborative effort between the librarians (Jim and Naomi) and the biologists (Sandi and Laurel).

Each of us is pleased to contribute another addition to the "Handy Answer" family. It is wonderful to see literary growth paralleling biological growth, from the parent of all of the titles in this series, *The Handy Science Answer Book* (first published in 1994) to this most recent offspring.

James Bobick, Naomi Balaban, Sandra Bobick, and Laurel Bridges Roberts

Acknowledgments

The authors thank the members of their families for ongoing encouragement and support while this volume was being transformed from a one page outline to the finished product. In particular, Jim and Sandi thank Andrew and Michael; Naomi thanks Carey, Shira, Avital, Devora, Ilana, Elli, and Reuven; and Laurel thanks Greg, Lark, and Mimi.

We also thank the staff at Visible Ink Press—Marty Connors, Roger Janecke, and Christa Gainor—for their guidance and encouragement; Neil Schlager and Vanessa Torrado-Caputo of Schlager Group for their editorial oversight and project management; Bob Huffman for cropping and sizing the illustrations; Mary Claire Krzewinski for page and cover design; Hans Neuhart of Electronic Illustrators Group for his creation of the original illustrations in the book; and Marco Di Vita at the Graphix Group for typesetting.

Photo and Illustration Credits

Milestones of
Major Events in Biology

Year	Event	Area of Biology
520 B.C.E.	Alcmaeon dissects animals and distinguishes between arteries and veins, describes the optic nerve, and intensively studies the brain.	Anatomy
350 B.C.E.	Aristotle attempts to classify animals, dividing the 500 known species into eight groups.	Taxonomy
350 B.C.E.	Theophrastus identifies plants and animals as fundamentally different.	Taxonomy
45 C.E.	Dioscorides' *De Materia Medica* describes 600 medicinal plants and nearly 1,000 other medicinals, including animal parts and minerals.	Botany and Zoology
50 C.E.	Pliny the Elder publishes *Naturalis Historia*, which summarizes all knowledge on astronomy, geography, and zoology.	General
900	Al-Razi (Rhazes) writes the first scientific paper on infectious diseases, describing smallpox and measles.	Microbiology
1538	William Turner publishes *New Letters on the Properties of Herbs*, the first English herbal to take a scientific approach to botany.	Botany
1542	Leonhard Fuchs publishes *On the History of Plants*, an early work on plant classification.	Botany
1543	Andreas Vesalius depicts human anatomy accurately in *De Humani Corporis Fabrica*.	Anatomy
1546	Georgius Agricola (Georg Bauer) first uses the term "fossil" in *De Natura Fossilium*.	Evolution
1546	Girolamo Fracastoro publishes *On Contagion and the Cure of Contagious Disease*, in which he proposes that contagious diseases are caused by invisible bodies transmitted from one person to another.	Microbiology
1551	Pierre Belon publishes *Natural History of Unusual Animals*, one of the earliest works of modern embryology.	Zoology
1551	Conrad Gesner publishes the first volume of *History of Animals*, an early illustrated treatise on animal classification.	Zoology
1558	Conrad Gesner uses magnification in scientific investigations.	Tools and Techniques

Year	Event	Area of Biology
1561	Gabriello Falloppio publishes *Anatomical Observations*, an early anatomy text that describes the female reproductive organs.	Anatomy
1583	Andreas Cesalpino writes *On Plants*, considered the first truly scientific textbook on botany since it defines and establishes the criteria for botanical taxonomy.	Botany
1590	Zacharias Janssen makes the first compound microscope.	Tools and Techniques
1597	John Gerhard publishes his *Generale Historie de Plantes*, the first catalogue of plants, which describes over 1,000 species.	Botany
1599	Ulisse Aldrovandi publishes the first three volumes of his *Natural History*, which lists and describes bird species and is recognized as the first serious zoological investigation.	Zoology
1603	Hieronymus Fabricius discovers that the veins contain valves, and he publishes *Concerning the Valves of the Veins*.	Anatomy
1604	Hieronymus Fabricius publishes *On the Formation of the Fetus*, the first important study of embryology, in which the placenta is first described.	Embryology
1621	The first botanical garden in Britain, the Oxford Physic Garden, is opened.	Botany
1628	William Harvey publishes *Exercitatio Anatomica de Motu Cordis et Sanguinis* (Anatomical Treatise on the Movement of the Heart and Blood), describing how blood circulates.	Anatomy
1637	René Descartes, in *Discourse on Method*, introduces the scientific method and advocates formulating results by using step-by-step deduction from self-evident truths rather than by inductive reasoning.	General
1658	Jan Swammardam observes and describes red blood cells.	Cytology
1661	Marcello Malpighi discovers blood capillaries.	Anatomy
1665	The first issue of the *Philosophical Transactions of the Royal Society of London*, the first scientific journal in England, is published.	General
1665	Robert Hooke coins the term "cell" and publishes *Micrographia*.	Cytology
1667	Robert Boyle studies bioluminescence of bacteria and fungi and observes that it will not take place unless air is present.	Biochemistry
1667	Nicolaus Steno coins the term "ovary" for the female reproductive gland.	Physiology
1668	Francisco Redi disproves the idea of spontaneous generation—that living organisms arise from nonliving matter, as in the case of maggots suddenly appearing on decaying meat.	General
1669	Marcello Malpighi describes metamorphosis in the silkworm and discovers the excretory tubules of insects—the first description of the anatomy of an invertebrate.	Zoology
1670	John Ray and Francis Willughby publish their *Catalogue of English Plants*.	Botany
1674	Anton von Leeuwenhoek uses a primitive microscope to study microorganisms described as "very little animalcules."	Microbiology

Year	Event	Area of Biology
1675	Marcello Malpighi publishes *Anatomy of Plants*, the first important work on plant anatomy.	Botany
1677	Anton van Leeuwenhoek discovers spermatozoa.	Cytology
1682	John Ray publishes *Methodus Plantarum*, in which he divides flowering plants into monocotyledons and dicotyledons.	Botany
1682	Nehemiah Grew publishes *Anatomy of Plants*, in which the male and female sex organs in plants (stamens and carpels) are first described.	Botany
1693	John Ray publishes *Synopsis of Animals*, which is the first important classification system for animals.	Zoology
1700	Joseph Pitton de Tournefort develops the binomial method of classification.	Taxonomy
1710	John Ray publishes *Study of Insects*, which is the first classification of insects.	Zoology
1727	Stephen Hales publishes *Vegetable Staticks*, which describes the uptake and movement of water through sunflowers and other plant physiology experiments.	Botany
1735	Carolus Linnaeus publishes his *Systema Naturae* and develops the modern classification system of plants and animals by genus and species.	Taxonomy
1740	Abraham Trembley discovers hydra and observes the phenomena of regeneration in that animal.	Zoology
1747	Albrecht von Haller publishes the first textbook of physiology.	Physiology
1748	Jean-Antoine Nollet discovers osmosis, the movement of a solvent such as water through a semipermeable membrane separating two solutions that have different concentrations.	Physiology
1749	Georges-Louis de Buffon writes *General and Particular Natural History*, a forty-four volume set that describes all that is known about animals and minerals at the time.	Zoology
1757	Alexander Monro distinguishes between the lymphatic system and the blood circulatory system.	Physiology
1759	Kaspar Wolff publishes *Theory of Generations*, which describes the development of plants from seeds and founding modern embryology.	Botany
1761	Joseph Gottlieb Kölreuter discovers the role of bees in pollination and realizes its importance for plant fertilization.	Botany
1766	Albrecht von Haller shows that nerves stimulate muscles to contract and that all nerves lead to the spinal cord and brain, laying the foundation of modern neurology.	Physiology
1768	Lazzaro Spallanzani studies regeneration in animals and concludes that the lower animals have a greater capacity to regenerate lost parts.	Zoology
1772	Joseph Priestley discovers that plants give off oxygen.	Botany
1774	Antoine-Laurent Lavoisier discovers that oxygen is consumed during respiration.	Physiology

Year	Event	Area of Biology
1777	Jan Ingenhousz studies the respiratory cycle in plants and concludes that sunlight is necessary for plants to produce oxygen—the first description of photosynthesis.	Botany
1786	The phenomena of contraction of muscle by electrical stimulation is discovered by Luigi Galvani.	Physiology
1789	Gilbert White publishes *The Natural History of the Antiquities of Selborne*, describing his observations of plant and animal life around his parish.	Environment
1793	Christian Sprengel describes the plant pollination process, noting the importance of wind and insects in cross-pollination.	Botany
1794	Erasmus Darwin, grandfather of Charles Darwin, publishes *Zoonomia or the Laws of Organic Life*, which contains his ideas on evolution.	Evolution
1796	Jan Ingenhousz concludes that plants use carbon dioxide.	Botany
1796	Edward Jenner develops a vaccine containing cowpox, which proves effective for preventing smallpox.	Microbiology
1798	Thomas Malthus publishes *Essay on the Principle of Population*, in which he proposes that population increases geometrically while food production increases only arithmetically, resulting in competition.	Environment
1799	The metric system, a uniform system of science weights and measures, is officially proposed by the French Academy and becomes a worldwide standard except in the United States.	Tools and Techniques
1800	Georges Cuvier publishes *Lessons on Comparative Anatomy*, establishing the discipline of comparative anatomy.	Anatomy
1801	Christiaan Hendrik Persoon publishes *An Overview of Fungi*, which was the founding treatise on modern mycology.	Mycology
1802	Jean-Baptiste de Lamarck proposes the term "biology" to describe a general science of living things.	General
1802	Gottfried Treveranus publishes a major work entitled *Biology*, which sought to summarize all basic knowledge of its time about the structure and function of living matter.	General
1802	The first dinosaur footprints are discovered in North America near South Hadley, Massachusetts.	Evolution
1803	John C. Otto is the first individual in the United States to band birds for scientific studies.	Zoology
1804	Alexander Brougniart classifies amphibians separately from reptiles.	Zoology
1804	Nicolas Théodore de Saussure demonstrates that water is necessary for photosynthesis.	Botany
1806	Louis-Nicolas Vauquelin isolates the first amino acid, asparagine.	Biochemistry
1809	Jean-Baptiste de Lamarck publishes *Zoological Philosophy*, in which he proposes his theories of evolution (Lamarckism).	Evolution

Year	Event	Area of Biology
1813	Augustin Pyrame de Candolle publishes *Elementary Theory of Botany*, in which he coins the word "taxonomy" and proposes that plant classification should be based on structure rather than function.	Taxonomy
1815-1818	Elias M. Fries, "father of mycology," publishes his first important work on mycology, *Observationes Mycologicae*.	Mycology
1817	Georges Cuvier divides the animal kingdom into four groups or Embranchments: 1) Vertebrata, 2) Mollusca, 3) Articulata, and 4) Radiata.	Taxonomy
1821	The first International Congress on Biology is held.	General
1822	William Beaumont begins the study of digestion when he treats Alexis St. Martin for a shotgun blast to the stomach.	Physiology
1826	Stamford Raffles founds the Royal Zoological Society in London.	Zoology
1826	Karl Ernst von Baer investigates the mammalian ovary and shows that the Graafian follicle contains the microscopic ovum (egg).	Embryology
1827–1838	John James Audubon publishes the first volume of his treatise entitled *Birds of America*.	Zoology
1828	Karl Ernst von Baer describes how fertilized cells develop into embryos in *Developmental History of Animals*.	Embryology
1830	A potato blight, caused by the fungus *Phytophthora infestans*, is first reported in the United States.	Mycology
1831	Robert Brown describes and names the nucleus in plant cells.	Cytology
1832	Henri Dutrochet discovers the small openings on leaves, which are stomata—later found to be involved in gas exchange.	Botany
1832–1836	Charles Darwin begins a five-year voyage to South America and the Pacific as naturalist on the HMS *Beagle*.	Evolution
1835	Johannes Purkinje observes ciliary motion in nerve cells using a compound microscope.	Anatomy
1835	The HMS *Beagle* visits the Galapagos Islands, and Charles Darwin observes the varieties of finches and tortoises.	Evolution
1836	Theodor Schwann discovers pepsin, the first animal enzyme to be isolated.	Physiology
1837	Henri Dutrochet proves that chlorophyll is essential for plant photosynthesis.	Botany
1838	Gerard Johann Mulder coins the word "protein."	Biochemistry
1838	Asa Gray and John Torrey publish the first part of *A Flora of North America*, which attempts to cover all species of plants in North America.	Botany
1839	Matthias Schleiden and Theodor Schwann propose the cell theory.	Cytology
1840	Rudolf Kölliker identifies spermatozoa as cells.	Cytology
1840	Charles Darwin's *Zoology of the Voyage of the Beagle* is published and describes the animals he saw or collected on his earlier scientific voyage.	Zoology
1842	Robert Remak names and describes the three layers of cells in an embryo as: ectoderm, mesoderm, and endoderm.	Embryology

Year	Event	Area of Biology
1842	Johann J. Steenstrup describes the alternation of sexual and asexual generations in both plants and animals.	Botany & Zoology
1842	Charles Darwin classifies coral reefs and explains how coral atolls are formed.	Environment
1844	John Dolland develops the immersion lens technique in microscopy.	Tools and Techniques
1844	The alkaloid cocaine, derived from coca leaves, is first isolated, but commercial production does not begin until 1862.	Botany
1845	Karl Theodor Ernst von Siebold recognizes protozoa as single-celled organisms and defines them as organisms.	Zoology
1846	Richard Owen publishes *Lectures on Comparative Anatomy and Physiology of the Vertebrate* Animals, one of the first texts on comparative vertebrate anatomy.	Anatomy
1851	Wilhelm Hofmeister establishes the alternation of generations in ferns and mosses and the differences between gymnosperms (conifers) and angiosperms (flowering plants).	Botany
1852	Claude Bernard proposes that living processes depend on a constant internal environment later called homeostasis.	Physiology
1855	Robert Remak describes cell division.	Cytology
1856	Louis Pasteur establishes that microorganisms are responsible for fermentation.	Microbiology
1856	The first identified human fossil, a Neanderthal skull, is discovered near Düsseldorf, Germany.	Evolution
1857	Rudolf Albert von Köllicker describes "sarcosomes" (now called "mitochondria") in muscle cells. The term "mitochondrion" is not introduced until 1897.	Cytology
1858	Rudolf Virchow publishes *Cellular Pathology as Based upon Physiological and Pathological Histology*, which states that "every cell comes from a cell" and establishing cellular pathology as essential in understanding disease.	Cytology
1858	Charles Darwin presents his theory of evolution at a meeting of the Linnaean Society of London.	Evolution
1859	Charles Darwin publishes *On the Origin of Species by Means of Natural Selection*.	Evolution
1859	Robert Bunsen and Gustav Kirchhoff invent the spectroscope, a device to chemically analyze and identify elements heated to incandescence.	Tools and Techniques
1861	Max Schultze describes the cell as consisting of protoplasm and a nucleus.	Cytology
1861	Louis Pasteur disproves the theory of spontaneous generation.	Microbiology
1862	Henry Walter Bates describes a type of biological mimicry in animals, known today as Batesian mimicry, in which a harmless and vulnerable species resembling a dangerous or unpalatable species is ignored by potential predators.	Evolution

Year	Event	Area of Biology
1862	Louis Pasteur publishes his *Germ Theory of Disease*.	Microbiology
1863	John Tyndall discovers what will become known as the greenhouse effect.	Environment
1864	Herbert Spencer coins the phrase "survival of the fittest" when discussing the process of natural selection in his work *Principles of Biology*.	Evolution
1864	Louis Pasteur invents the process of pasteurization, in which slow heating kills bacteria and other microorganisms and keeps wine and beer from turning sour.	Microbiology
1865	Gregor Mendel publishes *Experiments on Plant Hybridization* in the Proceedings of the Brunn Natural History Society, describing the laws of genetic inheritance.	Genetics
1865	Joseph Lister successfully uses his antiseptic system in surgery, initiating the development of sterile surgical technique.	Microbiology
1865	Julius von Sachs shows that chlorophyll is located in special bodies within plant cells—later called chloroplasts.	Botany
1865	Otto F. K. Deiters describes the structure of a nerve cell, including axons and dendrites.	Cytology
1866	Ernst Haeckel proposes the kingdom Protista for organisms that are not true plants or animals.	Taxonomy
1866	Ernst Haeckel coins the word "ecology" to identify the relationship between plants, animals, and their environment.	Environment
1868	Thomas H. Huxley proposes that birds are descendants of dinosaurs.	Evolution
1869	Johann Friedrich Miescher isolates and describes a substance in the nuclei of cells that he calls "nuclein"—today known as DNA.	Genetics
1871	Charles Darwin publishes *The Descent of Man*.	Evolution
1872	Angus Smith publishes *Air and Rain* and coins the phrase "acid rain."	Environment
1872	Ferdinand J. Cohn publishes *Researches on Bacteria*, the first major treatise on bacteria.	Microbiology
1872	The U.S. Congress passes legislation establishing Yellowstone as the world's first national park.	Environment
1873	Walther Flemming discovers "chromosomes," observes mitosis and provides the modern interpretation of nuclear division in *Cell Substance, Nucleus, and Cell Division*.	Cytology
1875	Oskar Hertwig discovers the process of fertilization when the nuclei of the egg and sperm unite.	Cytology
1875	Eduard Strasburger describes the process of mitosis (cell division) in *On Cell Formation and Cell Division*.	Cytology
1876	William Kuhne proposes the term "enzyme" to denote changes that occur during fermentation.	Biochemistry
1876	Robert Koch isolates the anthrax bacillus.	Microbiology
1877	Ilich Metchnikov describes the process of phagocytosis, whereby white blood cells engulf other cells and bacteria.	Microbiology

Year	Event	Area of Biology
1877	Ernst F. Hoppe-Seyler coins the term "biochemistry" for the study of chemical processes or metabolism in living organisms.	Biochemistry
1877	Robert Koch develops a method of obtaining pure cultures of bacteria.	Microbiology
1878	Fritz Müller describes Müllerian mimicry, in which several distasteful species evolve similar appearances.	Evolution
1880	The journal *Science* is founded and published by the American Association for the Advancement of Science.	General
1880	Charles Darwin and his son, Francis Darwin, publish *The Power of Movement in Plants*, which describes phototropism in plants.	Botany
1881	Wilhelm Pfeffer publishes *The Physiology of Plants: A Treatise upon the Metabolism and Sources of Energy in Plants*, which becomes the standard title in plant physiology.	Botany
1882	Robert Koch enunciates Koch's postulates.	Microbiology
1882	Wilhelm Engelmann demonstrates that photosynthesis is most effective in the red region of the color spectrum of visible light.	Botany
1883	Robert Koch isolates and identifies the bacillus that causes cholera.	Microbiology
1884	Hans Gram develops the Gram stain, a method for identifying bacteria.	Microbiology
1885	Louis Pasteur develops a rabies vaccine.	Microbiology
1887	Richard Petri develops the petri dish.	Microbiology
1887	Raphael Dubois is the first to explain the chemical nature of bioluminescence—the phenomenon of light production by living organisms such as bacteria, fungi, and insects.	Biochemistry
1888	Heinrich Wilhelm Gottfried von Waldeyer-Hartz observes and names the chromosomes in the nucleus of a cell.	Cytology
1888	Pierre Roux and Alexandre Yersin discover the first bacterial toxin, produced by *Corynebacterium diptheriae*, which causes diphtheria.	Microbiology
1890	Sergei Winogradsky isolates and identifies nitrifying soil bacteria and demonstrates their role in the nitrogen cycle.	Microbiology
1890	The U.S. Congress establishes Yosemite and General Grant national parks.	Environment
1891	Heinrich Wilhelm Gottfried von Waldeyer-Hartz coins the term "neuron."	Cytology
1892	The Sierra Club is founded by John Muir.	Environment
1892	Oscar Hertwig establishes cytology as a discipline of biology.	Cytology
1892	Dimitry Ivanovsky publishes his research on tobacco mosaic disease and concludes that the causative organism is smaller than bacteria.	Microbiology
1895	Wilhelm Röntgen accidentally discovers X-rays.	Physiology
1895	Martinus Beijerinck develops the technique for maintaining pure cultures of microorganisms.	Microbiology
1895	Emilie van Ermengem isolates the bacterium *Clostridium botulinum*, which causes botulism.	Microbiology

Year	Event	Area of Biology
1896	Niels Finsen discovers that ultraviolet (UV) light kills bacteria.	Microbiology
1897	C. Benda describes and names mitochondria in plant and animal cells.	Cytology
1898	Camillo Golgi describes the Golgi apparatus, an irregular network of small fibers, cavities, and granules in nerve cells. Its existence is confirmed in the 1940s with the use of the electron microscope.	Cytology
1898	Martinus Beijerinck establishes the field of virology by coining the word "virus" to describe a disease-producing microscopic agent that can pass through a bacteriological filter.	Microbiology
1898	Charles R. Barnes proposes the term "photosynthesis."	Botany
1899	Acetylsalicylic acid, aspirin, is produced by the company Bayer AG, using a process discovered by Adolph Kolbe.	Biochemistry
1900	Huge de Vries, Carl Correns, and Erich von Tschermak-Seysengg independently and simultaneously rediscover Mendel's finding on heredity.	Genetics
1901	Karl Landsteiner discovers three blood groups (A, B, and O) in humans. The fourth group, AB, is discovered one year later.	Physiology
1902	The hormone secretin, controller of the pancreas, is found to be secreted by the small intestine by William M. Bayliss and Ernest H. Starling.	Physiology
1902	Archibald Garrod identifies the first human genetic disease.	Genetics
1903	Walter Sutton discovers the mechanism of meiosis while studying the sperm cells of grasshoppers and proposes that heredity factors are located in chromosomes.	Genetics
1904	Herman W. Merkel notices the blights affecting American chestnut trees.	Botany
1904	Theodor Boveri concludes that a full set of chromosomes is necessary for the normal development of the sea urchin embryo. This discovery, together with Walter Sutton's proposal, will form the basis for the Sutton-Boveri chromosome theory of inheritance.	Genetics
1905	F. F. Blackman identifies a light-dependent stage and a light-independent stage in photosynthesis.	Botany
1905	The National Audubon Society is founded.	Environment
1905	Wilhelm Johannsen uses the terms "genotype" and "phenotype" to explain how plants that are genetically identical may have different physical characteristics.	Genetics
1905	The U.S. Forest Service is established.	Environment
1906	Richard Willstätter discovers that chlorophyll molecules contain magnesium.	Botany
1906	William Bateson coins the word "genetics" for the study of how physical, biochemical, and behavioral traits are transmitted from parent to offspring.	Genetics
1906	Charles S. Sherrington coins the term "synapse."	Physiology

Year	Event	Area of Biology
1907	Ivan Petrovich Pavlov investigates the conditioned reflex.	Behavior
1907	Ross Granville Harrison develops techniques for successfully culturing tissues.	Cytology
1908	Godfrey H. Hardy and Wilhelm Weinberg formulate the Hardy-Weinberg principle, relating mathematically the frequencies of genotypes to the frequencies of alleles in randomly mating populations.	Genetics
1908	Joseph Grinnell proposes the concept of ecological niche, stating that competition between species is a force that results in the adoption of similar yet separate habitats in the environment.	Environment
1908	The first biological autoradiograph is made when a whole frog is made radioactive and placed on a photographic plate.	Tools and Techniques
1909	Wilhelm Johannsen coins the word "gene" for the units of inheritance inside chromosomes.	Genetics
1909	Howard T. Ricketts identifies the microorganism (later named *Rickettsia*) that causes Rocky Mountain spotted fever.	Microbiology
1909	Charles D. Walcott discovers fossils of soft-bodied organisms in the Burgess Shale in British Columbia, Canada.	Evolution
1910	Paul Ehrlich introduces the technique of chemotherapy when he uses an arsenic compound (salvarsan) he synthesized to cure syphilis.	Microbiology
1910	F. Peyton Rous shows that when pieces of chicken tumor (sarcoma) are transplanted to other chickens, they also develop sarcomas.	Genetics
1910	Thomas Hunt Morgan, using the fruit fly (*Drosophila melanogaster*), for genetic research, discovers the concept of sex-linked inheritance for eye color.	Genetics
1911	Gene mapping is developed by Alfred H. Sturtevant.	Genetics
1911	Phoebus A. Levene discovers D-ribose, the sugar found in RNA.	Biochemistry
1913	Leonar Michaelis and Maud Leonora Menten propose the Michaelis-Menten equation to describe the rate at which enzyme-catalyzed reactions take place.	Biochemistry
1913	Charles Fabry proves the existence of an ozone layer in the upper atmosphere.	Environment
1914	Fritz Lipmann explains the role of ATP in cellular metabolism.	Cytology
1915	Frederick William Twort and Felix H. d'Herelle, independently in 1917, discover bacteriophages.	Microbiology
1915	Hermann Müller begins experiments with *Drosophila melanogaster* using X-rays as a means of inducing mutations.	Genetics
1915	Thomas Hunt Morgan, Alfred Sturtevant, Hermann Müller, and Calvin Bridges publish *The Mechanism of Mendelian Heredity*, claiming that invisible genes within the chromosomes of the cell nucleus determine the hereditary traits of the offspring.	Genetics

Year	Event	Area of Biology
1915	The Ecological Society of America is founded.	Environment
1916	Thomas Hunt Morgan prepares the first chromosome map for the four chromosomes of the fruit fly, *Drosophila melanogaster*.	Genetics
1916	The U.S. National Park Service is established.	Environment
1920	August Thienemann proposes the concept of trophic, or feeding, levels in an ecosystem with energy transferred from plants (the producers) to several levels of animals (the consumers).	Environment
1921	Otto Loewi demonstrates that nerves of the autonomic nervous system release a chemical substance when stimulated. He later identifies the substance as acetylcholine.	Physiology
1921	Frederick G. Banting and Charles H. Best isolate insulin from pancreatic tissue.	Physiology
1923	The first edition of *Bergey's Manual of Determinative Bacteriology* is published.	Microbiology
1923	Theodor Svedberg develops the ultracentrifuge.	Tools and Techniques
1923	R. Feulgen and H. Rossenbeck describe a DNA staining technique.	Cytology
1925	Tennessee passes the first state law prohibiting the teaching of Darwin's theory of evolution.	Evolution
1924	David Keilin discovers cytochrome, an iron-containing protein pigment important in cellular respiration.	Biochemistry
1924	Otto Warburg discovers cytochrome oxidase, an enzyme associated with biological oxidation.	Biochemistry
1926	Walter Bradford Cannon coins the term "homeostasis" to describe the body's ability to maintain a relative constancy in its internal environment.	Physiology
1926	James Sumner crystallizes the first pure protein, urease, and demonstrates that enzymes are proteins and act as catalysts.	Biochemistry
1926	Thomas Hunt Morgan publishes *The Theory of the Gene*, in which he proposes that all future research will center on the gene.	Genetics
1927	Thomas Hunt Morgan publishes *Experimental Embryology*.	Embryology
1927	Charles Elton publishes *Animal Ecology*, a landmark study that establishes ecology as a scientific field.	Environment
1927	John Belling introduces acetocarmine techniques for chromosome "squashes."	Cytology
1927	Henry H. Dale isolates acetylcholine, a neurotransmitter.	Physiology
1928	Alexander Fleming isolates and identifies the mold *Penicillium notatum*, which produces penicillin as an antibiotic, but is it not used therapeutically until 1940.	Microbiology
1928	Frederick Griffith discovers genetic transformation of a bacterium and calls the agent responsible the "transforming principle."	Genetics
1928	Albert Szent-Györgi isolates vitamin C.	Biochemistry

Year	Event	Area of Biology
1928	Louis J. Stadler shows that ultraviolet radiation can cause mutations.	Genetics
1929	Edward A. Doisy and Adolf Butenandt independently isolate estrogen, a female sex hormone.	Physiology
1929	Phoebus A. Levene discovers the sugar 2-deoxyribose, the five-carbon sugar found in DNA.	Biochemistry
1931	Adolf Butenandt isolates androsterone, a male sex hormone.	Physiology
1932	Hans Adolf Krebs describes the urea cycle in which ammonia is converted to urea in mammals.	Biochemistry
1932	Otto Warburg isolates flavoproteins, an important compound of cell metabolism.	Biochemistry
1932	Gerhard Domagk discovers that the azo dye Prontosil is effective against streptococcus infections. The active compound is sulfamilamide, which is a basic component of sulfa drugs.	Microbiology
1933	Ernst Ruska invents an electron microscope that magnifies objects 12,000 times.	Tools and Techniques
1933	Tadeus Reichstein successfully synthesizes vitamin C.	Biochemistry
1934	John D. Beunal takes the first X-ray diffraction photograph of a protein crystal, pepsin.	Tools and Techniques
1934	Roger Tory Peterson publishes the *Field Guide to Birds*, the first field guide.	Zoology
1934	Arnold O. Beckman invents the pH meter, which electronically measures acidity and alkalinity.	Tools and Techniques
1935	Hans Spemann is awarded the Nobel Prize in Physiology or Medicine for his discovery of the organizer effect in embryonic development.	Embryology
1935	Albert Szent-Györgyi isolates the proteins actin and myosin from muscle tissue and demonstrates that they combine to form actomyosin, which contracts when ATP is added.	Physiology
1935	Wendell Stanley crystallizes the tobacco mosaic virus and demonstrates that it is protein in nature.	Microbiology
1935	Fritz Zernicke invents the phase contrast microscope.	Tools and Techniques
1935	Arthur Tansley coins the term "ecosystem."	Environment
1937	William Gericke defines the term "hydroponics."	Botany
1937	Konrad Lorenz shows the importance of imprinting as a form of early learning.	Behavior
1937	Arne Tiselius invents electrophoresis, a technique to separate proteins based on their electrical charges.	Tools and Techniques
1937	Hans Adolf Krebs discovers the citric acid cycle, later called the Krebs cycle, a metabolic process that converts carbohydrates, fats, and proteins to energy.	Biochemistry
1937	G. Evelyn Hutchinson publishes important papers on the ecology of lakes.	Environment

Year	Event	Area of Biology
1938	Herbert F. Copeland proposes a forth kingdom of organisms, Monera, for all bacteria.	Environment
1939	Paul Muller develops dichlorodiphenyltrichloroethane (DDT) and discovers its effectiveness against insect populations. DDT is later found to be unsafe to humans.	Environment
1939	Robin Hill isolates chloroplasts and shows that when they are illuminated, oxygen is produced. The photolysis of water produces the oxygen and is the first stage of photosynthesis, commonly called the Hill reaction.	Botany
1939	Howard W. Florey and Ernst B. Chain produce pure penicillin, the first powerful antibiotic.	Microbiology
1940	Karl Landsteiner and Alexander Wiener discover the rhesus (Rh) factor in blood.	Physiology
1940	George W. Beadle and Edward L. Tatum establish the "one gene—one enzyme" hypothesis, currently referred to as the one gene—one polypeptide theory. In 1941 they demonstrate the "one gene—one enzyme" theory while working with *Neurospora crassa*.	Genetics
1942	Salvador E. Luria prepares and photographs bacteriophage with the electron microscope.	Microbiology
1942	Selman A. Waksman coins the term "antibiotic" for compounds produced by microorganisms that kill or inhibit the growth of other microorganisms.	Microbiology
1943	Selman A. Waksman discovers the antibiotic streptomycin, used to treat tuberculosis.	Microbiology
1944	Oswald T. Avery, Colin M. MacLeod, and Maclyn McCarthy demonstrate the role of DNA in genetic inheritance in studies on the *Pneumococcus* bacteria.	Genetics
1945	Salvador E. Luria and Alfred D. Hershey demonstrate that viruses can mutate.	Genetics
1945	Endoplasmic reticulum is identified in the first electron micrographs of an intact cell.	Cytology
1946	Melvin Calvin uses chromatography and radioisotopes to investigate carbohydrate biosynthesis during photosynthesis, which is referred to as the Calvin cycle and is not dependent upon light.	Botany
1946	Joshua Lederberg and Edward L. Tatum discover that conjugation or sexual reproduction occurs in *Escherichia coli*, and the field of bacterial genetics is established.	Microbiology
1947	David Lack publishes *Darwin's Finches: An Essay on the General Biological Theory of Evolution*, following his research on adaptation by natural selection and competitive exclusion.	Evolution
1947	Willard F. Libby develops the technique of radiocarbon (carbon-14) dating.	Tools and Techniques

Year	Event	Area of Biology
1947	Fritz Lipmann isolates coenzyme A.	Biochemistry
1948	Albert Lehninger and Eugene Kennedy discover that the synthesis of ATP occurs in the mitochondria.	Biochemistry
1948	Alfred Mirsky discovers RNA in chromosomes.	Genetics
1950	Albert Claude discovers the endoplasmic reticulum in cells with the aid of an electron microscope.	Cytology
1950	Barbara McClintock publishes her discovery of transposons—mobile genetic elements, popularly called jumping genes, in maize.	Genetics
1950	Erwin Chargaff establishes that A (adenine)=T (thymine) and G (guanine)= C (cytosine) in DNA.	Genetics
1951	Nikolaas Tinbergen writes *The Story of Instinct*, one of the first books on animal instinct and behavior.	Behavior
1951	The American Institute of Biological Sciences is founded.	General
1952	Maurice Wilkins and Rosalind Franklin use X-ray diffraction to study DNA. They show that DNA is coiled and has a spiral form with the sugar-phosphate backbone on the outside.	Genetics
1952	Rosalyn S. Yalow develops radioimmunoassay, a test that uses radioactive isotopes to measure concentrations of biological compounds.	Tools and Techniques
1952	Stanley L. Miller and Harold C. Urey demonstrate that amino acids can be formed from hydrogen, water, methane, and ammonia when subjected to electrical discharges—conditions that simulated the earth's early atmosphere, in which life may have originated from nonliving substances.	Evolution
1952	Alfred D. Hershey and Martha Chase use radioactive tracers to show that the DNA, not the protein, from bacteriophage T2 is the genetic material that infects bacteria.	Genetics
1952	Rita Levi-Montalcini discovers the first growth factor, a protein that stimulates cell division and growth in nerve cells.	Cytology
1953	James Watson and Francis Crick propose the double helix model of DNA.	Genetics
1953	Robert W. Briggs and Thomas J. King successfully grow tadpoles from eggs whose nucleus has been replaced with one from a partially differentiated cell of a developing embryo.	Embryology
1953	Jonas E. Salk begins testing a poliomyelitis vaccine that he developed in 1952.	Microbiology
1954	George Gamow proposes that there is a genetic code composed of triplet nucleotides in the DNA molecule.	Genetics
1954	Vincent du Vigneaud synthesizes the hormone oxytocin, the first naturally occurring protein to be synthesized in the laboratory.	Physiology
1954	J. H. Thio and Albert Levan show that humans have forty-six chromosomes.	Genetics

Year	Event	Area of Biology
1955	Severo Ochoa discovers polynucleotide phosphorylase and synthesizes ribonucleic acid (RNA).	Biochemistry
1955	Christian René de Duve discovers and names lysosomes.	Cytology
1955	Joshua Lederberg and Norton Zinder discover transduction, in which viruses carry part of the chromosome of one bacterial cell to another.	Genetics
1955	Fredrick Sanger determines the structure of insulin.	Biochemistry
1956	George Palade discovers ribosomes, where protein synthesis occurs.	Cytology
1956	Mahlon B. Hoagland discovers transfer RNA and shows that it can combine with a particular amino acid in the synthesis of proteins.	Genetics
1956	Arthur Kornberg synthesizes DNA in vitro.	Biochemistry
1956	Earl Sutherland Jr. isolates cyclic AMP.	Biochemistry
1957	Alick Isaacs and Jean Lindenmann discover interferon.	Cytology
1957	Heinz Fraenkel-Conrat and Bea Singer demonstrate that the genetic material of tobacco mosaic virus is RNA.	Genetics
1957	Melvin Calvin describes the metabolism of carbon in photosynthesis.	Botany
1957	Scientists discover a "hole" in the ozone layer over Antarctica.	Environment
1957	Albert B. Sabin develops an oral vaccine for polio.	Microbiology
1958	Botanist Frederick Campion Steward successfully regenerates an entire carrot plant from a tiny piece of phloem, thus showing that plants are totipotent.	Botany
1958	Arthur Kornberg isolates DNA polymerase I from *Escherichia coli*.	Biochemistry
1958	Edward O. Wilson discovers that ants communicate through compounds called "chemical releasers," later called pheromones.	Behavior
1959	Max Perutz determines the structure of hemoglobin using X-ray diffraction.	Biochemistry
1959	Severo Ochoa isolates the first RNA polymerase.	Biochemistry
1959	Adolf Butenandt coins the term "pheromone."	Physiology
1960	Kenneth Norris and John Prescott find that dolphins use echolocation to find the range and direction of objects in water.	Zoology
1960	Robert Woodward and Martin Strell independently synthesize chlorophyll.	Botany
1961	François Jacob and Jacques Monod discover that genetic expression is controlled by a regulatory gene, the operon.	Genetics
1961	Sidney Brenner, François Jacob, and Matthew Meselson discover messenger RNA (mRNA).	Genetics
1962	Marshall Nirenberg and coworkers discover the first three-letter code for the amino acid, phenylalanine.	Genetics
1962	Rachel Carson publishes *Silent Spring*, warning of impending environmental disaster and raising public awareness of dangerous substances in the environment.	Environment

Year	Event	Area of Biology
1963	John Carew Eccles suggests that the mind is separate from the brain. He states that the mind acts upon the brain by effecting subtle changes in the chemical signals that flow among brain cells. He is awarded the Nobel Prize for his work on the mechanisms of nerve-impulse transmission.	Physiology
1963	Robert Woodward synthesizes colchicine, the chemical compound that blocks the division of eukaryotic cells by interfering with the structure of the mitotic spindle.	Biochemistry
1963	Konrad Lorenz publishes *On Aggression*.	Behavior
1964	The U.S. Congress passes the Wilderness Act.	Environment
1965	Ruth Sager publishes investigations on nonchromosomal inheritance in chloroplasts of *Chlamydomonas reinhardtii*.	Genetics
1965	Robert Holley reports the first nucleotide sequence of a transfer RNA (tRNA).	Genetics
1965	Hans Ris and Walter Plaut discover DNA in chloroplasts.	Cytology
1966	Marshall Nirenberg and Har Gobind Khorana decipher the complete genetic code.	Genetics
1966	The Endangered Species Preservation Act is passed by Congress, allowing the listing of native U.S. animal species on the Endangered Species list.	Environment
1966	Mark Ptashre and Walter Gilbert independently identify the first repressor genes.	Genetics
1966	Francis Crick publishes *Of Molecules and Men*.	Biochemistry
1967	John B. Gurdon is the first individual to successfully clone a vertebrate, the South African clawed toad (*Xenopus laevis*), using the technique of nuclear transplantation.	Embryology
1968	Werner Arber discovers that bacteria defend themselves against viruses by producing DNA-cutting enzymes.	Genetics
1968	Paul R. Ehrlich publishes *The Population Bomb*.	Environment
1968	James Watson publishes *The Double Helix*.	Genetics
1969	R. H. Whitaker proposes a five-kingdom system of classification.	Taxonomy
1969	Jonathan R. Beckwith and coworkers isolate a single gene for the first time.	Genetics
1969	R. Bruce Merrifield produces the first synthetic enzyme, a copy of ribonuclease.	Biochemistry
1969	The U.S. Congress passes the Endangered Species Conservation Act, widening the scope of the Endangered Species Preservation Act to include species in danger of worldwide extinction.	Environment
1970	The Clean Air Act is passed by Congress.	Environment
1970	The first Earth Day (April 22) is held.	Environment
1970	The U.S. Environmental Protection Agency (EPA) is created.	Environment

Year	Event	Area of Biology
1970	Har Gobind Khorana prepares an artificial yeast gene from its chemical components.	Biochemistry
1970	Peter Duesberg and Peter Vogt discover the first oncogene in a virus.	Genetics
1970	Howard Temin and David Baltimore isolate reverse transcriptase, the enzyme required for the transcription of RNA to DNA.	Genetics
1971	The use of the insecticide DDT (dichlorodiphenyltrichloroethane) is banned in the United States.	Environment
1971	Choh Hao Li synthesizes human growth hormone.	Physiology
1971	Edward O. Wilson publishes *The Insect Societies*.	Behavior
1972	The Clean Water Act is passed by the U.S. Congress.	Environment
1972	DDT is banned.	Environment
1972	The Marine Protection, Research, and Sanctuaries Act (Ocean Dumping Act) is passed.	Environment
1972	The fluid mosaic model of the plasma membrane is proposed based on the work of Seymour J. Singer and Garth L. Nicholson.	Cytology
1972	Niles Eldridge and Stephen Jay Gould propose the theory of punctuated equilibrium.	Evolution
1972	Paul Berg assembles the first recombinant DNA molecule in vitro.	Genetics
1973	Representatives from eighty nations sign the convention on International Trade in Endangered Species (CITES), which prohibits trade in 375 endangered species of plants and animals and the products derived from them, such as ivory.	Environment
1973	Stanley Cohen and Herbert W. Boyer use a plasmid to clone DNA and establish the field of genetic engineering.	Biotechnology
1973	The U.S. Congress passes the Endangered Species Act combining U.S. and foreign species lists. In so doing, it establishes different categories for "endangered" and "threatened" species, including plants and invertebrates.	Environment
1974	Mario Molina and F. Sherwood Rowland suggest that chlorofluorocarbons (CFCs) used in refrigerators and as aerosol propellants may be damaging the ozone layer of the atmosphere.	Environment
1974	The skeletal structure of a cell (the cytoskeleton) is discovered using monoclonal antibodies and fluorescence.	Cytology
1975	Carl R. Woese divides the kingdom Monera into Archaeobacteria and Eubacteria.	Taxonomy
1975	César Milstein and Georges Köhler use genetic cloning to create monoclonal antibodies (MABs).	Genetics
1975	Edward M. Southern develops a method for transferring DNA fragments separated in a gel to a filter, preserving the relative positioning of the fragments and becoming an important technique for identifying cloned genes.	Biotechnology
1975	Edward O. Wilson publishes *Sociobiology: The New Synthesis*.	Behavior

Year	Event	Area of Biology
1976	The Toxic Substances Control Act is passed.	Environment
1976	R. A. Levin discovers Prochloran, a group of prokaryotes with a greater resemblance to chloroplasta than to cyanobacteria.	Taxonomy
1976	Amanda Chakrabarty develops an oil-eating strain of bacteria that consumes crude oil as a means of cleaning up oil spills.	Biotechnology
1976	The National Institutes of Health releases the first guidelines for recombinant DNA experimentation.	Genetics
1976	J. Michael Bishop and Harold Varmus discover proto-oncogenes.	Genetics
1977	Frederick Sanger publishes the complete nucleotide sequence of 5,386 bases for a virus, the bacteriophage φX174.	Genetics
1977	Phillip Sharp discovers introns in eukaryotic cells.	Genetics
1977	Walter Gilbert and Frederick Sanger develop methods for sequencing DNA.	Biotechnology
1978	Louise Brown is born as a result of in vitro fertilization.	Biotechnology
1979	James E. Lovelock publishes *Gaia: A New Look at Life on Earth*.	Environment
1980	The Superfund Act is passed.	Environment
1981	Karl Illmensee clones baby mice, and researchers in China clone a fish.	Biotechnology
1981	The first human retrovirus, which makes DNA from an RNA template, is discovered.	Genetics
1981	AIDS (acquired immunodeficiency syndrome) is reported for the first time.	Microbiology
1982	The first genetically engineered drug, Humulin, to treat diabetes, is approved by the U.S. Food and Drug Administration.	Biotechnology
1982	Thomas D. Seeley discovers how honeybees locate a site for a hive.	Behavior
1983	Kary Mullis develops polymerase chain reaction (PCR), a technique for quickly reproducing selected DNA segments.	Biotechnology
1984	Steen A. Willadsen clones sheep from embryo cells.	Biotechnology
1984	Alec Jefferys develops DNA fingerprinting.	Biotechnology
1986	The first gene known to inhibit cell growth is discovered.	Genetics
1986	A complete frog is found fossilized in amber in the Dominican Republic and estimated to be 35 to 40 million years old.	Evolution
1987	The United States Patent and Trademark Office announces that it will issue patents for animals produced by genetic engineering.	Biotechnology
1988	Harvard University receives a patent from the U.S. Patent and Trademark office for an "oncomouse," the first vertebrate to receive an oncogene in its genetic material.	Biochemistry
1989	The first approved transfer of a gene to a melanoma patient occurs.	Biotechnology
1990	The U.S. Congress passes the 1990 Clean Air Act.	Environment
1990	The Human Genome Project officially begins.	Genetics

Year	Event	Area of Biology
1990	In South Dakota, Sue Hendrickson discovers the largest, most complete fossil of *Tyrannosaurus rex* found to date.	Evolution
1991	Peter Goodfellow and Robin Lovell-Badge discover the location of the gene on the Y-chromosome that determines maleness in mice.	Genetics
1991	The Biosphere 2 Project begins in Oracle, Arizona.	Environment
1992	The first Earth Summit is held in Rio de Janeiro, Brazil, with officials from more than 150 countries.	Environment
1993	Scientists identify p53, a tumor-suppressor gene.	Genetics
1993	The U.S. Department of Interior announces plans for a National Biological Survey that would identify and catalog all living species of plants and animals in the United States.	Environment
1994	The first genetically engineered food product, the Flavr Savr tomato, receives FDA approval.	Biotechnology
1994	The first breast cancer gene is discovered	Genetics
1995	The bald eagle is downlisted from "endangered" to "threatened."	Environment
1995	Craig Venter deciphers for the first time the complete genome (DNA sequence) of an organism, the bacterium *Haemophilus influenzae*.	Genetics
1996	The first complete DNA sequence of a eukaryotic organism, the yeast *Saccharomyces cerevisiae*, is published.	Genetics
1997	Ian Wilmut clones a healthy sheep, named Dolly, from the cells of an adult ewe that continues to live through the adult stage.	Biotechnology
1997	The Kyoto Protocol is accepted by representatives from over 150 countries agreeing to reduce emissions of greenhouse gases.	Environment
1998	The first animal genome, consisting of 14,000 genes, to be mapped and sequenced is completed for *Caenorhabditis elegans*, a soil roundworm.	Genetics
1998	Human embryonic stem cells are grown successfully in tissue culture.	Cytology
1999	The first complete DNA sequence for a human chromosome (number 22) is reported.	Genetics
2000	David Ayares and his colleagues successfully clone five identical piglets.	Biotechnology
2000	The first draft of the complete structure of human DNA is completed, showing more than 30,000 genes.	Genetics
2000	Scientists at Cold Spring Harbor Laboratory in New York complete the genetic sequence of the first flowering plant, *Arabidopsis thaliana*, a member of the mustard family that has about 26,000 genes.	Genetics
2000	The first endangered condor, captured in the 1980s, is released in the wild.	Environment
2002	Scientists announce the discovery of a new order of insects, Mantophasmatodea, consisting of three members. All three members are fossil specimens in museums. Researchers expect to locate other specimens in museums as well as living specimens.	Zoology

Year	Event	Area of Biology
2003	Peter Schultz, a researcher at the Scripps Research Institute in California, announces the world's first, truly synthetic, man-made form of life—an autonomous bacterium—a new *Escherichia coli* that can manufacture its own twenty-first amino acid.	Genetics
2003	Researchers in Japan announce the creation of transgenic coffee plants, which have about 60 percent less caffeine. The first crop of beans (used to make coffee beverages) will mature and be harvested in 2007. It is expected to show a similar reduction in caffeine.	Biotechnology
2003	Hoping to reverse desertification of the Aral Sea, an eight-mile earthen dike is built to channel water from the Syr Dar'ya River into the Aral Sea. It is estimated that the water level will rise 13 ft (4 m) by 2007.	Environment

THE
HANDY
BIOLOGY
ANSWER
BOOK

CHEMISTRY FOR BIOLOGY

BASICS

What is **biochemistry**?

As a field of scientific study, chemistry may be divided into various subgroups. One major subgroup is organic chemistry, which refers to the study of carbon-based compounds, including carbohydrates and hydrocarbons such as methane and butane. Within the field of organic chemistry there is a discipline that focuses solely on the study of the organic molecules that are important to living organisms; this branch is known as biochemistry.

What is an **atom**?

An atom is the smallest unit of an element, containing the unique chemical properties of that element. Atoms are very small—several million atoms could fit in the period at the end of this sentence.

Parts of an Atom

Subatomic particle	Charge	Mass	Location
Proton	Positive	1.7×10^{-24} g	Nucleus
Neutron	Neutral	1.7×10^{-24} g	Nucleus
Electron	Negative	9.1×10^{-28} g	Orbitals around nucleus

How does the **nucleus of an atom** differ from the **nucleus of a cell**?

The English word nucleus—meaning "inside the shell"— is derived from the Latin word for nut. The nucleus of an atom is an enclosed space containing protons and neutrons. The nucleus of a cell is a membrane-enclosed organelle that contains the genetic material of a cell.

How do **elements differ** from one another?

Distinguishing one element from another requires a look at the subatomic particles of an atom—the protons, neutrons, and electrons. Each element has a specific number of protons; this number is used to ascribe an atomic number to an element. All atoms have an equal number of protons and electrons. For example, an atom of helium has an atomic number of 2 because it has two protons, and therefore two electrons. An atom with an equal number of protons and electrons, such as helium, has a net electrical charge of zero.

How does the **mass number** of an **element** differ from the **atomic number**?

The mass number is the sum of the number of protons and neutrons in the nucleus of an element. Using helium as an example, the mass number of helium is 4, as it has two protons and two neutrons in its nucleus. Since it has only two protons, the atomic number of helium is 2. When the atomic number changes (i.e., the number of protons changes), the result is a different element.

What are the most important **elements** in **living systems**?

The most important elements in living systems include oxygen, carbon, hydrogen, nitrogen, calcium, phosphorus, potassium, sulfur, sodium, chlorine, magnesium, and iron. These elements are essential to life due to their cellular function.

The Most Common and Important Elements in Living Organisms

Element	Percent of humans by weight	Functions in life
Oxygen	65	Part of water and most organic molecules; molecular oxygen
Carbon	18	Backbone of organic molecules
Hydrogen	10	Part of all organic molecules and water
Nitrogen	3	Component of proteins and nucleic acids
Calcium	2	Part of bone; essential for nerves and muscles
Phosphorus	1	Part of cell membranes and energy storage molecules; part of bone
Potassium	0.3	Important for nerve function
Sulfur	0.2	Structural component of some proteins
Sodium	0.1	Primary ion in body fluids; essential for nerve function

Element	Percent of humans by weight	Functions in life
Chlorine	0.1	Major ion in body fluids
Iron	Trace	Component of hemoglobin
Magnesium	Trace	Cofactor for enzymes; important to muscle function

What is a **molecule**?

Molecules are made of specific combinations of atoms. For example, carbon dioxide is made of one carbon atom and two oxygen atoms; water is made of two hydrogen atoms and one oxygen atom, and the atoms are joined by chemical bonds. Complex molecules such as starch may have hundreds of participating atoms linked in a specific pattern.

What is a **chemical bond**?

A chemical bond is an attraction between the electrons present in the outermost energy level or shell of a particular atom. This outermost energy level is known as the valence shell. Atoms with an unfilled outer shell are less stable and tend to share, accept, or donate electrons. When this happens, a chemical bond is formed. In living systems, chemical reactions, supervised by enzymes, link atoms together to form molecules.

What are the **major types** of **bonds**?

There are three major types of chemical bonds: covalent, ionic, and hydrogen. The form of bond that is established is determined by a specific arrangement between the electrons. Ionic bonds are formed when electrons are exchanged between two atoms, and the resulting bond is relatively weak. An ionic bond is found in salt. Covalent bonds occur when electrons are shared between atoms; this form of bond is strongest and is found in both energy-rich molecules and molecules essential to life. Hydrogen bonds are temporary, but they are important because they are crucial to the shape of a particular protein and have the ability to be rapidly formed and reformed, as in the case of muscle contraction. The following chart explains the types of bonds and their characteristics.

Summary of Three Types of Chemical Bonds

Type	Strength	Description	Examples
Covalent	Strong	Sharing of electrons results in each atom having a filled outermost shell of electrons	Bonds between hydrogen and oxygen in a molecule of water

3

Type	Strength	Description	Examples
Hydrogen	Weak	Bond between oppositely charged regions of molecules that have covalently bonded hydrogen atoms	Bonds between molecules of water
Ionic	Moderate	Bond between two oppositely charged atoms that were formed by the permanent transfer of one or more electrons	Bond between Na^+ and Cl^- in salt

What determines the **type of bond** that forms between **atoms**?

The electron structure of an atom is the best predictor of its chemical behavior. Atoms with filled outer shells tend not to form bonds. However, those atoms with 1, 2, 6, or 7 electrons in the outer shell tend to become ions and form ionic bonds. Atoms with greater than 2 or less than 6 electrons tend to form covalent bonds.

What is an **isotope**?

Atoms of an element that have different numbers of neutrons are isotopes of the same element. Isotopes of an element have the same atomic number but different mass numbers. Common examples are the isotopes of carbon: ^{12}C and ^{14}C. ^{12}C has 6 protons, 6 electrons, and 6 neutrons, while ^{14}C has 6 protons, 6 electrons, and 8 neutrons. Some isotopes are physically stable, while others, known as radioisotopes, are unstable. Radioisotopes undergo radioactive decay, emitting both particles and energy. If the decay leads to a change in the number of protons, the atomic number changes, transforming the isotopes into a different element.

What are **isomers**?

Isomers are compounds with the same molecular formula but differing atomic structure within their molecules. The atoms of structural isomers are connected in different ways. Geometric isomers differ in their symmetry about a double bond. Optical isomers are mirror images of each other.

How is the term "**mole**" used in **chemistry**?

A mole (mol) is a fundamental unit of measure; it refers to either the gram atomic weight or the gram molecular weight of a substance. A mole is equal to the quantity of a substance that contains 6.02×10^{23} atoms, molecules, or formula units. This number is also called Avogadro's number, named after Amedeo Avogadro (1776–1856), who is considered to be one of the founders of physical science.

Why is **carbon** an **important element**?

Carbon makes up 18 percent of the weight of the human body. Due to its unique electron configuration, carbon needs to share electrons. It can form four covalent bonds with other carbon atoms or a variety of other elements.

Why is **water** so important to **life**?

We are all aqueous creatures, whether because of living in a watery environment or because of the significant amount of water contained within living organisms. Therefore, all chemical reactions in living organisms take place in an aqueous environment. Water is important to all living organisms due to its unique molecular structure, which is v-shaped, with hydrogen atoms at the points of the v and an oxygen atom at the apex of the v. In the covalent bond between oxygen and hydrogen, the electrons spend more time closer to the oxygen nucleus than to the hydrogen nucleus. This uneven or unequal sharing of electrons results in a water molecule with a slightly negative pole and a slightly positive pole.

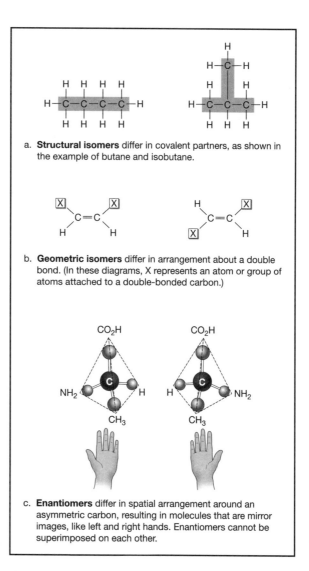

a. **Structural isomers** differ in covalent partners, as shown in the example of butane and isobutane.

b. **Geometric isomers** differ in arrangement about a double bond. (In these diagrams, X represents an atom or group of atoms attached to a double-bonded carbon.)

c. **Enantiomers** differ in spatial arrangement around an asymmetric carbon, resulting in molecules that are mirror images, like left and right hands. Enantiomers cannot be superimposed on each other.

The three types of isomers.

Water is the **universal solvent** in **biological systems**. What does this mean for **living organisms**?

A solvent is a substance that can dissolve other matter. Since all chemical reactions that support life occur in water, water is known as the universal solvent. It is the polar nature of water molecule (it contains both positive and negative poles) that causes it

5

to act as a solvent. Any substance with an electric charge will be attracted to one end of the molecule. If a molecule is attracted to water, it is termed hydrophilic; if it is repelled by water, it is termed hydrophobic.

Why is **liquid water** more **dense** than **ice**?

Pure, liquid water is most dense at 39.2°F (3.98°C), and decreases in density as it freezes. The water molecules in ice are held in a relatively rigid geometric pattern by their hydrogen bonds, producing an open, porous structure. Liquid water has fewer bonds; therefore, more molecules can occupy the same space, making liquid water more dense than ice.

What is an **ion**?

An ion is an atom that is charged by the loss or gain of electrons. For example, when an atom gains one or more electrons, it becomes negatively charged. When an atom loses one or more electrons, it becomes positively charged.

What is meant by **pH** and why is it so **important** to **living things**?

The term "pH" is taken from the French phrase *l'puissance d'hydrogen*, meaning "the power of hydrogen." The composition of water can be used to understand the concept of pH. Water is composed of two hydrogen atoms bonded covalently to an oxygen atom. In a solution of water, some water molecules will dissociate into the component ions: the H^+ ion and the OH^- ion. In chemistry the term "equilibrium reaction" is used to describe a situation in which there are an equal number of products and reactants. Water is an example of an equilibrium reaction: water contains H_2O, H^+, and OH^- at all times. It is the balance of the two ions that determines pH. When there are more H^+ ions than OH^- ions, the solution is referred to as an acid, and when there are more OH^- ions than H^+ ions, the solution is referred to as a base. The concentration of hydrogen ions in water influences the chemical reactions of other molecules.

Why is **pH** so **important** to **life**?

The concentration of hydrogen ions in water influences the chemical reactions of other molecules. An increase in the concentration of electrically charged ions

> ## How much water does the human body contain?
>
> The average human body is 50 to 65 percent water. Males on average have a higher percentage of body water than females because they have less body fat. The more fat you have, the less water you have.

interferes with or influences the ability of molecules (specifically proteins) to chemically interact. Most living systems function at an internal pH that is close to seven, but a biologically active molecule has an optimum pH at which it functions. The optimum pH level varies depending on the molecule and where it functions.

What is the **pH scale**?

The pH scale is the measurement of the H^+ concentration (hydrogen ions) in an aqueous solution. It is used to measure the acidity or alkalinity of a solution. The pH scale ranges from zero to fourteen. A neutral solution has a pH of seven; a solution with a pH greater than seven is basic (or alkaline); and a solution with a pH less than seven is acidic. The lower the pH, the more acidic the solution. As the pH scale is logarithmic, each whole number drop on the scale represents a tenfold increase in acidity (meaning the concentration of H^+ increases tenfold).

Scale of pH Values

pH Value	Examples of Solutions
0	hydrochloric acid (HCl), battery acid
1	stomach acid (1.0–3.0)
2	lemon juice (2.3)
3	vinegar, wine, soft drinks, beer, orange, juice, some acid rain
4	tomatoes, grapes, banana (4.6)
5	black coffee, most shaving lotions, bread, normal rainwater
6	urine (5–7), milk (6.6), saliva (6.2–7.4)
7	pure water, blood (7.3–7.5)
8	egg white (8), seawater (7.8–8.3)
9	baking soda, phosphate detergents, Clorox, Tums
10	soap solutions, milk of magnesia
11	household ammonia (10.5–11.9), nonphosphate detergents
12	washing soda (sodium carbonate)
13	hair remover, oven cleaner

What is **thermodynamics**?

Thermodynamics is the study of the relationships between energy and the activity of a cell. The laws of thermodynamics govern the way in which cells transform chemical compounds. The first law of thermodynamics states that the total energy of a system and its surroundings will always remain constant. The second law of thermodynamics states that systems (such as a cell) tend to become disordered. Disorder usually arises in the form of heat; entropy is the term that refers to the disorder that tends to disrupt cells, or even the universe.

MOLECULES

What **molecules** are most important to **living organisms**?

Four molecules are referred to as bio-organic because they are essential to living organisms and contain carbon: nucleic acids, proteins, carbohydrates, and lipids. These molecules are all large, and they are formed by a specific type of smaller molecule, known as a monomer.

What role do **bonds** have in **bio-organic molecules**?

Bonds are important to the structure of bio-organic molecules. Because chemical reactions actually involve electron activity at the subatomic level, shape determines function. For example, morphine has a shape similar to an endorphin, a natural molecule in the brain. Endorphins are pain suppressant molecules; thus, morphine essentially mimics the function of endorphins and can be used as a potent pain reliever.

What is meant by the term "**polar**" **molecule**?

Polar molecules have opposite charges at either end. "Polar" refers to the positive and negative sides of the molecule. If a molecule is polar, it will be attracted to other polar molecules. This can affect a wide range of chemical interactions, including whether a substance will dissolve in water, the shape of a protein, and the complex helical structure of DNA. Water is an example of a polar molecule.

What are **functional groups**?

There are numerous patterns of atoms and bonds that are found frequently in organic compounds. These configurations of atoms are called functional groups, as each has specific, predictable properties. For example, an amino acid contains both an amino and a carboxyl, while alcohols all have a hydroxyl group. This group results in the solubility of alcohol in water.

What is a **macromolecule**?

Macromolecules are literally "giant" polymers made from the chemical linkage of smaller units called monomers. To be considered a macromolecule, a molecule has to have a molecular weight greater than 1,000 daltons. A dalton is a standard unit of measurement that refers to the mass of a proton; it can be used interchangeably with atomic mass unit (amu).

How are **macromolecules built**?

There are four types of very large molecules that are important to life. These are carbohydrates, nucleic acids, lipids, and proteins. Although these are quite diverse in terms of structure, size, and function, the same mechanisms build and break them down:

- All are comprised of single units linked together to create a chain, similar to a freight train with many cars.

- All the monomers or single units contain carbon.

- All monomers are linked together through a process known as dehydration synthesis, which literally means "building by removing water." A hydrogen atom (H) is removed from one monomer, and a hydroxide (OH) group is removed from the next monomer in line. Atoms on the ends of the two monomers will then form a covalent bond to fill their electron shells, thereby building a polymer.

- All polymers are broken down by the same method, hydrolysis. Hydrolysis means "breaking with water." By adding H_2O, which contains hydrogen and hydroxide groups, back to the monomers, the bond is broken and the macromolecule separates into smaller pieces.

What are the most **common macromolecules** used as **energy sources** by cells?

Cells use a variety of macromolecules as energy sources. Carbohydrates, lipids, and even proteins can be metabolized for energy. ATP and related compounds are also used as temporary energy storage vehicles.

What is the comparative **value** of common **energy sources** for cells?

Energy source	Energy yield
Carbohydrate	4 kcal/g
Fat	9 kcal/g
Protein	4 kcal/g

What is **cholesterol**?

Cholesterol belongs to the subclass of lipids known as steroids. Steroids have a unique chemical structure. They are built from four carbon-laden ring structures that are fused together. The human body uses cholesterol to maintain the strength and flexibility of cell membranes. Cholesterol is also the molecule from which steroid hormones and bile acids are built.

Cholesterol and Some of Its Derivatives

Molecule	Function
Aldosterone	Maintains water and salt balance by the kidney, controls blood pressure
Bile acids	Produced by the liver, help in the digestion of dietary lipids
Cholesterol	Provides stability and flexibility to cell membranes
Cortisone	Carbohydrate metabolism
HDL (high density lipoproteins); LDL (low density lipoproteins)	Lipid-protein combinations that transport lipids in the blood
Testosterone, estrogens, progesterone	Maintain sex characteristics. Allow reproduction to occur

Why are some **fats** "**good**" and others "**bad**?"?

There are a number of ways to answer this question. One could argue that no fats are "bad," as fats are excellent sources of energy and help to maintain the health of the body. From this perspective, fat is only bad if one eats too much of it. Another way to answer would be to point out that there are several fats that are considered essential (the omega-6 and omega-3 fatty acids)—in other words, they are substances that our bodies require for maintenance but that we cannot manufacture. These are considered to be "good" fats, and comparatively, the fats we don't need to ingest are often dubbed as "bad." Finally, artificial fats should be mentioned. While these may have

What was the first amino acid to be discovered?

Asparagine, isolated from the asparagus plant, was discovered in 1806 by the French chemist Nicolas-Louis Vauquelin (1763–1829).

been created to maintain the flavor and texture of food while reducing the caloric content, they may be difficult to metabolize and therefore "bad" for us in the long run.

What is the **difference** between a **fat** and a **lipid**?

Lipids are bio-organic molecules that are hydrophobic. In other words, they do not mix with or dissolve in water. Among lipids there is a category known as "fats." Each fat molecule is comprised of a glycerol (alcohol) molecule and at least one fatty acid (hydrocarbon chain with an acid group attached).

How are **carbohydrates classified**?

Carbohydrates are classified in several ways. Monosaccharides (single unit sugars) are grouped by the number of carbon molecules they contain: triose has three, pentose has five, and hexose has six. Carbohydrates are also classified by their overall length (monosaccharide, disaccharide, polysaccharide) or function. Examples of functional definitions are storage polysaccharides (glycogen and starch), which store energy, and structural polysaccharides (cellulose and chitin), which provide support for organisms without a bony skeleton.

What do **proteins do**?

In a word, everything. Proteins allow life to exist as we know it. They are the enzymes that are required for all metabolic reactions. They are also important to structures like muscles, and act as both transporters and signal receptors.

Types and Functions of Proteins

Type of protein	Examples of functions
Defensive	Antibodies that respond to invasion
Enzymatic	Increase the rate of reactions, build and breakdown molecules
Hormonal	Insulin and glucagon, which control blood sugar
Receptor	Cell surface molecules that cause cells to respond to signals
Storage	Store amino acids for use in metabolic processes
Structural	Major components of muscles, skin, hair, horns, spiderwebs
Transport	Hemoglobin carries oxygen from lungs to cells

11

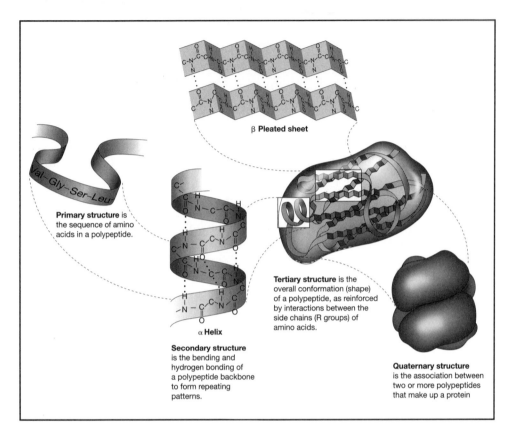

The four levels of protein structure.

Why are **proteins** among the most **complex molecules**?

Among the major bio-organic molecules, only proteins require up to four levels of structure in order to be functional. In addition, relatively slight environmental changes cause a shift in structural levels that may be sufficient to radically change the function of the protein.

The structural levels of protein and the functions they provide are:

- Primary: A polypeptide chain of up to 500 amino acids covalently bonded.

- Secondary: The formation of hydrogen bonds between nearby amino acids causes the polypeptide chain to twist and/or pleat.

- Tertiary: Distant amino acids form bonds and associations in reaction to changes that occur in the secondary level.

- Quaternary: Two separate polypeptide chains intermingle to form a molecule that has a larger, more complex structure than that found in the other protein levels.

What was one of the earliest diagnoses of a metabolic disease?

Alkaptonuria was identified in an Egyptian mummy from 1500 B.C.E.

What are **nucleic acids**?

DNA (deoxyribonucleic acid) and RNA (ribonucleic acid) are nucleic acids. Nucleic acids are molecules comprised of monomers known as nucleotides. These molecules may be relatively small (as in the case of certain kinds of RNA) or quite large (a single DNA strand may have millions of monomer units). Individual nucleotides and their derivatives are important in living organisms. ATP, the molecule that transfers energy in cells, is built from a nucleotide, as are a number of other molecules crucial to metabolism.

METABOLIC PATHWAYS

What is a **metabolic pathway**?

A metabolic pathway is a series of interconnected reactions that share common mechanisms. Each reaction is dependent on a specific precursor: a chemical, an enzyme, or the transfer of energy. One of the first studies of metabolic pathways was carried out in 1909 by the British physician Archibald Garrod (1857–1936). His study suggested a link between the inability to make a particular enzyme and inherited disease. The disease was alkaptonuria, a condition in which urine darkens upon exposure to air, due to the presence of the chemical alkapton. Garrod's discovery was one of the first incidents of a physical manifestation being tied to a specific metabolic disorder.

How do **cells store energy**?

Plants and animals use glucose as their main energy source, but the way this molecule is stored differs. Animals store their glucose subunits in the form of glycogen, a series of long, branched chains of glucose. Plants store their glucose as starch, formed by long, unbranched chains of glucose molecules. Both glycogen and starch are formed through the chemical reaction of dehydration synthesis and both are broken down through the process of hydrolysis.

How does the term "**negative feedback**" apply to **metabolism**?

Negative feedback is a cellular process that is similar to the manner in which an air conditioner operates: an air conditioner can be set to a certain temperature, and when the surrounding air reaches the set temperature, the air conditioner shuts off. Negative feedback is part of the homeostatic process through which cells conserve energy by synthesizing products only for their immediate needs.

13

How much ATP does the human body use?

Each cell in the human body is estimated to use between one billion and two billion ATP's per minute! With 100 trillion cells in the human body, how much ATP do we need? Roughly 1×10^{23} molecules. In the span of twenty-four hours, 100 trillion cells produce about 441 lbs (200 kg) of ATP.

What is **ATP**?

ATP (adenosine triphosphate) is the universal energy currency of a cell. Its secret lies in its structure. ATP contains three negatively charged phosphate groups. When the bond between the outermost two phosphate groups is broken, ATP becomes ADP (adenosine diphosphate). This reaction releases 7.3 kcal/mole of ATP, which is a great deal of energy by cell standards.

What is the **difference** between **catabolic** and **anabolic reactions**?

Catabolic and anabolic reactions are metabolic processes. Both the capture and use of energy by organisms involves a series of thousands of reactions (metabolism). A catabolic reaction is one that breaks down large molecules to produce energy; an example is digestion. An anabolic reaction is one that involves creating large molecules out of smaller molecules; an example is when your body makes fat out of extra nutrients you eat.

What is the **Krebs cycle**?

The Krebs cycle (also referred to as the citric acid cycle) is central to aerobic metabolism. It is an adaptation that allows cells to gain increased energy from glucose. The process is critical to the development of multicellular organisms, and is essential to the harvesting of high energy electrons during the final breakdown of the glucose molecule. By-products of this cycle are carbon dioxide and water. It is named in recognition of the German chemist Hans Krebs (1900–1981), who received the 1953 Nobel Prize in Physiology or Medicine.

What is the **Calvin cycle**?

Photosynthesis creates reactions that are both light dependent and light independent. The Calvin cycle is part of the reaction that is light independent and occurs in the chloroplast. This cycle is crucial to the capture of carbon dioxide, leading to the formation of sugar ($C_6H_{12}O_6$). It is named in honor of Melvin Calvin (1911–1997), who received the 1961 Nobel Prize in Chemistry for unraveling the process of glucose biosynthesis.

Did you know?

The efficiency of cellular respiration (the harvesting of energy from macromolecules) can be calculated by the following formula:

$$\frac{38 \text{ moles ATP/mole glucose} \times 7.3 \text{ kcal/mol ATP}}{686 \text{ kcal/mol glucose}} = 0.40 \text{ (40 percent efficiency)}$$

Why do we **die without oxygen**?

Most living organisms are aerobic; that is, they require oxygen to complete the total breakdown of glucose. As many as 36 ATP are produced through aerobic metabolism of one glucose molecule. Without oxygen, cells do not synthesize enough ATP to maintain a multicellular organism. Most people think that we need oxygen to breathe, but actually we need oxygen to recycle the spent electrons and hydrogen ions (H^+) produced as by-products of aerobic respiration. With these by-products, oxygen combines and forms "metabolic" water.

Where does the **carbon dioxide** that we **exhale** come from?

The carbon dioxide we exhale is a result of the breakdown of glucose during the energy harvesting phase of aerobic cell respiration. During this process, all of the carbon atoms (from the $C_6H_{12}O_6$) are released as carbon dioxide molecules. Of the six carbon dioxide molecules generated, four are released via the Krebs cycle.

What is a **redox reaction**?

A redox reaction is actually a series of chemical reactions involving the transfer of electrons from one atom or molecule to another. By transferring electrons, the cell is also transferring energy. The molecule that loses an electron is said to be oxidized, resulting in its charge becoming positive. The molecule that gains an electron is said to be reduced, and its charge becomes negative. Electrons do not travel alone in cells; rather, they are accompanied by a proton (H^+). A proton and an electron together make a hydrogen atom. Ultimately, a redox reaction involves the loss of hydrogen atoms from one molecule and the gain of hydrogen atoms by another molecule.

What is **fermentation**?

Scientists have theorized that fermentation was the process through which energy was first harvested from organic compounds. Fermentation evolved before Earth's atmosphere contained free oxygen. Aerobic respiration differs from fermentation in

15

that the products of glycolysis enter the Krebs cycle rather than being used to form lactic acid or alcohol. Fermentation must have predated the appearance of oxygen in the atmosphere, which occurred over 2.5 billion years ago. Fermentation is thus an ancient process and occurs normally in microorganisms that live in the absence of oxygen.

What **products** are produced by **fermentation**?

Fermentation is important in the production of wine, beer, soy sauce, baked products, and pickles.

ENZYMES

What is an **enzyme**?

Enzymes are proteins that act as biological catalysts. They decrease the amount of energy needed (activation energy) to start a metabolic reaction. Without enzymes, you would not be able to harvest energy

A wine fermentation tank. Fermentation is an ancient process and occurs normally in microorganisms that live in the absence of oxygen.

and nutrients from your food. As an example, lactose intolerance is an inability to produce lactase, the enzyme that breaks down milk sugar (lactose). While this condition is not life-threatening for adults, it can have severe consequences for infants and children.

What **roles do enzymes** actually play?

Enzymatic reactions can build up or break down specific molecules. The specific molecule an enzyme works on is the substrate. The molecule that results from the reaction is the product.

SUBSTRATE	ENZYME	PRODUCTS
Lactose	Lactase	glucose + galactose

Who **first used** the term **"enzyme,"** and how was it used?

In 1876 William Kuhne (1837–1900) proposed that the term "enzyme" be used to denote phenomena that occurred during fermentation. The word itself means "in yeast" and is derived from the Greek *en*, meaning "in," and *zyme*, meaning "yeast."

Why is the **shape** of an **enzyme** so important?

Shape is critical to the function of all molecules, but especially enzymes, which are three dimensional. The "active site" of an enzyme is the area where substrate binds and the reaction takes place. How an enzyme reacts with its substrate is similar to how a ship docks; there are minor bonds that form between the enzyme and substrate until docking is complete. Anything affecting a protein's shape would have an effect on its ability to react with the substrate.

How many enzymes are there?

Approximately 5,000 enzymes have been named, but there may be a total of 20,000 or more. A metabolic pathway may require a whole complex of enzymes to complete hundreds of reactions.

How are **enzymes named**?

Individual enzymes are named by adding the suffix "-ase" to the name of the substrate with which the enzyme reacts. An example of this method is the enzyme amylase, which controls the breakdown of amylose (starch). There are categories of enzymes that control certain reactions. Hydrolases control hydrolytic reactions; proteinases control protein breakdown; synthetases control synthesis reactions. There are exceptions: trypsin and pepsin, both digestive enzymes that breakdown protein, retain the names used before the modern form of nomenclature was adopted.

What are some of the most **common enzyme deficiencies**?

Lactose intolerance, a condition that results from the inability to digest the sugar present in milk—lactose—is one of the most common enzyme deficiencies. While nearly everyone is born with the ability to produce lactase—the enzyme responsible for lactose breakdown—many people lose this ability with age. Lactase production in adults varies by ethnic group.

Lactose Intolerance in Ethnic Groups

Ethnic Group	Estimated Prevalence of Lactose Intolerance
Asian Americans	>80 percent
Native Americans	80 percent
African Americans	75 percent
Mediterranean peoples	70 percent
Inuits	60 percent
Hispanics	50 percent
Caucasians	20 percent
Northern Europeans	<10 percent

Glucose-6-phosphate dehydrogenase deficiency is a more serious enzyme deficiency that is linked to the bursting of red blood cells (hemolysis). This deficiency is found in more than 200 million people, mainly Mediterranean, West African, Middle Eastern, and Southeast Asian populations.

How do I know my **enzymes are working**?

Obviously, it is difficult to track the activities of the 20,000 enzymes that are required by your body. However, in the case of the enzyme amylase, you can easily check to see if it is working. Amylase is an enzyme in your saliva that starts the breakdown of complex carbohydrates into simple sugars (glucose and maltose). A plain cracker held in the mouth long enough will begin this breakdown, and you will actually begin to taste the sweetness of the enzyme products. Also, lysozyme is an enzyme present in your respiratory tract secretions and tears. It prevents invasion by bacteria, which explains why the warm, moist, open environment of our eyes manages to remain relatively infection-free during a normal day.

Why do **enzymes** only work in **specific environments**?

Because changes in temperature and pH can cause the structure of a protein to change, every enzyme has criteria that must be met in order for it to perform its function. For example, the amylase that is active in the mouth cannot function in the acidic environment of the stomach; pepsin, which breaks down proteins in the stomach, cannot function in the mouth.

What **factors** affect **enzyme function**?

Enzymes may be controlled at a variety of levels:

- Since proteins are coded with DNA, a change in DNA will change the rate at which an enzyme is produced.
- Control molecules known as "competitive inhibitors" may prevent the substrate from reaching the enzyme.
- Allosteric or noncompetitive regulators may bind elsewhere on the enzyme, causing its shape to change and thereby altering its active site.
- Changes in environmental conditions such as pH, temperature, or salt concentration also affect shape.
- Absence of required enzyme "helpers" such as vitamins and minerals.

How can **mathematics** be used to **predict** an **enzyme's behavior**?

The calculation of enzyme activity is known as enzyme kinetics. One particularly useful model for making these predictions, the Michaelis-Menten equation, was formulated in 1913 by Leonor Michaelis (1875–1949) and Maud Menten (1879–1960). It is important for scientists to know how to control enzymatic reactions by varying the concentration of enzymes and substrates. After all, enzymes mediate all cellular pathways.

What are **vitamins** and **minerals**?

A vitamin is an organic, non-protein substance that is required by an organism for normal metabolic function but cannot be synthesized by that organism. In other words, vitamins are crucial molecules that must be acquired from outside sources. While most vitamins are present in food, vitamin D, for example, is produced as a precursor in our skin and converted to the active form by sunlight. Minerals, such as calcium and iron, are inorganic substances that also enhance cell metabolism. Vitamins may be fat- or water-soluble. Recommended amounts of vitamins are to ensure normal enzymatic function, and excessive intake can be toxic.

Vitamin	Major Sources	Major Function
A	Animal products; plants contain only vitamin A building blocks.	Aids normal cell division and development. Particularly helpful in the maintenance of visual health.
B-complex	Fruits and vegetables (folate); meat (thiamine, niacin, vitamin B_6, and B_{12}); milk (riboflavin, B_{12})	Energy metabolism; promotes harvesting energy from food.

19

Vitamin	Major Sources	Major Function
C	Fruits and vegetables, particularly citrus, strawberries, spinach, and broccoli.	Collagen synthesis; antioxidant benefits; promotes resistance to infection.
D	Egg yolks; liver; fatty fish; sunlight.	Supports bone growth; maintenance of muscular structure and digestive function.
E	Vegetable oils; spinach; avocado; shrimp; cashews.	Antioxidant.
K	Leafy, green vegetables; cabbage.	Blood clotting.

APPLICATIONS

What is the **difference** between **saturated** and **unsaturated fat**?

Fat is a type of lipid molecule constructed by glycerol and three fatty acids. The molecular structure of the fatty acids determines whether the fat is saturated or unsaturated. Fats with hydrogen atoms but without double bonds are "saturated." Unsaturated fatty acids have double bonds, and therefore have fewer hydrogen atoms.

What is a **trans fat**?

The term "trans" fat refers to the arrangement of hydrogen atoms around the carbon backbone of the fatty acid. A trans-fatty acid is a molecule that has a carbon backbone with hydrogen atoms attached in a manner that is not normally found in nature. Most naturally occurring fatty acids have their hydrogen arranged in the "cis" form. In trans fats, some of these hydrogens are attached on opposite sides of the fatty acid molecule in what is known as a "trans" (as opposed to "cis") formation.

Can one **type of fat** be **changed** into **another type**?

The process of hydrogenation can convert an unsaturated fatty acid into a hydrogenated fatty acid. This process, which is achieved by adding extra hydrogen atoms to unsaturated fat, has become both a bane and a blessing. Hydrogenation is the process that allows unsaturated vegetable oils to be turned into margarine. This method prevents oxidation, and thus rancidity, and has allowed for the development of foods with less animal and saturated fats. However, the consumption of hydrogenated fatty acids may be linked to increased risk of heart disease, because the fats cause a change in the structure of targeted unsaturated fatty acids. The consumption of trans fats has been shown to slightly increase the levels of bad cholesterol (LDL) in the blood.

A laboratory worker tests for steroids. Many athletic competitions now test athletes for steroid use.

What kinds of **molecules** are **hormones**?

There are two basic types of hormones, those that are water soluble (hydrophilic) and those that are not (hydrophobic). Hydrophilic hormones include those derived from amino acids, peptides, and proteins. The hydrophobic group includes steroids, which are derived from cholesterol. Steroid hormones include testosterone, estrogen, cortisol, and aldosterone.

Can all **anabolic steroid** use be **detected**?

One of the newest "designer steroids" is tetrahydrogestrinone (THG). This chemical is not detected by routine urine tests and has been classified as a nutritional supple-

What are anabolic steroids?

Anabolic steroids are hormones that work to increase synthesis reactions, particularly in muscle. These steroids are also known as the androgens (a precursor of testosterone, the primary male hormone). They are commonly used by athletes to enhance performance, although they do have clinical applications for certain medical conditions. Major health hazards caused by anabolic steroid abuse or unsupervised use include liver tumors, cancer, jaundice, fluid retention, high blood pressure, an increase in bad cholesterol (LDL), a decrease in good cholesterol (HDL), testicular shrinking, reduced sperm count, baldness, and breast development (in men).

ment. However, according to the Food and Drug Administration (FDA), THG is related to two other synthetic anabolic steroids, gestrinone and trenbolone. Tetrahydrogestrinone is designed to boost muscle mass and strength; however, dangerous side effects include benign and malignant liver tumors, hepatitis, aggressive mood swings, decreased fertility, acne, and an increased risk of heart attacks. A new test has been developed that can detect THG in urine samples.

How do we know the **energy needs** of different **organisms**?

The basal metabolic rate (BMR) of an organism refers to the amount of energy needed, while resting, to maintain normal life processes. BMR is determined by measuring an organism's production of heat or carbon dioxide, or by analyzing oxygen consumption. An organism's BMR can vary according to a variety of factors, including age, gender, size, body temperature, the surrounding environmental temperature, available food quality, hormonal levels, the time of day, activity levels, and the amount of oxygen available.

What is a **calorie**?

There are actually two kinds of calories. Ask a chemist and you will learn that a calorie (with a lowercase "c") is the amount of energy (heat) required to raise 1 gr (1 ml) of water by 1°C. A nutritionist, on the other hand, would describe a "big C" or kilocalorie (kcal)—the amount of energy required to raise 1 kg (1 L) of water by 1°C. The kcal is the unit used to describe the energy value in food. For example, if a chocolate chip cookie is completely incinerated, the amount of heat energy released would be enough to raise the temperature of 1 L of water by approximately 300°C! However, as this system adheres to the laws of thermodynamics, the conversion is not totally efficient (only about 25 percent of energy actually performs cellular work).

What is the average BMR of humans?

The average male has a BMR of 1,600–1,800 kilocalories (kcal) per day; the BMR of the average female is 1,300–1,500 kcal per day. This is the same energy consumption as a 100 watt light bulb.

Can **humans make** all the **lipids** they need?

No. There are two types of fatty acids that we need but can't make: linoleic acid, an omega-6 fatty acid, and linolenic acid, which is also known as an omega-3 fatty acid. These acids are used to maintain cell membranes and to build hormone-like messengers known as eicosanoids. Eicosanoids are twenty carbon fatty acids—examples include leukotrienes and prostaglandins. Good sources of omega-3 and omega-6 lipids are nuts, grains, and both fats and oils from vegetables or fish.

Why are **carbohydrates** a major part of the **human diet**?

Our cells have evolved in a manner that has made carbohydrates the human body's primary energy source; in fact, the entire metabolic system begins with glucose. While most of the cells in our bodies can adapt, at least temporarily, to using lipids and proteins as energy sources, our brain cells require glucose. This means that when blood glucose levels fall too low, the brain cells shut down and we faint. By fainting, the flow of blood and glucose to the brain is increased.

What is so **good** about **fish oil**?

Fish oils, extracted from species such as mackerel, salmon, anchovy, sardines, and tuna, are excellent sources of the essential omega-3 fatty acids—however, they are not the only possible source. Nuts and seeds also provide omega-3 fatty acids. Nutritionists recommend that these fatty acids be acquired through a healthy and diverse diet rather than by taking nutrient supplements.

Why do you need **protein** in your **diet**?

Of the twenty amino acids used by humans as building blocks for proteins, eight are essential. In other words, our bodies cannot synthesize them, nor can we survive without them. Luckily, all of the essential amino acids can be acquired by eating animal meat (complete proteins) and/or certain plant sources (complementary proteins). By combining grains, such as rice, and legumes, such as beans and peas, one is able to consume adequate daily amounts of protein. The recommended daily allowance (RDA) of protein for healthy adults is 0.4 g/lb (0.8 g/ kg) of body weight.

What are some **diseases** associated with **inadequate dietary intake** of **protein**?

There are several diseases associated with insufficient protein intake. Marasmus, a condition caused by chronic malnutrition, occurs most often among children six to eighteen months of age. The arms and legs of children with marasmus are described as having a "matchstick-like" appearance. Kwashiorkor, another disease caused by malnutrition, derives its name from a Ghanian phrase that describes the change a first-born child often undergoes when a second child is born. Kwashiorkor sufferers typically consume enough calories but are protein deficient—usually a result of being weaned at too early an age. Kwashiorkor causes an enlargement of the liver and a buildup of fluid within the body, particularly in the belly. Children with kwashiorkor tend to have plump-looking limbs and distended bellies, but the deceptive enlargement is caused by swelling, not body fat.

Fish oil from salmon and other fishes is an excellent source of omega-3 fatty acids.

What is a **better source of energy**—fat or sugar?

Fats contain 9 kcal/g, while carbohydrates average about 4 kcal/g. Fats are useful for storing energy, but sugars are easily broken down during metabolism. Nutritionists and diet gurus are debating which source of energy is better. The absolute answer lies in individual situations. For example, babies require fat in their diets for both energy and to build a healthy nervous system. Middle-aged adults quite often take in too much fat for their metabolic requirements. As one ages, average energy needs decrease by 5 percent per decade.

How **much cholesterol** is **needed** in your diet?

None. The body is capable of making all the cholesterol it needs. Almost all cells can synthesize some cholesterol, but most require more to maintain plasma membranes. This additional cholesterol, which is found in the body, is synthesized by the liver. All cell membranes contain cholesterol, and it is also used in producing bile salts and steroid hormones.

What is **dietary fiber**?

Dietary fiber is a type of carbohydrate that cannot be broken down by digestive enzymes. Because of this, the fiber passes through the digestive tract more quickly, aiding in elimination. The

This child's distended belly is evidence of malnutrition.

term "dietary fiber" includes the cellulose found in plant cell walls and the chitin that makes up the support tissues of fungi (mushrooms), crustaceans, and insects.

Who was **Linus Pauling**?

Linus Carl Pauling (1901–1994)—the only person to win two, unshared Nobel Prizes (1954 prize in Chemistry; 1962 Peace Prize)—revolutionized the study of chemistry, aided the development of molecular biology, and made important advances in medical research. He was responsible for determining the molecular basis of sickle-cell anemia; was one of the scientists engaged in the race to determine the molecular structure of DNA; and became a strident advocate for the use of large doses of vitamin C to prevent illness.

25

How are **fats used** in the **body**?

Function	Specific structures
Cell structure	Cell membranes use phospholipids and cholesterol for membrane structure and stability.
Energy storage	Triglycerides stored in adipose tissue.
Insulation	Protects body from changes in ambient temperature. Promotes signal conduction in the nervous system (similar to electrical wire insulation).
Messengers	Steroid hormones. HDL and LDL carry cholesterol through blood stream.

What happens to the **proteins** present in an **egg** when it is **cooked**?

The "white" of an egg is rich in the protein albumin. When subjected to high heat, the bonds that form the three-dimensional structure of albumin are irreversibly changed. This causes the clear, jellylike consistency of the egg to become firm and white. This process is known as protein denaturation. Proteins may also be denatured by a change in the pH value of the fluid surrounding them or by the addition of chemical detergents. Most denaturation is irreversible; however, scientists can use partial or temporary denaturation to control the action of enzymes used in experiments.

What happens to **proteins** during a **hair perm**?

During a hair perm, the bonds that form the secondary and tertiary levels of structure in the protein keratin (the major protein of hair) are chemically broken and then reformed; hydrogen bonds are primarily involved in this process. The same principle, breaking and forming bonds, is used to make hair either very curly or very straight.

Why are **fevers dangerous**?

Fever is usually an indication of bacterial or viral infection. While there is still debate about whether fevers actually speed up the body's inflammatory response to infection, there appears to be no clinical evidence that reducing fevers slows healing. Approximately 4 percent of children between the ages of three months and five

Linus Pauling made important contributions in chemistry and molecular biology.

years experience febrile seizures. It has been commonly thought that high fevers in children could cause denaturation of central nervous proteins, leading to seizures and possibly brain damage. However, the truth of that theory has not been fully determined.

What happens to **proteins** as we **age**?

A variety of molecular changes typify aging. For example, there is a slowdown in production of the proteins (collectively known as collagen) that maintain the skin integrity. As the amount of collagen in skin decreases, the skin loses elasticity and begins to sag, bag, and wrinkle. The loss of elasticity can be demonstrated by gently pinching a fold of skin on the back of the hand of differently aged people. The skin of young children will immediately return to form when released, while that of an elderly person will take longer.

How is **energy stored** in the **body**?

Source	Storage site
Carbohydrates	Glycogen (approx 500 gms in average human) in liver and skeletal muscles.
Lipids	Adipose tissue. Healthy adult males have 12–18 percent body fat; healthy adult females have 12–25 percent body fat.
Protein	Throughout the body; last choice as energy source.

What is the **relationship** between **metabolism** and **life span**?

Several studies have concluded that a program maintaining proportions of carbohydrates, lipids, and proteins, but reducing the consumption of calories, can increase longevity. However, these conclusions have only been demonstrated in tropical guppyfish and the worm *Caenorhabditis elegans*. To date, there have been no large clinical studies that have demonstrated the value of this regimen for humans.

How is **alcohol produced**?

The type of alcohol found in alcoholic beverages is known as ethanol. Ethanol is produced by yeast through the process known as alcoholic fermentation. During the breakdown of glucose to harvest energy, cells generate ethanol as a way to recycle a molecule that is crucial to metabolism. Interestingly, yeast can only survive in media (like beer or wine) that contain less than 10 percent alcohol by volume. This means that beverages with an alcohol content greater than 10 percent have been fortified with additional alcohol or have been distilled. Distillation involves the boiling off and then the condensation of the ethanol, which evaporates at a lower temperature than water.

What effect does **aspirin** have on **enzymes**?

Aspirin is probably the most common over-the-counter medication. Aspirin blocks the production of cyclooxygenase (COX) enzymes. COX-1 and COX-2 are two important enzymes with different functions: COX-1 catalyzes the biosynthesis of hormones important in maintaining the stomach lining; COX-2 catalyzes the biosynthesis of chemicals that promote inflammation, fever, and pain when an injury occurs. The positive effects of aspirin (pain and inflammation reduction) are due to the blocking of COX-2 enzymes. However, the negative aspects of aspirin (stomach problems) are due to the blocking of COX-1 enzymes.

How can **spinach neutralize the explosive trinitrotoluene (TNT)**?

TNT is a dangerous explosive. There are more than half a million tons of TNT stockpiled throughout the United States. Spinach contains a powerful enzyme called nitroreductase that is able to neutralize TNT by converting it to other compounds that are less dangerous. Through additional reactions, these less-harmful compounds can be converted to carbon dioxide gas. This enzymatic process is of great interest to the United States military.

Is there a **chemistry** to **love**?

Neurochemicals are released in the physical bodies of two people that are attracted to each other, so one can say that there is indeed a chemical reaction between two peo-

How is chocolate related to marijuana?

In 1992 a chemical called anandamide was discovered in chocolate. Anandamide is a messenger molecule that binds to the same receptors that bind marijuana. This action helps alleviate pain and aids in relaxation.

Researchers think that the presence of anandamide in chocolate may explain why it is by far the most-craved food.

ple. The neurochemicals involved are phenylethylamine (also found in chocolate), which speeds up the flow of information between nerve cells; dopamine, which makes us feel good; and norepinephrine, which makes the heart race.

Why does **eating chocolate** make some people **feel good**?

Chocolate contains over 300 known chemicals, some of which can alter mood, such as caffeine. Chocolate also contains a small amount of the chemical phenylethylamine (PEA), a stimulant to the nervous system that makes people feel more alert and gives a sense of overall well-being.

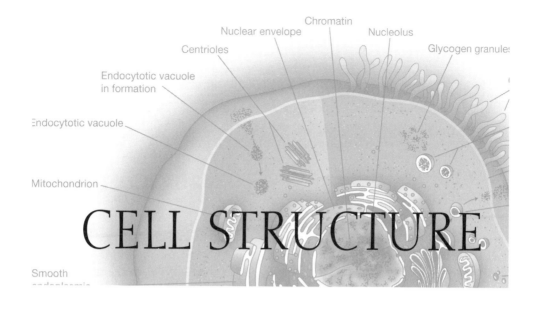

Chromatin

Nuclear envelope

Nucleolus

Centrioles

Glycogen granules

Endocytotic vacuole
in formation

Endocytotic vacuole

Mitochondrion

Smooth

CELL STRUCTURE

INTRODUCTION AND
HISTORICAL BACKGROUND

What is a **cell**?

A cell is a membrane-bound unit that contains hereditary material (DNA) and cyto-plasm; it is the basic structural and functional unit of life.

What is the **cell theory**?

The cell theory is the concept that as all living things are made up of essential units called cells, they are the fundamental components of all life. The cell is the simplest collection of matter that can live. There are diverse forms of life existing as single-celled organisms. More complex organisms, including plants and animals, are multi-cellular cooperatives composed of diverse specialized cells that could not survive for long on their own. All cells come from preexisting cells and are related by division to earlier cells that have been modified in various ways during the long evolutionary history of life on Earth. Everything an organism does occurs fundamentally at the cellular level.

When were **cells first studied**?

The scientific study of cells developed gradually from the first description of cells in the seventeenth century. During this early period, investigations of cells were done using the primitive light microscopes available at the time and consisted almost entirely of morphological descriptions of cell structure. In the eighteenth and nine-teenth centuries research expanded to include the study of cell chemistry and physiol-

Is a chicken egg a single cell?

There is only one set of cellular structures (nucleus, cytoplasm, and cellular membrane) within the shell of a chicken egg. Therefore, it is referred to as a single cell. However, because of the abundance of yolk and albumin, which are nutrients for the developing embryo (the chick), this cell is far larger than a normally functioning cell within a chicken or any other organism could be.

ogy, efforts that proceeded independently from morphological studies. The study of cell structure, cell chemistry, and cell physiology continued as separate fields of experimentation until the beginning of the twentieth century, when the rapidly developing field of biochemistry began to influence cell biology. At the same time, genetics became established as a field of investigation.

What is **cytology**?

Cytology is the study of cellular structure based on microscopic techniques. Cytology became a separate branch of biology in 1892, when the German embryologist Oscar Hertwig (1849–1922) proposed that organismic processes are reflections of cellular processes.

Can **cells** be **seen without** a **microscope**?

Yes, but not many. Most cells are far smaller than the period at the end of this sentence. Bird and frog egg cells may be somewhat larger and can in fact be observed by the unaided eye.

What **scientists** made **important early discoveries** associated with the **cell**?

In 1665 Robert Hooke (1635–1703), curator of instruments for the Royal Society of London, was the first to see a cell, initially in a section of cork, and then in bones and plants. In 1824 Henri Dutrochet (1776–1847) proposed that animals and plants had similar cell structures. Robert Brown (1773–1858) discovered the cell nucleus in 1831, and Matthias Schleiden (1804–1881) named the nucleolus (the structure within the nucleus now known to be involved in the production of ribosomes) around that same time. Working independently, Schleiden and Theodor Schwann (1810–1882) described preliminary forms of the general cell theory in 1839, the former stating that cells were the basic unit of plants and Schwann extending the idea to animals. In 1855 Robert Remak (1815–1865) became the first to describe cell division. Shortly after Remak's discovery, Rudolph Virchow (1821–1902) stated that all cells come from preexisting cells. The work of Schleiden, Schwann, and Virchow firmly established the cell theory. In 1868 Ernst Haeckel (1834–1919) proposed that the nucleus was responsible for heredity. Chromosomes were named and observed in the nucleus of a cell in 1888 by Wilhelm von Waldeyen-Hartz (1836–1921). Walther Flemming

> ## Who was the first to propose the terms "prokaryotic" and "eukaryotic" to describe cells?
>
> The French marine biologist Edouard Chatton (1883–1947) proposed the terms *procariotique* (prokaryotic) and *eucariotique* (eukaryotic) in 1937. Prokaryotic, meaning "before nucleus" was used to describe bacteria and eukaryotic meaning "true nucleus" was used to describe all other cells.

(1843–1905) was the first individual to follow chromosomes through the entire process of cell division.

What is the **origin** of the **term "cell"**?

The term "cell" was first used by Robert Hooke (1635–1703), an English scientist who described cells he observed in a slice of cork in 1665. Using a microscope that magnified thirty times, Hooke identified little chambers or compartments in the cork that he called *cellulae*, a Latin term meaning "little rooms" because they reminded him of the cells inhabited by monks. It is from this word that we got the modern term "cell." He calculated that one square inch of cork would contain 1,259,712,000 of these tiny chambers or "cells"!

PROKARYOTIC AND EUKARYOTIC CELLS

What are the **differences** between **prokaryotic** and **eukaryotic cells**?

Eukaryotic cells are much more complex than prokaryotic cells, having compartmentalized interiors and membrane-bound organelles within their cytoplasm. The major feature of a eukaryotic cell is its membrane-bound nucleus, which compartmentalizes the activities of the genetic information from other types of cellular metabolism.

Comparison of Prokaryotic and Eukaryotic Cells

Characteristic	Prokaryotic Cells	Eukaryotic Cells
Organisms	Eubacteria and Archaebacteria	Protista, Fungi, Plants, Animals
Cell size	Usually 1–10 µm across	Usually 10–100 µm across
Membrane-bound organelles	No	Yes
Ribosomes	Yes	Yes
Mode of cell division	Cell fission	Mitosis and meiosis
DNA location	Nucleoid	Nucleus

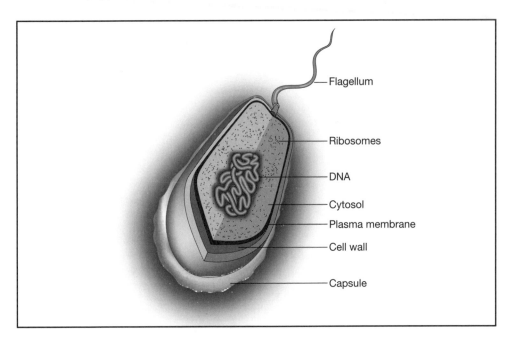

Generalized prokaryotic cell.

Characteristic	Prokaryotic Cells	Eukaryotic Cells
Membranes	Some	Many
Cytoskeleton	No	Yes

What groups of **organisms** have **prokaryotic cells** and which ones have **eukaryotic cells**?

All living organisms are grouped into three large groups called domains. They are: Bacteria, Archae, and Eukarya. The domains Bacteria (eubacteria or "true" bacteria) and Archae (archaebacteria or "ancient" bacteria) consist of unicellular organisms with prokaryotic cells. The domain Eukarya consists of four kingdoms: Protista, Fungi, Plantae, and Animalia. Organisms in these groups have eukaryotic cells. Eukaryon means "true nucleus."

How do the **cells** of **bacteria, plants** and **animals** compare to each other?

	Bacterium (Prokaryote)	Plant (Eukaryote)	Animal (Eukaryote)
Exterior structures			
Cell wall	Present (protein polysaccharide)	Present (cellulose)	Absent
Plasma membrane	Present	Present	Present

How do prokaryotic cells compare in size to eukaryotic cells?

Eukaryotic cells are generally much larger and more complex than prokaryotic cells. Most eukaryotic cells are 100 to 1,000 times the volume of typical prokaryotic cells.

	Bacterium (Prokaryote)	Plant (Eukaryote)	Animal (Eukaryote)
Flagella and cilia	May be present	Absent except in sperm of a few species	Frequently present
Interior structures			
Endoplasmic reticulum	Absent	Usually present	Usually present
Ribosome	Present	Present	Present
Microtubule	Absent	Present	Present
Centriole	Absent	Absent	Present
Golgi apparatus	Absent	Present	Present
Cytoskeleton	Absent	Present	Present
Other organelles			
Nucleus	Absent	Present	Present
Mitochondrion	Absent	Present	Present
Chloroplast	Absent	Present	Absent
Nucleolus	Absent	Present	Present
Chromosone	A single circle of naked DNA	Multiple; DNA-protein complex	Multiple: DNA-protein complex
Microbody	Absent	Present	Present
Lysosome	Absent	Absent	Present
Vacuole	Absent	Usually a large single vacuole	Absent

What is the **chemical composition** of a typical **mammalian cell**?

Molecular component	Percent of total cell weight
Water	70
Proteins	18
Phospholipids and other lipids	5
Miscellaneous small metabolites	3

Molecular component	Percent of total cell weight
Polysaccharides	2
Inorganic ions (sodium, potassium, magnesium, calcium, chlorine, etc.)	1
RNA	1.1
DNA	0.25

ORGANIZATION WITHIN CELLS

How many classes of **internal membrane-bound structures** are found in **eukaryotic** cells?

There are two classes of internal membrane-bound structures in eukaryotic cells. There are discrete organelles such as mitochondria, chloroplasts, and peroxisomes; then there is the dynamic endomembrane system—nuclear membrane, endoplasmic reticulum, Golgi apparatus, lysosomes, and vacuoles.

What are **organelles**?

Organelles—frequently called "little organs"— are found in all eukaryotic cells; they are specialized, membrane-bound, cellular structures that perform a specific function. Eukaryotic cells contain several kinds of organelles, including the nucleus, mitochondria, chloroplasts, endoplasmic reticulum, and Golgi apparatus.

What are the major **components** of the **eukaryotic cell**?

Structure	Description
Cell Nucleus	
Nucleus	Large structure surrounded by double membrane
Nucleolus	Special body within nucleus; consists of RNA and protein
Chromosones	Composed of a complex of DNA and protein known as chromatin; resemble rodlike structures after cell division
Cytoplasmic Organelles	
Plasma membrane	Membrane boundary of living cell
Endoplasmic reticulum (ER)	Network of internal membranes extending through cytoplasm

What are the three major parts of all cells?

All cells have three major features in common: a plasma membrane, cytoplasm, and a nucleoid or nucleus.

Structure	Description
—Smooth endoplasmic reticulum	Lacks ribosomes on the outer surface
—Rough endoplasmic reticulum	Ribosomes stud outer surface
Ribosomes	Granules composed of RNA and protein; some attached to ER and some are free in cytosol
Golgi complex	Stacks of flattened membrane sacs
Lysosomes	Membranous sacs (in animals)
Vacuoles	Membranous sacs (mostly in plants, fungi, and algae)
Microbodies (e.g., peroxisomes)	Membranous sacs containing a variety of enzymes
Mitochondria	Sacs consisting of two membranes; inner membrane is folded to form cristae and encloses matrix
Plastids (e.g., chloroplasts)	Double membrane structure enclosing internal thylakoid membranes; chloroplasts contain chlorophyll in thylakoid membranes
The Cytoskeleton	
Microtubules	Hollow tubes made of subunits of tubulin protein
Microfilaments	Solid, rod-like structures consisting of actin protein
Centrioles	Pair of hollow cylinders located near center of cell; each centriole consists of nine microtubule triplets (9 x 3 structure)
Cilia	Relatively short projections extending from surface of cell; covered by plasma membrane; made of two central and nine peripheral microtubules (9 + 2 structure)
Flagella	Long projections made of two central and nine peripheral microtubules (9 + 2 structure); extend from surface of cell; covered by plasma membrane

What is the **cytosol**?

The cytosol is the semi-fluid substance of the cell in which the organelles are suspended.

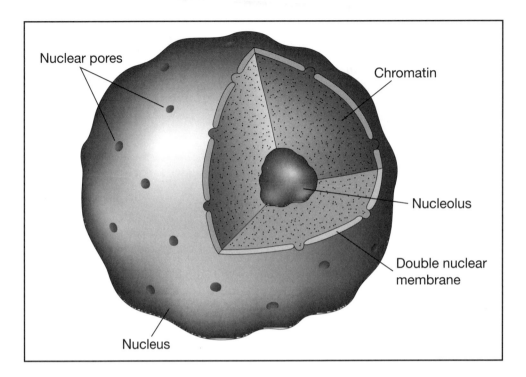

The nucleus of a cell.

NUCLEUS

What is the **major function** of the **nucleus**?

The nucleus is the information center for the cell and the storehouse of the genetic information (DNA) that directs all of the activities of a living eukaryotic cell. It is usually the largest organelle in a eukaryotic cell and contains the chromosomes.

When was the **nucleus first described**?

The Scottish botanist Robert Brown (1773–1858) first named and described the nucleus in 1831, while studying orchids. Brown called this structure the nucleus, from the Latin *nucula*, meaning "little nut" or "kernel."

What are the main **components** of the **nucleus**?

The boundary around the nucleus consists of two membranes (an inner one and an outer one) that form the nuclear envelope. Nuclear pores are small openings in the nuclear envelope that permit molecules to move between the nucleus and the cytoplasm. The nucleolus is a prominent structure within the nucleus. The nucleoplasm

How much DNA is in a typical human cell?

If the DNA in a single human cell were stretched out and laid end-to-end, it would measure approximately 6.5 ft (2 m). The average human body contains 10 to 20 billion mi (16 to32 billion km) of DNA distributed among trillions of cells. If the total DNA in all the cells from one human was unraveled, it would stretch to the Sun and back more than 600 times.

is the interior space of the nucleus. In addition, the DNA-bearing chromosomes of the cell are found in the nucleus. The nucleus is the repository for the cell's genetic information and the control center for the expression of that information.

Why is **DNA important**?

DNA is the chemical inside the nucleus of a cell that carries the genetic instructions for making living organisms.

When was **DNA first discovered**?

DNA was first discovered in 1869 by Johann Friedrich Miescher (1844–1895). He isolated and described what he called "nuclein," which eventually became known as deoxyribonucleic acid (DNA), the genetic information of the cell.

How is **DNA organized** in the **nucleus**?

Within the nucleus DNA is organized with proteins into a fibrous material called chromatin. As a cell prepares to divide or reproduce, the thin chromatin fibers condense, becoming thick enough to be seen as separate structures, which are called chromosomes.

How is **DNA related** to the **size** of a **cell**?

The amount of DNA that a cell must accommodate is significant, even for organisms with genomes of modest size. There is enough DNA in the typical *E. coli* cell to encircle it more than 400 times. A typical human cell contains enough DNA to wrap around the cell more than 15,000 times.

Which **organelles** contain **DNA**?

DNA is found in nuclei, mitochondria, chloroplasts, and some centrioles.

Do **all cells** have a **nucleus**?

Prokaryotic cells do not have an organized nucleus. Most eukaryotic cells have a single organized nucleus. Red blood cells are the only mammalian cells that do not have a nucleus.

Why does the **size of the nucleolus vary** in cells?

Cells with a high rate of protein synthesis have a large number of ribosomes. In these active cells, nucleoli tend to be large and can account for 20 to 25 percent of the nuclear volume.

What is the **function** of the **nucleolus**?

The nucleolus is a large, spherical structure present in the nucleus of a eukaryotic cell. It is the site where ribosome subunits are assembled, and where both ribosomal RNA synthesis and processing occur.

How many **nuclear pores** are there in a nucleus?

A typical nucleus in a mammalian cell has 3,000 to 4,000 pores or 10 to 20 pores per square micrometer (μm). Nuclear pores are not merely perforations, but channels composed of more than one hundred different proteins.

What is a **chromosome**?

A chromosome is the threadlike part of a cell that contains DNA and carries the genetic material of a cell. In prokaryotic cells chromosomes consist entirely of DNA and are not enclosed in a nuclear membrane. In eukaryotic cells the chromosomes are found within the nucleus and contain both DNA and RNA.

When were **chromosomes first observed**?

Chromosomes were observed as early as 1872, when Edmund Russow (1841–1897) described seeing items that resembled small rods during cell division; he named the rods "Stäbchen." Edouard van Beneden (1846–1910) used the term *bâtonnet* in 1875 to describe nuclear duplication. The following year, 1876, Edouard Balbiani

> ## Which organism has the largest number of chromosomes?
>
> **O**phioglossum reticulatum, a species of fern, has the largest number of chromosomes with more than 1,260 (630 pairs).

(1825–1899) described that at the time of cell division the nucleus dissolved into a collection of *bâtonnets étroits* ("narrow little rods"). Walther Flemming (1843–1905) discovered that the chromosomal "threads" or *Fäden* split longitudinally during mitosis.

Which organism has the **least number** of **chromosomes**?

The organism with the least number of chromosomes is the male Australian ant, *Myrmecia pilosula*, with one chromosome per cell. Male ants are generally haploid—that is, they have half the number of normal chromosomes—while the female ant has two chromosomes per cell. Bacteria have one circular chromosome consisting of DNA and associated proteins.

What are **telomeres** and where are they located?

Telomeres are protective structures composed of DNA that contain multiple repetitions of one short nucleotide sequence instead of genes. They are found at each end of a eukaryotic chromosome.

CYTOPLASM

Which **organelle divides** and **compartmentalizes** the cell?

The endomembrane system, which fills the cell and divides it into compartments, and is visible only through electron microscopy. The endoplasmic reticulum (ER), a series of interconnected membranous tubes and channels in the cytoplasm, is the largest and most extensive system of the internal membranes.

What is the **function** of the **endomembrane system**?

The endomembrane system allows macromolecules to diffuse or be transferred from one of the components of the system to another.

What is the **difference** between **rough** and **smooth endoplasmic reticulum**?

Rough endoplasmic reticulum are regions rich in ribosomes that manufacture proteins. These regions appear to have a pebbly surface that is somewhat similar to sandpaper. Regions with no ribosomes are referred to as smooth endoplasmic reticulum. The smooth endoplasmic reticulum is involved in the synthesis of lipids and steroids, carbohydrate metabolism, and the inactivation and detoxification of drugs and other components that may be toxic or harmful to the cell.

What is the **function** of the **Golgi apparatus**?

The Golgi apparatus (frequently called the Golgi body) is a collection of flattened stacks of membranes. It serves as the packaging center for cell products. It collects materials at one place in the cell, and packages them into vesicles for use elsewhere in the cell or transportation out of the cell.

Who **discovered** the **Golgi apparatus**?

In 1898 Camillo Golgi (1843–1926), an Italian physician, first described an irregular network of small fibers, cavities, and granules in nerve cells. It was not until the 1940s, and the invention of the electron microscope, that the existence of the Golgi apparatus was confirmed. In 1906 Golgi and Santiago Ramón y Cajal (1852–1934) were awarded the Nobel Prize for Physiology or Medicine for their investigations into the fine structure of the nervous system.

How **many Golgi bodies** are in a **cell**?

Protists contain one or a small number of Golgi bodies. Animal cells may contain twenty or more Golgi bodies, while plant cells may contain several hundred Golgi bodies.

What are the two major **types** of **vesicles** in **eukaryotic** cells?

Vesicles are small, intracellular membrane sacs usually formed from the Golgi apparatus. Lysosomes and microbodies are examples of vesicles found in eukaryotic cells.

What is the **function** of **lysosomes**?

Lysosomes are single, membrane-bound sacs that contain digestive enzymes. The digestive enzymes break down all the major classes of macromolecules including proteins, carbohydrates, fats, and nucleic acids. Throughout a cell's lifetime, the lysosomal enzymes digest old organelles to make room for newly formed organelles. The lysosomes allow cells to continually renew themselves and prevent the accumulation of cellular toxins.

Who **discovered lysosomes**?

Lysosomes are a relatively modern discovery in cell biology; they were observed by
Christian de Duve (1917–)in the early 1950s. In 1955, after six years of experi-

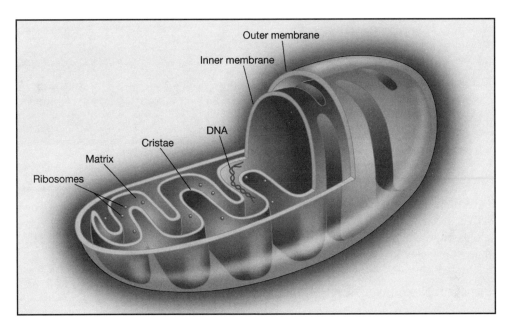

The structure of a mitochondrion.

ments, de Duve was convinced that he had found an organelle that had not been previously described and was involved in intracellular lysis (digestion). He named the organelle a lysosome. This organelle was the first to be described entirely on biochemical criteria. The results were later verified using electron microscopy. In 1974 De Duve, Albert Claude (1898–1983), and George Palade (1912–) shared the Nobel Prize in Physiology or Medicine for their work detailing the functions of the lysosome.

What is a **peroxisome** and what is its **function** in a cell?

Peroxisomes, also discovered by Christian de Duve (1917–), are surrounded by a single membrane and are the most common type of microbody in cells. They are especially prominent in algae, the photosynthetic cells of plants, and both mammalian kidney and liver cells. Peroxisomes contain detoxifying enzymes and produce catalase, which breaks down hydrogen peroxide into hydrogen and water.

Why are **ribosomes** an **important** organelle?

Ribosomes, one of the most complex aspects of the molecular machine, are the site of protein synthesis in a cell. They consist of a large and small subunit composed of ribosomal RNA and protein. However, compared with membrane-bound organelles, ribosomes are tiny structures!

Where are **ribosomes** found in a **cell**?

Ribosomes are found in the cytoplasm of both prokaryotic and eukaryotic cells, as well as in the matrix of mitochondria and the stroma of chloroplasts. In eukaryotic cytoplasm, ribosomes are found in the cystol, and are bound to the endoplasmic reticulum as well as the outer membrane of the nuclear envelope.

How are **ribosomes different** from other **organelles**?

Ribosomes differ from most other organelles in not being bound by a membrane.

Are there **differences** between **prokaryotic** and **eukaryotic ribosomes**?

Prokaryotic and eukaryotic ribosomes resemble each other structurally, but they are not identical. Prokaryotic ribosomes are smaller in size, contain fewer proteins, have smaller RNA molecules, and are sensitive to different inhibitors of protein synthesis.

What are **mitochondria**?

A mitochondrion (singular) is a self-replicating, double-membraned body found in the cytoplasm of all eukaryotic cells. The outer membrane of a mitochondrion is smooth, while the inner membrane is folded into numerous layers that are called cristae. Mitochondria are the location for much of the metabolism necessary for protein synthesis, and the production of both ATP and lipids.

When were **mitochondria discovered**?

In 1857 Rudolf Albert von Kölliker (1817–1905), histologist and embryologist, first described "sarcosomes" (now called mitochondria) in muscle cells. The term "mitochondrion" (meaning "threadlike granule") was first used in 1898. Functionally active mitochondria were first isolated in 1948. Kölliker was among the first biologists to interpret tissue structure in terms of cellular elements.

How **many mitochondria** are there in a cell?

The number of mitochondria varies according to the type of cell. The number ranges between one and 10,000, but averages about 200. Each cell in the human liver has

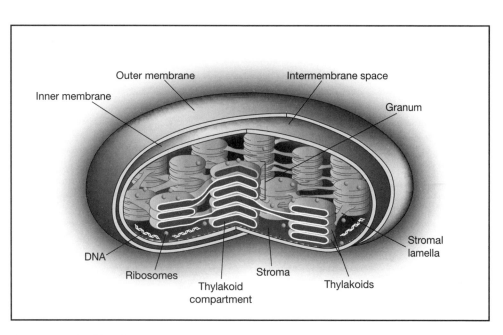

The structure of a chloroplast.

over 1,000 mitochondria. Cells with high energy requirements, such as muscle cells, may have many more mitochondria.

What is the **function** of **chloroplasts**?

Chloroplasts are the structural and functional units where photosynthesis takes place—the process whereby green plants use light energy for the synthesis of organic molecules from carbon dioxide and water, with oxygen released as a by-product. They contain the green pigment chlorophyll, which traps light energy for photosynthesis.

When were **chloroplasts first described**?

Because chloroplasts are a large cell structure (bigger than any other organelle except the nucleus), they were described and studied early in the history of cytology. Anton van Leeuwenhoek (1632–1723) and Nehemiah Grew (1641–1712) described these organelles in the seventeenth century, in the early stages of the microscopic study of cells.

Do all **plant cells** contain **chloroplasts**?

No, not all plant cells contain chloroplasts. The various types of plant cells arise from meristem, rapidly dividing and undifferentiated tissue. Meristem cells do not contain chloroplasts, but have smaller organelles called proplastids. Depending on their location in a plant and how much light they receive, proplastids develop into one of sever-

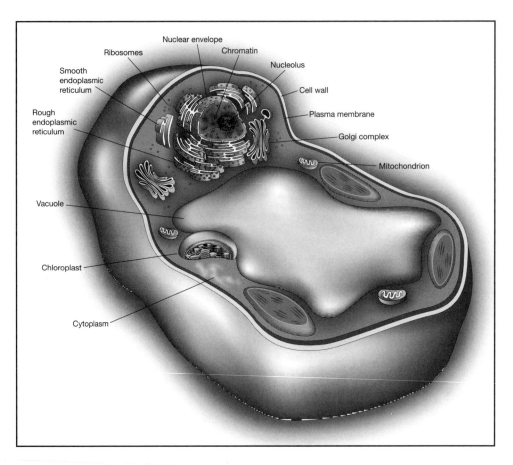

A generalized plant cell.

al kinds of plastids with different functions. Chloroplasts are one example of a plastid that converts light energy to chemical energy, subsequently used in synthesizing organic molecules—the process of photosynthesis.

What are the **different types** of **plastids?**

Proplastids can differentiate into several types of plastids that are involved in cellular storage. Amyloplasts store starch, proteinoplasts store proteins, and elaioplasts store lipids. In addition, some proplastids develop into chromoplasts.

What are the main **components** of **chloroplasts**?

Chloroplasts have outer and inner membranes, which are in close association with each other. They also have a closed compartment of stacked membranes—called grana—that lie inside the inner membrane. A chloroplast may contain one hundred or more grana, and each granum may contain a few to several dozen disk-shaped structures called thylakoids that contain chlorophyll on their surface. The fluid that surrounds the thylakoid is called the stroma.

How many **chloroplasts** are in **plant cells**?

Unicellular algae may only have one large chloroplast, whereas a plant leaf cell may have between 20 and 100.

How do **chromoplasts differ** from **chloroplasts**?

Chromoplasts are pigment-containing plastids found in plants. They are responsible for the characteristic red, orange, or yellow coloration visible in some flowers, fruits, and plants.

What is the **cytoskeleton** and what is its **function**?

The cytoskeleton is a structural feature of eukaryotic cells that was revealed by advanced microscopy. It consists of an extensive three-dimensional network of interconnected filaments and tubules that extends throughout the cytosol, from the nucleus to the inner surface of the cell membrane. These filaments and tubules determine cell shape and facilitate a variety of cell movements.

What are the three **types of fibers** found in the **cytoskeleton** of **eukaryotic** cells?

The cytoskeleton is a network of fibers that maintains the shape of cells. The three types of fibers are actin filaments, microtubules, and intermediate filaments. Actin filaments are long fibers composed of two protein chains. They are responsible for cellular movements, such as contraction, crawling, "pinching" during division, and formation of cellular extensions. Microtubules are hollow tubes composed of a ring of thirteen protein filaments. They are responsible for moving materials within the cell. Intermediate filaments are tough, fibrous protein molecules structured in an overlapping arrangement. They are intermediate in size when compared to actin filaments and microtubules, and provide structural stability to cells.

CELL WALL AND PLASMA MEMBRANE

What groups of **organisms** have a **cell wall**?

A cell wall is present in organisms in the kingdoms Archaebacteria, Eubacteria, Protista, Fungi, and Plantae. Animals are the only organisms that do not have a cell wall.

What are the main components of the **cell wall** in **plants**?

The cell wall is one of the features of plant cells that distinguishes them from animal cells. The wall protects the plant cell and maintains its shape. The cell wall consists mainly of cellulose microfibrils embedded in a matrix of protein and sugar polymers.

How does a **primary cell wall** differ from a **secondary cell wall**?

A primary cell wall is laid down during cell division, and is relatively thin and flexible in order to accommodate cell enlargement and elongation. It is strengthened when the cell matures and stops growing. A secondary cell wall is present between the plasma membrane and primary cell wall in some cells. The secondary wall is often deposited in several laminated layers. It is strong and durable, and provides both cell protection and support. Wood consists mainly of secondary cell walls.

Are **bacterial cell walls** different from **plant cell walls**?

Cell walls in prokaryotes (e.g., bacteria) and plants define the cell's shape and give rigidity to the cell. Unlike plant cell walls, bacterial cell walls consist mainly of peptidoglycans and not cellulose. Peptidoglycans are polysaccharide chains (amino sugars) cross-linked by small peptides.

What are the **functions** of the **cell membranes**?

Cell membranes define and compartmentalize space, regulate the flow of materials, detect external signals, and mediate interactions between cells.

What is the **structure** of the **plasma membrane**?

The plasma membrane is a thin membrane that surrounds and defines the boundaries all living cells. It consists of a double layer (bilayer) of phospholipids with various proteins attached to or embedded in it.

What **contributions** did **Overton** and **Langmuir** make to the study of **membranes**?

As early as the 1890s Charles Overton (1865–1933) was aware that cells seemed to be enveloped by a selectively permeable layer that allowed different substances to enter and leave cells at significantly different rates. He recognized that lipid-soluble substances penetrated readily into cells, whereas water-soluble substances did not. He concluded that lipids were present on the cell surface as some sort of a "coat." Irving Langmuir (1881–1957) proposed that that cells were covered by a lipid monolayer.

His work became the basis for further investigation into the membrane structure. Langmuir was awarded the Nobel Prize for Chemistry in 1932.

Who first proposed that the **cell membrane** is composed of a **bilayer structure of lipids**?

In 1925 Evert Gorter (1881–1954) and F. Grendel, two Dutch physiologists, hypothesized that there is a bilayer structure of lipids on the cell surface. Their work was significant because it was the first attempt to understand membranes at the molecular level.

Who first proposed a **model** for the **plasma membrane**?

Following the earlier work of Gorter and Grendel on cell membranes, Hugh Davson (1909–1996) and James F. Danielli (1911–1984) proposed a sandwich model for the structure of cell membranes in 1935. This model was a phospholipid bilayer between two layers of globular proteins. Since cell membranes are so fragile in vivo, one can only propose theoretical models for their structure. Current techniques still do not permit direct observation of the functioning of plasma membranes.

What is the **current model** of the **plasma membrane**?

The current model, frequently referred to as the fluid mosaic model, is based on work completed by Seymour J. Singer (1924–) and Garth L. Nicholson (1943–) in 1972. Their research revealed that the plasma membrane is a mosaic of integral proteins bobbing in a fluid bilayer of phospholipids. This pattern is not static because the positions of the proteins are constantly changing, moving about like icebergs in a sea of lipids. Peripheral proteins are not embedded in the lipid bilyaer but are appendages loosely bound to the membrane surface. Membrane carbohydrates on the surface function as cell markers to distinguish one cell from another. This model has been tested repeatedly and has been shown to accurately predict the properties of many kinds of cellular membranes; this structure has also been confirmed using a technique known as freeze-fracture electron microscopy.

What are the main **functions** of the **plasma membrane**?

The main purpose of the plasma membrane is to provide a barrier that keeps cellular components inside the cell while simultaneously keeping unwanted substances from entering the cell. The membrane allows essential nutrients to be transported into the cell and aids in the removal of waste products from the cell. The specific functions of a membrane depend on the kinds of phospholipids and proteins present in the plasma membrane.

What are the main **components** of the **plasma membrane**?

Component	Function
Cell surface markers	"Self"-recognition; tissue recognition
Interior protein network	Determines shape of cell
Phospholipid molecules	Provide permeability barrier, matrix for proteins
Transmembrane proteins	Transport molecules across membrane and against gradient.

What are **cell junctions**?

Cell junctions are the specialized connections between the plasma membranes of adjoining cells. The three general types of cell junctions are tight junctions, anchoring junctions, and communicating junctions. Tight junctions bind cells together, forming a barrier that is leak-proof. For example, tight junctions form the lining of the digestive tract, preventing the contents of the intestine from entering the body. Anchoring (or adhering) junctions link cells together, enabling them to function as a unit and forming tissue, such as heart muscle or the epithelium that comprises skin. Communicating (or gap) junctions allow rapid chemical and electrical communication between cells. They consist of channels that connect the cytoplasm of adjacent cells.

CILIA AND FLAGELLA

How are **cilia** and **flagella** similar?

Cilia and flagella are the motile appendages of eukaryotic cells. They are thick, flexible structures that exhibit a beating motion and project from the surfaces of many cells. A cell having one or a small number of appendages can be identified having flagella (singular, flagellum) if they are relatively long in proportion to the size of the cell. If the cell has many short appendages, they are called cilia (singular, cilium). Cilia and flagella have the same internal structure, but differ in their length, the number occurring per cell, and mode of beating. They are axonemal in shape, formed by a main cylinder of tubules that are approximately 0.25 μm in diameter. The axoneme has a pattern of "9 + 2" consisting of nine outer doublets of tubules and two additional central microtubules (the central pair). Cilia are about 2–10 μm long, while flagella are much longer, ranging from 1 μm to several millimeters, although they are most commonly 10–200 μm. They are both intracellular structures, bound by an extension

> ## What are the largest and smallest organelles in a cell?
>
> The largest organelle in a cell is the nucleus. The next largest organelle would be the chloroplast, which is substantially bigger than a mitochondrion. The smallest organelle in a cell is the ribosome.

of the plasma membrane. Both cilia and flagella are used by cells to move through watery environments or to move materials across cell surfaces.

How does the **motion** exhibited by **cilia** and **flagella differ** from one another?

Cilia move back-and-forth, causing their motion to be perpendicular to their axis of direction. Flagella undulate in a whip-like motion, moving in the same direction as their axis.

MISCELLANEOUS

Which **cell structures** are unique to **plant cells**?

The chloroplast, central vacuole, tonoplast, cell wall, and plasmodesmata are only found to occur in plant cells.

What are the **functions** of the **central vacuole**?

The central vacuole can encompass 80 percent or more of the cell. It is usually the largest compartment in a mature plant cell, and is surrounded by the tonoplast. The important functions it provides include storage, waste disposal, protection, and growth.

What are **plasmodesmata**?

Plasmodesmata are present in plant cells; they are channels or canals that occur in the cell wall, connecting the cytoplasms of adjacent cells. They allow molecules direct communication through adjacent cells. Plant cells are united into functioning tissues by plasmodesmata.

What **cell structures** are unique to **animal cells**?

Lysosomes and centrioles are found only in animal cells.

What is the **function** of **centrioles** in animal cells?

Centrioles both assemble and organize long, hollow cylinders of protein known as microtubules. They are also known as microtubule-organizing centers (MTOCs).

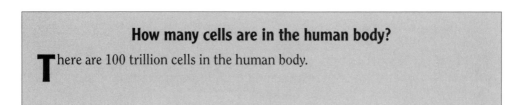

How many cells are in the human body?

There are 100 trillion cells in the human body.

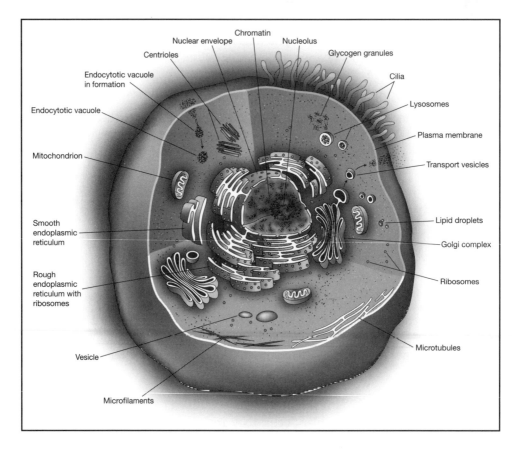

A generalized animal cell.

What are the **functions** of **microtubules**?

Microtubules help determine the cell shape, move chromosomes during cell division, and provide the internal structure of cilia and flagella.

Is there an average **life span** of a **cell**?

No! Most living cells do not live longer than a month. Even cells that live longer, such as liver and brain cells, constantly renew their components so that no part of the cell is really more than a month old. One example of a cell with a longer life span is the memory B cell. Memory B cells can survive for several decades—even a lifetime—providing an immune response to pathogens invading for a second time.

What is the **evolutionary relationship** between simple **prokaryotic** cells and the complex cells of **eukaryotes**?

Mitochondria and chloroplasts have prokaryote-like features. For example, although most of the DNA in eukaryotic cells resides in the nucleus, both mitochondria and chloroplasts have DNA molecules in their inner compartments. Mitochondrial and chloroplastic ribosomes are similar to the ribosomes of prokaryotes. The endosymbiont theory proposes that eukaryotic organisms evolved from prokaryotic ancestors; this idea supports the notion that organelles evolved from prokaryotic organisms that originally lived inside larger cells, eventually losing the ability to function as autonomous organisms.

CELL
FUNCTION

BASIC FUNCTIONS

Why are **cells** so **small**?

Cells come in a variety of sizes and shapes. Bacteria are among the smallest (0.2–0.3 μm in diameter), while cells of plants and animals are generally larger (10–50 μm in diameter). Cellular size is determined by the surface/volume ratio required to maintain access to the substances and enzymes that cells need to complete their functions. For further understanding of the surface/volume ratio, imagine that a cell is the shape of a cube: the volume of a cube increases as does the length or diameter, but the surface area only increases as the cube does. Once the upper limits of surface area/volume ratio are reached, the cell generates more cytoplasmic volume than can be met by the small increases in membrane surface area.

How do substances **move in** and **out** of **cells**?

Cells constantly transport substances across their cell membranes. Endocytosis (from the Greek *endo*, meaning "in ," and *cytosis*, meaning "cell") is the process by which cells bring molecules into their structure. Exocytosis (from the Greek *exo*, meaning "out," and *cytosis*) is the process by which cells transport materials out of their structure, across their own cell membrane. There are two different types of endocytosis: pinocytosis (from the Greek *pino*, meaning "to drink," and *cytosis*) and phagocytosis (from the Greek *phago*, meaning "to eat," and *cytosis*).

How do **cells drink**?

Cells drink by a process called pinocytosis, which is a form of endocytosis. During pinocytosis the cell membrane folds inward, forming a small pocket (vesicle) around fluid that is directly outside the cell membrane. Fluid consumed by cells may contain small molecules, such as lipids. Endothelial cells, which line capillaries, are constantly undergoing the process of pinocytosis, "drinking" from the blood within the capillaries.

How do **cells eat**?

Cells eat by a process known as phagocytosis, which is important to two types of cells: amoebas (unicellular protozoans) and mammalian white blood cells (macrophages, neutrophils). Phagocytosis is very important because it provides food for the protozoan. It also plays a critical role in defense systems by eliminating microorganisms or damaged cells in mammals. Once a particle (or microorganism) is ingested, it is wrapped within a vesicle, and then the vesicle fuses with a lysosome. The digestive enzymes of the lysosome then digest the contents of the vesicles.

How do **cells secrete substances**?

The release of material from a cell is known as exocytosis. First the cell forms the cell product and then packages it. The package, or vesicle, is comprised of the same material that makes up the cell membrane. When the vesicle reaches the membrane, the two structures merge together much like air bubbles do in liquid. The contents of the vesicle are then expelled from the cell. For example, secretory cells that manufacture specific proteins, such as the pancreatic cells that manufacture insulin, use the process of exocytosis to secrete insulin into the blood.

How do **cells move**?

Not all cells move; those that do move exhibit distinct body features and methods for inducing movement. Cells with flagella (such as sperm) move by whipping the flagellum back and forth. Several protists (e.g., paramecia) have cilia—which are much smaller than flagella—that produce movement by shifting through liquid in a way that is similar to how an oar can be used to propel a boat through water. Pseudopodia, often referred to as "fake feet," are cellular extensions caused by a stretching of the cell membrane. Pseudopodia are used by amoebae (protists) for movement and acquiring

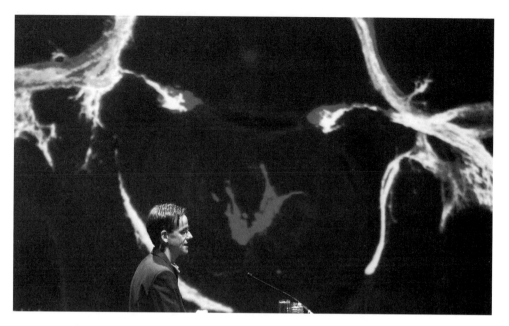

A stem cell researcher. The blobs on the image behind him are stem cells.

prey. Macrophages, white blood cells that play an important role in the immune system of human beings, use pseudopodia to attack and devour invading microbes.

What is a **stem cell**?

Stem cells are undifferentiated cells—meaning that they do not have a specific function—that are capable, under certain conditions, of producing cells that can become a specific type of tissue. Stem cells present in adult humans are found in bone marrow and other tissues (such as fat). However, most research interest is focused on stem cells present in fetal tissue, which, in a laboratory setting, can reproductively divide indefinitely, and be stimulated into becoming a variety of different cell types. The potential benefits of stem cells have made the research of these cells an exciting research topic.

What are the potential **uses** of **stem cells**?

Stem cells could be used to grow new hearts that could be transplanted without fear of rejection. They could be used to renew the function of injured structures like the spinal cord. They could be used as cell models for drug testing, thereby increasing the speed for finding cures.

What is the **cell cycle** and how is it **controlled**?

The life cycle of a single eukaryotic cell is known as the cell cycle. The cycle has two major phases: interphase and mitosis. When a cell is not dividing, it is in interphase;

57

for example, a mature neuron conducting an impulse is in interphase. Some cells remain in interphase almost indefinitely. Interphase includes the G_1, S, and G_2 phases, which are periods of growth during which a cell increases in size, complexity, and protein content. The G_1 phase prepares for DNA synthesis (known as the S phase), and G_2 prepares for both mitosis and the synthesis of proteins.

Cells that undergo mitotic phases of division and renewal are cyclically predictable; an example of this would be an epithelial cell, which divides about every fifteen days. The phase of mitosis (also known as M phase) may require a great deal of preparation time.

What type of **signals** control **cell reproduction**?

Cell reproduction is controlled by both external and internal signals. External signals include the availability of nutrients and space for growth. An example of an internal signal is the rise and fall of protein levels at specific points of the cell cycle, which is maintained by checkpoints and feedback controls. Cyclins and cyclin-dependent kinases are protein regulators that activate cell cycles and stimulate the synthesis of proteins.

What are examples of **growth factors** that **affect cell division**?

In order for a cell to grow, it must have specific nutrients. Some cells may require a "messenger"—referred to as a growth factor—to stimulate growth. There are several classes of proteins known as growth factors, which may also influence other cell activities, including embryonic development and response to tissue injury.

Examples of Growth Factor Families

Growth Factor	Target Cells
Epidermal Growth Factor	Epithelial cells

Can a cell ever repeat an S phase without dividing?

In some cells, the S phase (during which DNA is synthesized) can take place many times before mitosis occurs, resulting in an astronomical amount of DNA! For example, hair and glandular cells may have from 16 to 4,000 times the haploid amount of DNA. The record for animals is the giant neuron of the snail *Aplysia*, which has at least 75,000 times the haploid amount of DNA.

Growth Factor	Target Cells
Transforming Growth Factor	Epithelial cells
Platelet derived growth factor	Smooth muscle
Transforming growth factor	Fibroblast (connective tissue cells)
Fibroblast growth factor	Fibroblasts, many other cell types
Interleukin-2	Cytotoxic T(thymus-derived) lymphocytes
Colony stimulating factor	Macrophage precursors

What are the **stages of mitosis**?

Mitosis involves the replication of DNA and its separation into two new daughter cells. While only four phases of mitosis are often listed, the entire process is actually comprised of six phases:

- Interphase: Involves extensive preparation for the division process.
- Prophase: The condensation of chromosomes; the nuclear membrane disappears; formation of the spindle apparatus; chromosomes attach to spindle fibers.
- Metaphase: Chromosomes, attached by spindle fibers, align along the midline of a cell.
- Anaphase: The centromere splits and chromatids move apart.
- Telophase: The nuclear membrane reforms around newly divided chromosomes.
- Cytokinesis: The division of cytoplasm, cell membranes, and organelles occur. In plants, a new cell wall forms.

Do all **cells divide** at the **same rate**?

No, all cells do not divide at the same rate. Cells that require frequent replenishing, such as skin or intestinal cells, may only take roughly twelve hours to complete a cell

59

cycle. Other cells, such as liver cells, remain in a resting state (interphase) for up to a year before undergoing division. There are also cells that exist for a lifetime in a non-dividing state, an example of which would be human brain cells.

How are **organelles partitioned** during **mitosis**?

Following the telophase of mitosis, cytokinesis—physical separation of the daughter cells—occurs. Although the exact mechanisms are not clear, it appears that larger organelles, such as the endoplasmic reticulum and Golgi complex, undergo fragmentation into small vesicles during mitosis and later reassemble in daughter cells.

What are the major differences between **cell division in plants and cell division in animals**?

The major differences in plant and animal cell division are in the assembly of the spindle apparatus. The site of spindle apparatus assembly is the centrosome. In animal cells, a pair of centrioles is at the center of the centrosome. In contrast, most plants lack centrioles, but they do have a centrosome. In animal cells, a cleavage furrow forms during cytokinesis, which deepens and then pinches the parent cells in two. Plant cells, which have cell walls, do not have a cleavage furrow. Instead, a cell plate is produced in the middle of the parent cell, which grows toward the perimeter of the cell until it reaches the plasma membrane, dividing the cell in two. A new cell wall then forms from the cell plate.

What is **meiosis**?

Meiosis is often referred to as reduction division, meaning that the number of chromosomes present is reduced from 2N (diploid) to N (haploid). The meiotic process consists of two separate cell divisions and occurs in the gonads (ovaries and testis). It is most important to sexual reproduction because of genetic variation that occurs as a result of this process.

How are **cell structures divided** during **meiosis**?

Meiosis is just part of the larger process of gametogenic production that occurs in either the ovaries or testes. During sperm formation (spermatogenesis), the cells that eventually become mature sperm go through two successive meiotic divisions. This results in four haploid spermatids that will develop into mature sperm cells. The sperm are then reconfigured into a cell that is specialized for one thing: fertilizing an egg. The mature sperm is basically a nucleus with one set of chromosomes, mitochondria, and flagellum for propulsion.

In humans, the cells that will develop into mature eggs are present in the ovary before the oogenesis (egg formation) ever occurs. Immature eggs (oogonia) remain in a stage referred to as "meiosis I" until they mature during puberty. During the stage of "meiosis II," oocytes are ready to be released, but will not fully complete the meiotic process until after fertilization takes place. During the progression from one diploid oocyte to four haploid cells, cytoplasm is shunted unequally to only one cell. The end result is one large mature ovum and two or three very small haploid cells called polar bodies.

What happens in a **cell** if **mitochondria** are **defective**?

Since mitochondria are the energy producers of a cell, if a cell's mitochondria are defective, any cells that have a high metabolic rate will also be affected. There are metabolic poisons that affect specific aspects of mitochondrial function. These include cyanide, dinitrophenol (an ingredient of early diet drugs), valinomycin, and gramicidin. If mitochondria are damaged, they begin to leak free radicals that can alter DNA.

Because mitochondria have their own DNA, it is also possible for them to be altered by genetic mutations. Mutations in mitochondrial genes are thought to play a role in degenerative neurological diseases such as Parkinson's and Alzheimer's.

Is there **movement** among the **molecules** that comprise a **cell membrane**?

A cell membrane is mainly composed of phospholipids and proteins, which are two types of bio-organic molecules. Within the membrane, phospholipids are able to move laterally. Depending on temperature and fatty acid composition, phospholipids generally move faster than proteins. The proteins drift about slowly in a body of lipids, much like icebergs in an ocean. Proteins are able to change shape (also known as conformation). For example, carrier proteins are able to bind specific molecules such as glucose in order to provide transportation for the molecule. Once glucose is attached to the carrier protein, the protein changes shape and ferries the glucose inside of the cell.

Are all **cell membranes** alike?

Although all cell membranes have the same general structure, the membrane composition is different for each species . Depending on their function, membranes vary in the amount of protein they contain or the type of membrane receptors they contain. For example, plants that can survive frigid temperatures, such as winter wheat, are able to increase their concentration of unsaturated phospholipids during the winter to prevent membrane solidification. Another example is muscle cells, which have plasma membrane receptors for the neurotransmitter acetylcholine that tells them when to contract.

Do all **cells** use the same **source of energy**?

Most cells use glucose as their primary energy source, but lipids and proteins can be broken down to provide energy as well. Lipids are catabolized to their monomers, glycerol and fatty acids, which are then metabolized within the pathways of cell respiration. Proteins are also catabolized to their amino acid building blocks, which are then fed into the process of glycolysis, also known as the Krebs cycle.

Do all **cells require oxygen**?

Not all cells require oxygen; there are cells that use a metabolic pathway called fermentation to produce energy. Examples of anaerobic (non–oxygen dependent) organisms include yeast and bacteria that are able to thrive in an environment with low levels of oxygen. However, the majority of organisms are aerobic (oxygen dependent) because of the high ATP yield that oxygen provides. ATP is the main energy source for cells. Under some circumstances, oxygen-dependent cells can harvest energy from fermentation for short periods of time. However, this shortcut eventually results in a buildup of a lactic acid, a toxic waste product, and produces little ATP.

Why do **cells die**?

Cells die for a variety of reasons, many of which are not deliberate. For example, cells can starve to death, asphyxiate, or die from trauma. Cells that sustain some sort of damage, such as DNA alteration or viral infection, frequently undergo programmed cell death. This process eliminates cells with a potentially lethal mutation or limits the spread of the virus. Programmed cell death can also be a normal part of embryonic development. Frogs undergo cell death that results in the elimination of tissues, allowing a tadpole to morph into an adult frog.

What is **programmed cell death**?

Apoptosis, or programmed cell death, is a process by which cells deliberately destroy themselves. The process follows a sequence of events controlled by nuclear genes. First, the chromosomal DNA breaks into fragments, and this is followed by breakdown of the nucleus. The cell then shrinks and breaks up into vesicles that are phagocytosed by macrophages and neighboring cells.

While programmed cell death may seem counterproductive at first glance, it plays an important role in maintaining the life and health of organisms. During human embryonic development, apoptosis removes the webbing between the fingers and toes; it is also vital to the development and organization of both the immune and nervous systems.

Why is **light** important for **living organisms**?

Virtually all life depends on the availability of light, which powers photosynthesis (the

process of synthesizing energy). Light travels in waves, and its energy is contained in

> ## What color wavelengths are the most effective for photosynthesis?
>
> The most effective wavelengths of light for photosynthesis are blue (430 nm) and red (670 nm). Curiously, green plants photosynthesize least effectively in green light.

packets called photons. The energy of a photon is inversely proportional to the wavelength of the light—the longer the wavelength, the less energy per photon. Sunlight consists of a spectrum of colors present in light.

What is **photosynthesis**?

Photosynthesis (from the Greek word *photo*, meaning "light," and *synthesis*, from the Greek work *syntithenai*, which means "to put together") is the process by which plants use energy derived from light in order to make food molecules from carbon dioxide and water. Photosynthesis is a dual-staged process with multiple components. Light reactions or light-dependent reactions compose the first steps of photosynthesis. During the light reactions, light energy derived from sunlight is converted to chemical energy. Oxygen (O_2) is produced as a waste product of this process. The steps of the second stage are the carbon-fixation reactions known as the Calvin cycle. The Calvin cycle is a series of reactions that assemble sugar molecules from carbon dioxide (CO_2) and the energy-containing products of the light reactions. Carbon fixation is the conversion of carbon dioxide (CO_2) into organic compounds.

How do **chloroplasts** work?

Chloroplasts are able to capture solar energy to perform photosynthesis, the reduction of carbon dioxide to simple carbohydrates. This process entails a series of reactions that result in the chemical splitting of water and the release of oxygen into the environment. During the light phase, chlorophyll molecules absorb energy from light and their electrons become energized. These excited electrons pass energy from one chlorophyll molecule to another, resulting in the production of ATP and a special nucleic acid-type carrier known as NADPH. This molecule carries the electrons to the next stage of photosynthesis, the dark phase. The dark reactions manufacture sugars using the energy stored during NADPH and ATP. The conversion of carbon dioxide present in the atmosphere into carbon atoms in living organisms is called carbon fixation.

Why is **photosynthesis important**?

Ultimately, photosynthesis is the process that provides food for the entire world. Each year more than 250 billion metric tons of sugar are created through photosynthesis.

Photosynthesis is a source of food not only for plants, but also all organisms that are not capable of internally producing their own food, including humans.

Which **scientists** made **significant discoveries** toward the theory of **photosynthesis**?

The ancient Greeks and Romans believed that plants derived their food from the soil. The earliest experiment to test this hypothesis was performed by the Belgian scientist Jan Baptista van Helmont (1577–1644), who grew a willow tree in a container of soil and fed it only water. At the end of five years, the weight of the willow tree had increased by 164 lbs (74.4 kg), while the weight of the soil had decreased by 2 oz (57 gr). Van Helmont concluded the plant had received all its nourishment from the water and none from the soil. Joseph Priestley (1733–1804) demonstrated that air was "restored" by plants. In 1771 Priestly conducted an experiment in which he placed a lighted candle in a glass container and allowed it to burn until extinguished by lack of oxygen. He then put a plant into the same chamber and allowed it to grow for a month. Repeating the candle experiment a month later, he found that the candle would now burn. Priestley's experiments showed that plants release oxygen (O_2) and take in carbon dioxide (CO_2) produced by combustion. The Dutch physician Jan Ingenhousz (1730–1799) confirmed Priestley's ideas, emphasizing that air is "restored" only by green plants in the presence of sunlight. Evidence of photosynthesis's two-stage process was first presented by F. F. Blackman (1866–1947) in 1905. Blackman had identified that both a light-dependent stage and a light-independent stage occur during photosynthesis. In 1930 C. B. van Niel (1923–1977) became the first person to propose that water, rather than carbon dioxide, was the source of the oxygen that resulted from photosynthesis. In 1937 Robert Hill (1899–1991) discovered that chloroplasts are capable of producing oxygen in the absence of carbon dioxide only when the chloroplasts are illuminated and provided with an artificial electron acceptor.

How are **plant cells** able to use **light** to **produce sugars**?

Plant cells use chloroplasts to convert light energy to chemical energy (sugar). These chloroplasts (which are likely to have originated as free-living bacteria) use the energy contained in UV radiation to raise electrons to higher energy states through the process of photosynthesis. The energized electrons are used to build and rearrange a number of different molecules. Some of these eventually become glucose, which in turn can be used to produce sucrose, also known as table sugar.

Do **plant cells** really **produce oxygen**?

Yes, plant cells produce oxygen through the process of photosynthesis. Splitting water molecules to harvest their electrons causes the release of oxygen. By submerging a small piece of an aquatic plant in a beaker containing water, one can actually see the oxygen bubbles produced.

How do **plant** cells **use carbon dioxide**?

Plant cells reduce carbon dioxide to sugar by using the electrons that are produced when chlorophyll absorbs light. Six carbon dioxide (CO_2) molecules, along with six water (H_2O) molecules, can be converted into a simple sugar ($C_6H_{12}O_6$).

How do the two forms of chlorophyll, **chlorophyll** *a* and **chlorophyll** *b*, **participate** in **photosynthesis**?

During photosynthesis, light is absorbed by pigments present in organisms. Chlorophyll *a* is the primary pigment required for photosynthesis and occurs in all photosynthetic organisms except photosynthetic bacteria. Accessory pigments such as carotenoids and chlorophyll *b* absorb light that chlorophyll *a* cannot absorb; these pigments extend the range of visible light useful for photosynthesis.

Do humans really need **cholesterol**?

Cholesterol is an important component in the cell membranes of animals. Cholesterol, a lipid-based molecule (steroid), actually has two functions: 1) To help stabilize the membrane. 2) To maintain membrane flexibility as temperature changes. Normally the human body is capable of producing all the cholesterol it needs. Dietary ingestion of excess saturated fats and cholesterol is currently thought to be the source of the plaque that builds up in arteries and can cause heart attacks and strokes. Dietary cholesterol can be found in all animal sources, including shellfish. Plant phytosterols have the same function as cholesterol found in animals, but they do not affect the human body in the same way.

Can **cells** conduct **electricity**?

Yes. All living cells have membranes that allow them to maintain a difference in concentration of atoms located on the inside and outside of the cell. Some of these atoms are ions that have a charge. This ability to maintain an imbalance of ions is called cell membrane potential and is analogous to the electrical potential of a battery.

CELL JOBS

How do **cells communicate** with each other?

Cells communicate with each other via small, signaling molecules that are produced by specific cells and received by target cells. This communication system operates on both a local and long-distance level. The signaling molecules can be proteins, fatty acid derivatives, or gases. Nitric oxide is an example of a gas that is part of a locally based signaling system and is able to signal for a human's blood pressure to be lowered. Hormones are long-distance signaling molecules that must be transported via the circulatory system from their production site to their target cells. Plant cells, because of their rigid cell walls, have cytoplasmic bridges called plasmodesmata that allow cell-to-cell communication. Animals use gap junctions to transfer material between adjacent cells.

How do **cells recognize** each other?

Cells recognize each other through the molecules that are attached to their extracellular matrix, which fills the space between cells, binding cells and tissues together.

Process	Recognition Type
Embryonic development	Cell-cell recognition to form tissues
Tissue matching	Organ rejection, including blood groups
Immune surveillance	Recognizes invaders and cells that have been invaded, including cancer cells
Viral infections	Viruses target specific cells based on membrane markers (HIV virus targets specific CD4 markers)
Hormones	Target specific cells or organs
Neurotransmitters	Stimulated neural function, particularly important to mood disturbances (lack of serotonin receptors can cause depression)

Process	Recognition Type
Cancer	Cancer cells develop abnormal cell membrane markers
Fertilization	Based on specific markers on egg and sperm cell membranes

How do cells **respond** to **cellular signals**?

In order to respond to a signal, a cell needs a receptor molecule that recognizes the signal. A cell's response to a specific signal varies according to the signal. Some signals are local signals (e.g., growth factors), while others act as distance signals (e.g., hormones). There are two basic types of hormones, those that bind to receptors on the cell surface and those whose receptors are found within cytoplasm. Both types cause the cellular machinery to change its activities.

How do **cells** respond to **steroid hormones**?

Progesterone, estrogen, testosterone, and glucocorticoids are steroid hormones that are lipid-based signaling molecules. After entering a target cell, the steroid hormone binds to a receptor protein and starts a cascade of events that ultimately activates ("turns on") or inhibits ("turns off") a specific set of genes.

How do **cells** respond to **insulin**?

Protein-based hormones such as insulin bind to cell-surface receptors. While they do not enter the cell, they cause changes in the cell's metabolism. Specific cells in the pancreas secrete insulin, a hormone that regulates the concentration of glucose in the blood. Skeletal muscles and the liver are targets for insulin. Insulin deficiency is responsible for Type 1 diabetes. In contrast, type II diabetes, also known as adult onset diabetes mellitus, is not the result of insulin deficiency but is rather due to insulin resistance. Cells with insulin resistance do not respond to increasing insulin levels by transporting glucose into cells.

Can a **cell survive** in **complete darkness**?

Yes, cells can survive in complete darkness. A new theory on the origin of life argues that living systems arose in small compartments of total darkness located within iron sulfide rocks that were formed by hot springs on the sea floor. Cells of plant roots live in total darkness and carry out all normal plant cell activities, with the exclusion of photosynthesis. And, if you think about it, cells located in the very middle of your body also live in a pretty dark environment!

How do **plant** cells **store energy**?

Plant cells store energy in complex carbohydrates such as starch, disaccharides, and lipids. These energy sources may be used to fuel cellular metabolism or to provide energy for the germination of seeds.

What is **cell fluorescence**?

Fluorescence is luminescence caused by a pigment molecule. Pigments are molecules that absorb some colors of light while reflecting others. For example, green pigments (like those in leaves) absorb red and blue wavelengths of light but reflect green wavelengths. After being energized by photons (units of light energy), the electrons of some pigment molecules actually give off light as they fall back to their normal state. Chlorophyll molecules that play a role in photosynthesis have this ability, as do molecules found in several organisms, such as jellyfish. The body of the insect *Photinus pyralis,* better known as a lightning bug, has the enzyme luciferase, which generates the chemical reaction that leads to a drop in the energy state of electrons—a reaction similar to that which occurs in a chlorophyll molecule—allowing the insect to "glow."

What is the **most-common blood cell**?

Red blood cells, also known as erythrocytes, are the most-common blood cells. A milliliter (ml) of blood contains approximately 5 billion red blood cells. The average person has a total of 25 trillion red blood cells in their blood stream! Red blood cells are very small; it would take a string of 2,000 red blood cells to circle a pencil.

How does a **red blood cell** carry **oxygen**?

Cells use oxygen to harvest the energy stored in molecules like sugars and fats. Red blood cells use a protein called hemoglobin to carry oxygen to all the cells of the body. Hemoglobin is actually made of four separate protein strands; each one surrounds a central, iron-containing molecule. The hemoglobin group carries the oxygen molecule; each molecule is made up of two atoms of oxygen bonded together.

Do **red blood cells function** the same way in **humans** and **insects**?

Insects don't have blood like ours; instead, their bodies contain a fluid known as hemolymph. Their hemoglobin is not concentrated in cells *within* the hemolymph, but rather is found *floating* in the hemolymph. An insect's hemoglobin still carries oxygen, but it is somewhat smaller than that found in mammals like humans. Squids, octopi, and crustaceans also have oxygen-carrying molecules in their plasma, but their bodies use a copper-based molecule known as hemocyanin to carry oxygen instead.

How does **sickle cell anemia** affect the function of **red blood cells**?

The genetic defect in sickle cell anemia is a mutation in one of the polypeptide chains that comprise the hemoglobin molecule. The abnormal molecule produced in sickle cell anemia causes a change in the shape of affected red blood cells. The disease morphs a typically rounded red blood cell disk into a crescent or sickle shape. The abnormal red blood cells affected by sickle cell anemia are rigid and are more likely to clump together. Because of their tendency to group, they are more prone to sticking to the walls of blood vessels; they can even block the blood vessels themselves. Because of this, the hemoglobin molecule isn't able to carry oxygen as well as the cells need it to.

Who **discovered** how **muscles work**?

Hugh Huxley (1924–) and Andrew Huxley (1917–) (the scientists were unrelated) researched theories regarding muscle contraction. Hugh Huxley was initially a nuclear physicist who entered the field of biology at the end of World War II. He used both X-ray diffraction and electron microscopy to study muscle contraction. Andrew Huxley was a muscle biochemist who obtained data similar to Hugh's, indicating that the contractile proteins thought to be present in muscles are not contractile at all, but rather slide past each other to shorten a muscle. This theory is called the sliding filament theory of muscle contraction.

How do **muscle** cells **work**?

Muscle cells—whether the skeletal muscles in the arms or legs, the smooth muscles that line the digestive tract and other organs, or the cardiac muscle cells in the heart—work by contracting. Skeletal muscle cells are comprised of thousands of contracting units known as sarcomeres. The proteins actin (thin filament) and myosin (thick filament) are the major components of the sarcomere. These units perform work by moving structures closer together through space. Sarcomeres in the skeletal muscles pull parts of the body through space relative to each other (e.g., walking or swinging your arms).

To visualize how a sarcomere works:

• Interlace the fingers of your two hands with the palms facing toward you (represents actin, myosin).

• Push the fingers together so that the overall length from one thumb to the other is decreased (sarcomere length decreases).

• Any object attached to either thumb would be pulled through space as the fingers move together (sliding filament theory).

What **sources** do **muscle cells** use for **energy**?

Muscle cells use a variety of energy sources to power their contractions. For quick energy, the cells utilize their stores of ATP and creatine phosphate, which is another phosphate-containing compound. These stored molecules are usually depleted within the first twenty seconds of activity. The cells then switch to other sources, most notably glycogen, a carbohydrate that is made of glucose molecules strung together in long-branching chains.

Do all **muscle cells** work the **same** way?

Although all muscles work by contracting, not all muscle types have sarcomeres, the muscle contraction units. Cardiac muscle cells have sarcomeres but use different sup-

Lifting weights will cause muscles to grow in size but does not actually increase the number of muscle cells present.

How many connections are in a very small piece of brain tissue?

A thin piece of brain tissue 3 mm (0.12 in) in diameter could have one billion connections between neurons.

port structures during contraction than those found in skeletal muscles. Smooth muscle cells don't use sarcomeres at all.

How do **muscle cells** use **calcium**?

Calcium ions are stored inside muscle cells. The calcium ions are released when a muscle cell gets a signal to induce contraction, which initiates the movement of the contractile proteins within muscles. When calcium concentrations fall, muscle contractions stop.

How do **muscle cells** respond **to activities like** weight lifting?

Weight lifting will cause muscles to grow—the size of muscle cells increases, but the body does not actually grow more muscle cells. Rather, weight lifting causes the body to grow more of the thick (myosin) and thin (actin) proteins that aid muscle contraction. This process makes the muscle not only bigger, but stronger as well. Some muscles gain strength faster than others. In general, large muscles, like those present in your chest and back, grow faster than smaller ones, like those in your arms and shoulders. Most people can increase their strength between 7 and 10 percent after ten weeks of training each muscle group at least twice a week.

Do all **brain cells** work the same way?

There are two basic types of cells in the nervous system: neurons that carry messages and the support cells that both feed and protect them. In the central nervous system—comprised of the brain and spinal cord—these support cells are known as neuroglial or glial cells. These cells perform a variety of functions in maintaining the health of the neurons of the brain. It is estimated that glial cells actually comprise about 90 percent of the cells in the nervous system.

How do **neurons** carry **messages**?

Imagine a long trail of dominoes running from your finger to the spinal cord in your back. When you touch a hot object, the messages this action produces in neurons is similar to the reaction that would happen if you knocked over the first domino of the set. After the first domino falls, each domino that follows it would be knocked over in

71

turn. This process is similar to what happens between neurons. The message is carried from the finger tip all the way to the spinal cord. Neurons maintain a concentration of sodium and potassium ions (Na^+ and K^+, respectively). The openings present in membranes allow these ions to pass through. As the sodium and potassium gates open and then close, the electrical charge of the nerve cell changes and messages are transmitted along the neuron membrane.

How do **brain cells** store **memories**?

The part of your brain responsible for processing memory is the hippocampus. It is believed that memories are formed at the level of individual nerve cells. The synapse is the point at which adjoining nerve cells touch, and it is this juncture that is the building block of memory systems. Information moves across the synapse—the information signal is carried inside a cell by a second messenger (known as cyclic AMP), which then activates other cell machinery. The end result is the switching on of a gene that regulates memory. The product of the gene, a protein, promotes synaptic growth and can convert short term memory to long term memory.

How do **brain cells function** differently in people with **Alzheimer's disease**?

The build up of toxic proteins (plaques) in the brains of Alzheimer's patients is linked to changes in several genes that regulate memory and learning. The protein responsible for this effect is a small peptide, beta-amyloid. This protein is made by the body throughout the entire lifetime, but in Alzheimer's patients, either too much is synthesized or too little is broken down. The buildup of plaque has been implicated in neuron death, which eventually leads to dementia. Researchers have studied mice that were genetically engineered with genes to cause them to accumulate beta-amyloid deposits in their brain. The mice developed neurological problems consistent with Alzheimer's disease that occurs in humans. Results from the tests performed on mice indicated that the functions of at least six genes were suppressed in the presence of the renegade protein.

Can **cells** ever **change jobs**?

The more specialized of a function a cell performs, the less likely it is for the cell to change jobs within an organism. However, there are some cells that have unspecialized functions and are able to adapt to the changing needs of the body. In mammals, a

How fast do neurons carry messages?

Neurons can transmit messages at speeds of up to 350 km per hour (217.5 mi per hour), at a rate of 500 impulses per second.

good example of cells with adapting functions would be bone marrow cells, which are responsible for producing different types of cells in the blood. Bone marrow cells produce red blood cells and five types of white blood cells. Slime molds of the kingdom Protista have cells that are capable of drastically changing their function. The cellular adaptations that can occur in slime molds allow them to change from single-celled amoebas to multicellular, reproductive spore producers.

CELLULAR SPECIALIZATION

Do **cells** have specific **shapes**?

While it doesn't seem that animal cells have a specific shape, they all have a cytoskeleton. The cytoskeleton enables cells to assume and maintain complex shapes, such as the star-like shape of a neuron or the biconcave shape of a red blood cell. Epithelial and connective tissues also have very particular shapes—the columnar epithelium resembles a wall of bricks. Plant cells usually have a specific shape due to their rigid cell wall, which decreases the elasticity of the cell.

How do **cells** become **tissues**?

Tissues are comprised of different types of cells that have a common purpose or function. Multicellular organisms are so complex that tissues are needed to perform vital functions such as support, transportation, and the movement of animal cells. In animals, the groups of cells that form tissue may become structurally linked via their cytoskeletons and membranes. As an example, epithelial cells are anchored to underlying connective tissue by a basement membrane. These cells are often exposed to mechanical stress, and a system of cell junctions between epithelial cells helps them respond when stretched.

How does a **cell** respond to **injury**?

In most tissues, injured cells die and are subsequently replaced. However, in the case of nervous tissue, where dead cells cannot be replaced, a number of events can occur. Nerve growth factor is produced by adjoining neurons; this can induce sprouting and growth of previously dormant neurons.

73

How do **skin cells** synthesize **vitamin D**?

Vitamin D is crucial to normal bone growth and development. When UV light shines on a lipid present in skin cells, the compound is transformed into vitamin D. People native to equatorial and low latitude regions of the earth have dark skin pigmentation as a protection against strong, nearly constant exposure to UV radiation. Most people native to countries that exist at higher latitudes—where UV radiation is weaker and less constant—have lighter skin, allowing them to maximize their vitamin D synthesis. During the shorter days of winter, the vitamin D synthesis that occurs in people that live in higher latitudes is limited to small areas of skin exposed to sunlight.

Increased melanin pigmentation, present in people native to lower latitudes, reduces the production of vitamin D. Susceptibility to vitamin D deficiency is increased in these populations by the traditional clothing of many cultural groups native to low latitudes, which attempts to cover the body completely to protect the skin from overexposure to UV radiation. Most clothing effectively absorbs irradiation produced by ultraviolet B rays. The dose of ultraviolet light required to stimulate skin synthesis of vitamin D is about six times higher in African Americans than in people of European descent. The presence of darker pigmentation and/or veiling may significantly impair sun-derived vitamin D production, even in sunny regions like Australia.

Why don't **organisms** that **produce toxins die** during the synthesis of toxic material?

Organisms produce toxins as a defense mechanism; therefore, most organisms are immune to their own toxins as well as that of other members of the same species. Toxins are a cell product packaged into a membranous sac known as a vesicle. The contents of the vesicle are secreted into the salivary gland, skin surface, or fangs of the organism during the process of exocytosis. The toxin is then injected, ingested, or absorbed by other organisms—often a predator. Each toxin creates a specific effect. For example, the snake venom known as bungarotoxin deactivates acetylcholine receptors on muscles. Since it is secreted externally, it is unlikely that the snake producing the toxin would be affected.

Venom Toxins by Molecule and Action

Type of Toxin	Physiological Effect
Neurotoxin	Nervous system damage
Mytotoxin	Muscle damage
Nephrotoxin	Kidney damage
Cytotoxin	Damages cell membrane
Cardiotoxin	Damages heart
Hemotoxin	Damages circulatory system

Can humans ever grow horns?

Yes, there are documented cases of humans having horns. A horn is actually a projection that forms above the surface of the skin that is composed of compacted keratin.

How are **drugs detoxified** by **cells**?

Drugs are detoxified by the smooth endoplasmic reticulum of liver cells. Detoxification usually involves changing the molecular structure of a toxin, a modification that increases the toxin's solubility, allowing it to be safely carried away by the blood and excreted via urine. Cells are able to increase their detoxification efforts when drug levels increase. Investigations conducted with rats that have been injected with a sedative known as phenobarbital have shown a striking increase in the amount of smooth endoplasmic reticulum of the liver cells.

What is the most **specialized cell** in **mammals**?

Depending on the criteria you choose, several types of cells could be nominated as the most specialized cell in the mammalian body. The two top candidates for this honor would probably be: 1) the cells that produce gametes (sperm and eggs); 2) the red blood cells that carry oxygen and carbon dioxide in the blood. Red blood cells are perhaps the most highly specialized. They only live for roughly 120 days, but during that time they may travel more than 500 mi (800 km) through various organs and blood vessels! Red blood cells lack a nucleus, so they are unable to reproduce; new cells are formed in bone marrow.

How do some **cells**, such as those found in **nails** and **horns**, change shape and become **hard**?

All of the surface cells of the body (except those in eyes) contain a molecule called keratin, a fibrous protein that is particularly well suited to withstand abrasions. Certain cells, like those of the nails and hair, have increased amounts of keratin that provide extra strength, helping them maintain their shape. Whether it is in the claws on the toes of your cat, the horns on the head of your favorite cow, or the 5 million hairs that cover your body, keratin strengthens and protects organisms from everyday wear and tear.

How do **stomach cells survive** the **acid** in the stomach?

The cells that line your stomach produce a basic, alkaline mucus. The mucus has a higher pH that counteracts the acid produced by the stomach—which has a pH of 2.0.

Because of this, cells are protected from the digestive enzymes at work in the stomach. If stomach acid manages to reach the tissue below the protective mucus layer, ulcers can result.

What **prevents urine** from **leaking out** of the **bladder** and getting into the body?

The cells that form your bladder are held together by tight junctions, which are connections between cells that hold them together so closely that urine can't slip through to reach the rest of the body. These connections, formed by protein strands that bind the cell membranes, also play an important role in keeping food in the digestive tract until it has been completely processed.

How do **skin cells** keep your **blood** from **leaking** out?

The skin is comprised of multiple layers of cells. The outermost layers are made of dead cells full of keratin. Sebaceous glands coat these dead cells with an oily secretion that makes them water resistant. However, they are not waterproof; about one pint of fluid from deeper tissues leaks through the skin's surface every day and evaporates. This excretion is in addition to perspiration produced by excessive heat or strenuous activity. The very strong junctions holding the cells together, known as desmosomes, prevent large of amounts of fluid leakage across the skin barrier. The linkages are so efficient that the epidermal skin cells tend to slough off in sheets rather than individually.

How do **sperm** work?

Cells called spermatogonium become sperm through the process of meiosis. A mature sperm has only half the DNA required for a functional cell, so it can't survive on its own. However, they do have flagella and mitochondria that power their journey through the reproductive tract in search of an egg. Human sperm typically die within forty-eight hours if they have not fertilized an egg.

Can **unfertilized eggs function** like other **cells** in the body?

A human egg is about 2,000 times the size of a sperm and has all the organelles and proteins required of a living cell but only one half of the necessary DNA. Because of this, the egg cannot be considered a functional cell. Without fertilization, human egg cells survive about twenty-four hours after ovulation. Prior to that, the egg is maintained by the support structures known as follicular cells, which transfer nutrients to the egg. However, in a process called parthenogenesis, an egg can develop without being fertilized; this type of fertilization occurs in rotifers, aphids, and whiptailed lizards.

How can **bacteria survive** inside our **gut**?

Bacteria known as intestinal flora survive in both the small and large intestine because
of the neutral pH of the intestinal environment. Because these bacteria have low

requirements for oxygen and sunlight exposure, they are well suited to life in this environment. They use our digested food as a source of nutrition and even provide a benefit to us by synthesizing several of the vitamins (biotin, as well as vitamins K and B5) we need. If the flora are thriving, it is more difficult for disease-causing microbes to establish themselves in the intestines and attack our bodies.

What happens if a cell's **DNA** is **mutated**?

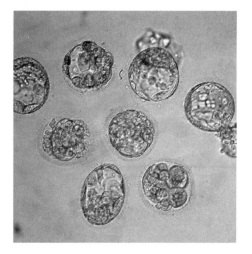
Cloned embryos at the eight-cell stage.

DNA contains the code for building everything within the cell. If the DNA code is changed, or mutated, the cell will build things differently. Mutations can change the way that a cell produces substances, such as proteins and carbohydrates; builds organelles; or responds to messages. These changes can enhance a cell's ability to survive, although it is more likely that drastic changes will result in a decrease of efficiency or even the death of the cell.

What is a **malignant cell**?

Cancerous cells that reproduce quickly and expand beyond the tissue or organ where they originated are described as malignant. Malignant tumors are more difficult to eradicate because they may colonize organs far removed from their origin. For example, cancers that originate in the lung can quickly spread to the brain and other organs through the circulatory system.

APPLICATIONS

How is **cell cloning** used in **scientific research**?

Cell cloning is the process by which an exact copy is made of a cell. This cellular process is known as mitosis and is required for the growth and repair of multicellular organisms. Different types of cells in the body differ in their ability to perform mitosis. Some cells, like skin cells, produce clones quite often. Others, like those of the nervous system, will not reproduce after they have reached maturity and have differentiated. The scientific purpose of cloning is to produce many copies of certain types of cells that can then be used for a variety of purposes, like basic research or the growth of replacement organs.

Other than fetal cells, what are **human sources** of **stem cells**?

Many types of cells have been found that can be used as stem cell sources.

Cell Source	Possible Use
Brain	Neurodegenerative diseases; spinal cord injury
Hair and skin	Burn healing
Breast (from cosmetic surgery)	Breast duct regeneration
Fatty tissue (liposuction leftovers)	Cartilage, bone, fat
Bone marrow	Almost any tissue; embryonic healing capacity
Pancreas	Diabetes treatment
Heart	Healing following myocardial infarction
Baby teeth (with associated tissue)	Can be used similarly to bone marrow

Can **cells** be used as **factories**?

Humans have been using cells as factories for millennia. The production of cheese, yogurt, beer, and wine relies on the ability of individual cells to produce specific products, such as lactic acid and ethanol. More recently, scientists have been able to manipulate cell genes, so that cells will produce substances that have little relation to their normal function. Examples of products from bioengineered cell factories include the production of human insulin to treat diabetics and Factor VIII (a naturally produced clotting factor in humans) for hemophiliacs.

How do **cells** become **cancerous**?

Cancer is caused by the unrestrained growth of cells. Cells that do not "follow the rules" of normal cell cycling may eventually become cancerous. This means that the cells reproduce more often than normal, creating tumors. Usually this happens over an extended period of time and begins with changes at the molecular level. There are more than one hundred distinct types of cancer, each of which behaves in a specific fashion and responds to treatment differently.

Where do **cancer cells** come from?

When the reproductive rate of cells exceeds their death rate, the tissue becomes enlarged, forming a tumor. Although these cells are initially identical to the others in the tissue, they gradually take on characteristics of malignancy. The cancer cells reproduce rapidly and tend to be abnormally large or small. Malignant tumors grow very quickly and invade other tissues. Cancer types are named for the location of the tissue that gives rise to the tumor and the organs involved. Genetics, viruses, or even environmental exposure to substances like those in cigarette smoke may cause tumor formation. However, not all tumors are malignant; tumors that grow within a well-defined capsule are benign and unlikely to be life-threatening.

Are some forms of **cancer communicable**?

Yes and no. Cancer cells arise from within the body. However, certain types of cancer may actually be caused by viruses, which may be transferred from one organism to another. Currently, it is estimated that viruses may play a role in as much as 15 percent of human cancers worldwide.

Viruses and Human Cancers

Virus family	Human tumors
Hepatitis B	Liver cancer
Epstein Barr	Nasopharyngeal cancer; Burkitt's lymphoma
Herpes	Kaposi sarcoma
Papillomavirus	Cervical cancer
HIV	Kaposi sarcoma; cervical cancer; non-Hodgkin's lymphoma
SV40	Mesothelioma

How do **cancer** cells **feed** themselves?

In the 1960s Dr. Judah Folkman (1933–) realized that malignant tumors could not grow without nourishment, which is delivered by the blood. Rapidly growing tumors actually cause the formation of new blood vessels in a process known as angiogenesis. Folkman's hypothesis was that by identifying the substances used to cause angiogenesis, drugs could be formulated to prevent new vessel formation, thus starving the tumors. This work has led to the identification of at least two substances that inhibit angiogenesis: endostatin and angiostatin. These drugs hold promise as new therapies to combat aggressive tumors.

How do **genes control cancer** cells?

In normal cells, there are two types of genes that are important in determining whether or not cancerous tissue can form. These genes control the production of proteins that affect the cell cycle. Proto-oncogenes are DNA sequences that promote normal cell division. By mutation, these genes may be converted into oncogenes, which promote the overproduction of cells. Another class of genes, known as tumor-suppressor genes, prevents excess reproduction of cells. Mutation in these genes can also allow cells to became cancerous.

What aspects of a **cancer cell** do most **anticancer treatments target**?

Anticancer drugs attempt to slow down or stop the ongoing cell division that occurs in cancerous tissues. Treatment protocols include radiation, heat exposure, freezing, surgery, and/or drug therapy.

Why don't all types of **cancers respond** the same way to **anticancer drugs**?

The purpose of most anticancer drugs is to target the overproduction of malignant cells. Different types of cancers arise from different types of tissue. Since every tissue is made up of cells specialized for a certain function, it is not surprising that different forms of cancers will have different responses to the same drug. For example, a drug that targets the overproduction of liver cells (which are adapted to filtering and monitoring the blood supply) might have little effect on nerve cells that specialize in carrying messages.

Can **human cells grow outside the body**?

Human cells can be cultured *in vitro* (a Latin term meaning "outside the body") if the cells are given oxygen and appropriate nutrients. However, these cells usually pass through a limited number of cell divisions (about fifty) before they begin to die.

Can we **make** artificial **cells**?

Research in progress at the National Aeronautical and Space Administration (NASA) is focused on artificial cells as a means to deliver medicine in outer space; these cells are able to withstand dehydration and thus can be safely stored for long periods. Artificial cells are made of a polymer that acts like a cell membrane, but the polymer is stronger and more manageable than real membranes. These polymers are called polymersomes and can be made to cross-link with other polymers. Researchers feel that many different kinds of molecules can be encapsulated within these polymersomes and then delivered to specific target organs. An example would be an artificial blood cell that not only delivers oxygen but also medication as it travels through the body.

How does **cyanide** affect **cells**?

Cyanide acts by inhibiting the enzymes cells need for oxygen utilization. Without these enzymes, a cell cannot produce ATP and will die. People can be accidentally exposed to cyanide. Very small amounts of cyanide naturally occur in some foods and plants. For example, cyanide is present in cigarettes and in the smoke produced by burning plastics. Cyanide is also used to make paper and textiles, clean metals, separate gold from its ore, and in the chemicals used to develop photographs. Pesticides used in ships and buildings may also contain cyanide.

What are **beta-blockers** and how do they affect **cell function**?

In order to understand how beta-blockers work, one must understand that most drugs work by binding to receptors on the cell membrane. Drugs called agonists activate cell receptors by causing either an increase or decrease of cell function. Beta-blockers are a class of drugs called antagonists because they block agonists from binding to cell receptors. An example of a beta-blocker is propranolol, which is used to treat high blood pressure and angina. Propranolol works by protecting the heart against sudden surges of stress hormones like adrenaline.

How do **statin** drugs affect **cell function**?

Statins are a group of drugs that work to lower cholesterol levels, particularly the "bad cholesterol", low-density lipoprotein known as LDL. The drugs work in two ways: 1) They block an enzyme that is needed for cholesterol production. 2) They increase LDL membrane receptors in the liver.

Cholesterol can only get into cells by binding to specific receptors that remove the LDL from blood. The extra receptors that statins create help decrease the cholesterol levels. As Americans become more aware that high cholesterol is a major risk factor for heart disease, statins are becoming increasingly popular.

How does **carbon monoxide** affect **cell function**?

Carbon monoxide is a highly poisonous gas. Because of its molecular similarity to oxygen, hemoglobin can bind to carbon monoxide instead of oxygen, and this subsequently disrupts hemoglobin's efficiency as an oxygen carrier. Carbon monoxide actually has a much greater affinity (about 300 times more!) for hemoglobin than oxygen. When carbon monoxide replaces oxygen, this causes cell respiration to stop, leading to death. The particular danger of carbon monoxide poisoning lies in the fact that a person exposed to high levels of this toxin cannot be saved by being transporting to an environment free of the poison and rich with oxygen. Since the hemoglobin remains blocked, artificial respiration with overpressurized, pure oxygen must first be performed to return the hemoglobin to its original function and the body to normal cell respiration.

How does **alcohol** affect **cell function**?

Alcohol causes varying effects on different cells. In general, alcohol increases tissue sensitivity to injury and prevents post-injury recovery. Alcohol stimulates brain cells by disrupting calcium channels within the cell membranes. It is thought that alcohol affects the fluidity of the membrane phospholipids. Alcohol also causes mitochondrial damage, depressed platelet function, decreased synthesis and transportation of proteins from the liver, and the activation of pancreatic enzymes that may subsequently damage the lining of the lung.

How do **diet drugs** affect **cell metabolism**?

Diet drugs target the areas of the brain that control food intake and produce neurotransmitters. Drugs that attempt to control food intake are adrenergic; their role is to enhance metabolism and increase the efficiency of calorie burning. The second type of drug targets the brain chemical serotonin. Serotonin is linked to appetite, particularly cravings for carbohydrates. Higher levels of serotonin lead to less cravings for carbohydrates. For example, the diet drug dexfenfluramine inhibits the removal of serotonin in the brain. However, there are side effects to the use of diet drugs, and there is insufficient data regarding the safety and effectiveness of their long-term use.

Is it possible to **eliminate fat cells** permanently?

Current evidence suggests that the original number of fat cells in any area of the body is controlled by one's genetic make up. People who lose fat mass through diet and exercise do not lose fat cells. Rather, the fat cells atrophy. Although fat cells may be removed by liposuction, studies have suggested that liposuction does not effectively control weight. Surgical removal of fat cells only removes about 10 percent or less of fat cells in the body.

How does **caffeine** affect **cells**?

Caffeine is probably the most-common drug ingested by people worldwide. Caffeine affects cells by stimulating lipid metabolism and slowing the use of glycogen as an energy source. As a whole, the body responds to caffeine by extending endurance, allowing you to stay awake for longer periods of time or perform extra activities. Adverse effects of excess caffeine intake include stomach upset, headaches, irritability, and diarrhea.

Can caffeine affect athletic performance?

Athletic competitions at collegiate, national, and international levels prohibit the intake of caffeine in amounts greater than five to six cups of coffee during the two hours prior to competition.

Caffeine Source and Average Dose

Source	Average Dose (mg)
Coffee (12 oz) brewed	300
Coffee (12 oz) decaffeinated	7
Tea (12 oz)	100
Tea (12 oz) iced	70
Soft drinks (12 oz)	30–46
Dark chocolate (1 oz)	20
Milk chocolate (1 oz)	6
Cold remedies	0–30
Pain relievers	0 (aspirin)–130 (Excedrin)
Diet pills	200–280

BACTERIA, VIRUSES, AND PROTISTS

INTRODUCTION AND HISTORICAL BACKGROUND

How has the **classification of organisms** changed throughout history?

From Aristotle (384–322 B.C.E.) to Carolus Linnaeus (1707–1778), scientists who proposed the earliest classification systems divided living organisms into two kingdoms—plants and animals. During the nineteenth century, Ernst Haeckel (1834–1919) proposed establishing a third kingdom—Protista—for simple organisms that did not appear to fit in either the plant or animal kingdom. In 1969 R. H. Whitaker (1920–1980) proposed a system of classification based on five different kingdoms. The groups Whitaker suggested were the bacteria group Prokaryotae (originally called Monera), Protista, Fungi (for multicellular forms of nonphotosynthetic heterotrophs and single-celled yeasts), Plantae, and Animalia. This classification system is still widely accepted; however a six-kingdom system of classification was proposed in 1977 by Carl Woese (1928–). The groups proposed in the six-kingdom approach are Archaebacteria and Eubacteria (both for bacteria), Protista, Fungi, Plantae, and Animalia. In 1981 Woese proposed a classification system based on three domains (a level of classification higher than kingdom): Bacteria, Archaea, and Eukarya. The domain Eukarya is subdivided into four kingdoms: Protista, Fungi, Plantae, and Animalia.

What are the **major characteristics** of each **kingdom** of living **organisms**?

Kingdom	Cell Type	Characteristics
Monera (Bacterial and Archaean Kingdoms)	Prokaryotic	Single cells lacking distinct nuclei and other membranous organelles

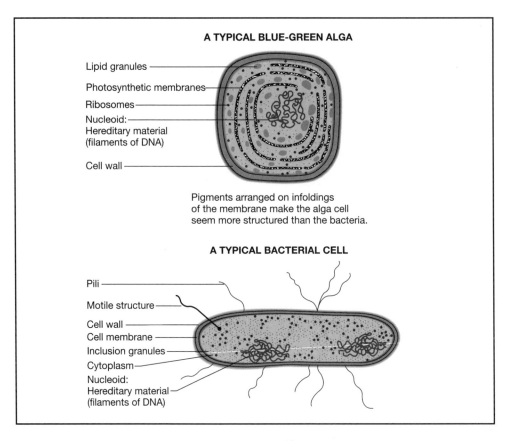

A TYPICAL BLUE-GREEN ALGA

Lipid granules

Photosynthetic membranes

Ribosomes

Nucleoid:
Hereditary material
(filaments of DNA)

Cell wall

Pigments arranged on infoldings
of the membrane make the alga cell
seem more structured than the bacteria.

A TYPICAL BACTERIAL CELL

Pili

Motile structure

Cell wall

Cell membrane

Inclusion granules

Cytoplasm

Nucleoid:
Hereditary material
(filaments of DNA)

Two prokaryotic cells.

Kingdom	Cell Type	Characteristics
Protista	Eukaryotic	Mainly unicellular or simple multicellular, some containing chloroplasts. Includes protozoa, algae, and slime molds.
Fungi	Eukaryotic	Single-celled or multicellular, yeasts, not capable of photosynthesis
Plantae	Eukaryotic	Multicellular organisms with chloroplasts capable of photosynthesis
Animalia	Eukaryotic	Multicellular organisms, many with complex organ systems

How many different organisms have been identified by biologists?

Approximately 1.5 million different species of microorganisms, plants, and animals have been described and formally named. Some biologists believe this is only a fraction of species that exist, estimating that there are more than 10 million species waiting to be discovered, classified, and named. It is estimated that 15 percent of all species are marine organisms. Most scientists agree that only 5 percent of bacteria, fungi, nematode, and mite species have been discovered.

BACTERIA

How has the **classification** of **bacteria evolved**?

Early systems for classifying bacteria were based on structural and morphological elements displayed by the organisms (i.e., shape, size, and the presence or absence of several elements [capsules, photosynthetic or non-photosynthetic capabilities, flagella, endospores, etc.]). Other approaches relied on differential stains such as the Gram stain. More recently, genetic and molecular characteristics have been employed to reflect true evolutionary relatedness.

Who **developed** the **Gram stain** and why is it **important**?

Staining is a process in which a scientist colors a microorganism with a dye that emphasizes certain structure. The Gram stain was developed in 1884 by Hans Christian Gram (1853–1938), a Danish physician who worked at a hospital in Berlin, Germany. The Gram stain is one of the first steps in the identification of bacteria, and is used to categorize bacteria into two major groups: gram-positive and gram-negative. When the Gram stain is combined with other information on cellular morphology and biochemical characteristics, conclusive identification of an unknown type of bacteria can usually be made. The Gram stain is the most widely used stain in microbiology.

What is the most **important structural difference** between **gram-negative bacteria** and **gram-positive bacteria**?

The most important structural difference between gram-negative and gram-positive bacteria is that gram-negative bacteria are enclosed by two bilayers—a cytoplasmic membrane and an outer membrane. Between the two membranes lies a thin peptidoglycan layer that is linked to the outer membrane. In contrast, gram-positive bacteria have a thick layer of peptidoglycan. The cell wall of gram-positive bacteria is between two to eight times as thick as the cell wall of gram-negative bacteria. The thinness of the cell wall of gram-negative bacteria affects the bacteria's ability to retain the crystal violet-iodine complex of the Gram stain.

Which **criteria** were used to form what is referred to as the **classical approach** of **bacterial taxonomy**?

In this approach bacteria are grouped into genera and species on the basis of the following characteristics: 1) Structural and morphological characteristics including shape, size, arrangement, capsules, flagella, endospores, and Gram stain. 2) Biochemical and physiological traits such as optimum temperature and pH ranges for growth, oxygen requirements, growth factor requirements, respiration and the fermentation of end products, antibiotic sensitivities, and types of carbohydrates used as an energy source.

What is *Bergey's Manual of Determinative Bacteriology*?

Bergey's Manual of Determinative Bacteriology is an extensive reference manual used for bacterial classification. The first edition was published in 1923 under the sponsorship of the Society of American Bacteriologists (organized in 1899 and now known as the American Society for Microbiology). This reference work was first conceived by David H. Bergey (1860–1937) with the assistance of a special committee of the Society of American Bacteriologists chaired by Francis C. Harrison. The most recent edition was published in four volumes in 2001 under the name *Bergey's Manual of Systematic Bacteriology*.

What are **Archaebacteria**?

Archaebacteria (domain Archaea) are primitive bacteria that often live in extreme environments. This domain includes the following: 1) Thermophiles ("heat lovers"), which live in very hot environments, including the hot sulfur springs of Yellowstone National Park, which reach temperatures ranging from 140 to 176°F (60 to 80°C). 2) Halophiles ("salt lovers"), which live in locations with high concentrations of salinity, such as the Great Salt Lake in Utah, which has salinity levels that range from 15 to 20 percent. Seawater normally has a level of salinity of 3 percent. 3) Methanogens, which obtain their energy by using hydrogen gas (H_2) to reduce carbon dioxide (CO_2) to methane gas (CH_4).

How many **groups** are identified in the **domain Bacteria**?

Biologists recognize at least a dozen different groups of bacteria.

Major Group	Gram Reaction	Characteristics	Examples
Actinomycetes	Positive	Produce spores and antibiotics; live in soil environment.	*Streptomyces*
Chemoautotrophs	Negative	Live in soil environment; important in the nitrogen cycle.	*Nitrosomonas*
Cyanobacteria	Negative	Contain chlorophyll and are capable of photosynthesis; live in aquatic environment.	*Anabaena*

Major Group	Gram Reaction	Characteristics	Examples
Enterobacteria	Negative	Live in intestinal and respiratory tracts; ability to decompose materials; do not form spores; pathogenic.	*Escherichia, Salmonella, Vibrio*
Gram-positive cocci	Positive	Live in soil environment; inhabit the skin and mucous membranes of animals; pathogenic to humans.	*Streptococcus, Staphylococcus*
Gram-positive rods	Positive	Live in soil environment or animal intestinal tracts; anaerobic; disease-causing.	*Clostridia, Bacillus*
Lactic acid bacteria	Positive	Important in food production, especially dairy products; pathogenic to animals.	*Lactobacillus, Listeria*
Myxobacteria	Negative	Move by secreting slime and gliding; ability to decompose materials.	*Chondromyces*
Pseudomonads	Negative	Aerobic rods and cocci; live in soil environment.	*Pseudomonas*
Rickettsias and chlamydias	Negative	Very small, intracellular parasites; pathogenic to humans.	*Rickettsia, Chlamydia*
Spirochetes	Negative	Spiral-shaped, live in aquatic environment.	*Treponema, Borrelia*

What is the **most abundant group** of **organisms**?

The eubacteria are the most abundant group of organisms on Earth. More living eubacteria inhabit the human mouth than the total number of mammals living on Earth.

When were **bacteria discovered**?

Antoni van Leeuwenhoek (1632–1723), a Dutch fabric merchant and civil servant, discovered bacteria and other microorganisms in 1674 when he looked at a drop of pond water through a glass lens. Early, single-lens instruments produced magnifications of 50 to 300 times real size (approximately one-third of the magnification produced by modern light microscopes). Primitive microscopes provided a perspective into the previously unknown world of small organisms, which von Leeuwenhoek called "animalcules" in a letter he wrote to the Royal Society of London. Because of these early investigations, von Leeuwenhoek is considered to be the "father of microbiology."

How did the **discovery** of **bacteria** impact the theory of **spontaneous generation**?

The theory of spontaneous generation proposes that life can arise spontaneously from nonliving matter. One of the first scientists to challenge the theory of spontaneous generation was the Italian physician Francesco Redi (1626–1698). In 1668 Redi performed an experiment to show that meat placed in covered containers (either glass-covered or gauze-covered) remained free of maggots, while meat left in an uncovered container eventually became infested with maggots from flies laying their eggs on the meat. After the discovery of microorganisms by Antoni van Leeuwenhoek (1632–1723), the controversy surrounding spontaneous generation was renewed, as it had been assumed that food became spoiled by organisms arising spontaneously within food. In 1776 Lazzaro Spallanzani (1729–1799) showed that no growth occurred in flasks that were boiled after sealing. The controversy over the theory of spontaneous generation was finally solved in 1861 by Louis Pasteur (1822–1895). He showed that the microorganisms found in spoiled food were similar to those found in the air. He concluded that the microorganisms that caused food to spoil were from the air and did not spontaneously arise.

Who were the **founders of** modern bacteriology?

German bacteriologist Robert Koch (1843–1910) and French chemist Louis Pasteur (1822–1895) are considered the founders of bacteriology. In 1864 Pasteur devised a way to slowly heat foods and beverages to a temperature that was high enough to kill most of the microorganisms that would cause spoilage and disease, but would not ruin or curdle the food. This process is called pasteurization.

By demonstrating that tuberculosis was an infectious disease caused by a specific species of *Bacillus*, Koch in 1882 set the groundwork for public-health measures that would go on to significantly reduce the occurrences of many diseases. His laboratory procedures, methodologies for isolating microorganisms, and four postulates for determining agents of disease gave medical investigators valuable insights into the control of bacterial infections.

What period of time has come to be known as the **golden age** of **microbiology**?

The era known as the "golden age" of microbiology began in 1857 with the work of Louis Pasteur (1822–1895) and lasted about sixty years. During this period of time, there were many important scientific discoveries. Joseph Lister's (1827–1912) practice of treating surgical wounds with a phenol solution led to the advent of aseptic surgery. The advancements Paul Ehrlich (1854–1915) made to the theory of immunity synthesized the "magic bullet", an arsenic compound that proved effective in treating syphilis in humans. In 1884 Elie Metchnikoff (1845–1916), an associate of Pasteur, published a report on phagocytosis. The report explained the defensive process in which the body's white blood cells engulf and destroy microorganisms. In 1897 Masaki Ogata reported that rat fleas transmitted bubonic plague, ending the centuries-old mystery of how plague was transmitted. The following year, Kiyoshi Shiga (1871–1957) isolated the bacterium responsible for bacterial dysentery. This organism was eventually named *Shigella dysenteriae*. During the "golden age" of microbiology, researchers identified the specific microorganism responsible for numerous infectious diseases. The following chart identifies many of these diseases, their infectious agent, who discovered them, and the year they were discovered.

Diseases and Discoverers

Disease	Infectious Agent	Discoverer	Year Discovered
Anthrax	*Bacillus anthracis*	Robert Koch	1876
Gonorrhea	*Neisseria gonorrhoeae*	Albert L. S. Neisser	1879
Malaria	*Plasmodium malariae*	Charles-Louis Alphonse Laveran	1880
Wound infections	*Staphylococcus aureus*	Sir Alexander Ogston	1881
Tuberculosis	*Mycobacterium tuberculosis*	Robert Koch	1882
Erysipelas	*Streptococcus pyogenes*	Friedrich Fehleisen	1882
Cholera	*Vibrio cholerae*	Robert Koch	1883
Diphtheria	*Corynebacterium diphtheriae*	Edwin Klebs and Friedrich Löffler	1883–1884

Disease	Infectious Agent	Discoverer	Year Discovered
Typhoid fever	*Salmonella typhi*	Karl Eberth and Georg Gaffky	1884
Bladder infections	*Escherichia coli*	Theodor Escherich	1885
Salmonellosis	*Salmonella enteritidis*	August Gaertner	1888
Tetanus	*Clostridium tetani*	Shibasaburo Kitasato	1889
Gas gangrene	*Clostridium perfringens*	William Henry Welch and George Henry Falkiner Nuttall	1892
Plague	*Yersinia pestis*	Alexandre Yersin and Shibasaburo Kitasato	1894
Botulism	*Clostridium botulinum*	Emile Van Ermengem	1897
Shigellosis	*Shigella dysenteriae*	Kiyoshi Shiga	1898
Syphilis	*Treponema pallidum*	Fritz R. Schaudinn and P. Erich Hoffman	1905
Whooping cough	*Bordetella pertussis*	Jules Bordet and Octave Gengou	1906

How did **Louis Pasteur's theory of fermentation** differ from the accepted concept of fermentation?

Louis Pasteur (1822–1895) proposed that fermentation is a process carried out by what he referred to as "living ferments." The other renowned chemists of the time believed that fermentation was a purely chemical process in which microorganisms were a byproduct, not the cause.

What are the **main components** of a **bacterial cell**?

The major components of a bacterial cell are the plasma membrane, cell wall, and a nuclear region containing a single, circular DNA molecule. Plasmids—small circular pieces of DNA that exist independently of the bacterial chromosome—are also present in a bacterial cell. In addition, some bacteria may have flagella, which aids in movement; pili or fimbriae, which are short, hairlike appendages that help bacteria adhere to various surfaces, including the cells that they infect; or a capsule of slime around the cell wall that protects it from other microorganisms.

Do **bacteria** all have the **same shape**?

Bacteria have three main shapes—spherical, rod-shaped, and spiral. Spherical bacteria, known as cocci, occur singularly in some species and as groups in other species. Cocci have the ability to stick together and form a pair (diplococci); when they stick together in long chains, they are called streptococci. Irregularly-shaped clumps or clusters of bacteria are called staphylococci. Rod-shaped bacteria, called bacilli, occur as single rods or as long chains of rods. Spiral- or helical-shaped bacteria are called spirilla.

Is the **color** of **bacteria** of **significance**?

The pigments that occur in bacteria are helpful in the identification of bacteria. Bacteria can be red, purple, green, or yellow. Some bacteria produce pigments only under certain environmental conditions, such as particular temperatures.

What is a **petri dish**, and **who developed** it?

A petri dish is a shallow dish consisting of two round, overlapping halves. The petri dish is used to grow bacteria and other microorganisms on a solid culture medium, usually nutrient agar. The top of the dish is larger than the bottom so that when the dish is closed, a strong seal is created, preventing contamination of the culture. This device was developed in 1887 by Julius Richard Petri (1852–1921), a member of Robert Koch's laboratory. Petri dishes are very easy to use, can be stacked on each other to save space, and are one of the most common items in a microbiology laboratory.

Are any **bacteria visible** to the **naked eye**?

Epulopiscium fishelsoni, which lives in the gut of the brown surgeonfish (*Acanthurus nigrofuscus*), is visible to the naked eye. It was first identified in 1985 and mistakenly classified as a protozoan. Later studies analyzed the organism's genetic material and proved it to be a bacterium of unprecedented size: 0.015 in (0.38 mm) in diameter, or about the size of a period in a small-print book.

What **organism** has the **smallest cellular genome**?

The microorganism *Mycoplasma genitalium* has the smallest known cellular genome, consisting of 580,070 "base pairs." A base pair is a couple of hydrogen-bonded, nitrogenous bases—one purine and one pyrimidine—that join the component strands of the DNA double helix. By comparison, *Escherichia coli* has 4,639,221 base pairs.

What are the **four phases** of the **bacterial population growth curve**?

The four phases of the bacterial population growth curve are: lag phase, exponential phase (also called logarithmic), stationary phase, and death phase (also called decline). During the lag phase there is no increase in cell numbers, although the bacteria are synthesizing enzymes present in their environment in preparation for the exponential phase. During the exponential or logarithmic phase, the bacterial population grows at a rate that doubles the population during the generation time. The stationary phase incurs neither an increase nor a decrease in the cell population. The population growth cannot continue at the exponential rate since the nutrient supplies have been depleted and waste products have accumulated. The final phase of the bacterial population growth curve is the death phase, during which more cells die than are replaced by new cells.

What is the **generation time** for various **bacteria**?

Generation time is defined as the time required for a bacterial population to double in its number. If a culture tube is inoculated with one cell that divides every 20 minutes, the total cell population will grow to two cells after a period of 20 minutes, and four cells after 40 minutes; the growth will continue at this rate.

Generation Time for Selected Bacteria

Bacterium	Temperature (°C)	Generation time (minutes)
Escherichia coli	37	17
Shigella dysenteriae	37	23
Salmonella typhimurium	37	24
Pseudomonas aeruginosa	37	31
Staphylococcus aureus	37	32
Bacillus subtilis	36	35
Clostridium botulinum	37	35
Streptococcus lactis	30	48
Lactobacillus acidophilus	37	66
Mycobacterium tuberculosis	37	792

What **effect** do **pH** levels have on the **growth** of **bacteria**?

pH is the measure of the hydrogen ion activity of a solution. The pH scale ranges from 0 (very acidic) to 14 (extremely alkaline or basic). The pH, or concentration of hydrogen ions (H$^+$) in an environment, is critical to bacterial growth because it can affect enzyme activity. An extremely high or low pH can denature and inactivate enzymes, or disrupt cell processes. An environment's pH level dramatically affects the growth of bacteria and other microorganisms. Each species has an optimum pH level to sustain growth, as well as a range of pH levels in which they are able to survive. Acidophiles have their growth optimum between pH 0 and 5.5; neutrophiles, between pH 5.5 and 8.0; and alkalophiles prefer the pH range of 8.5 to 11.5. Extreme alkalophiles have optimum growth at pH 10 or higher. The following table shows the pH ranges and the optimum pH that several different organisms require for growth.

Organism	pH Range for Growth	pH Optimum for Growth
Thiabacillus thiooxidans	1.0–6.0	2.0–3.5
Lactobacillus acidophilus	4.0–6.8	5.8–6.6
Escherichia coli	4.4–9.0	6.0–7.0
Clostridium sporogenes	5.0–9.0	6.0–7.6
Nitrobacter sp.	6.6–10.0	7.6–8.6
Nitrosomonas sp.	7.0–9.4	8.0–8.8

How does a **bacterium's requirement for oxygen** act as a basis for **classification**?

Bacteria are divided into four major groups on the basis of their response to oxygen. Aerobic bacteria grow in the presence of oxygen. Microaerophilic bacteria grow best at oxygen concentrations lower than those present in air —less than 20 percent of oxygen. Anaerobic bacteria grow best in the absence of oxygen. Facultative anaerobes can grow in the presence or absence of oxygen, with more growth when oxygen is present.

How do **bacteria reproduce**?

Bacteria reproduce asexually, by binary fission—a process in which one cell divides into two similar cells. First the circular, bacterial DNA replicates, and then a transverse wall is formed by an ingrowth of both plasma membrane and the cell wall.

Do **bacteria** ever **reproduce sexually**?

Although sexual reproduction involving the fusion of gametes does not occur in bacteria, genetic material is sometimes exchanged between bacteria. This is done in three different ways. The first method is a transformation in which fragments of DNA are released by a broken cell and taken in by another bacterial cell. The second possibility is a transduction in which a bacteriophage carries genetic material from one bacterial cell to another. Lastly, conjugation in which two cells of different mating types come together and exchange genetic material is also possible.

How **quickly** do **bacteria reproduce**?

In a favorable environment, bacteria can reproduce very rapidly. Favorable circumstances include laboratory cultures or the natural habitat. For example, under optimal conditions, *Escherichia coli* can divide every twenty minutes. A laboratory culture started with a single cell can produce a colony of 10^7 to 10^8 bacteria in about twelve hours.

What is the **relationship** between **bacteria** and **temperature**?

All microorganisms have temperature ranges that determine growth. Overall, microorganisms are unique in their ability to exist and grow at temperatures ranging from 14°F to 230°F (-10°C to 110°C). Temperature restrictions are due to limitations in cell metabolism. A microorganism's maximum temperature is the highest temperature at which growth can occur; minimum temperature is the lowest temperature at which growth can occur. A microorganism's optimum temperature is the temperature at which the growth rate is the fastest. The maximum, minimum, and optimum temperatures define the range of growth for each microorganism and are collectively referred to as the cardinal temperatures. Bacteria are divided into four groups on the basis of their cardinal temperatures for growth: psychrophiles, mesophiles, thermophiles, and extreme thermophiles. The following chart lists the temperature ranges at which these groups can grow.

Bacteria Group	Possible Temperature	Optimum Temperature
Psychrophiles	14°F to 77°F (-10°C to 25°C)	50°F to 68°F (10°C to 20°C)
Mesophiles	50°F to 113°F (10°C to 45°C)	68°F to 104°F (20°C to 40°C)
Thermophiles	86°F to 176°F (30°C to 80°C)	104°F to 158°F (40°C to 70°C)
Extreme thermophiles	176°F (80°C) and above	

What is the **role** of **plasmids**?

Plasmids are small, circular molecules of DNA containing genetic information. They contain about 2 percent of the genetic information of a cell and are separate from

> ## How many bacteria are found in soil?
>
> **A** teaspoon of agricultural soil contains approximately 10 million bacteria. There are between 200 and 500 lbs (91 and 27 kg) of bacteria in one acre of soil.

chromosomes. Although plasmids are not essential to the life of a bacterium, they determine a cell's resistance to antibiotics, commonly referred to as the R factor or resistance factor. Certain plasmids allow the transfer of genetic material, which is essential in genetic engineering.

How are **bacteria classified** on the basis of **metabolic activity**?

Bacteria are either heterotrophic or autotrophic. Heterotrophs rely on organic compounds for carbon and energy needs, while autotrophs require inorganic nutrients and carbon dioxide as their sole source of carbon. Most bacteria are heterotrophs, and must obtain organic compounds from other organisms. The majority of heterotrophs are free-living saprobes (also known as saprophytes or saprotrophs), and obtain their nourishment from dead, organic matter. Autotrophs can be photosynthetic or chemosynthetic. Photosynthetic autotrophs obtain their energy from light, while chemosynthetic autotrophs obtain their energy by oxidizing inorganic chemicals.

Where are **bacteria found**?

Bacteria inhabit every place on Earth—including places where no other organism can survive. Bacteria have been detected as high as 20 mi (32 km) above the Earth and 7 mi (11 km) deep in the waters of the Pacific Ocean. They are found in extreme environments, such as the Arctic tundra, boiling hot springs, and our bodies.

How do bacteria **survive without nutrients**?

When a population of bacteria loses its food supply, many bacteria dehydrate. During this process bacteria produce a thick, tough spore coat. As spores, bacteria are able to rest for long periods of time. When conditions become favorable, the spores become active again. The bacteria absorb water, break down their thick, tough spore coats, and begin to form new cell walls.

How **long** can **bacterial spores** remain **dormant**?

In 1995 scientists revived *Bacillus* spores from the digestive tract of an amber-encased, stingless Dominican bee, *Proplebeia dominicana*, which lived from 25 million to 40 million years ago. Previously documented records based on ampules that Louis Pasteur collected indicated that spores were able to survive for only seventy years.

What are **bioluminescent bacteria**?

Bioluminescence is the production of light, with very little heat, by some organisms. The light-emitting substance (luciferin) in most species is an organic molecule that emits light when it is oxidized by molecular oxygen in the presence of an enzyme (luciferase). The enzyme picks up electrons from flavoproteins in the electron transport chain and then emits some of the electron's energy as a photon of light. Bioluminescence is primarily a marine phenomenon occurring in many regions of oceans or seas. One example is the "milky sea" found in the Indian Ocean, where the sea appears to have a soft, white glow.

Who was the first to postulate that **microorganisms** are the **cause** of **disease**?

Robert Koch (1843–1910) was the first to identify that various microorganisms are the cause of disease. His four basic criteria of bacteriology, known as Koch's postulates, are still considered fundamental principles of bacteriology. Koch was awarded the Nobel Prize in Physiology or Medicine in 1905 for his research on tuberculosis.

What are **Koch's postulates**?

Koch's postulates are the four basic criteria an organism must meet in order to be identified as pathogenic (capable of causing disease). The characteristics are as follows: 1) The organism must be found in tissues of animals that have been infected with the disease, rather than in disease-free animals. 2) The organism must be isolated from the diseased animal and grown in a pure culture or *in vitro*. 3) The cultured organism must be able to be transferred to a healthy animal, which will show signs of the disease after having been exposed to the organism. 4) The organism must be able to be isolated from the infected animal.

Why are some **strains** of **bacteria pathogenic** and other strains not?

Strains of bacteria possess genetic differences; these differences are not sufficient for them to be considered as separate species, but each strain is distinctive. For example, there are many different strains of *Escherichia coli* (*E. coli*). Some, such as *E. coli*

0157:H7, cause serious diseases, while others live in the intestine and can be considered beneficial, as they aid digestion.

How **dangerous** is *Clostridium botulinum*?

The bacterium *Clostridium botulinum* can grow in food products and produce a toxin called botulinum, the most toxic substance known. Microbiologists estimate that one gram of this toxin can kill 14 million adults! This bacterium can withstand boiling water (212°F or (100°C) but is killed in five minutes at 248°F (120°C). This tolerance makes *Clostridium botulinum* a serious concern for people who can vegetables at home. If home canning is not done properly, this bacterium will grow in the anaerobic conditions of the sealed container and create extremely poisonous food. The endospores of *Clostridium botulinum* can germinate in poorly prepared canned goods, so individuals should never eat food from a can that appears swollen, as it is a sign that the can has become filled with gas released during germination. Consuming food from a can containing endospores that have undergone germination can lead to nerve paralysis, severe vomiting, and even death.

What is **Botox**?

Botox, the trade name for botulinum toxin type A, is a protein produced by the bacterium *Clostridium botulinum*. Although it is the same toxin that causes food poisoning, purified botulinum toxin that is sterile and has been converted to a form that can be injected and used in a medical setting. Botox was first approved by the Food and Drug Administration (FDA) in December 1989 to treat two eye muscle disorders, uncontrollable blinking (blepharospasm) and misaligned eyes (strabismus). In 2000 the toxin was approved to treat cervical dystonia, a neurological movement disorder that causes severe neck and shoulder contractions. In small doses it is able to block nerve cells from releasing a chemical called acetylcholine, which signals muscle contractions. By selectively interfering with a muscle's ability to contract, existing frown lines are smoothed out, improving the appearance of the surrounding skin.

What **diseases** are **caused** by **rickettsiae** and **chlamydiae**?

Organism	Disease
Chlamydia trachomatis	Trachoma, Lymphogranuloma venereum (LGV), Nongonococcal urethritis (NGU)

99

Organism	Disease
Coxiella bumeti	Q fever
Rickettsia prowazeki	Epidemic typhus
Rickettsia rickettsii	Rocky Mountain spotted fever
Rickettsia typhi	Endemic typhus

Are **rickettsiae** and **chlamydiae** bacteria or viruses?

For many years, rickettsiae and chlamydiae were thought to be viruses because they are very small and are intracellular parasites. They are now known to be bacteria because they possess both DNA and RNA, have cell walls similar to those found in gram-negative bacteria, divide by binary fission, and are susceptible to antibiotics that produce an effect in most bacteria.

What are **mycoplasmas**?

Mycoplasmas are the smallest, free-living organisms, and the only bacteria that exist without a cell wall. Some mycoplasmas have sterols (a type of lipid lacking fatty acids) in their plasma membranes, which provide the strength a membrane needs to maintain cellular integrity without a wall. Since mycoplasmas lack cell walls that provide shape and rigidity, they have no definitive forms and are pleomorphic. *Mycoplasma pneumoniae* causes a disease known as primary atypical pneumonia (PAP), a mild form of pneumonia confined to the lower respiratory tract. Because mycoplasmas do not have cell walls, penicillin is ineffective in stopping their growth. Tetracycline, which inhibits protein synthesis, is recommended as the antibiotic of choice for treatment of PAP.

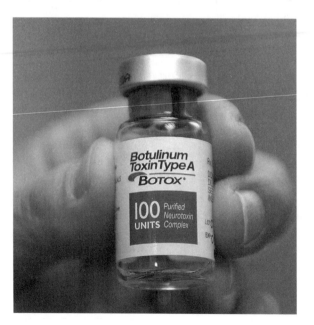

Botox, the trade name for botulinum toxin type A, can smooth frown lines and wrinkles.

What is **anthrax**?

Bacillus anthracis, the etiologic agent of anthrax, is a large, gram-positive, nonmotile, spore-forming, bacterial rod. The three virulence factors of *B. anthracis* are edematous toxins, lethal toxins, and capsular antigens. There are three major, clinical forms of human anthrax: cutaneous, inhalational, and gastrointestinal. If left untreated, anthrax—in all forms—can lead to septicemia and death.

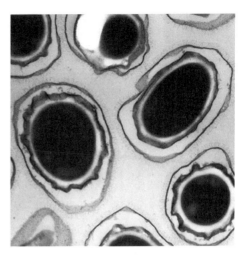

Anthrax spores. There are three major, clinical forms of human anthrax: cutaneous, inhalational, and gastrointestinal.

VIRUSES

What is a **virus**?

A virus is an infectious, protein-coated fragment of DNA or RNA. Viruses replicate by invading host cells and taking over the cell's "machinery" for DNA replication. Viral particles can then break out of the cells, spreading disease.

Are **viruses living** organisms?

Viruses cannot grow or replicate on their own and are inert outside their living host cell. Once they enter a host cell they become active. As such, they are between life and nonlife and are not considered living organisms.

What is the **structure** of a **virus**?

Viruses consist of strands of the genetic material nucleic acid, the basis of a genome, which is surrounded by a protein coat called a capsid. The capsid protects the genome and gives the virus its shape. Viruses may be either helical or icosahedral. Some viruses display a combination of helical and icosahedral symmetry, known as complex symmetry. The capsid is often subdivided into individual protein subunits called capsomeres. The organization of the capsomeres yields the symmetry of the virus. Animal viruses often form an envelope around the capsid. This envelope is rich in proteins, lipids, and glycoprotein molecules.

How do **viruses compare** to **bacteria**?

Characteristic	Bacteria	Viruses
Able to pass through bacteriological filters	No	Yes
Contains a plasma membrane	Yes	No
Contains ribosomes	Yes	No

101

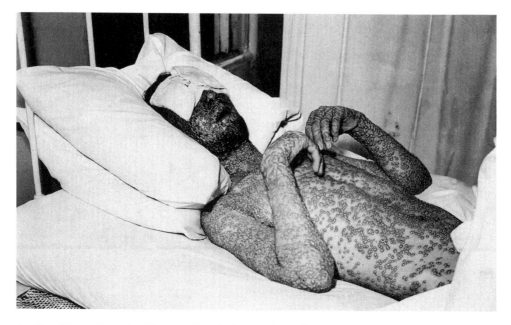

A victim of the smallpox virus. Viruses replicate by invading host cells and taking over the cell's "machinery" for DNA replication.

Characteristic	Bacteria	Viruses
Possesses genetic material	Yes	Yes
Requires a living host to multiply	No	Yes
Sensitive to antibiotics	Yes	No
Sensitive to interferon	No	Yes

What is the **average size** of a **virus**?

Viruses are much smaller than bacteria. The smallest viruses are about 17 nm in diameter, and the largest viruses are up to 1000 nm (1 μm) in length. By comparison, the bacterium *Escherichia coli* is 2000 nm in length, a cell nucleus is 2800 nm in diameter, and an average eukaryotic cell is 10,000 nm in length.

Average Size of Common Viruses

Virus	Size (in nanometers)
Smallpox	250
Tobacco mosaic	240
Rabies	150
Influenza	100

A Chinese health worker wears protective garments to protect him against the SARS virus.

Virus	Size (in nanometers)
Bacteriophage	95
Common cold	70
Polio	27
Parvovirus	20

How do **viruses enter** their **host cells** to **reproduce**?

A virus is able to enter a host cell by either tricking the host cell to pull it inside, as the cell would do to a nutrient molecule, or by fusing its viral coat with either the host cell wall or membrane and then releasing its genes into the host. Some viruses inject their genetic material into the host cell, leaving their empty viral coats outside of the host cell.

Where did **viruses originate**?

The most widely accepted hypothesis is that viruses are bits of nucleic acid that "escaped" from cells. According to this view, some viruses trace their origin to animal cells, some to plant cells, and others to bacterial cells. The variety of origins may explain why viruses are species-specific—that is, why some viruses only infect species that they are closely related to, or the organisms from which they originated. **103**

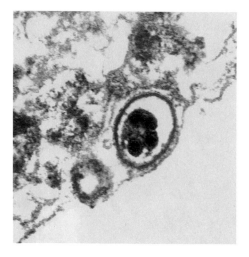

The various herpesviruses, like the one shown here, are DNA viruses.

This hypothesis is supported by the genetic similarity between a virus and its host cell.

Where are **viruses found**?

Viruses lie dormant in any environment (land, soil, air) and on any material. They infect every type of cell—plant, animal, bacterial, and fungal.

What was the **first virus** to be **isolated in a laboratory**?

In 1935 Wendell Stanley (1904-1971) of the Rockefeller Institute (known today as Rockefeller University) prepared an extract of the tobacco mosaic virus and purified it. The purified virus precipitated in the form of crystals. During this investigation Stanley was able to demonstrate that viruses can be regarded as chemical matter rather than as living organisms. The purified crystals retained the ability to infect healthy tobacco plants, thus characterizing them as viruses, not merely chemical compounds derived from a virus. Subsequent studies showed that the tobacco mosaic virus consisted of a protein and a nucleic acid. Further studies showed that this virus consisted of RNA (ribonucleic acid) surrounded by a protein coat. Stanley was awarded the Nobel Prize in Chemistry in 1946 for his discovery.

Do **viruses contain** both **DNA** and **RNA**?

Viruses have either DNA or RNA as their genomic material, whereas cells—including bacteria—have both.

What is the **difference** between **DNA** and **RNA viruses**?

In DNA viruses, the synthesis of viral DNA is similar to how the host cell would normally carry out DNA synthesis. The virus inserts its genetic material into the host's DNA. In RNA viruses, transcription (the first stage of protein synthesis, in which messenger RNA is produced from the DNA) takes place with the help of RNA polymerase.

Is **influenza** a **serious viral disease**?

The influenza virus is probably the most lethal virus in human history. It is an acute respiratory disease characterized by fever, chills, headache, generalized muscular aches, and a frequent cough. Influenza affects people of all ages, but can be particularly severe for the very young, the very old, and people with complications due to other diseases. It is estimated that within 18 months during 1918–1919, there were 200 million cases of influenza, causing 21 million deaths worldwide.

Which animal viruses are DNA viruses and which are RNA viruses?

Viruses	Diseases Caused
DNA Viruses	
Adenoviruses	Approximately forty types of viruses infecting human respiratory and intestinal tracts, causing sore throats, tonsillitis, and conjunctivitis
Herpesviruses	Herpes simplex type 1 (cold sores), herpes simplex type 2 (genital herpes), varicella-zoster (causing chickenpox and shingles)
Papovaviruses	Human warts, degenerative brain diseases, polyomas (tumors)
Parvoviruses	Infections in dogs, swine, arthropods, and rodents; causes gastroenteritis in humans after eating infected shellfish
Poxviruses	Smallpox, cowpox
RNA Viruses	
Paramyxoviruses	Rubeola (measles), mumps, distemper (in dogs)
Orthomyxoviruses	Influenza in humans and other animals
Picornaviruses	Polioviruses, Hepatitis A, human colds; coxsackievirus and echovirus cause aseptic meningitis; enteroviruses infect the intestine; rhinoviruses infect the respiratory tract
Reoviruses	Vomiting and diarrhea
Retroviruses	AIDS, some types of cancer
Rhabdoviruses	Rabies
Togaviruses	Rubella, yellow fever, encephalitis

> **When a virus invades a host cell, how many new viruses are produced?**
>
> **V**iruses can be very prolific. As one example, a poliovirus may produce 100,000 new viruses within a single host cell.

Why is it **difficult** to **treat viral infections** with medications?

Antibiotics are ineffective against viral infections because viruses lack the structures (e.g., a cell wall) with which antibiotics interfere. In general, it is difficult to treat viral infections with medications without affecting the host cell, as viruses use the host cell's machinery during replication. Several antiviral drugs have been developed that are effective against certain viruses.

Disease	Viral pathogen	Antiviral drug
AIDS	Human immunodeficiency virus	Azidathymidine (AZT), didanosine, dideoxycytosine
Chronic hepatitis	Hepatitis B or C	α-interferon
Genital herpes, shingles, chickenpox	Herpesvirus	Acyclovir, idoxuridine, trifluridine, vidarabine
Influenza A	Influenza	Amatadine

What **naturally occurring substance** provides **protection** against **viral infections?**

Interferons protect the adjacent cells against viral penetration. Interferons are glycoproteins produced by body cells upon exposure to a virus. In 1957 Alick Isaacs (1921–1967) and Jean Lindenmann (1924–) identified a group of over twenty substances that were later designated as alpha, beta, and gamma interferons.

What is a **bacteriophage?**

A bacteriophage, also called a phage, is a virus that infects bacteria. The term "bacteriophage" means "bacteria eater" (from the Greek word *phagein*, which means "to devour"). Phages consist of a long nucleic acid molecule (usually DNA) coiled within a polyhedral-shaped protein head. Many phages have a tail attached to the head. Fibers extending from the tail may be used to attach the virus to the bacterium.

When were **bacteriophages first discovered**?

Bacteriophages were discovered in the early 1900s by Frederick W. Twort (1877–1950), a British scientist, and Felix d'Hérelle (1873–1949), a French scientist. In 1915 Twort reported observing a filterable agent that destroyed bacteria growing on solid media; d'Hérelle independently confirmed the discovery in 1917. It was actually d'Hérelle who named the agent "bacteriophage." However, very few scientists accepted these findings and the work on the growth and infectious nature of bacteriophages. It was not until the 1930s that Martin Schlesinger, a German biochemist, characterized bacteriophages, establishing their own unique place in the microbial world.

How are **bacteriophages classified**?

Bacteriophages are classified as either lytic or temperate. Lytic phages destroy the host cell. When a lytic virus infects a susceptible host cell, it uses the host's metabolic machinery to replicate viral nucleic acid and produce viral proteins. There are five steps in this process: attachment, penetration, replication, assembly, and release. During this process, which takes approximately 30 minutes, almost 100 phages are released.

Temperate phages do not always destroy their host cell. After attachment and penetration, the DNA from a temperate phage becomes incorporated into the host bacterial DNA; it is then referred to as a prophage. The prophage replicates at the same time as the bacterial DNA. The viral genes may be repressed indefinitely. Bacterial cells carrying prophages are known as lysogenic cells.

What is the **difference between** a **virus** and a **viroid**?

Viroids are small fragments of nucleic acid (RNA) without a protein coat. They are usually associated with plant diseases and are several thousand times smaller than a virus.

What is a **prion**?

Prions are abnormal forms of natural proteins. Current research indicates that a prion is composed of about 250 amino acids. Despite extensive and continuing investigations, no nucleic acid component has been found. Like viruses, prions are infectious agents.

Who was the **first** to use the **word prion**?

In a paper Stanley Prusiner (1942–) wrote in 1982, he used the term "prion" in place of the expression "proteinaceous infectious particle" when describing an infectious agent. Prusiner was awarded the Nobel Prize in Physiology or Medicine in 1997.

How do **prions work**?

Scientists have not discovered exactly how prions work. Current research shows that prions accumulate in lysosomes. In the brain, it is possible that the filled lysosomes burst and damage cells. As diseased cells die, the prions contained in the cells are released and are able to attack other cells.

What **diseases** have been **linked** to **prions**?

It is thought that prions are responsible for the group of brain diseases known as transmissible spongiform encephalopathies (TSEs). This group includes the disease that is referred to as bovine spongiform encephalopathy (mad cow disease) when it occurs in cattle and Creutzfeldt-Jakob disease when it occurs in humans.

SINGLE-CELLED ORGANISMS

Who **first proposed** the **kingdom Protista**?

The German zoologist Ernst Haeckel (1834–1919) first proposed the kingdom Protista in 1866, for the newly discovered organisms that were neither plant nor animal. The term "protist" is derived from the Greek term *protistos*, meaning "the very first."

What are the **characteristics** of the **protists**?

Protists are a diverse group of organisms. All protists are eukaryotic. Many are unicellular, but they may be multicellular, multinucleate, or exhibit a colonial organization. Although most are microscopic, some are much larger, reaching lengths of nearly 200 ft (60 m). In early, traditional taxonomic schemes, they were united on the basis of being neither plant nor animal nor fungus. Current evidence suggests that protists exhibit characteristics of the plant, animal, and fungi kingdoms.

What are the **major groups** of organisms in the **kingdom Protista**?

There is little agreement among taxonomists on how to classify the protists, but they may be conveniently divided into seven general groups that share certain characteristics of locomotion, nutrition, and reproduction. The following chart exhibits the general groupings:

Group	Characteristics
Sarcodinas	Amoebas and related organisms that have no permanent locomotive structure
Algae	Single-celled and multicellular organisms that are photosynthetic
Diatoms	Photosynthetic organisms with hard shells formed of silica
Flagellates	Organisms that propel themselves through water with flagella

Group	Characteristics
Sporozoans	Nonmotile parasites that spread by forming spores
Ciliates	Organisms that have many short, hairlike structures on their cell surface associated with locomotion
Molds	Heterotrophs with restricted mobility that have cell walls composed of carbohydrates

What are the **differences** between the **fifteen major phyla** of **protists**?

The great diversity among the protists has made their grouping and classification difficult. Characteristics used to classify protists include mode of locomotion, presence or absence of flagella and cilia, body form and coverings, pigmentation and the ability to conduct photosynthesis, mode of nutrition, and whether the organism is unicellular or multicellular.

Characteristics of the Major Phyla in the Kingdom Protista

Phylum	Morphology	Body form/ covering	Locomotion	Pigmentation/ photosynthesis
Rhizopoda	Single cell	No definite shape; shells in some	Pseudopodia (means "false foot")	None
Foraminifera	Single cell	Shells	Cytoplasmic projections	None
Actinopoda	Single cell	Skeletons	Pseudopods reinforced with microtubules	None
Chlorophyta	Single cell; colonial and multicellular	Cellulose in cell wall	Flagella; some species are nonmotile	Chlorophyll
Rhodophyta	Mostly multicellular	Cellulose	Nonmotile	Chlorophyll and phyco-erythrin
Phaeophyta	Multicellular	Cellulose	Flagella	Chlorophyll
Chrysophyta	Single cell; some colonial	Shells with silica	No flagella; move by gliding over secreted slime	Chlorophyll
Pyrrhophyta	Single cell; some colonial	Mostly cellulose plates	Flagella	Chlorophyll; carotenoids
Euglenophyta	Single cell allowing	Flexible pellicle carotenoids euglenoids to change their shape	Flagella	Chlorophyll;

Phylum	Morphology	Body form/ covering	Locomotion	Pigmentation/ photosynthesis
Zoomastigophora	Single cell	None	Flagella	None
Apicomplexa	Single cell	Spores	Nonmotile	None
Ciliophora	Single cell	None	Cilia	None
Acrasiomycota	Single cell during most of the life cycle; multicellular during reproductive stages	Cellulose (spores)	Pseudopods for single cells	None
Myxomycota	Multinucleate mass of cytoplasm	None	May have flagella	None
Oomycota	Coenocytic mycelium	Cellulose	Flagella	None

How many **different species** of **protists** are known?

Biologists estimate there may be as many as 200,000 species of protists. The estimated number of species identified per phyla as of the early twenty-first century is as follows:

Phylum	Common Name	Number of Species
Rhizopoda	Amoebas	Hundreds
Foraminifera	Foraminiferas	Hundreds
Actinopoda	Actinopods	Hundreds
Chlorophyta	Green algae	7,000
Rhodophyta	Red algae	4,000
Phaeophyta	Brown algae	1,500
Chrysophyta	Diatoms	11,500
Pyrrhophyta	Dinoflagellates	2,100
Euglenophyta	Euglenoids	1,000
Zoomastigophora	Zoomastigotes; flagellates	Thousands
Apicomplexa	Sporozoans	3,900
Ciliophora	Ciliates	8,000
Acrasiomycota	Cellular slime molds	70
Myxomycota	Plasmodial slime molds	500
Oomycota	Water molds	580

Where are **amoebas** found?

Amoebas are found in soil, freshwater, and saltwater. They have no definite body shape and continually change form (the term "amoeba" is derived from the Greek word meaning change) as they move using their pseudopods. Pseudopods, meaning "false feet," are projections of cytoplasm. As the cytoplasm extends and fills, the amoeba are moved. The pseudopods also surround and capture food in a vacuole.

What **disease** is caused by the protist *Entamoeba histolytica*?

Entamoeba histolytica, a parasitic amoeba, causes amoebic dysentery, an intestinal disorder. Estimates indicate that up to 10 million individuals in the United States have parasitic amoebas, but only 2 million exhibit symptoms of the disease. In tropical areas, up to half the population may be infected.

What **disease** is caused by protists of the genus *Trypanosoma*?

Species of *Trypanosoma* are the cause of "sleeping sickness," East Coast fever, and Chagas's disease. All of these diseases are common in tropical areas. The trypanosomes that cause these diseases are transmitted by biting insects, such as the tsetse fly.

What **disease** is caused by protists of the genus *Plasmodium*?

The sporozoan *Plasmodium* causes malaria. An individual of the genus *Plasmodium* enters the human body through a bite of a mosquito of the genus *Anopheles* that has been infected by the protist.

Which **protist** is an **indicator** of **polluted water**?

Euglenoids are unicellular flagellates; many euglenoids are capable of photosynthesis and are autotrophic. They are commonly found in freshwater ponds and puddles. Others do not carry on photosynthesis and are heterotrophic, often found in water with large amounts of organic material. Euglenoids frequently serve as bioindicators and are found in large numbers in polluted waters.

What are **diatoms**?

Diatoms are microscopic algae of the phylum Chrysophyta in the kingdom Protista. Almost all diatoms are single-celled algae and dwell in both fresh- and saltwater; they are abundant in the cold waters of the northern Pacific Ocean and the Antarctic. Diatoms are yellow or brown in color, and are an important food source for marine plankton and many small animals.

Diatoms have hard cell walls; these "shells" are made from silica that has been extracted from the water. It is unclear how the extraction of silica from water is accomplished. When they die, their glassy shells—called frustules—sink to the bottom of the sea and harden into rock called diatomite. One of the most famous and accessible diatomites is the Monterrey Formation along the coast of central and southern California.

What **ciliate characteristics** give the impression they are **complex organisms**, although they are in fact **unicellular**?

Ciliates have many specialized organelles within their single cells. These organelles include cilia for locomotion, two types of nuclei (micronuclei and macronuclei), an oral groove to ingest food, food vacuoles for digestion, a contractile vacuole to regulate water balance, and a cytoproct to expel solid waste particles. Species of the genus *Paramacium,* which are found in freshwater, respond to their environment as if controlled by a nervous system. For example, when they encounter physical barriers, they react by reversing the beating of their cilia in order to proceed in a new direction.

What **differences** are there between **cellular slime molds** and **plasmodial slime molds**?

Although both are called slime molds, cellular and plasmodial slime molds have few features in common and differ in their life cycles. Cellular slime molds are similar to amoebas. They move, feed, and reproduce as a single cell. Plasmodial slime molds (plasmodium) consist of a multinucleate mass of cytoplasm lacking cell walls.

What **slime mold** serves as a **model organism** in **developmental biology**?

Dictyostelium discoideum has been studied as a model for the developmental biology of complex organisms. Under optimum conditions, this organism lives as individual, amoeboid cells. When food is scarce, the cells stream together into a moving mass, resembling a slug, that differentiates into a stalk with a spore-bearing body at its top. This structure releases spores that can grow into a new amoeboid cell. The development from identical, free-living cells to a multicellular organism simulates many of the properties of more complex and complicated organisms.

How did the protist *Phytophthora infestans* influence **Irish history**?

Phytophthora infestans, one of the potato's most lethal pathogens, causes the late blight of potato disease. This pathogen was responsible for the Irish potato famine of 1845–1849. *P. infestans* causes the leaves and stem of the potato plant to decay, eventually causing the tuber to stop growing. In addition, the tubers are attacked by the pathogen and rot. It has been estimated that 1.5 million Irish people emigrated from their country and moved to various parts of the world, but most immigrated to the United States. An estimated 400,000 people perished during the famine due to malnutrition.

What **evidence** has led scientists to believe **land plants evolved** from **green algae**?

Many scientists believe that ancient green algae evolved into land plants. The chloroplasts present in green algae are the same as those of land plants. In addition, green algae have cell walls of similar composition to land plants; both store food, such as starch, in the same manner. Most green algae live in freshwater habitats with highly variable conditions. The ongoing changes in their environment have made them highly adaptable.

APPLICATIONS

What is **pasteurization**?

Pasteurization is the process of heating liquids, such as milk, to destroy microorganisms that can cause spoilage and disease. This process was developed by Louis Pasteur (1822–1895) as a method to control the microbial contamination of wine. Pasteurization is commonly used to kill pathogenic bacteria, such as *Mycobacterium*, *Brucella*, *Salmonella*, and *Streptococcus*, common to milk and other beverages.

There are three methods for pasteurizing milk. In the first method, low-temperature holding (LTH), milk is heated to 145°F (62.8°C) for thirty minutes. In the second method, high-temperature short-time (HTST), milk is exposed to a temperature of 161°F (71.7°C) for fifteen seconds. This technique is also known as flash pasteurization. The most recent method allows milk to be treated at 286°F (141°C) for two seconds; this approach is referred to as ultra-high temperature (UHT) processing. Shorter-term processing results in improved flavor and extended product shelf life.

How were **bacteria involved** in **World War I**?

During World War I the British needed the organic solvents acetone and butanol. Butanol was required for the production of artificial rubber, and acetone was used as a solvent in the manufacture of the smokeless explosive powder cordite. Prior to 1914, acetone was made by the dry heating (pyrolysis) of wood. Between 80 and 100 tons of birch, beech, or maple wood were required to make one ton of acetone. When the war broke out, the demand for acetone quickly exceeded the existing world supply. However, by 1915 Chaim Weizmann (1874—1952) had developed a fermentation process by which the anaerobic bacterium *Clostridium acetobutylicum* converted 100 tons of molasses or grain into 12 tons of acetone and 24 tons of butanol. British and Canadian breweries were converted into acetone and butanol factories until new fermentation facilities could be constructed. Commercial acetone and butanol were made by this fermentation process until they were replaced by cheaper petrochemicals in the late 1940s and 1950s.

When was the term **"antibiotic" first used**?

The term "antibiotic" means "against life." In 1889 Paul Vuillemin used the term to describe the substance pyocyanin, which he had isolated several years earlier. Pyocyanin inhibited the growth of bacteria in test tubes but was too lethal to be used in disease therapy. Antibiotics are chemical products or derivatives of certain organisms that inhibit the growth of other organisms.

How do **antibiotics destroy** an **infection**?

Antibiotics function by weakening the cell wall, or interfering with the protein synthesis or RNA synthesis of the bacterial cell. For example, penicillin weakens the cell wall to the point that the internal pressure causes the cell to swell and eventually burst. Certain antibiotics are more effective against gram-negative bacteria, while others are more effective against gram-positive bacteria.

What part of the **cell** is **affected by** different groups of **antibiotics**?

Antibiotic Antibiotic	Site of Activity in the Bacterial Cell	Method of Interfering with Bacterial Cell
Aminoglycosides, Chloramphenicol, Clindamycin, Erythromycin, Spectinomycin, Tetracyclines.	Protein synthesis	Inhibit protein synthesis; often bind to ribosomes
Bacitracin, Cefoxitin, Cephalosporins, Penicillin, Vancomycin	Cell wall	Cell wall inhibitors; interfere with cell-wall synthesis
Polymyxins	Cell membrane	Injure the plasma membrane
Metronidazole, Nalidixic acid, Rifampin	Nucleic acid	Inhibit RNA or DNA synthesis
Isoniazid, Sulfa drugs, Trimethoprim	Metabolic reaction	Inhibit the synthesis of certain metabolites; precursors of proteins, DNA, and RNA

Which **antibiotics** are most **effective** against **gram-positive** bacteria and which are most effective against **gram-negative** bacteria?

Penicillin and its semisynthetic derivatives are effective against gram-positive bacteria. The aminoglycosides, such as Gentamicin, neomycin, and kanamycin, are effective against gram-negative bacteria. Antibiotics that are effective against a broad range of gram-positive and gram-negative bacteria are called broad-spectrum antibiotics.

Which **pathogens** are **antibiotic resistant**?

Pathogens that are antibiotic or antimicrobial resistant produce diseases and infections that cannot be treated with standard antibiotics. These microbes have changed and mutated in ways that greatly reduce or eliminate the effectiveness of antibiotic drugs in curing or preventing infections.

What is an example of an **antibiotic-resistant** pathogen?

Staphylococcus aureus, a bacterium that causes a variety of infections, including urinary tract infections and bacterial pneumonia, became resistant to penicillin fifty years ago. Stronger and more effective antibiotics have been developed to treat staph infections caused by the bacterium.

What **factors** have contributed to an **increase** in the number of **resistant bacteria**?

Bacteria mutate in order to adapt to new conditions. A mutation that enables a microbe to survive in the presence of an antibiotic quickly spreads throughout a microbial population. Since bacteria replicate very rapidly, a mutation can swiftly become prevalent.

The overuse of antibiotics promotes the emergence of resistant bacteria. Antibiotics may also be prescribed for viral infections that they are not effective against. Furthermore, patients often fail to follow the directions for taking antibiotics precisely. A prescribed dose of antibiotics should be taken until it is completed. Although an individual can feel better shortly after starting a treatment, not completing the full course of antibiotics often destroys only the most vulnerable bacteria. Relatively resistant bacteria are able to survive and prosper in a human's body. Because antibiotic-resistant strains of bacteria do not respond to standard treatments, illnesses are able to last for longer periods of time and can result in death. The proliferation of resistant bacteria has made it more difficult to establish effective treatments.

Which **antibiotic** is considered the "last resort" for treating resistant strains of **staphylococcal** and **enterococcal** infections?

Staphylococcal and enterococcal infections often have to be treated with Vancomycin, as this antibiotic is lethal to these resistant strains of bacteria often found in hospi-

What percentage of antimicrobial drugs produced in the United States are used in animal feedstocks?

An estimated 40 to 50 percent of the antimicrobial drugs produced annually in the United States are used in animal feed.

tals. Until recently, Vancomycin was effective against these resistant pathogens. However, in the late 1980s a Vancomycin-resistant strain of enterococcus developed and began posing a threat to hospital patients and the health care community. Researchers continue to chemically alter the Vancomycin molecule to help it remain effective as the antibiotic of last resort.

What are the **agricultural uses** of **antibiotics**?

Antibiotics are sprayed onto fruit trees and other food-bearing plants to control disease. In addition, antibiotics are added to feed stocks to prevent disease and improve the growth rate among food-producing animals.

Is there a **link** between **antibiotic use in animal feed** and the increase in bacterial infections becoming **resistant** to antibiotics in humans?

Scientists have discovered a link between agricultural use of antibiotics, particularly in animal feed, and the increase of food-borne infections in humans who consume products derived from these animals. Resistant bacteria present in animals can survive the slaughtering and meat-packaging process. Undercooked meat will harbor these bacteria and when eaten can cause illness in humans. To further complicate the situation, the antibiotics used to treat humans made ill by infections may be similar to those used routinely in animals, rendering them less effective.

How does the use of **antibiotics** in **animal feed promote animal growth**?

Farmers introduced the use of antibiotics to animal feed more than forty years ago. The main reason for antibiotic use was, and continues to be, to keep the animals healthy. Animals are often kept closely together in pens, a condition that promotes the spread of bacterial infections. The use of antibiotics inhibited the incidence and spread of infection among livestock. An unplanned side effect of this treatment was accelerated animal growth. Scientists believe antibiotics suppress the intestinal bacteria *Clostridium perfringens*, which produces toxins that may retard animal growth. After this discovery, farmers began to give their animals antibiotics to promote weight gain as well as to treat infection.

How do **antibiotics differ** from **antibacterials**?

Antibiotics and antibacterials both interfere with the growth and reproduction of bacteria. Antibiotics are medications for humans and animals. Antibacterials, found in soaps, detergents, health and skincare products, and household cleaners, are used to disinfect surfaces and eliminate potentially harmful bacteria.

What are the **differences** between **non-residue-producing antibacterials** and **residue-producing antibacterials**?

Non-residue-producing antibacterials quickly destroy bacteria, rapidly disappear from the surface they were applied due to evaporation or chemical breakdown, and leave no active residue behind. Examples of non-residue-producing antibacterials are alcohols, aldehydes, gaseous substances such as formaldehyde, and halogen-releasing compounds such as chlorine and peroxides.

In contrast, the disinfecting action of residue-producing antibacterials is prolonged because they leave long-acting residues on the surfaces to which they are applied. Examples of residue-producing antibacterials are anilides such as triclocarban, bisphenols such as triclosan, heavy metal compounds such as silver and mercury, and quaternary ammonium compounds such as benzalkonium chloride.

Which **chemical agents** are useful to **control microorganisms**?

Chemical Agents Used to Control Microorganisms

Chemical agent (Category of chemical)	Antiseptic or disinfectant	Antimicrobial spectrum	Applications
Acids	Benzoic acid, lactic, salicylic acid, undecylinic acid, and propionic acids	Many bacteria and fungi	Skin infections, food preservatives
Acridine dyes	Acriflavine, proflavine	Staphylococci, Gram-positive bacteria	Skin infections
Alcohol	70 percent ethyl alcohol	Vegetative bacterial cells, fungi, protozoa, viruses	Instrument disinfectant, skin antiseptic
Carionic detergents	Commercial detergents	Broad variety of microorganisms	Industrial sanitation, skin antiseptic, disinfectant

117

Chemical agent (Category of chemical)	Antiseptic or disinfectant	Antimicrobial spectrum	Applications
Chlorine (Halogen)	Chlorine gas, sodium hypochlorite, chloramines	Broad variety of bacteria, fungi, protozoa, viruses	Water treatment, skin antisepsis, equipment spraying, food processing
Copper (Heavy metal)	Copper sulfate	Algae, some fungi	Algicide in swimming pools, municipal water supplies
Ethylene oxide	Ethylene oxide gas	All microorganisms, including spores	Sterilization of instruments, equipment, heat-sensitive objects (plastics)
Formaldehyde	Formaldehyde	Broad variety of bacteria, fungi, protozoa, viruses	Embalming, vaccine production, gaseous sterilant
Glutaraldehyde	Glutaraldehyde gas, formalin	All microorganisms, including spores	Sterilization of surgical supplies
Hydrogen peroxide	Hydrogen peroxide	Anaerobic bacteria	Wound treatment
Iodine (Halogen)	Tincture of iodine, Iodophors	Broad variety of bacteria, fungi, protozoa, viruses	Skin antisepsis, food processing, preoperative preparation
Mercury (Heavy metal)	Mercuric chloride, merthiolate, metaphen	Broad variety of bacteria, fungi, protozoa, viruses	Skin antiseptics, disinfectants
Phenols	Cresols, hexachlorophene, hexylresorcinol, chlorhexidine	Gram-positive bacteria, some fungi	General preservatives, skin antisepsis with detergent
Silver (Heavy metal)	Silver nitrate	Organisms in burned tissue, gonococci	Skin antiseptic, eyes of newborns
Triphenyl-methane dyes	Malachite green; crystal violet	Staphylococci, some fungi, Gram-positive bacteria	Wounds, skin infections

Do **bacteriostats, sanitizers, disinfectants,** and **sterilizers** control microorganisms the same way?

The Environmental Protection Agency (EPA) classifies antimicrobial agents as non–public health products and public health products. Non–public health products are products used to control the growth of algae, odor-causing bacteria, microorganisms infectious only to animals, and bacteria that cause spoilage, deterioration, or fouling of materials. Examples of non–public health products are jet fuel, paints, and treatments for both textile and paper products. Public health products control microorganisms infectious to humans in any inanimate environment. Bacteriostats, sanitizers, disinfectants, and sterilizers are all public health products.

Sanitizers are used to reduce, not eliminate, microorganisms from inanimate environments to levels considered safe as determined by public health codes or regulations. Non–food contact sanitizers are used on dishes, utensils, and equipment found in dairies and food-processing plants. Non–food contact sanitizers include carpet sanitizers, air sanitizers, laundry additives, and in-tank toilet sanitizers.

Disinfectants are used on hard, inanimate surfaces and objects in order to destroy or irreversibly inactivate infectious bacteria, but not necessarily their spores. Disinfectants are classified according to whether they will be used in hospital or general environments. Hospital disinfectants are critical to infection control and are used on both medical and dental instruments, floors, walls, bed linens, and toilet seats.

Sterilizers, also called sporicides, destroy all forms of microbial life and their spores. Sterilization is crucial to infection control, and the process is widely used in hospitals as well as on medical instruments and equipment. Examples of sterilizers are autoclaves, dry-heat ovens, low-temperature gas, and liquid-chemical sterilants.

In the home setting, does the use of **antibacterial soap reduce** the risk of **infection**?

Research has uncovered little evidence to support claims that the use of antibacterial soap in a domestic setting reduces the risk of infection or prevents infections. These residue-producing antibacterials help control the spread of infection in health-care settings, such as hospitals and nursing homes.

When was a **chemical disinfectant first used** during a **surgical procedure**?

The use of a chemical disinfectant during a surgical procedure was first documented at the Glasgow Royal Infirmary in March 1865. Before surgery on a patient, Joseph Lister (1827–1912) sprayed the air with a fine mist of the phenol carbolic acid and soaked his surgical instruments in a carbolic acid solution. Although the patient later died, Lister continued his experiments. In an 1867 article in the *Lancet*, it was reported that Lister's postoperative surgery mortality rate dropped from 45 percent to 9 percent after he started to use chemical disinfectants in operative settings.

How many **antimicrobial products** are available?

More than 8,000 antimicrobial products are registered with the U.S. Environmental Protection Agency (EPA). The EPA regulates products that kill microbes on inanimate surfaces. The various products contain more than 300 different active ingredients and are available as sprays, liquids, concentrated powders, and gases. More than 50 percent of the 8,000 antimicrobial products are used to control infectious microorganisms in health-care settings.

Which **antibacterial products** are the most **effective** for **general use**?

Non-residue-producing agents, such as hydrogen peroxide and bleach, are effective agents for controlling microbes. Several consumer products with residue-producing agents have proved to be effective for specific conditions. For example, antibacterial toothpaste helps control periodontal disease; antibacterial deodorants suppress odor-causing bacteria; and antidandruff shampoos help control dandruff.

What **criteria** are used to **identify** emerging infectious diseases?

Emerging infectious diseases are ones that are new, changing, have shown an increase in incidence, or have the potential to increase in the near future. There are several criteria used to identify an emerging infectious disease, including: 1) A disease causing symptoms that are clearly different from all other diseases. 2) Diagnostic techniques that allow the identification of a new pathogen. 3) A locally defined disease that becomes widespread. 4) A rare disease that increases in occurrence and becomes common. 5) A mild disease that becomes severe. 6) A slowly developing disease that increases in life span.

What **factors contribute** to the **spread** of emerging **infectious diseases**?

Some of the factors contributing to the spread of emerging infectious diseases include environmental changes, unwarranted use and abuse of antibiotics, and the prevalence of modern transportation, which allows diseases to spread in wide geographic areas very rapidly.

Name of Infectious Agent	Type of Micro-organism	Disease	Year of Emergence	Factors Contributing to Emergence
Borrelia burgdorferi	Bacteria	Lyme disease	1975	Reforestation near homes; increase in deer population.
Legionella pneumophila	Bacteria	Legionnaires' disease	1976	Cooling and plumbing systems.
Staphylococcus aureus	Bacteria	Toxic shock syndrome	1978	Ultra-absorbent tampons.
Escherichia coli O157:H7	Bacteria	Hemolytic-uremic syndrome	1982	Mass food processing, allowing contamination of meat.
Human immuno-deficiency virus (HIV)	Virus	AIDS	1983	Migration to cities, sexual transmission, intravenous drug use, transfusions, organ transplants.
Flavivirus	Virus	Dengue fever	1984	Transportation and travel.
Hepatitis C virus (HCV)	Virus	Hepatitis C	1989	Transfusions, organ transplants, contaminated hypodermic needles, sexual transmission.
Sin nombre virus (SNV)	Virus	Hantavirus pulmonary syndrome	1993	Environmental changes.
Filovirus	Virus	Ebola hemorrhagic fever	1995 (earlier outbreaks in 1975 and 1979)	Unknown.
Coronavirus	Virus	Severe acute respiratory syndrome	2003	Close person-to-person contact.

What are the most **prevalent microbial diseases**?

According to the Centers for Disease Control, the most common microbial diseases in the United States are:

Disease	Number of cases per year (as of 2002)
Gonorrhea	351,852
Salmonellosis	44,264

Disease	Number of cases (as of 2002)
AIDS	42,745
Syphilis	32,871
Lyme disease	23,763
Shigellosis	23,541
Varicella (chickenpox)	22,841
Hepatitis (all types)	18,626
Tuberculosis	15,075
Pertusis (whooping cough)	9,771

Why are bacteria known as pollution fighters?

Through a process known as bioremediation, bacteria are able to degrade many pollutants found in soil and water. They are also capable of altering a harmful substance so it becomes harmless or even beneficial to its environment. Scientists are working to improve bacteria's efficiency as a natural pollution fighter.

Which bacteria were used to clean the beach after the Exxon *Valdez* oil spill?

Bacteria from the genus *Pseudomonas* are capable of degrading oil as their source of nutrition. This microorganism was used to clean up the spill that emerged from the Exxon *Valdez*. In order to improve the efficiency of this bacteria, nitrogen and phosphorus were introduced to beaches polluted by the spill. The number of oil-degrading bacteria on the beaches increased sufficiently enough to remove much of the oil from the beaches.

What are some examples of fermented foods?

Fermented foods are foods produced by microorganisms. A combination of bacteria and fungi are used in the fermentation process of many food products. Examples of fermented foods are cheese, sour cream, yogurt, sauerkraut, vinegar, olives, pickles, soy sauce, miso, chocolate, coffee, and most alcoholic beverages.

Which bacteria are essential in the production of various food products?

Food	Microorganism
Buttermilk and sour cream	*Streptococcus cremoris, Leuconostoc citrovorum*
Pickles	*Enterobacter aerogenes, Leuconostoc spp., Lactobacillus brevis*
Sauerkraut	*Leuconostoc spp., Lactobacillus brevis*
Swiss cheese	*Lactobacillus spp., Propionibacterium spp.*
Vinegar	*Acetobacter aceti*
Yogurt	*Streptococcus thermophilus, Lactobacillus bulgaricus*

spp. = unspecified species

What is the **difference** between **regular acidophilus milk** and **sweet acidophilus milk**?

Both acidophilus milk and sweet acidophilus milk are inoculated with the bacterium *Lactobacillus acidophilus*. Many health practitioners believe that this bacterium, a normal member of the human intestinal flora, aids in digestion. Regular acidophilus milk is produced by adding the bacterium to vats of skim milk as part of the fermentation process; it has a characteristic sour taste. When the bacterium is added to pasteurized milk and packaged without fermentation, sweet acidophilus milk, which lacks the characteristic sour taste, is produced.

After the Exxon *Valdez* oil spill, bacteria from the genus *Pseudomonas* were used to clean up the spill.

How were the **white cliffs** of **Dover, England,** formed?

The white cliffs of Dover are composed of a variety of protist fossil shells, including coccolithophores (a type of algae) and foraminiferans. The shells of dead foraminiferans are deposited on the bottom of the ocean and go on to form gray mud. The gray mud is gradually transformed into limestone (chalk). Geologic uplifting brings the formations of limestone up to land.

What are some **commercial uses** of **red algae**?

The cell walls of red algae contain a mucilaginous outer component usually composed of agar and carrageenan. Agar is used for making gel capsules, a material that is used to make dental impressions and as a base for cosmetics. It is also the basis of the scientific laboratory media used for growing bacteria, fungi, and other organisms. Agar is also used to prevent baked goods from drying out and for rapid-setting jellies. Carrageenan is used as a stabilizer for emulsions such as paints, cosmetics, ice cream,

The white cliffs of Dover, England, are made up of various protist fossil shells.

and many dairy products. The genus *Porphyra*, commonly called "nori," is cultivated as a food and is specifically popular in Japan.

What is **agar**?

Agar is a polysaccharide extract of red algae that is used as a solidifying material in microbiological media. Agar was developed as a culture media for bacteria by Robert Koch (1843–1910). Koch was interested in the isolation of bacteria in pure culture. Because isolation was difficult in liquid media, he began to study ways in which bacteria could be grown on solid media. After sterile, boiled potatoes proved unsatisfactory, a better alternative was suggested by Fannie E. Hesse (1850–1934), the wife of Walther Hesse (1846–1911), who was one of Koch's assistants. She suggested that agar, which she had used to thicken sauces, jams, and jellies, be used to solidify liquid nutrient broth. Agar is generally inexpensive and, once jelled, does not melt until reaching a temperature of 212°F (100°C). If 1–2 g of agar are added to 100 ml of nutrient broth, it produces a solid medium that is not degraded by most bacteria.

What are the **commercial uses** of **brown algae**?

Kelp, one of the species of brown algae, is harvested from ocean water for the sodium, potassium salts, iodine, and alginates that it contains. The alginates are carbohydrates used in the formation of gels.

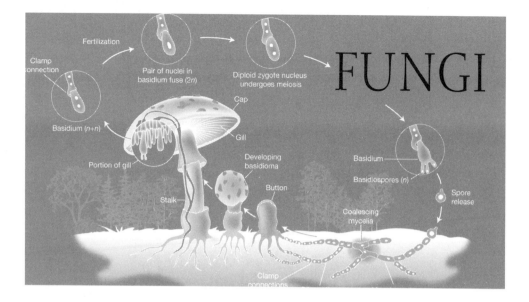

FUNGI

INTRODUCTION AND HISTORICAL BACKGROUND

What **characteristics** do all **fungi** share?

In the earliest classification systems, fungi were classified as plants. The first classification system to recognize fungi as a separate kingdom was proposed in 1784. Researchers identified four characteristics shared by all fungi: fungi lack chlorophyll; the cell walls of fungi contain the carbohydrate chitin (the same tough material a crab shell is made of); fungi are not truly multicellular since the cytoplasm of one fungal cell mingles with the cytoplasm of adjacent cells; fungi are heterotrophic eukaryotes (unable to produce their own food from inorganic matter), while plants are autotrophic eukaryotes.

What is the **scientific study of fungi** called?

The scientific study of fungi is called mycology, which is derived from the Greek word *mycote*, meaning fungus. The first extensive taxonomical study of fungi, *Systemia Mycologicum*, was published by Elias Fries (1794–1878) and Christian Hendrick Persoon (1761–1836) in 1821–32. The book is still considered the authoritative resource on fungal taxonomy.

Who is the "**father of mycology**"?

Elias Fries (1794–1878) is considered the "father of mycology." Born in Sweden, he received a degree in philosophy from Lund University in 1814. His first important work on mycology, *Observationes mycologicae*, was published between 1815 and 1818. He devoted his career to the study of botany, concentrating specifically on fungi and lichens.

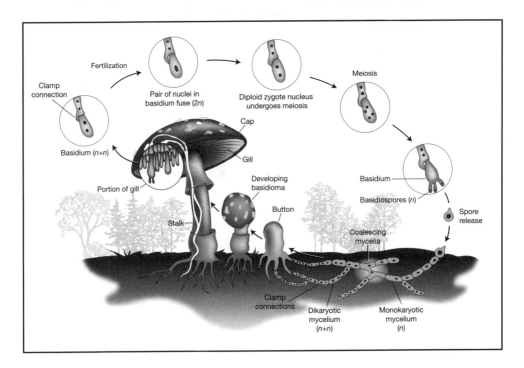

The life cycle of a mushroom.

Which **children's author** studied and **drew illustrations** of fungi?

Beatrix Potter (1866–1943), perhaps best known for having written *The Tale of Peter Rabbit* in 1902, began drawing and painting fungi in 1888. She eventually completed a collection of almost 300 detailed watercolors, which are now in the Armitt Library in Ambleside, England. In 1897 she prepared a scientific paper on the germination of *Agaricineae* spores for a meeting of the Linnean Society of London. Although her findings were originally rejected, experts now consider her ideas correct.

What **organisms** are included in the **kingdom Fungi**?

Members of the kingdom Fungi range from single-celled yeasts to *Armillaria ostoyea*, a species that covers 2,200 acres (890 hectares)! Also included are mushrooms that are commonly consumed, the black mold that forms on stale bread, the mildew that grows on damp shower curtains, rusts, smuts, puffballs, toadstools, shelf fungi, and the death cap mushroom, *Amanita phalloides*. Of the bewildering variety of organisms that live on the planet Earth, perhaps the most unusual and peculiarly different from human beings are fungi. Fungi are able to rot timber, attack living plants, spoil food, and afflict humans with athlete's foot and even worse maladies. Fungi also decompose dead organisms, fallen leaves, and other organic materials. In addition, they produce antibiotics and other drugs, make bread rise, and ferment beer and wine.

How are **different species** of fungi **classified**?

Fungi classification is based mainly on the type of reproductive spore a fungi produces. If the reproductive, sexual spores of a species have not been identified or examined, the fungi are placed in the phyla Deuteromycota. The three phyla of fungi are Ascomycota, Basidiomycota, and Zygomycota. Each phyla has unique reproductive spores: ascospores, basidiospores, and zygospores. A zygospore is a large spore enclosed in a thick wall; it is the result of the fusion of the nuclei of two cells that are morphologically similar to each other. An ascospore is the fusion of two cells that are either morphologically similar or dissimilar. Ascospores are produced in a saclike structure called an ascus. Members of the phyla Ascomycota are known as sac fungi. Basidiospores are formed on a base pedestal called a basidium. Members of the phyla Basidiomycota are known as club fungi due to the shape of the pedestal. The phylum Chytridiomycota has at times been classified in the kingdom Fungi and at other times in the kingdom Protista. Recent evidence obtained from comparisons of protein and nucleic-acid sequences has suggested they belong in the kingdom Fungi.

Characteristics of Fungal Phyla

Phylum	Representative Species	Type of Sexual Spore	Method of Asexual Reproduction
Chytridiomycota (Chytrids)	*Allomyces, Coelomomyces*	None	Zoospores
Zygomycota (Zygote fungi)	*Rhizopus* (black bread mold), *Glomus* (endomycorrhizal fungus)	Zygospore	Nonmotile spores
Ascomycota (Sac fungi)	*Neurospora, Marchella* (edible morels), powdery mildews, *Tuber* (truffles), yeasts	Ascospore	Budding, conidia (nonmotile spores), fragmentation
Basidiomycota (Club fungi)	Poisonous (*Amanita*) and edible (*Agaricus*) mushrooms, bracket fungi, stinkhorns, puffballs, rusts, smuts	Basidiospore	Uncommon
Deuteromycetes (Imperfect fungi)	*Aspergillus, Penicillium*	Sexual reproduction has not been observed	Spores; conidia

How many **species** of fungi have been **identified**?

Scientists have identified between 70,000 and 80,000 species of fungi; almost 2,000 new species are discovered each year. Some mycologists estimate that there are about 1.5 million species worldwide, placing the fungi second only to insects in the number of different species.

What are some of the **best-known deuteromycetes**?

Deuteromycetes are mostly free-living and terrestrial, but some are pathogenic. The best-known pathogenic deuteromycetes include the causal agent (*Aspergillus niger*) of a respiratory disease called aspergillosus, athlete's foot (*Epidermophyton flocco-sum*), ringworm (*Microsporum canis*), and candida "yeast" infection (*Candida albicans*). Some other famous deuteromycetes are species of the genus *Penicillium*, particularly *P. notatum*, for the role it played in the discovery of penicillin; *P. chrysogenum*, for the commercial production of penicillin; *P. griesofulvum*, for the production of griesofulvin (an effective antibiotic against ringworm and athlete's foot); and *P. roquefortii* and *P. camembertii*, which are used to make Roquefort and Camembert cheeses, respectively.

Where are **fungi found**?

Fungi grow best in dark, moist habitats, but they can be found wherever organic material is available. Moisture is necessary for their growth, and they can obtain water from the atmosphere as well as from the medium upon which they live. When the environment becomes very dry, fungi survive by going into a resting stage or by producing spores that are resistant to drying. The optimum pH for most species is 5.6, but some fungi can tolerate and grow in pH levels ranging from 2 to 9. Certain fungi can grow in concentrated salt solutions or sugar solutions such as jelly or jam, which cannot sustain bacterial growth. Fungi also thrive in a wide range of temperatures. Even refrigerated food may be susceptible to fungal invasion.

Since **fungi lack the chlorophyll** necessary to produce their own food, how do they **obtain food**?

Fungi are saprobes that absorb nutrients from waste and dead organisms. Instead of taking food inside its body and then digesting it, as an animal would, a fungus digests

> ## Which organism is considered the largest living organism on Earth?
>
> One specimen of the tree-root fungus *Armillaria ostoyae* is considered the largest living organism on Earth. The species can grow to cover 3.4 mi (5.5 km) in diameter; a specimen in Oregon is known to cover more than 2,200 acres (890 hectares) of forest, comprising hundreds of tons in weight. The specimen is estimated to be 2,400 years old.

food outside its body by secreting strong, hydrolytic enzymes onto the food. In this way, complex organic compounds are broken down into simpler compounds that a fungus can absorb through the cell wall and cell membrane.

Why are **fungi disappearing**?

Gourmets with a taste for the subtle flavors of fresh, wild mushrooms are discovering that those delicacies are increasingly hard to find. In Europe, just a few years ago, it was easy to pick a basket of one of the most-prized fungi, the apricot chanterelle. However, not only are these mushrooms becoming scarce, the ones that do grow wildly do not grow to the same size they once did. In 1975 it took 50 times more chanterelles to make up 2.2 lb (1 kg) than it did in 1958. Other fungi are also becoming rare. For example, the average number of fungal species per 120 square yards (1,000 square meters) in the Netherlands has dropped from 37 to 12. One reason for this decline has been an increase in air pollution. Fungi are more sensitive to air pollution than plants because fungi have no protective covering, whereas the aerial parts of plants are protected by cuticles and bark. In addition, while plants extract water from soil using their roots, some fungi absorb water directly from the atmosphere, along with the pollution that may be present in the air. Poor air quality contributes to the decline of fungi.

Which **organism** is known as the "**humongous fungus**"?

The "humongous fungus" is an enormous, underground fungus that grows in northern Michigan. The fungus, *Armillaria gallica*, was discovered in 1992. It encompasses 35 acres (15 hectares) and is thought to be at least 1,500 years old. Scientists proved that the organism was a single fungus by taking twenty samples of the fungus and performing DNA analyses on sixteen fragments from each sample. The genetic material of each sample was identical.

What is the **relationship** between **fungi** and **ants**?

The leaf-cutting ants, found in Central and South America as well as in the southern United States, have a mutually beneficial, or symbiotic, relationship with certain fungi of the genus *Septobasidium*. The ants are not able to digest the cellulose found in leaves. The fungus breaks down the cellulose—a food source for the fungus—and converts it into carbohydrates and proteins, which the ants can digest. The ants then eat the fungus. The ants provide the fungus with a guaranteed food supply and eliminate competing fungi. The ants and fungus are not known to occur independently from each other.

STRUCTURE

What are the **structures** present in **typical fungi**?

Most fungi are a mass of intertwined filaments known as hyphae that are surrounded by a rigid cell wall. Each hypha cell has a distinct nucleus. Individual cells may be separated by walls called septa. The hyphae form a radially expanding network called the mycelium. Cytoplasm flows freely throughout hyphae, passing through major pores present in the septa. Because of this streaming, proteins synthesized throughout the hyphae may be carried to their tips, which are actively growing. As a result, fungal hyphae may grow very rapidly when food and water are abundant and the temperature is optimum.

Which **organelle** is **absent** in **fungi**?

Centrioles, which divide and organize the spindle filters during mitosis and meiosis, are lacking in all fungi. The nuclear envelope does not break down and reform; instead, a spindle apparatus is formed within the envelope. Fungi regulate the formation of microtubules during mitosis with small, amorphous structures called spindle plaques.

What main **carbohydrate** is **stored** by fungi?

The main carbohydrate stored by fungi is glycogen, which is also the main storage carbohydrate of animals. This fact suggests that fungi are more closely related to animals than plants, which store starch as their main carbohydrate.

What are **imperfect fungi**?

Imperfect fungi are also called deuteromycetes or conidial fungi. They are an assemblage of distinct fungal species that are known to reproduce only asexually; the sexual reproductive features have not been identified and are not used as the basis for classification. In this group, sexual reproduction has not been observed. Most imperfect fungi are thought to be ascomycetes that have lost the ability to reproduce sexually. The best-known members of the deuteromycetes are the genera *Penicillium* and *Aspergillus*. Whenever a mycologist discovers a sexual stage in one of these fungi, the species is reclassified from the imperfect category to a particular phylum; the phylum selected depends on the type of sexual structures.

What is a **fruiting body**?

Macrofungi such as mushrooms and toadstools produce fruiting bodies. A fruiting body is a structure that enables the dispersal of spores for reproduction. It is the structure of a fungus that is visible above the ground. Fruiting bodies are found in a variety of shapes, ranging from the common cap-and-stem mushrooms to the more exotic, antler-like, coral-like, cage-like, trumpet-shaped, or club-shaped mushrooms. The method of spore dispersal for the various types of macrofungi is related to the shape of the fruiting body.

Are all **large fungi** shaped like **mushrooms**?

The fruiting bodies of fungi come in a seemingly endless array of forms and colors. Many are variations on the familiar stalk-and-cap pattern of the common mushrooms sold in stores, although some have minute spore-bearing pores instead of gills on the undersides of their caps. Many fungi do not resemble mushrooms at all. Puffballs are solid, fleshy spheres. Bird's nest fungi form little cups containing "eggs" packed with spores. One kind of fungus looks like a head of cauliflower, and others resemble upright, branching clumps of coral. Some protrude from tree trunks like shelves, while others look like glistening blobs of jelly.

How do **bacterial** and **fungal spores differ**?

The main purpose of bacterial spores—known as endospores—is to protect bacterial cells so they can survive extreme, harsh conditions. Fungi reproduce both sexually and asexually through the formation of spores. Asexual spores are formed by the hyphae of one organism; the organisms that form from these spores are identical to their parents. Sexual spores result from the fusion of nuclei from two strains of the same species of fungus. Organisms from sexual spores derive characteristics from each parent.

What are the main **types** of **asexual spores**?

The main types of asexual spores among the fungi are arthrospores, chlamydospores, sporangiospores, and conidia (from the Greek word *conidios*, meaning "dust"). Conidia and sporangiospores are produced from a fruiting body. Neither arthrospores nor chlamydospores involve a fruiting body. Arthrospores (from the Greek term *arthro*, meaining "joint") are formed by fragmentation of the hyphae. Chlamydospores are formed along the margin of the hyphae. They are thick-walled spores.

What do **spores look like**?

Spores vary greatly in size, shape, color, and surface texture. They are generally small. On average, they are usually less than 20 μm, and rarely exceed more than 100 μm—approximately one-tenth the thickness of a dime!

What are **sclerotia**?

Sclerotia are the aggregate of hyphae enclosed in thick walls that form a protective covering when conditions (temperature and water) are not conducive for growth. When conditions improve, the sclerotia germinate and produce stalks containing spore-bearing bodies. Ascospores are embedded in the tips of the stalks. When transported by wind, ascospores may land on grasses or grains, especially rye. The sclerotia continue to grow after they have landed on the host plants.

What are **dimorphic fungi**?

Many fungi, particularly those that cause disease in humans, are dimorphic—that is, they have two forms. In response to changes in temperature, nutrients, or other environmental factors, they can change from a yeast form to a mold form.

FUNCTION

How do **fungi reproduce**?

Fungal reproduction occurs in two different ways—sexually or asexually. Asexual reproduction occurs through fission, budding, or—most commonly—by spore formation. Sexual reproduction occurs by means that are characteristic for each group. Two types of reproductive structures are found in fungi. Sporangia produce spores, whereas gametangia produce gametes. In order to sexually reproduce, fungi often carry out some type of conjugation. Hyphae of two genetically different mating types come together and fuse, forming a diploid zygote. Most fungi reproduce sexually with nuclear exchange rather than gametes.

Do **fungi** exhibit an **alternation of generations** similar to plants?

Organisms with life cycles that exhibit an alternation of generations have both a haploid phase and a diploid phase. During the haploid phase, reproduction is conducted through gametes, which fuse together to form a zygote. During the diploid phase, reproduction is conducted by spores. Spores develop individually and divide through mitosis, producing the gametophytes of the next generation. Because fungi have both a haploid and diploid phase, they exhibit an alternation of generations.

A bread mold.

What is the **life cycle** of a **typical fungus** such as the **black bread mold,** *Rhizopus nigricans*?

The life cycle of black bread mold, *Rhizopus nigricans*, is characteristic of the members of the phylum Zygomycota. This fungus has a period of sexual reproduction and a period of asexual reproduction that occurs more frequently. During sexual reproduction, there is a fusion of gametangia. The resulting zygosporangium forms a thick coat that awaits favorable conditions in order to conduct further development. When conditions are favorable, the zygosporangium germinates into a sporangium. During asexual reproduction, spores are produced in the sporangium and then dispersed.

What is the **dikaryotic phase** of **fungal life cycles**?

The dikaryotic phase of the fungal life cycle is unique. During this unusual phase, which is common in many species of fungi, cells contain two distinct nuclei. These two nuclei divide simultaneously as the mycelium grows; growth continues until fusion occurs during karyogamy.

What is **mycorrhiza**?

Symbiosis is the close association of two or more different organisms. One type of symbiosis is known as mutualism, defined as an association that is advantageous to both parties. The most common—and possibly the most important—mutualistic, symbiotic relationship in the plant kingdom is known as mycorrhiza. The word myc-

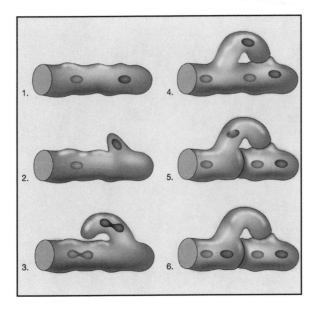

The dikaryotic phase of fungal life.

orrhiza is derived from the Greek words *mykes*, meaning "fungus," and *rhiza*, meaning "root." Mycorrhiza is a specialized, symbiotic association between the roots of plants and fungi that occurs in the vast majority of plants—both wild and cultivated. In a mycorrhizal relationship, the fungi assist their host plants by increasing the plants' ability to capture water and essential elements such as phosphorus, zinc, manganese, and copper from the soil, and transfer them into the plant's roots. The fungi also provide protection against attack by pathogens and nematodes. In return for these benefits, the fungal partner receives carbohydrates, amino acids, and vitamins essential for its growth directly from the host plant. Basidiomycetes (mushrooms, bracket fungi, etc.) are the fungal, mycorrhizal partners of trees and other woody plants. Zygomycetes (molds, etc.) are the fungal partners of nonwoody plants. It has been estimated that mycorrhizal fungi amount to 15 percent of the total weight of the world's plant roots.

Which is more **common—endomycorrhiza** or **ectomycorrhiza**?

Both endomycorrhiza and ectomycorrhiza have the ability to be mycorrhizal fungi. In endomycorrhiza, the hyphae of the fungus penetrate the outer cells of the plant root and extend into the surrounding soil. In ectomycorrhiza, the hyphae surround but do not penetrate the roots. Endomycorrhiza are much more common than ectomycorrhiza. The fungal component of endomycorrhiza is a zygomycete. While only about thirty species of zygomycetes are known to be involved in endomycorrhizal relationships, the zygomycetes are associated with more than 200,000 species of plants. Basidiomycetes are the most common fungal component of ectomycorrhiza, although some ascomycetes also form ectomycorrhizal relationships. More species of fungi are involved in ectomycorrhiza (at least 5,000), but most are only associated with a single species of plant. Furthermore, the total number of plants involved in ectomycorrhiza is limited to a few thousand.

Which **plants** most commonly **form ectomycorrhizal relationships**?

The most common plants associated with ectomycorrhiza are trees and shrubs growing in temperate regions. These trees include pines, firs, oaks, beeches, and willows. These plants tend to be more resistant to extreme temperatures, drought, and other harsh environmental conditions. Some ectomycorrhizal fungi may provide protection from acidic precipitation.

What **roles** do fungi play in **recycling**?

Fungi play a key role in the recycling of many elements. As the primary decomposers in the biosphere, they break down organic matter, including dead plants and other vegetation. As fungi actively decompose materials, carbon, nitrogen, and the mineral components present in organic compounds are released; these elements can all be recycled. During decomposition, carbon dioxide is released into the atmosphere and minerals are returned to the soil. It is estimated that, on average, the top 8 in (20 cm) of fertile soil contain nearly 5 metric tons of fungi and bacteria per 2.47 acres (1 hectare)! Without fungi acting as decomposers, dead, organic matter would overpower the world and life on Earth would eventually become impossible.

Do fungi only **decompose dead and decaying organic matter**?

Not only do fungi consume dead and decaying organic matter, but some attack living plants and animals as they serve as a source for necessary organic molecules. Fungi often cause diseases among plants and animals. They are some of the most-harmful pests to living plants and are responsible for billions of dollars in agricultural losses each year. Food products that have been harvested and stored are not immune to fungal decay. Fungi often secrete substances into the foods they attack, making the foods unpalatable or even poisonous.

MUSHROOMS AND EDIBLE FUNGI

Which **mushrooms** and other **edible fungi** are the most popular?

Category	Common Name	General Characteristics	Examples
Agarics	Gilled mushrooms	Cap, stem, and gills.	Common white button mushroom
Boletes	Sponge mushrooms	Stem; pores on the underside of the cap. Form mycorrhizal associations with trees.	King bolete (also called cepe, porcini, and steinpilz).
Clavarias	Coral mushrooms	Fruiting bodies are club shaped. Usually branched. Often found growing on the ground. Resemble marine corals.	Yellow spindle coral.

Category	Common Name	General Characteristics	Examples
Gasteromycetes	Puffballs	Generally spherically shaped; stem-like base with a rounded top. Spores are released through a pore present in the top or when the outer skin is broken.	Common puffballs, giant puffballs, stinkhorns.
Hydnums	Toothed mushrooms	Cap and stem with "teeth" present on the underside of the cap.	Hedgehog tooth mushroom.
Tremallales	Jelly fungi	Irregularly shaped; has gelatinous flesh.	Wood ears.
Tuber	Truffles	Live below ground. The spore-bearing asci are enclosed in the fruiting bodies.	Perigord truffle.

What **purpose** do the **gills** of mushrooms serve?

Gills, which can be present on the undersurface of a mushroom's cap, serve two main purposes. The first is to maximize the surface area on which spores are produced, allowing a very large number of spores to be produced. The second purpose is to help hold up the cap of the mushroom. Spores are produced in the basidia—specialized cells that line the surface of the gills. It has been estimated that a mushroom with a cap that is 3 in (7.5 cm) in diameter can produce as many as 40 million spores per hour!

How many kinds of **mushrooms** are **edible**?

Among the basidiomycetes, there are approximately 200 varieties of edible mushrooms and about 70 species of poisonous ones. Some edible mushrooms are cultivated commercially; more than 844 million lb (382,832 metric tons) are produced in the United States each year.

How are **mushrooms grown commercially**?

The most common, commercially grown mushroom is the white button mushroom, *Agaricus bisporus*. Mushroom farms consist of special planting beds in buildings with temperature and humidity conditions that are controlled. The beds contain soil mixed with a material that is rich in organic matter. The beds are inoculated with mushroom spawn—a pure culture of the mushroom fungus grown in large bottles on an organic-rich medium. The mycelium grows and spreads throughout the soil mixture for several weeks.

> ## What is a spore print?
>
> **A** spore print is an important tool used in the species identification of a mushroom. Mature mushrooms yield the best spore prints. To make a spore print, remove the stalk from the mushroom and place the cap gill-side-down on a piece of paper. Cover the mushroom cap with a glass and leave it undisturbed for several hours or overnight. The spores present in the mushroom will drop onto the paper. The color of the spore print, the size and shape of the spores, and the pattern of the gills can be used to identify the mushroom.

Mushroom formation is induced by adding a layer of casing soil to the surface of the bed. Mushrooms appear on the surface of the bed through a process known as a "flash." Mushrooms must be collected immediately after flashing, while they are still fresh.

Which common, **edible mushrooms** are grown in **North America** and **Europe**?

Common Name	Scientific Name
American matsutake	*Tricholoma magnivelare*
Blewit	*Clitocybe nuda*
Chanterelle	*Cantharellus cibarius*
Chicken mushroom	*Laetiporus sulphureus*
Hen-of-the-woods	*Grifola frondosa*
Honey mushroom	*Armillaria mellea*
Horn of plenty or black trumpet	*Craterellus cornucopioides, Cruterellus fallax*
King bolete	*Boletus edulis*
Meadow mushroom	*Agaricus campestris* (and others)
Morels	*Morchella esculenta* (and others)
Oyster mushroom	*Pleurotus ostreatus*
Parasol	*Lepiota procera, Lepiota americana, Lepiota rhacodes*

Are there reliable **rules** to **identify poisonous mushrooms**?

There are no general, reliable rules to identify poisonous mushrooms. Some of the edible varieties are quite easily recognized, but some edible varieties closely resemble poisonous mushrooms and can only be distinguished by an expert. The common lore that poisonous mushrooms make silver spoons turn black, while mushrooms that can be peeled are edible, is not true. Some of the deadliest mushrooms, amanitas, do not

Hunting for morel mushrooms. Morels are a favored edible mushroom.

turn silver spoons black and can be peeled! The only rule to follow is that one must be able to identify a mushroom *with certainty* prior to eating it.

What is **unusual** about *Amanita* **mushrooms**?

Some of the most poisonous mushrooms belong to the genus *Amanita*. Toxic species of this genus have been known by the names "death angel" (*Amanita phalloides*) and "destroying angel" (*Amanita virosa*). Ingestion of a single cap of either of these species can kill a healthy, adult human! Even ingesting a tiny bit of amatoxin—the toxin present in species of this genus—may result in liver ailments that will last the rest of a person's life.

What **toxic substances** are **produced** by **mushrooms**?

The most toxic substances produced by mushrooms are amatoxins and phallotoxins (both cyclopeptides). These toxins act by interfering with RNA and DNA transcription, inhibiting the formation of new cells. The toxins collect in the liver, ultimately leading to liver failure.

What **antidote** is available for **mushroom poisoning**?

No effective antidote for mushroom poisoning in humans has been discovered. The toxins produced by mushrooms accumulate in the liver and lead to irreversible liver

---

What are portobellos?

Portobellos are extremely large, dark-brown mushrooms that are the fully mature form of the crimino mushroom, which is a variation of the commonly cultivated white mushroom. The name "portobello" was introduced in the 1980s as a marketing ploy to popularize an unglamorous mushroom that could not be sold.

---

damage. Unfortunately, there may be no indication of poisoning for several hours after ingesting a toxic mushroom. When the symptoms do present, they often resemble typical food poisoning. Liver failure becomes apparent three to six days after ingesting a poisonous mushroom. Oftentimes a liver transplant may be the only possible form of treatment.

Can **mushrooms grow** up **overnight**?

A mushroom is only the fruiting body—that is, reproductive structure—of a much larger fungus body that grows unseen in rotting logs, rich humus, and dark, damp places. Many familiar mushrooms have fruiting bodies that are fleshy and umbrella shaped. Warm, damp weather triggers their sudden appearance. Usually first to be noticed are small, round "button caps" composed of densely packed hyphae. Soon after the outer covering ruptures, the stem elongates, and the cap enlarges to its full size. This entire process can indeed happen overnight!

What is unusual about the **stinkhorn fungus** *Dictyophora*?

The stinkhorn fungus *Dictyophora indusiata* is one of the world's fastest-growing organisms. It pushes out of the ground at a rate of about 0.2 in (0.5 cm) per minute. The growth rate is so fast that a crackling sound can be heard as the tissues of the fungus swell and stretch. During growth, a delicate, net-like veil forms around the fungus, giving this fungus its common name, "the lady of the veil". The fungus then decomposes and in the process, produces a strong odor that is similar to the smell of decaying flesh. This odor attracts flies that crawl over the fungus and collect its spores on their feet. This process ensures that the spores are carried to new areas. Although the odor produced by species of *Dictyophora* is quite unpleasant, members of this genus are considered delicacies in China, where they are marketed as aphrodisiacs.

How is a **fairy ring** formed?

Long ago it was believed that the circles of mushrooms that sometimes form in meadows marked the locations where fairies gathered at night to dance. Fairy rings, or fungus rings, are frequently found in grassy areas. There are three types of rings: those that do not affect their surrounding vegetation, those that cause increased vegetational growth, and those that damage their surrounding environment. The rings are started from a mycelium (the underground, food-absorbing part of a fungus). The fungus growths are circular because a round inner band of decaying mycelium forms underground. This band uses up the resources present in the soil that is directly above it. When the fungus forms caps that present above ground, the mushrooms grow around the mycelium, creating a ring effect. Each succeeding generation grows further from the center.

What are **truffles** and where do they come from?

Truffles, a delight of gourmets, are arguably the most-prized edible fungi. Found mainly in western Europe, they grow near the roots of trees (particularly oak, but also chestnut, hazel, and beech) in open woodlands. Unlike typical mushrooms, truffles develop 3 to12 in (7.6 to 30.5 cm) underground, making them difficult to find. Truffle hunters use dogs and pigs that have been specially trained to find the flavorful morsels. Both animals have a keen sense of smell and are attracted to the strong, nut-like aroma of truffles. In fact, trained pigs are able to pick up the scent of a truffle from 20 ft (6.1 m) away. After catching a whiff of a truffle's scent, the animals rush to the origin of the aroma and quickly root out the precious prize. Once the truffle is found, the truffle hunter (referred to in French as a *trufficulteur*) carefully scrapes back the earth to reveal the fungus. Truffles should not be touched by human skin, as doing so can cause the fungus to rot.

What do **truffles look like**?

Truffles have a rather unappealing appearance—they are somewhat round, but irregularly shaped, and have thick, rough, wrinkled skin that varies from off-white to almost black in color. The fruiting bodies present on truffles are fragrant, fleshy structures that usually grow to about the size of a golf ball; they range from white, gray, or brown to nearly black in color. There are nearly seventy known varieties of truffles, but the most desirable is the black truffle—also known as black diamond—that grows in France's Perigord and Quercy regions as well as Italy's Umbria region. The flesh of the black diamond appears to be black, but it is actually dark brown, and contains white striations. The flesh has an aroma that is extremely pungent. The second-most popular is the white truffle (actually off-white or beige) of Italy's Piedmont region. Both the aroma and flavor of this truffle are earthy and garlicky. Fresh truffles are available from late fall to midwinter and can be stored in the refrigerator for up to three days.

Truffles are arguably the most prized edible fungi.

How are **truffles** used in **cooking**?

Dark truffles are generally used to flavor foods such as omelets, polentas, risottos, and sauces. White truffles are usually served raw; they are often grated over foods such as pasta or dishes containing cheese, as their flavors are complementary. They are also added at the last minute to cooked dishes.

What **mushroom** has been identified as a **fossil**?

Mushrooms are rarely found in fossil records since there is little in the structure of a mushroom that can be fossilized. In 1990 two researchers in the Dominican Republic discovered a fossil of a fleshy, gilled mushroom. The mushroom, *Coprinites dominicana,* is believed to be 35 million to 40 million years old and is the only known fossil mushroom from the tropics. Microscopic examination suggests it is related to modern-day inky-cap mushrooms.

Mushrooms of the genera *Conocybe* and *Psilocybe*, both of which have hallucinogenic properties, were considered sacred by the Aztecs. These mushrooms are still used in religious ceremonies held by the descendents of the Aztecs. Psilocybin, which is chemically related to lysengic acid diethylamide (LSD), is a component to both genera and is responsible for the trance-like state and colorful visions experienced by those who eat these mushrooms.

LICHENS

What are **lichens**?

Lichens are organisms that grow on rocks, tree branches, and bare ground. They are composed of two different entities living together in a symbiotic relationship: 1) a population of either algal or cyanobacterial cells that are single or filamentous; and 2) fungi. Lichens do not have roots, stems, flowers, or leaves. The fungal component of a lichen is called the mycobiont (from the Greek terms *mykes*, which means "fungus," and *bios*, meaning "life"), and the photosynthetic component is called the photobiont (from the Greek terms *photo*, meaning "light," and *bios*, meaning "life"). The scientific name given to the lichen is the name of the fungus and is most often an ascomycete. As the fungus has no chlorophyll, it cannot manufacture its own food, but it can absorb food from algae. Lichens and algae enjoy a symbiotic relationship. Lichens can often be found growing around and on top of algae, providing the algae with protection from the sun, thus decreasing the loss of moisture. Fungi and algae were the first organisms recognized as having a symbiotic relationship. A unique feature of this relationship is that it is so perfectly developed and balanced that the two organisms behave as a single organism.

Who is considered the **"father of lichenology"**?

Erik Acharius (1757–1819) is considered the father of lichenology. He was the founder of modern lichen taxonomy, having described and arranged lichens into forty distinct genera. His four major works of research—*Lichenographiæ suecicæ prodromus* (1798), *Methodus lichenum* (1803), *Lichenographia universalis* (1810), and *Synopsis methodica Lichenum* (1814)—formed the foundations of modern lichenology.

What is unusual about the **natural distribution** of **lichens**?

Lichens are widespread because they are able to live and grow in some of the harshest environments on Earth. They occur from arid desert regions to the Arctic; they grow on bare soil, tree trunks, rocks, fence posts, and alpine peaks all over the world. Some lichens are so tiny that they are almost invisible to the naked eye. Others, like reindeer "mosses," may cover acres of land with ankle-deep growth. One species of the genus *Verrucaria* grows underwater as a submerged marine lichen. Lichens are often the first inhabitants of newly exposed rocky areas. In Antarctica there are more than 350 species of lichens but only two species of plants.

What is the **ecological role** of **lichens**?

Lichens account for approximately 8 percent of the vegetation covering Earth's surface. In certain environments, such as regions of tundra, they cover vast areas of land. Lichens delay global warming by consuming significant amounts of carbon dioxide (CO_2) during photosynthesis. When they cover the ground, they prevent soil from drying out. In desert areas they are able to capture and conserve the moisture present in fog and dew. Lichens release nutrients, such as nitrogen and phosphorus, which are important in regions with nutrient-poor soils as the nutrients aid tree growth. Lichens are also an important food source for many species of animals, including wild turkeys and reindeer of the Arctic tundra. Birds such as the olive-headed weaver of Madagascar and the goldfinch of Europe use lichens to build their nests.

What is the **relationship** between **lichens** and **air pollution**?

Lichens are extremely sensitive to pollutants in the atmosphere and can be used as bioindicators of air quality. They absorb minerals from the air, from rainwater, and directly from their substrate. Lichen growth has been used as an indicator of air pollution, especially sulfur dioxide. Pollutants are absorbed by lichens, causing the destruction of their chlorophyll, which leads to a decrease in the occurrence of photosynthesis and changes in membrane permeability. Pollutants upset the physiological balance between the fungus and the alga or cyanobacterium, and degradation or destruction of the lichens results. Lichens are generally absent in and around cities, even though suitable substrates exist; the reason for this is the polluted exhaust from automobiles and industrial activity. They are beginning to disappear from national parks and other relatively remote areas that are becoming increasingly contaminated by industrial pollution. The return of lichens to an area frequently indicates a reduction in air pollution.

What are other examples of **lichens assessing pollution**?

Lichens are used to assess radioactive pollution levels in the vicinity of uranium mines, environments where nuclear-powered satellites have crashed, former nuclear bomb testing sites, and power stations that have incurred accidents. Following the Chernobyl nuclear power station disaster in 1986, arctic lichens as far away as Lapland were tested and showed levels of radioactive dust that were as much as 165 times higher than had been previously recorded.

143

How did the **increased level** of **radioactive dust** (cesium 137) in **lichens affect the food chain** following the **Chernobyl** nuclear power station disaster?

Lichens are a primary source of food for reindeer; reindeer is commonly consumed by humans that live in regions of tundra. When the accumulated level of radioactive dust present in lichens became so high in the reindeer that fed off of them, the reindeer meat became unsuitable for human consumption. Hundreds of tons of reindeer carcasses were disposed of as toxic waste.

YEASTS

How do **yeasts differ** from other **fungi**?

Yeasts remain unicellular—that is, as single cells—throughout their life. Most species reproduce by budding, others through binary fission or spore formation. Each bud that separates from its mother yeast cell can grow into a new yeast cell. Some yeast cells group together to form colonies.

How is **yeast** utilized in **food** and **beverage manufacturing**?

Yeast is used in winemaking, beer making, and bread making. Yeast converts food into alcohol and carbon dioxide (CO_2) during fermentation. In the manufacture of wine and beer, the yeast's manufacture of alcohol is a desired and necessary component of the final product. The CO_2 is what gives beer and champagne their bubbly effect. Bread making requires the CO_2 produced by yeast for certain doughs to rise. Yeasts used in brewing and baking are cultivated strains carefully kept to prevent contamination.

What is the difference between **active dry yeast** and **compressed fresh yeast**?

Both active dry yeast and compressed fresh yeast are leavening agents. Active dry yeast comprises tiny, dehydrated granules of yeast. Although the granules are alive, the yeast cells are dormant due to their lack of moisture; because the cells are dormant, dry yeast has a long shelf life. Active dry yeast becomes active when mixed with warm liquid. Compressed fresh yeast is moist and extremely perishable. It must be stored under refrigeration and used within one to two weeks.

What is the **role** of **yeast** in **beer production**?

Beer is made by fermenting water, malt, sugar, hops, yeast (species *Saccaromyces* spp.), salt, and citric acid. Each ingredient has a specific role in the creation of beer. Malt is produced from a grain—usually barley—that has sprouted, been dried in a kiln, and ground into a powder. Malt gives beer its characteristic body and flavor. Hops is made from the fruit that grows on the herb *Humulus lupulus* (a member of the mulberry family). The fruit is picked when ripe and is then dried; this ingredient gives beer a slightly bitter flavor. Yeast is used for the fermentation process.

> ## What is the difference between baker's yeast and brewer's yeast?
>
> **B**aker's yeast is used as a leavening agent. Leavening agents are used to increase the volume of baked goods. Brewer's yeast is a special non-leavening agent used in beer making. It is a rich source of B vitamins and is also used as a food supplement.

Making beer is a complex process. One method begins by mixing and mashing malted barley with a cooked cereal grain such as corn. This mixture, called "wort," is filtered before hops is added to it. The wort is then heated until it is completely soluble. The hops is removed, and after the mixture has cooled, yeast is added. The beer ferments for 8 to 11 days at temperatures that range between 50° and 70°F (10° and 21°C). The beer is then stored and kept at a state that is close to freezing. During the next few months the liquid takes on its final character before carbon dioxide is added for effervescence. The beer is then refrigerated, filtered, and pasteurized in preparation for bottling or canning.

Is the **same strain** of **yeast** used to make **lager** and **ale** beers?

Two common strains of yeast are used to ferment beer: *Saccharomyces carlsbergensis* and *Saccharomyces cerevisiae*. *Saccharomyces carlsbergensis*, also known as bottom yeast, sinks to the bottom of the fermentation vat. Strains of bottom yeast ferment best at 42.8–53.6°F (6–12°C) and take 8 to 14 days to produce *lager* beer. *Saccharomyces cerevisiae*, also known as top-fermenting yeast, is distributed throughout wort and is carried to the top of the fermenting vat by carbon dioxide (CO_2). Top-fermenting yeast ferments at a higher temperature (57.2–73.4°F [14–23°C]) over only 5 to 7 days. Top-fermenting yeasts produce ales, porter, and stout beers.

Why is the yeast *Saccharomyces cerevisiae* important in **genetic research**?

Biologists have studied *Saccharomyces cerevisiae*, a yeast used by bakers and brewers, for many decades because it offers valuable clues to aid in the understanding of how more-advanced organisms work. For example, humans and yeast share a number of similarities in their genetic make up. The DNA present in certain regions of yeast contain stretches of DNA subunits that are nearly identical to those in human DNA. These similarities indicate that humans and yeast both have similar genes that play a critical role in cell function. In 1996 an international consortium of scientists from the United States, Canada, Europe, and Japan completed the genome sequence (all 12,057,500 subunits contained in the nuclear DNA) of *S. cerevisiae*. It is the first eukaryotic organism to be completely sequenced. With their rapid generation time, yeasts continue to be the organism of choice to provide significant insights into the functioning of eukaryotic systems.

APPLICATIONS

What is the **economic impact** of **fungi**?

Fungi produce gallic acid, which is used in photographic developers, dyes, indelible ink, as well as in the production of artificial flavoring, perfumes, chlorine, alcohols, and several acids. Fungi are also used to make plastics, toothpaste, soap, and in the silvering of mirrors. In Japan almost 500,000 metric tons of fungus-fermented soybean curd (tofu and miso) are consumed annually. Different strains of the rust fungus *Puccinia graminis* cause billions of dollars of damage annually to food and timber crops throughout the world.

What are some **beneficial uses** of the **sclerotia** of the fungus *Claviceps purpurea*?

The sclerotia of *Claviceps purpurea* are known as the plant disease ergot. Ergot is used pharmaceutically to produce drugs used to induce labor in pregnant women and to control bleeding after childbirth. Ergotamine, an ergot alkaloid, is used to treat migraine headaches.

How does **ergot** affect **humans** and **cattle**?

Eating bread and other grain products contaminated with ergot causes the disease called St. Anthony's fire. Common during the Middle Ages, this disease, which causes sensations of intense heat followed by a complete loss of sensation in an infected person's limbs, no longer occurs very frequently due to improved techniques of grain production and milling. Cattle that graze on grains infected with ergot are able to ingest enough to cause death or the spontaneous abortion of fetuses.

Which **fungus** may have played a role in the **Salem witch trials**?

The Salem witch hunts of 1692 may have initially been caused by an infestation of a microbiological poison. The fungus *Claviceps purpurea*, commonly known as rye smut, produces the poison ergot. When ingested, this poison produces symptoms similar to the ones that presented in the girls who accused others of being witches in Salem. Historians and biologists have reviewed environmental conditions in New England from 1690 to 1692 and have found that conditions were perfect for an occurrence of rye smut overgrowth. The weather conditions during those years were particularly wet and cool. Rye grass had replaced wheat as the principal grain because wheat had become seriously infected with wheat rust during long periods of cold and damp weather. The symptoms of ergot poisoning include convulsions, pinching or biting sensations, stomach ailments, as well as temporary blindness, deafness, and muteness.

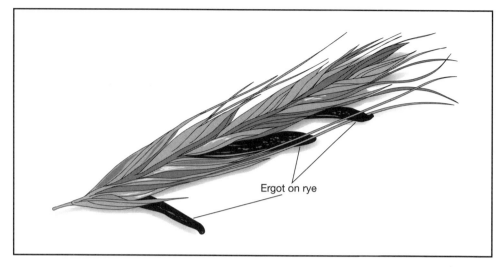

Ergot on rye. Eating bread and other grain products contaminated with ergot causes the disease called St. Anthony's fire.

Which **tree** that is native to the **United States** has become **extinct due to fungus**?

The American chestnut (*Castenea dentate marsh*) was widespread across eastern North America until the early 1900s. This type of chestnut tree made up almost half of the population of hardwood forests in central and southern Pennsylvania, New Jersey, and southern New England. In its entire range, the species dominated deciduous forests, making up almost one-quarter of the trees. The fungus *Cryphonectria parasitica*, commonly known as chestnut blight, destroyed nearly every specimen of the American chestnut tree.

What part of the American chestnut tree does the fungus *Cryphonectria parasitica* attack?

Cryphonectria parasitica, or chestnut blight, attacks a tree's layers of living bark and the adjacent layers of wood. The fungus kills the cells present in the bark that serve the function of carrying the food made in the leaves of a tree to other parts of a tree. As such, nutrients are not able to reach various parts of a tree. The fungus also clogs the cells present in the wood of a tree's trunk that serve to carry water and nutrients through the body of a tree. This fungus leaves the roots of a tree unaffected, allowing a tree to send up new sprouts. However, within a number of years, the bark and wood of new sprouts also become affected.

What other **tree species** has been **adversely affected** by a **fungus**?

Elm trees are susceptible to the fungus *Ophiostoma ulmi*, which causes Dutch elm disease. The fungus lives in the tubular cells present in the outermost wood of trees. **147**

As the cells become plugged, water and nutrients are not able to move from the roots to the top of a tree, and eventually the tree dies.

When was **Dutch elm disease first identified** in **North America?**

Dutch elm disease was first identified in 1930 in Cincinnati, Ohio. The source of the fungus was shown to be elm logs imported from Europe. By 1940 the disease had spread to nine states; by 1950 it was found in seventeen states and had spread into southern Canada. Today it is found wherever elm trees grow throughout North America.

How many species of **fungi** are **plant pathogens?**

Elm trees are susceptible to the fungus *Ophiostoma ulmi*, which causes Dutch elm disease.

More than 8,000 species of fungi cause disease in plants. Most diseases found among both cultivated and wild plants are caused by fungi. Some pathogenic fungi grow and multiply in their host plants; others grow and multiply on dead organic matter and host plants. Fungi that are pathogenic to plants can occur below the soil surface, at the soil surface, and throughout the body of a plant. Fungi are responsible for leaf spots, blights, rusts, smuts, mildews, cankers, scabs, fruit rots, galls, wilts, tree diebacks and declines, as well as root, stem, and seed rots.

What are **rusts** and **smuts** and what **effect** do they produce in **crops?**

Rusts and smuts are very small fungi responsible for many serious plant diseases. Cereals and other grains are highly susceptible to attack by rusts and smuts. Many rusts and smuts have complicated life cycles as they are known to use more than one

The unique flavor of the blue cheese known as Roquefort is produced by the fungi genus *Penicillium*.

plant species as a host during their lifetime. For example, wheat rust spends a portion of its life in barberry plants and a portion in wheat.

How are **fungi related** to **soy sauce**?

Aspergillus tamari and other deuteromycetes are used to produce soy sauce by slowly fermenting boiled soybeans. Soy sauce provides foods with more than its special flavor; the soybeans and fungi give soy sauce amino acids that are vital to human life. Fungi have been used in many cultures to improve the nutrient quality of the diet.

What **cheeses** are **associated** with **fungi**?

The unique flavor of cheeses such as Roquefort, Camembert, and Brie is produced by members of the genus *Penicillium*. Roquefort is often referred to as "the king of cheeses"; it is one of the oldest and best-known cheeses in the world. This "blue cheese" has been enjoyed since Roman times and was a favorite of Charlemagne, king of the Franks and emperor of the Holy Roman Empire (742–814). Roquefort is made from sheep's milk that has been exposed to the mold *Penicillium roqueforti* and aged for three months or more in the limestone caverns of Mount Combalou, near the village of Roquefort in southwestern France. This is the only place true Roquefort can be aged. It has a creamy, rich texture and a flavor that is simultaneously pungent, piquant, and salty. It has a creamy white interior marked by blue veins; the cheese is held together with a snowy white rind. True Roquefort is authenticated by the presence of a red sheep on the emblem present on the cheese's wrapper.

Penicillium camemberti gives Camembert and Brie cheeses their special qualities. Napoleon is said to have christened Camembert cheese with its name; supposedly the name comes from the Norman village where a farmer's wife first served it to Napoleon. This cheese is formed of cow's milk and has a white, downy rind and a smooth, creamy interior. When perfectly ripe and served at room temperature, the cheese should ooze thickly. Although Brie is made in many places, Brie from the region of the same name east of Paris is considered one of the world's finest cheeses by connoisseurs. Similar to Camembert, it has a white, surface-ripened rind and smooth, buttery interior.

What **human diseases** are caused by **fungal pathogens**?

Disease	Fungal Pathogen	Phylum	Organs Affected
Aspergillosis, Otomycosis	*Aspergillus fumigatus*	Ascomycetes	Lungs, ears
Blastomycosis	*Blastomyces dermatidis*	Ascomycetes	Lungs
Candidiasis (Vaginitis, Thrush and Onychia)	*Candida albicans*	Deuteromycetes	Intestine, vagina, skin, mouth
Coccidiodomycosis	*Coddidiodes immitis*	Deuteromycetes	Lungs
Cryptococcosis	*Cryptococcus neoformans*	Basidiomycetes	Lungs, spinal cord, meninges
Histoplasmosis	*Histoplasma capsulatum*	Ascomycetes	Lungs
Sporotrichosis	*Sporothrix schenkii*	Deuteromycetes	Skin
Tinea capitis (ringworm of the head); Tinea corporis (ringworm of the body); Tinea favosa (ringworm of the scalp); Tinea pedis (athlete's foot)	Species of the genera *Epidermophyton, Microsporum,* and *Trichophyton.*	The genera *Trichophyton* and *Microsporum* are of Ascomycetes. The genus *Epidermophyton* is of Deuteromycetes.	Skin

What are the **special challenges** of producing **effective antifungal drugs**?

Since fungi are eukaryotic, their cellular structure is similar to—if not the same as—that of animals and humans. Drugs that affect metabolic pathways in fungi often affect corresponding pathways in host cells, which can result in the host sustaining drug toxicity. Many antifungal drugs can only be used topically. Very few drugs have been found to be selectively toxic—that is, toxic to fungi and not their human hosts.

What are some important **antifungal drugs** and their **mechanism** of **operation**?

Many antifungal drugs work by interfering with the function or synthesis of ergosterol. Ergosterol is found in the cytoplasmic membrane of fungal cells but is not found in human cells. Some antifungal drugs interfere with fungus-specific structures and functions, such as the cell wall.

Category of Drug	Mechanism of Operation	Example of Drug
Allylamines	Ergosterol synthesis	Terbenafine
Azoles	Ergosterol synthesis	Fluconazole, Itraconazole, Ketoconazole, Clotrimazole, Miconazole, Voriconazole
Nucleic acid analogs	DNA synthesis	5-Fluorocytosine
Polyenes	Ergosterol synthesis	Amphotericin B
Polyoxins	Chitin synthesis	Polyoxin A, Polyoxin B

How was it **discovered** that **microorganisms** are effective against **bacterial infections**?

British microbiologist Alexander Fleming (1881–1955) happened upon the discovery of penicillin's use as a antibacterial agent. In 1928 Fleming was researching staphylococci at St. Mary's Hospital in London. As part of his investigation, he had spread staphylococci on several petri dishes before going on vacation. Upon his return he noticed a green-yellow mold contaminating one of the petri dishes. The staphylococci had failed to grow near the mold. He identified the mold as being of the species *Penicillium notatum*. Further investigation proved that staphylococci and other gram-positive organisms are killed by *P. notatum*. It was not until the 1940s that Howard Florey (1898–1968) and Ernst Boris Chain (1906–1979) rediscovered the benefits of penicillin and were able to isolate it for medical use. In 1945 Fleming, Florey, and Chain shared the Nobel Prize in Physiology or Medicine for their work on penicillin.

How are **antibiotics produced** today?

Until the mid-1950s all antibiotics were products of microorganisms. In the late 1950s researchers succeeded in synthesizing the nucleus of *Penicillium notatum*. This achievement allowed for various new groups to be attached to the synthesized nucleus, which paved the way for new forms of penicillin to be created. Synthesized

151

Alexander Fleming was the first to discover that penicillin could be used as an antibacterial agent.

antibiotics use natural molecules and add side chains that create a drug more effective than penicillin.

What **fungus** plays an important **role** in **human organ transplantation**?

The soil-inhabiting fungus *Tolypocladium inflatum* is the source of cyclosporin, a medication that suppresses the immune reactions that cause organ transplant rejections. Cyclosporin does not cause the undesirable side effects that other immune-suppressing medications do. This remarkable drug became available in 1979, making it possible to resume organ transplants, which had essentially been abandoned. As a result of cyclosporin, successful organ transplants are almost commonplace today.

Are there other **medicinal uses** for **fungi**?

Two species of fungi have been popular in China and Japan for their medicinal value—*Lentinula edodes*, a species of shiitake, and *Ganoderma lucidum*, a reishi mushroom. Lentinan and *L. edodes* mycelium extract (LEM) are two extracts from *L. edodes* that are being studied for their medicinal properties. Lentinan has been found to enhance the function of the immune system and slow tumor growth. LEM appears to improve liver function and may have potential as an antiviral agent. Asian herbalists consider reishi to be the fungus of choice for treating of a variety of illnesses.

Which **fungi** commonly found in **North America** have demonstrated **beneficial medicinal effects**?

Clinical trials have been conducted to test potential health benefits of fungi commonly found in North America. The following chart reveals some of the benefits that several common fungi may provide.

Fungus (Mushroom)	Common Name	Medicinal Use
Armillaria mellea	Honey mushroom	Has been shown to reduce hypertension.
Grifola frondosa	Maitake	Encouraging results for treatment of many cancers, hypertension, and hepatitis B.

What was the first semisynthetic penicillin?

The group known as the aminopenicillins, which include the drugs ampicillin and amoxicillin, were the first semisynthetic penicillins.

Fungus (Mushroom)	Common Name	Medicinal Use
Grifola umbellate	Zhu ling	When combined with other herbal medications and/or conventional treatments, it may aid cancer treatments.
Inonotus obliquus	Chaga	Success seen for managing lung, breast, and genital cancers.
Schizophyllum commune	Split gill	Significantly prolonged life for patients with various cancers when used in conjunction with conventional therapies.
Trametes versicolor	Turkey tail	Improved survival rates for patients with a number of different cancers.
Wolfiporia cocos	Tuckahoe	When used with metronidazole, the cure rate for patients with viral hepatitis was nearly double that of patients taking either medication by itself.

Why are species of the genus *Neurospora* important?

Pink bread molds of the genus *Neurospora* have long served as powerful laboratory models used to study genetics, biochemistry, and molecular biology. Scientists first demonstrated the concept that one gene produces a corresponding protein by studying *Neurospora*. Its ease of growth and the extensive genetic information available for this organism make it a convenient model for the study of many processes found in higher plants and animals. Among the fungi, it is second only to yeast as a basic model organism.

What fungus is associated with chemical warfare?

Fusarium, a genus that is included in the group considered "imperfect fungi," produces trichothecenes, which are highly toxic substances that have been tested as agents of chemical warfare. Trichothecenes are very hardy compounds that can survive autoclaving and do not degrade when exposed to light. They are relatively easy to produce and have been prepared as biological weapons. The toxin causes the chronic disease alimentary toxic aleukia (ATA).

153

Armillaria mushrooms have been shown to reduce hypertension.

Which **mycotoxin** is found in certain **agricultural products** and is believed to be **carcinogenic** to humans?

The fungus *Aspergillus flavus* produces toxic aflatoxins. This mold is found in soil, decaying vegetation, grains, cereals, hay, corn, peanuts, sweet potatoes, rice, and animal feed. It is believed that products contaminated with aflatoxins are carcinogenic to humans.

How were **fungi involved** in **World War I**?

During World War I the Germans needed glycerol to make nitroglycerin, which is used in the production of explosives such as dynamite. Before the war, the Germans had imported their glycerol, but the British naval blockade during the war prevented such imports. The German scientist Carl Neuberg (1877–1956) knew that trace levels of glycerol are produced when *Saccharomyces cerevisiae* is used during the alcoholic fermentation of sugar. He sought and developed a modified fermentation process in which the yeast would produce significant quantities of glycerol and less ethanol. The production of glycerol was improved by adding 3.5 percent sodium sulfite at a pH of 7.0 to the fermentation process, which blocked one chemical reaction in the metabolic pathway. Neuberg's procedure was implemented with the conversion of German beer breweries to glycerol plants. The plants produced 1,000 tons of glycerol per month. After the war ended, the production of glycerol was not in demand, so it was suspended.

What fungi have traveled in space?

Colonies of luminescent fungi are sometimes maintained aboard space flights. Sensitive to escaped fuel and other noxious fumes, the fungi exhibit dulled luminescence when exposed to as little as 0.02 part per million of fuel. Thus, the fungi serve as an early warning system for the presence of noxious fumes, much as canaries warned underground miners of a lack of oxygen or the presence of dangerous gases such as methane.

What **colored pigments** are **produced** by **lichens**?

Lichens are often strikingly colored because of pigments that play a role in protecting the photosynthetic partner from the destructive action of the sun's rays. These pigments can be extracted from lichens and used as natural dyes. The traditional method of manufacturing Scotland's famous tweeds makes use of fungal dyes; synthetic dyes are now commonly used. Orchil, used to dye woolens, is a pigment produced specifically by lichens. The lichen *Rocella tinctoria* is used to produce litmus, a widely used acid-base indicator.

Which **lichens** are **most often used** by the **perfume** and **cosmetic industries**?

The musk-like fragrance and fixative properties of *Evernia prunastri* and *Pseudevernia furfurnacea* make them popular components for perfumes and cosmetics. The essential oils of these lichens are extracted with solvents. *E. prunastri* and *P. furfurnacea* are common to southern France, Morocco, and the Serbo-Croatian peninsula.

What is the **difference** between **white rot fungi** and **brown rot fungi**?

White rot fungi, found in the wood of deciduous trees, first attack the lignin of wood. Once the lignin is digested, the fungi destroy cellulose and other major parts of cells. The partially decayed wood with residual cellulose is off-white in color, hence the name "white rot fungi." Brown rot fungi, found in conifers, damage the cellulose first but do very little, if any, damage to the lignin. The name "brown rot fungi" came about because infected wood becomes dark reddish-brown to golden in color.

How do **fungi cause trees** to **become hollow**?

Trees with hollow interiors are a familiar sight in parks, forests, and throughout urban areas. Fungal decay is often the cause of the hollowing of the trees. Wood-decaying fungi can enter a tree following a wound to the tree. The tree will produce a band of cells that tries to resist the formation of decay directly around the wound. The fungus that has entered the tree through the wood will continue to decay the inner

Lichens, like the yellow lichen shown here, are often used as natural dyes and in the perfume and cosmetic industries.

wood while the wood protected by the band of cells will continue to be sound. Hollow trees are also formed by fungi that solely attack the heartwood of a tree; this is most commonly observed in older, coniferous trees of the Pacific coast of North America, reaching from California to Alaska. Heart-rot fungi can enter a tree either through a wound or through small branch stubs. It decays the inner wood, resulting in a weak tree that is susceptible to damage by strong winds.

Do **different fungi rot different parts** of **trees**?

Most fungi that rot wood in standing trees are basidiomycetes, the group that includes mushrooms, fleshy shelves, and rigid brackets. Most wood rotters attack only one or two related species of trees. Conifers and deciduous trees are more prone to fungi damage. Many wood-rot fungi damage only specific parts of a tree. For example, *Ganoderma lucidum* and *Heterobasidion annosum* specifically rot roots and are rarely found in higher sections of trees. *Cerenna unicolor* and *Climacodon septentrionale* are common in high sections of trees and are rarely found in the roots. *Laetiporus sulfurous* and *Fomitopsis pinicola* are found in all sections of a tree except the smallest branches of their host.

Which **species** of **tree** are **highly resistant** to **fungal decay** and which are **highly susceptible**?

In general, black locust, walnut, white oak, cedar, and black cherry trees are highly resistant to fungal decay. Species that are highly susceptible to fungal decay include aspen, willow, silver maple, and American beech trees.

What are the **ideal conditions** for **fungi** to **attack wood**?

Fungi tend to attack woods when temperatures range between 50° and 90°F (10° and 32°C). Wood needs to be moist for fungi to grow. The most serious decay occurs when the moisture content of the wood is approximately 30 percent. Wood with a moisture content of less than 20 percent will usually not decay, and any infection will not progress. Wood that is too wet will not decay because the excess moisture does not allow fungi sufficient access to air, thus impeding their growth.

Which **woods** are **recommended** for **construction** when the wood will be constantly **exposed to moisture**?

Decay-resistant woods such as redwood and cedar or wood treated with preservatives are recommended for use in construction when the wood will be exposed to moisture. The most effective and least toxic preservative is chromated copper arsenate (CCA). Lumber treated with CCA has a characteristic light-green stain.

What is **dry rot**?

Dry rot is a misleading common name for this form of decay since one of the primary requirements for fungal decay is moisture. This form of decay is referred to as dry rot because it is found in wood that is not visibly moist or damp. The fungus *Serpula lacrymans* produces specialized mycelia that enable it to carry water and nutrients from a location with the conditions necessary for decay to locations where these conditions are not met. Mycelia can carry water and nutrients up to 15 ft (4.5 m), across materials such as stone and concrete. The environmental requirements for the fungus are humid air with a relative humidity greater than 95 percent and temperatures between 32° and 82°F (0° and 28°C). *Serpula lacrymans* is common in wooden buildings of Europe, Asia, Australia, and Japan. It is uncommon in North America, although a related fungus, *Meruliporia incrassate*, causes similar damage that is not as extensive.

Have **fungi** been **effective** in **biocontrol**?

Biocontrol is defined as the use of one living organism to kill or control another organism. Fungi that parasitize insects are a valuable weapon for biocontrol. The spores of a parasitic fungus are sprayed on pest insects. The fungus then attacks and controls its host. A fungus was identified as killing populations of silkworms in as early as 1834. The spores of the same fungus are now used as a mycoinsecticide—a parasitic fungus used to kill insects—to control Colorado potato beetles. The spores of other fungi are used to control spittlebugs, leaf hoppers, citrus rust mites, and other insect pests.

PLANT DIVERSITY

INTRODUCTION
AND HISTORICAL BACKGROUND

What are the **major subdisciplines** of **botany**?

The major subdisciplines of botany are:

Subdiscipline	Description
Agronomy	Application of plant science to crop production.
Bryology	The study of mosses and liverworts.
Economic botany	The study of the utilization of plants by humans.
Ethnobotany	The study of the use of plants by indigenous peoples.
Forestry	The study of forest management and the utilization of forest products.
Horticulture	The study of ornamental plants, vegetables, and fruit trees.
Paleobotany	The study of fossil plants.
Palynology	The study of pollen and spores.
Phytochemistry	The study of plant chemistry, including the chemical processes that take place in plants.
Plant anatomy	The study of plant cells and tissues.
Plant ecology	The study of the role plants play in the environment.
Plant genetics	The study of genetic inheritance in plants.
Plant morphology	The study of plant forms and life cycles.
Plant pathology	The study of plant diseases.
Plant physiology	The study of plant function and development.
Plant systematics	The study of the classification and naming of plants.

If you were to compress the history of Earth into a single year, what would be some major "dates" in plant evolution?

Time (million years)	Event	Date
3,600	First algae* appear	Mar. 21
433	Land plants appear	Nov. 27
400	Ferns and gymnosperms appear	Nov. 30
300	Major coal deposits formed	Dec. 8
65	Flowering plants appear	Dec. 26

*According to current plant classification systems, algae are no longer considered plants. They are Protista.

Who is known as the **founder of botany**?

The ancient Greek scientist Theophrastus (ca. 372–ca. 287 B.C.E.) is known as the father of botany. His two works on botany, *On the History of Plants* and *On the Causes of Plants*, were so comprehensive that 1,800 years passed before any new significant botanical information that had not been covered by Theophrastus was discovered. He integrated the practice of agriculture into botany and established theories regarding plant growth and the analysis of plant structure. He related plants to their natural environment and identified, classified, and described 550 different plants.

What is *Gray's Manual*?

Gray's Manual of Botany, first published in 1848 by Asa Gray (1810–1888) under the title *Manual of the Botany of Northern United States*, was the first accurate and modern guide to the plants of eastern North America. The publication contained keys and thorough descriptions of plants. The eighth, and centennial, edition was largely rewritten and expanded by Merritt Lyndon Fernald (1873–1950) and published in 1950. This edition was corrected and updated by R. C. Rollins and reprinted in 1987 by Dioscorides Press.

What **contributions** did **John and William Bartram** make to **botany**?

John Bartram (1699–1777) was the first American-born botanist. He and his son, William Bartram (1739–1823), traveled throughout the American colonies observing the flora and fauna of the colonies. Although John Bartram never published his observations, he was considered the authority on American plants. In 1791 his son William published his notes on American plants and animals as *Bartram's Travels*.

How are **plants identified** based on their **growth patterns**?

Herbaceous or nonwoody plants die at the end of each growing season. Woody plants add a new layer of wood each year.

What are the **four major groups** of **plants**?

Plants are divided into phyla based on whether they are vascular (containing vascular tissue consisting of cells joined into tubes that transport water and nutrients) or nonvascular. The phyla of vascular plants are then further divided into seedless plants and those that contain seeds. Plants with seeds are divided into flowering and nonflowering groups. Nonvascular plants have traditionally been called bryophytes. Because bryophytes lack a system for conducting water and nutrients, they are restricted in size and live in moist areas close to the ground. Examples of bryophytes are mosses, liverworts, and hornworts. Examples of seedless, vascular plants are ferns, horsetails, and club mosses. The conifers, which are cone-bearing, are seed-bearing, nonflowering vascular plants. The majority of plants are seed-bearing, flowering, vascular plants known as angiosperms.

What are the **phyla** of **plants**?

Phyla	Number of Species	Characteristics	Example
Bryophyta	12,000	Nonvascular	Mosses
Hepaticophyta	6,500	Nonvascular	Liverworts
Anthocerotophyta	100	Nonvascular	Hornworts
Psilophyta	6	Vascular, homosporous, no differentiation between root and shoot	Whisk ferns
Lycophyta	1,000	Vascular, homosporous or heterosporous	Club mosses
Arthrophyta	15	Vascular, homosporous	Horsetails
Pterophyta	12,000	Vascular, homosporous	Ferns
Cycadophyta	100	Vascular, heterosporous, seed-forming	Cycads (commonly known as "sago palms")
Ginkgophyta	1	Vascular, heterosporous, seed-forming, deciduous tree	Ginkgo
Gnetophyta	70	Vascular, heterosporous, seed-forming	Ephedra, shrubs, vines
Coniferophyta	550	Vascular, heterosporous, seed-forming	Conifers (pines, spruces, firs, yews, and redwoods)
Anthophyta	240,000	Vascular, heterosporous, seed-forming	Flowering plants

How has **plant classification changed** over the years?

The earliest classifications of plants were based on whether the plant was considered medicinal or was shown to have other uses. *De re Rustica* by Cato the Censor (234–149 B.C.E.) lists 125 plants and was one of the earliest catalogs of Roman plants. Gaius Plinius Secundus (23–79), known as Pliny the Elder, wrote *Historia Naturalis*, which was published in the first century. The book was one of the earliest catalogs of significant plants in the ancient world, describing more than 1,000 plants. Plant classification became more complicated as more and more plants were discovered. One of the earliest plant taxonomists was the Italian botanist Caesalpinus (1519–1603). In 1583 he classified more than 1,500 plants according to various attributes, including leaf formation and the presence of seeds or fruit.

John Ray (1627–1705) was the first botanist to base plant classification on the presence of multiple similarities and features. His *Historia Plantarum Generalis*, published between 1686 and 1704, was a detailed classification of more than 18,000 plants. The book included a distinction between monocotyledon and dicotyledon flowering plants. The French botanist J. P. de Tournefort (1656–1708) was the first to characterize genus as a taxonomic rank that falls between the ranks of family and species. De Tournefort's classification system included 9,000 species in 700 genera. The Swedish naturalist Carolus Linnaeus (1707–1778) published *Species Plantarum* in 1753. It organized plants into twenty-four classes based on reproductive features. The Linnaean system of binomial nomenclature remains the most widely used system for classifying plants and animals. It is considered an artificial system since it often does not reflect natural relationships.

During the late-eighteenth century several natural systems of classification were proposed. The French botanist Antoine Laurent de Jussieu (1686–1758) published *Genera Plantarum*. The tome *Prodromus Systematis Naturalis Regni Vegetabilis* was started in 1824 by the Swiss botanist Augustin Pyrame de Candolle (1778–1841) and completed 50 years later. Another *Genera Plantarum* was published between 1862 and 1883 by the English botanists George Bentham (1800–1884) and Sir Joseph Dalton Hooker (1817–1911).

Charles Darwin's (1809–1882) ideas on evolution began to influence systems of classification during the late-nineteenth century. The first major phylogenetic system of plant classification was proposed around the close of the nineteenth century. *Die natürlichen Pflanzenfamilien* (*The Natural Plant Families*), one of the most complete phylogenetic systems of classification and still in use through the twentieth century, was published between 1887–1915 by the German botanists Adolf Engler (1844–1930) and Karl Prantl (1849–1893). Their system recognizes about 100,000 species of plants, organized by their presumed evolutionary sequence.

Systems of classifications were also developed during the twentieth century. Some works focused on groups of plants, especially flowering plants, rather than all

An ant preserved in amber. Amber is the fossilized resin of trees. The two major deposits of amber are in the Baltic region and the Dominican Republic.

plants. Charles Bessey (1845–1915) was the first American scientist to publish a system of classification in the early-twentieth century. Cladistics is one of the newest approaches to classification. It is often defined as a set of concepts and methods for determining cladograms, which portray branching patterns of evolution.

What is the **origin** of **land plants**?

Many scientists believe land plants evolved from green algae. Green algae, especially the charaophytes, share a number of biochemical and metabolic traits with plants. Both contain the same photosynthetic pigments—carotenes, xanthophylls, as well as chlorophylls *a* and *b*. Cellulose is a major component of the cell walls of plants and algae, and both store their excess carbohydrates as starch. In addition, some aspects of cell division, particularly the formation of new cross-walls, only occurs in plants and certain charaophytes, such as species of the genera *Cerara* and *Colechaete*.

How does a **plant become** a **fossil**?

Fossilization is dependent upon where organisms grow and how quickly they are covered by sediment . Rarely do paleobotanists find the fossil remains of whole plants. Usually only fossilized parts of plants are found. Fossilization occurs in many different ways. Three common methods of fossilization are compression, impression, and molding or casting.

163

Compression fossils are often formed in water, where heavy sediment flattens leaves or other plant parts. The weight of the sediment squeezes out water present in the plant tissue, leaving only a thin film of tissue. An impression fossil is an imprint of an organism that is left behind when the organism's remains have been completely destroyed, leaving only the contour of the plant. Fossil molds and casts are formed when animal or plant tissues become surrounded by hardened sediment; the tissue then decays. The hollow negative created by the tissue is called a mold. When fossil molds fill with sediment over time, the sediment often conforms to the contours of the mold, resulting in a fossil called a cast.

How is **petrified wood** formed?

Petrified wood is formed when water containing dissolved minerals, such as calcium carbonate ($CaCO_3$) and silicate, infiltrate wood or plants. The foreign materials either replace or enclose the organism, so the structural details of the plant are retained. The process takes thousands of years. Botanists find these types of fossils to be very important since they allow for the study of the internal structure of extinct plants. After time passes, the plant or wood seems to have turned to stone because the original form and structure are retained, but it does not actually turn into stone.

What is **amber**?

Amber is the fossilized resin of trees. Resin is the sticky material often seen oozing from the trunk of a pine tree. Resin hardens as it dries and is the source of both turpentine and rosin. Resin comes primarily from kauri pine forests. The translucent material emerges from trees and forms lumps that appear deep orange or yellow in color. These lumps may weigh up to 99 lb (45 kg). The two major deposits of amber are in the Baltic region and the Dominican Republic. Amber has also been found in the central Appalachian region of the United States. Prehistoric insects have been so remarkably preserved in amber that the pieces still contain the insect's intact DNA. Amber is the only jewel of plant origin.

What is the **alternation of generations** in **plants**?

All plants exhibit an alternation of generations between diploid sporophytes and haploid gametophytes. Sporophytes produce haploid spores as a result of meiosis. The spores grow into multicellular, haploid individuals known as gametophytes. Spores are the first cells of the gametophyte generation. Gametophytes produce gametes as a result of mitosis. Male and female gametes fuse to form a zygote, which grows into a sporophyte. The zygote is the first cell of the following sporophyte generation.

What are **heterosporous plants**?

Heterosporous plants produce two types of spores—microspores and megaspores—which develop into the male gametophyte and female gametophyte, respectively. In 1580 the physician Prospero Alpini (1553–1616) identified that plants exist in male and female forms.

BRYOPHYTES

Where are **bryophytes found**?

Mosses, liverworts, and hornworts—collectively known as bryophytes—are often found in moist environments. However, there are species that inhabit almost all environments, from hot, dry deserts to the coldest regions of Antarctica. They are most noticeable when they grow in a dense mass.

What are the main **features** of **bryophytes**?

Bryophytes are generally small, compact plants that rarely grow to more than 8 in (20 cm) tall. They have parts that appear leaflike, stemlike, and rootlike, and lack vascular tissue (xylem and phloem). Most species have rhizoids, a cuticle, a cellular jacket to retain moisture around sperm-producing and egg-producing structures, and large gametophytes that hold onto sporophytes. They require water to reproduce sexually. In nature, they are noted for their intense shades of green.

In the life cycles of **bryophytes**, which is **dominant**, the **sporophyte** or **gametophyte**?

In all of the bryophytes—mosses, liverworts, and hornworts—gametophytes are the most conspicuous, dominant phase. A mat of moss consists of haploid gametophytes. Sporophytes are typically smaller and present only part of the time.

Which **bryophytes** are most closely **related** to **green algae**?

Hornworts are more closely related to green algae than to any other group of plants. Hornwort cells usually have a single, large chloroplast with a pyrenoid (granular, starch-containing body) similar to those of green algae. Mosses and liverworts are like all other plants because they have many dish-shaped chloroplasts per cell.

What is the **purpose** of **rhizoids**?

Rhizoids are a characteristic feature of mosses, liverworts, and hornworts. Rhizoids are slender, usually colorless projections that consist of a single cell or a few cells. They serve to anchor mosses, liverworts, and hornworts to their substrate and absorb water.

What **feature** of **liverworts** hints to their name?

Liverworts were named during the Middle Ages, when herbalists followed the theoretical approach known as the *Doctrine of Signatures*. The core philosophy of this perspective was that if a plant part resembled a part of the human body, it would be useful in treating ailments of that organ or part. The thallus of thalloid liverworts resembles a lobed liver. Therefore, in line with the philosophy presented by the doctrine, the plant was used to treat liver ailments. The word "liver" was combined with "wort," which means herb, to form the name "liverwort."

What **value** do **liverworts** have **ecologically**?

Liverworts provide food for animals. Due to their ability to retain moisture, they also assist in the decay of logs and aid in the disintegration of rocks into soil.

What **plants** are **erroneously called mosses**?

Not all plants called mosses are bryophytes. Irish moss (*Chondrus crispus*) and related species are actually red algae. Iceland moss (*Cetraria islandica*) and reindeer moss (*Cladonia rangiferina*) are lichens. Club mosses (genus *Lycopodium*) are seedless, vascular plants, and Spanish moss (*Tillandsia usneoides*) is a flowering plant in the pineapple family.

Why are **mosses important**?

Some mosses are decomposers that break down the substrata and release nutrients for the use of more complex plants. Mosses play an important role in controlling soil erosion. They perform this function by providing ground cover and absorbing water. Mosses are also indicators of air pollution. Under conditions of poor air quality, few mosses will exist. Peat is used as fuel to heat homes and generate electricity. Bryophytes are among the first organisms to invade areas that have been destroyed by a fire or volcanic eruption.

What is **unusual** about **cave moss**?

Cave moss (*Schistostega pennata*) is a small plant with reflective, sub-spherical cells at its tips. These cells give off an eerie glow that is gold and green in color. In Japan the plant has been the subject of numerous books, television shows, newspaper and magazine articles, and even an opera! There is a national monument to this species near the coast of Hokkaido, where it grows near a small cave.

How can **bryophytes** be **used** as **bioindicators**?

Bioindicators are physiological, chemical, or behavioral changes that occur in organisms as a result of changes in the environment. Bryophytes of the genus *Hypnum* are particularly sensitive to pollutants, especially sulfur dioxide. As a result, most bryophytes are not found in cities and industrial areas. Mosses and liverworts, especially *Hypnum cupressiforme* and *Homalotecium serieceum*, were used as bioindicators to monitor radioactive fallout from the Chernobyl reactor accident in 1986.

The Hawley bog. Peat moss (genus *Sphagnum*) grows mostly in bogs and is favored by gardeners for its ability to increase the water-holding capacity of soils.

What **economic importance** do **bryophytes** have?

Bryophytes are used in a variety of industrial applications. Different species of mosses are used as furniture stuffing, as soil conditioner, as cushioning, and as material used to absorb oil after spills.

What are the **uses** of **peat moss**?

Peat moss (genus *Sphagnum*) grows mostly in bogs. Peat moss is favored by gardeners for its ability to increase the water-holding capacity of soils. Due to large, dead cells in the leaf-like parts, it is able to absorb five times as much water as cotton plants. Peat moss is also used as damp cushions by florists to keep other plants and

What is the oldest group of seedless, vascular plants?

According to the fossil record, the club mosses (lycopods) are the most-ancient group of seedless, vascular plants.

flowers damp. Species of *Sphagnum* also have medicinal purposes. Certain aboriginal people use peat moss as disinfectants and, due to its absorbency, as diapers. Peat moss is acidic and is an ideal dressing for wounds. During World War I the British used more than one million wound dressings made of peat moss. Native North Americans used species of the genera *Mnium* and *Bryum* to treat burns. In Europe species of the genus *Dicranoweisia* have been used to waterproof roofs.

FERNS AND RELATED PLANTS

What were the **first vascular plants**?

The word "vascular" comes from the Latin word *vasculum*, meaning "vessel" or "duct." It is believed that the first vascular plants were members of the division Rhyniophyta, which flourished about 400 million years ago but are now extinct. Members of the extinct genus *Cooksonia*, named for the paleobotanist Isabel Cookson (1893–1973), were the first vascular plants to be identified. These plants were only a few centimeters tall, with water-conducting cells present in their stems but without roots and leaves.

What are the **four groups** of **seedless, vascular plants**?

The seedless, vascular plants include: ferns of the genus *Pterophyta*, which is the largest group; the whisk ferns of the genus *Psilophyta*; the club mosses of the genus *Lycophyta*; and the horsetails of the genus *Arthrophyta*.

What are the main **features** of **vascular plants**?

Vascular plants have leaves, roots, cuticles, stomata, specialized stems, tissues that conduct efficiently, and, in most cases, seeds. Their sporophytes are large, dominant, and nutritionally independent.

What is the **relationship** between **ancient plants** and **coal formation**?

Coal, formed from ancient plant material, is organic. Most of the coal mined today was formed from prehistoric remains of primitive land plants, particularly those of the Carboniferous period, which occurred approximately 300 million years ago. Five main

Most of the coal mined today was formed from prehistoric remains of primitive land plants, particularly those of the Carboniferous period, which occurred approximately 300 million years ago.

groups of plants contributed to the formation of coal. The first three groups were all seedless, vascular plants: ferns, club mosses, and horsetails. The last two groups were the now-extinct seed ferns and the primitive gymnosperms. Forests of these plant groups were in low-lying, swampy areas that periodically flooded. When these plants died, they decomposed, but as they were covered by water, they did not decompose completely. Over a period of time the decomposed plant material accumulated and consolidated. Layers of sediment formed over the plant material during each flood cycle. Heat and pressure built up in these accumulated layers and converted the plant material to coal. The various types of coal (lignite, bituminous, and anthracite) were formed as a result of the different temperatures and pressures to which the layers were exposed.

In the **life cycles** of seedless, vascular plants such as ferns, is the **sporophyte** or **gametophyte dominant**?

The seedless, vascular plants exhibit an alternation of generations consisting of heteromorphic diploid and haploid phases, as found in all plants. The life cycle of seedless, vascular plants is dominated by the sporophyte. The sporophyte is the "plant"—for example, trees and flowers—that everyone visualizes when they think of plants.

How does **light affect** the **growth** of **fern gametophytes?**

Light controls spore germination in ferns. Wavelengths in the red range of the spectrum (about 700 nm) induce spore germination, while wavelengths in the blue light

of the spectrum (about 400 nm) prevent spore germination. Red light also induces
apical growth and positive phototropism, increases the gap phase in mitosis, and
delays the formation of cell plates during cytokinesis. On the other hand, blue light
inhibits these phenomena.

What is special about the **spores** of **leptosporangiate ferns**?

Leptosporangiate ferns are the most-common ferns in North America. The sporangia
of leptosporangiate ferns arise from a single surface cell, are relatively small, and have
a delicate stalk and a thin sporangial wall. The small number of spores per leptospo-
rangium is a multiple of four, varying between 16 and 512, most often 16 or 32, in
homosporous species. Each plant is able to produce millions of spores because of the
large number of sporangia per sorus and the enormous number of sori per leaf. One
mature plant of the species *Thelypteris dentate* can produce more than 50 million
spores each season!

What is a **fiddlehead**?

The type of fern typically grown as a houseplant is of the diploid, or sporophyte, generation. It is composed of a rhizome, an underground stem that occurs horizontally, which
produces roots and leaves called fronds. As each young frond first emerges from the
ground, it is tightly coiled and resembles the top of a violin, hence the name fiddlehead.

Why are **horsetails** called "**scouring rushes**"?

The epidermal tissue of horsetails contains abrasive particles of silica. Scouring rush-
es were used by Native Americans to polish bows and arrows. Early North American
settlers, who cleaned their pots and pans along stream banks, used horsetails—found
in abundance in such areas—to scrub out their dishes.

What **club moss** is called the "**resurrection plant**"?

Selaginella lepidophylla, found in the deserts of the southwestern United States and
Mexico, is called the resurrection plant because of its ability to defy severe drought

conditions. During periods of drought, this plant forms a tight, dried-up ball. When rain comes, its branches expand, become green, and carry out photosynthesis.

Which **club moss** is threatened with **extinction**?

Many club mosses are attractive plants often used as Christmas decorations or in floral wreaths. The species *Lycopodium nutans,* which is native to Hawaii and is known commonly known as Wawae'ioie, is threatened with extinction and has been placed on the endangered species list.

What **seedless, vascular plant** played a **role in early photography**?

Prior to the invention of flashbulbs, photographers used flash powder that consisted almost entirely of dried spores from club mosses of the genus *Lycopodium*.

GYMNOSPERMS

What are **gymnosperms** and which **plants** are included in this **group**?

Gymnosperms (from the Greek terms *gymnos,* which means "naked," and *sperma,* meaning "seed") produce seeds that are totally exposed or borne on the scales of cones. The four phyla of gymnosperms are: Coniferophyta, conifers including pine, spruce, hemlock and fir; Ginkgophyta, consisting of one species, the ginkgo or maidenhair tree; Cycladophyta, the cycads or ornamental plants; and Gnetophyta, a collection of very unusual vines and trees.

What is the **oldest genus** of **living trees**?

The genus *Ginkgo,* commonly known as maidenhair trees, comprises the oldest living trees. This genus is native to China, where it has been cultivated for centuries. It has not been found in the wild and it is likely that it would have become extinct had it not been cultivated. Fossils of 200-million-year-old ginkgoes show that the modern-day ginkgo is nearly identical to its forerunner. As of the early twenty-first century, only one living species of ginkgo remains, *Ginkgo biloba*. The fleshy coverings of the seeds produced by females of the species *G. biloba* have a distinctly foul odor. Horticulturists prefer to cultivate the male plant from shoots to avoid the odor and mess created by the female tree.

What plant produces the **largest seed cones**?

The largest seed cones are produced by cycads. They may be up to 1 yd (1 m) in length and weigh more that 3.3 lb (15 kg).

What is **taxol**?

Taxol is a drug used to treat ovarian cancer. It "freezes" cancer cells early in the process of cell division. When the cells are unable to divide, they eventually die. Taxol is obtained from the bark of the Pacific yew, *Taxus brevifolia*, a gymnosperm that grows in the Pacific Northwest. Because the Pacific yew is a small tree that grows slowly and is not found in abundance, researchers have synthesized taxol.

What are the **distinguishing characteristics** of **fir, pine, and spruce trees**?

The best way to tell the difference between the three trees is by their cones and needles:

Species	Needles	Cones
Balsam fir	Needles are 1–1.5 in (2.54–3.81 cm) long, flat, and arranged in pairs opposite each other.	Upright, cylindrical, and 2–4 in (5–10 cm) long.
Blue spruce	Needles are roughly 1 in (2.54 cm) long, grow from all sides of the branch, are silvery blue in color, and are very stiff and prickly.	3.5 in (8.89 cm) long.
Douglas fir	Needles are 1–1.5 in (2.54–3.81 cm) long, occur singularly, and are very soft.	Cone scales have bristles that stick out.
Fraser fir	Similar to Balsam fir, but needles are smaller and more rounded.	Upright, 1.6 – 2.4 inches (4-6 cm) long
Scotch pine	Two needles in each bundle; needles are stiff, yellow green, and 1.5–3 in (3.81–7.62 cm) long.	2–5 in (5–12.7 cm) long.
White pine	Five needles in each bundle; needles are soft and 3–5 in (7.62–12.7 cm) long.	4–8 in (10–20.3 cm) long.
White spruce	Dark-green needles are rigid but not prickly; needles grow from all sides of the twig and are less than an inch (2.54 cm) long.	1–2.5 in (2.54–6.35 cm) long and hang downward.

Which **conifers** in North America **lose** their **leaves** in **winter**?

Dawn redwood trees (*Metasequoia*) are deciduous. Their leaves are bright green in summer and turn coppery red in the fall before they drop. Previously known only as a fossil, the tree was found in China in 1941 and has been growing in the United States since the 1940s. The U.S. Department of Agriculture distributed seeds to experimen-

> ## In the life cycle of a pine tree, is the sporophyte or gametophyte dominant?
>
> In the evolution of seed plants, one of the key terrestrial adaptations in reproduction was the increasing dominance of the sporophyte generation. The mature pine tree is a sporophyte.

tal growers in the United States, and the dawn redwood tree now grows all over the country. The only native conifers that shed all of their leaves in the fall are the bald cypress (*Taxodium distichum*) and the Larch (*Larix larcina*).

Do **pine trees** keep their **needles forever**?

Pine needles occur in groups, called fascicles, of two to five needles. A few species have only one needle per fascicle, while others have as many as eight. Regardless of the number of needles, a fascicle forms a cylinder of short shoots that are surrounded at their base by small, scalelike leaves that usually fall off after one year of growth. The needle-bearing fascicles are also shed a few at a time, usually every two to ten years, so that any pine tree, while appearing evergreen, has a complete change of needles every five years or less.

How long does it take to **produce** a **mature pine cone**?

From the time young cones appear on the tree, it takes nearly three years for them to mature. The sporangia of a pine tree are located on scalelike sporophylls that are densely packed in structures called cones. Conifers, like all seed plants, are heterosporous, meaning that male and female gametophytes develop from spores produced by separate cones. Small pollen cones produce microspores that develop into the male gametophytes or pollen grains. Larger, ovulate cones make megaspores that develop into female gametophytes. Each tree usually has both types of cones. This three-year process culminates in the production of male and female gametophytes, brought together through pollination, and the formation of mature seeds from the fertilized ovules. The scales of ovulate cones then separate, and the seeds are scattered by wind. A seed that lands on a habitable place germinates, its embryo emerging as a pine seedling.

Is **hemlock** poisonous?

There are two species known commonly as hemlock: *Conium maculatum* and *Tsuga canadensis*. *Conium maculatum* is a weedy plant, and all parts of it are poisonous. In ancient times minimal doses of the plant were used to relieve pain, although there was

While the size of giant redwood trees implies that they are composed of very strong wood, the opposite is true. The wood is useless as timber because it is brittle and shatters into splintery, irregular pieces when struck.

a great risk of poisoning from this form of treatment. *Conium maculatum* was also used to carry out death sentences in ancient times. The Greek philosopher Socrates was condemned to death and sentenced to drink a potion made from hemlock. The poisonous species should not be confused with *Tsuga canadensis*, a member of the evergreen family. The leaves of *T. canadensis* are not poisonous and are often used to make tea.

Are **giant redwood trees** found only in **California**?

Although redwoods extend somewhat into southern Oregon, the vast majority of giant redwoods are found in California. The closest relative to this form of redwood is the Japanese cedar found in regions of Asia. This tree grows to a height of 150 ft (45.7 m), with a circumference of 25 ft (7.6 m). There are two species of the genus *Sequoia*, which are commonly known as the redwood and big tree. Both can be seen in either Redwood National Park or Sequoia National Park. At the latter park, the most impressive tree is known as the General Sherman Tree. It is 272 ft (83 m) tall, has a diameter of 32 ft (9.75 m), and a circumference of 101 ft (30.8 m). The weight of the tree is estimated to be more than 6,000 tons. Other trees found in Sequoia National Park exceed 300 ft (91.4 m) in height but are more slender. The General Sherman Tree is about 4,000 years old, the oldest living thing next to the bristlecone pine. Approximately 150 million years ago these giant trees were widespread across the Northern Hemisphere. While the size of these giant trees implies that they are composed of very strong wood, the opposite is true. The wood is useless as timber because it is brittle and shatters into splintery, irreg-

What is the origin of the word "sequoia"?

The word "sequoia" was proposed by Austrian botanist Stephen Endlicher (1804–1849) to commemorate the eighteenth-century Cherokee leader Sequoyah (*ca.* 1770–1843), remembered for developing an eighty-three-letter alphabet for the Cherokee language.

ular pieces when struck. Perhaps the weakness of the wood is why so many of these giant trees still survive and have not been harvested by the logging industry.

In what ways are **gymnosperms economically important**?

Gymnosperms account for approximately 75 percent of the world's timber and a large amount of the wood pulp used to make paper. In North America the white spruce, *Picea glauca*, is the main source of pulpwood used for newsprint and other paper. Other spruce wood is used by to manufacture violins and similar string instruments because the wood produces a desired resonance. The Douglas fir, *Pseudotsuga menziesii*, provides more timber than any other North American tree species and produces some of the most-desirable lumber in the world. The wood is strong and relatively free of knots. Uses for the wood include house framing, plywood production, structural beams, pulpwood, railroad ties, boxes, and crates. Since most naturally occurring areas of growth have been harvested, the Douglas fir is being grown in managed forests. The wood from the redwood *Sequoia sempervirens* is used for furniture, fences, posts, some construction, and has various garden uses.

In addition to the wood and paper industry, gymnosperms are important in making resin and turpentine. Resin, the sticky substance in the resin canals of conifers, is a combination of turpentine, a solvent, and a waxy substance called rosin. Turpentine is an excellent paint and varnish solvent but is also used to make deodorants, shaving lotions, medications, and limonene—a lemon flavoring used in the food industry. Resin has many uses; it is used by baseball pitchers to improve their grip on the ball and by batters to improve their grip on the bat; violinists apply resin to their bows to increase friction with the strings; dancers apply resin to their shoes to improve their grip on the stage.

ANGIOSPERMS (FLOWERING PLANTS)

What **factors** have contributed to the **success** of **seed plants**?

Seed plants do not require water for sperm to swim to an egg during reproduction. Pollen and seeds have allowed them to grow in almost all terrestrial habitats. The sperm

175

of seed plants is carried to eggs in pollen grains by the wind or animal pollinators such as insects. Seeds are fertilized eggs that are protected by a seed coat until conditions are proper for germination and growth. Seeds are also dispersed by wind or animals.

What is the **oldest-known fossil flower**?

The fossil of the world's earliest known flower was discovered in 1986. This flowering plant is from the 120-million-year-old Koonwarra fossil beds found near Melbourne, Australia. The fossil flower was an important find because all the parts of the flower were attached to an intact plant. The fossil resembles a small black pepper plant, less than 1 in (3 cm) long. Paleobotanists believe that this plant represents an ancestral type of flower.

What was one of the **earliest flowering plants**?

Scientists do not know for certain which plant was the world's first flowering plant, but many surmise that it was the cattail *Typha latifolia*, a species still found today. Although it looks like a reed, it is actually a flowering plant. The flowers are tiny, and the petals and sepals are made up of a few bristles.

What are the **two major groups** of **angiosperms**?

Angiosperms—made up of the largest number of plant species (240,000)—are classified into two major groups, monocots and dicots. The description of monocots and dicots is based on the first leaves that appear on the plant embryo. Monocots have one seed leaf, while dicots have two seed leaves. There are approximately 65,000 species of monocots and 175,000 species of dicots. Orchids, bamboo, palms, lilies, grains, and many grasses are examples of monocots. Dicots include most trees that are nonconiferous, shrubs, ornamental plants, and many food crops.

What are the **major differences** between **monocots** and **dicots**?

The seed leaves, also called cotyledons, are different in monocots and dicots. Monocots have one cotyledon, while dicots have two cotyledons. Other differences are illustrated in the following chart:

	Seed Leaves	Leaf Veins	Stems	Flowers	Roots	Examples
Monocots	One cotyledon	Usually parallel	Vascular bundles in scattered arrangement	Floral parts usually in multiples of three	Fibrous root system	Corn, lily, pineapple, banana
Dicots	Two cotyledons	Usually netlike	Vascular bundles arranged in a ring	Floral parts usually in multiples of four or five	Usually a taproot	Pea, rose, sunflower, ash

Wheat being harvested. Wheat is the most widely cultivated cereal in the world

Among **angiosperms**, what is the **most important family**?

Angiosperms, commonly known as flowering plants, include the grass family. This family is of greater importance than any other family of flowering plants. The edible grains of cultivated grasses, known as cereals, are the basic foods of most civilizations. Wheat, rice, and corn are the most extensively grown of all food crops. Other important cereals are barley, sorghum, oats, millet, and rye.

What is the most **widely cultivated cereal** in the world?

Wheat is the most widely cultivated cereal in the world; the grain supplies a major percentage of the nutrients needed by the world's population. Wheat is one of the oldest domesticated plants, and it has been argued that it laid the foundation for Western civilization. Domesticated wheat had its origins in the Near East at least 9,000 years ago. Wheat grows best in temperate grassland biomes that receive 12 to 36 in (30 to 90 centimeters) of rain per year and have relatively cool temperatures. Some of the top wheat-producing countries are Argentina, Canada, China, India, the Ukraine, and the United States.

What are some **economically important angiosperms**?

Angiosperms produce lumber, ornamental plants, and a variety of foods. Some examples of economically important angiosperms are:

Common Family Name	Genus Name	Economic Importance
Gourd	*Cucurbitaceae*	Food (melons and squashes)

177

Common Family Name	Genus Name	Economic Importance
Grass	*Poaceae*	Cereals, forage, ornamentals
Lily	*Liliaceae*	Ornamentals and food (onions)
Maple	*Aceraceae*	Lumber and maple sugar
Mustard	*Brassicaceae*	Food (cabbage and broccoli)
Olive	*Oleaceae*	Lumber, oil and food
Palm	*Arecaceae*	Food (coconut), fiber, oils, waxes, furniture
Rose	*Rosaceae*	Fruits (apple and cherry), ornamentals (roses)
Spurge	*Euphorbiaceae*	Rubber, medicinals (castor oil), food (cassava), ornamentals (poinsettia)

What are **Joshua trees**?

Yucca brevifolia, a large shrub found in the southwestern region of the United States, received its common name from Mormon pioneers. They named the tree after the prophet Joshua because its greatly extended branches resemble how Joshua used his outstretched arms to point his spear toward the city of Ai.

What is a **banyan tree**?

The banyan tree, *Ficus benghalensis*, native to tropical regions of Southeast Asia, is a member of the genus *Ficus*. It is a magnificent evergreen that can reach 100 ft (30.48 m) in height. As the massive limbs spread horizontally, the tree sends down roots that develop into secondary, pillar-like supporting trunks. Over a period of years a single tree may spread to occupy as much as 2,000 ft (610 m) around its periphery.

What is a **monkey ball tree**?

The osage orange tree, *Maclura pomifera*, produces large, green, orangelike fruits. The fruit is roughly spherical, 3.5 to 5 in (8.8 to 12.7 cm) in diameter, and have a coarse, pebbly surface.

Why do the leaves of the **mimosa plant close** in response to **touch**?

When a leaf of the mimosa plant—also known as the "sensitive plant"—is touched, a minute electric current is generated that is quickly transmitted to the cells at the base of each leaflet. As soon as the signal arrives to the cells, the water contained in the cells is released. Due to loss of water, the leaves collapse downward.

What is the **fastest-growing land plant**?

Bamboo (*Bambusa* spp.), native to tropical and subtropical regions of Southeast Asia and islands of the Pacific and Indian Oceans, is the plant that gains height most quickly. Bamboo can grow almost 3 ft (1 m) in 24 hours. This rapid growth is produced partly by cell division and partly by cell enlargement.

How is it possible to **identify the genus and species** of a plant based on its **wood**?

The arrangement of plant cells present in the xylem of wood allows for the identification of a plant's genus. For example, the wood of oak trees (genus *Quercus*) is dense and heavy because it has abundant fibers.

Which **tree species** from the **United States** have **lived the longest**?

Of the 850 different species of trees in the United States, the oldest species is the bristlecone pine, *Pinus longaeva*. This species grows in the deserts of Nevada and southern California, particularly in the White Mountains. Some of these trees are believed to be over 4,600 years old. The potential life span of these pines is estimated to be 5,500 years. But potential age of the bristlecone pine is very young when compared to the oldest surviving species in the world, the maidenhair tree (*Ginkgo biloba*) of China. This species of tree first appeared during the Jurassic era, some 160 million years ago. Also called icho, or the ginkyo (meaning "silver apricot"), this species has been cultivated in Japan since 1100 B.C.E.

Longest-Living Tree Species in the United States

Name of tree	Average lifespan (years)
Bristlecone pine (*Pinus longaeva*)	3,000–4,700
Giant sequoia (*Sequoiadendron giganteum*)	2,500
Redwood (*Sequoia sempervirens*)	1,000–3,500
Douglas fir (*Pseudotsuga menziesii*)	750
Bald cypress (*Taxodium distichum*)	600

The Venus fly trap imprisons its victims when trigger hairs on its leaves are touched.

How are **carnivorous plants** categorized?

Carnivorous plants are plants that attract, catch, and digest animal prey, absorbing the bodily juices of prey for the nutrient content. There are more than 400 species of carnivorous plants. The species are classified according to the nature of their trapping mechanism. All carnivorous plants have traps made of modified leaves with various incentives or attractants, such as nectar or an enticing color, that can lure prey. Active traps display rapid motion in their capture of prey. The Venus fly trap, *Dionaea muscipula*, and the bladderwort, *Utricularia vulgaris*, have active traps that imprison victims. Each leaf is a two-sided trap with trigger hairs on each side. When the trigger hairs are touched, the trap shuts tightly around the prey. Semi-active traps employ a two-stage trap in which the prey is caught in the trap's adhesive fluid. As prey struggles in the fluid, the plant is triggered to slowly tighten its grip. The sundew (*Drosera capensis*) and butterwort (*Pinguicula vulgaris*) have semi-active traps. Passive traps entice insects using nectar. The passive-trap leaf has evolved into a shape resembling a vase or pitcher. Once lured to the leaf, the prey falls into a reservoir of accumulated rainwater and drowns. An example of the passive trap is the pitcher plant (*Sarracenia purpurea*). The Green Swamp Nature Preserve in southeastern North Carolina has the most numerous types of carnivorous plants.

What is **unique** about the **water lily** *Victoria amazonica*?

It is very big! Found only on the Amazon River, this water lily has leaves that can reach up to 6 ft (1.8 m) in diameter. The flowers of *Victoria amazonica* reach up to 12 in (30 cm) in height and open at dusk, but only open on two successive nights.

How did the **navel orange originate**?

Navel oranges are oranges without seeds. In the early-nineteenth century an orange tree in a Brazilian orchard produced seedless fruit although the rest of the trees in the orchard produced seeded oranges. This naturally occurring mutation gave rise to what we now refer to as the navel orange. A bud from the mutant tree was grafted onto another orange tree; branches that resulted were then grafted onto other trees, soon creating orchards of navel orange trees. Every navel orange tree is derived from pieces of the tree that first produced the mutated fruit.

How long can an orange **tree produce oranges**?

An average orange tree will produce fruit for fifty years, but eighty years of productivity is not uncommon, and a few trees are known to be still producing fruit after more than a century. An orange tree may attain a height of 20 ft (6.1 m), but some trees are as much as 30 ft (9.1 m) high. Orange trees grow well in a variety of soils but prefer subtropical settings.

How are **seedless grapes** grown?

Since seedless grapes cannot reproduce in the manner that grapes usually do (i.e., dropping seeds), growers have to take cuttings from the plants, root them, and then plant the plant cuttings. Seedless grapes come from a naturally occurring mutation in which the hard seed casing fails to develop. Although the exact origin of seedless grapes is unknown, they might have been first cultivated thousands of years ago in present-day Iran or Afghanistan. Currently, 90 percent of all raisins are made from Thompson seedless grapes.

Do **seedless watermelons** occur naturally?

Seedless watermelon was first introduced in 1988 after fifty years of research. A seedless watermelon plant requires pollen from a seeded watermelon plant. Farmers frequently plant seeded and seedless plants close together and depend on bees to pollinate the seedless plants. The white "seeds," also known as pods, found in seedless watermelons serve to hold a fertilized egg and embryo. Because a seedless melon is sterile and fertilization cannot take place, pods do not harden and become a black seed, as occurs in seeded watermelons.

What is the **difference** between **poison ivy, oak, and sumac**?

These North American woody plants grow in almost any habitat and are quite similar in appearance. Each variety of plant has three-leaf compounds that alternate, berry-like fruits, and rusty brown stems. Poison ivy (*Rhus radicans*) grows like a vine rather than a shrub and can grow very high, covering tall, stationary items such as trees.

The fruit of *R. radicans* is gray in color and is without "hair," and the leaves of the plant are slightly lobed.

Rhus toxicodendron, commonly known as poison oak, usually grows as a shrub, but it can also climb. Its leaflets are lobed and resemble the leaves of oak trees, and its fruit is hairy. Poison sumac (*Rhus vernix*) grows only in acidic, wet swamps of North America. This shrub can grow as high as 12 ft (3.6 m). The fruit it produces hangs in a cluster and ranges from gray to brown in color. Poison sumac has dark-green leaves that are sharply pointed, compound, and alternating; it also has inconspicuous flowers that are yellowish green. All parts of poison ivy, poison oak, and poison sumac can cause serious dermatitis.

What is **kudzu**?

Kudzu (*Pueraria lobata*) is a vine that was brought from Japan for the 1876 Centennial Exposition in Philadelphia. It was intentionally planted throughout the southern United States during the 1930s in an attempt to control erosion. In fact, the federal government paid farmers as much as eight dollars an acre to plant it. In 1997, however, the government reversed its position on kudzu and referred to it as a "noxious weed." Kudzu grows over everything that it encounters, draping itself across power poles and pine trees like a shawl. The plant is responsible for more than $50 million in lost farm and timber production each year. It grows at a rate of 120,000 acres per year. As of the early twenty-first century it covers between 2 and 4 million acres of land throughout the United States, occurring from Connecticut in the East, to Missouri and Oklahoma in the West, and south to Florida and Texas. Kudzu grows as fast as 1 ft (30 cm) per day. The latest approach to controlling the growth of kudzu is to have goats chew on it, devouring the leaves, stems, and roots.

What are **succulents**?

A group of more than thirty plant families including the amaryllis, lily, and cactus families form what is known as the succulents (from the Latin term *succulentis*, meaning "fleshy" or

The noxious weed kudzu, imported to the United States from Japan in the 1890s, is responsible for more than $50 million in lost farm and timber production each year.

"juicy"). Most members of the group are resistant to droughts as they are dry-weather plants. Even when they live in moist, rainy environments, these plants need very little water.

What is a **stone plant**?

The species *Lithops turbuniformis* is known by several common names, including stone plant, flowering stone, and living stone. The genus name *Lithops* comes from the Greek words *lithos*, which means "stone," and *opsis*, meaning "face" or "appearance." The plant is native to South Africa, Namibia, and the West Coast of Africa. The plants are very difficult to see since their appearance mimics the rocks and stones of their natural habitat. They have two thick leaves and a single flower that is either yellow or white in color. Stone plants thrive in very arid conditions with minimal rainfall.

APPLICATIONS

What are some specific **examples** of how **plants** are **economically important**?

Materials of plant origin are found in a wide variety of industries including paper, food, textile, and construction. Chocolate is made from cocoa seeds, specifically seeds of the species *Theobroma cacao*. Foxglove (*Digitalis purpurea*) contains cardiac glycosides used to treat congestive heart failure. The berries obtained from the plant

Piper nigrum produce black pepper. The berries are dried, resulting in black peppercorns, which can then be cracked or ground. Tea can be made from the leaves of *Camellia sinensis*. Fiber taken from the stem of flax plants (*Linum usitatissimum*) have been used to make linen, while the flax seeds are commonly consumed and are a source of linseed oil. Paper money is even made from flax fibers!

What products does **one acre** of **trees yield** when cut and processed?

There are approximately 660 trees on one acre of forest; this number of trees can yield approximately 105,000 ft (32,004 m) of lumber, more than 30 tons (30,000 kg) of paper, or 16 cords of firewood.

How much **wood** is needed to make **one ton of paper**?

In the United States, wood pulp is usually used in paper manufacturing. Pulp is usually measured by cord or weight. Although the fiber used in making paper is derived overwhelmingly from wood, many other ingredients are needed as well. One ton of paper typically requires two cords of wood; 55,000 gal (208,000 L) of water; 102 lb (46 kg) of sulfur; 350 lb (159 kg) of lime; 289 lb (131 kg) of clay; 1.2 tons of coal; 112 kilowatt hours of power; 20 lb (9 g) of dye and pigment; and 108 lb (49 kg) of starch. Other ingredients may also be necessary.

How much **wood** is used for each **Sunday edition** of the **New York Times**?

More than 150 acres of forest are cut down for each Sunday edition of the *New York Times*. Most of the world's paper comes from wood pulp. In the United States each person uses an average of 731 lb (322 kg) of paper per year, or 2 lb (910 g) per day. Less than 50 percent of paper is recycled. Recycling 4 ft (1.2 m) of newspapers would save a 40 ft (12 m) tall tree.

Which **woods** are used for **telephone poles**?

The principal woods used for telephone poles are southern pine, Douglas fir, western red cedar, and lodgepole pine. Ponderosa pine, red pine, jack pine, northern white cedar, and western larch are also used.

What **wood** is the favorite for **butcher's blocks**?

Because of its resilience, the preferred wood for making butcher's blocks is derived from the American sycamore (*Platanus occidentalis*), also known as the American planetree, buttonball, buttonwood, and water beech. The wood of *Platanus occidentalis* is also used as for veneers for decorative surfaces, fence posts, and fuel.

What **wood** is used to make **baseball bats**?

Wooden baseball bats are made from white ash (*Fraxinus Americana*). This wood is ideal for producing bats because it is tough and light, and can thus help drive a ball a great distance. A tree roughly seventy-five years old and 15.7 in (40 cm) in diameter can produce approximately sixty bats.

From **where** do **frankincense** and **myrrh** originate?

Frankincense is an aromatic gum resin obtained by tapping the trunks of trees belonging to the genus *Boswellia*. The milky resin hardens when exposed to the air, forming irregularly shaped granules—the form in which frankincense is usually marketed and sold. Also called olibanum, frankincense is used as an ingredient in many different products, including pharmaceuticals, perfumes, fixatives, fumigants, and incense. Myrrh comes from a tree of the genus *Commiphora*, native to the northeastern region of Africa and the Middle East. Myrrh is also a resin obtained from trees; it is used in pharmaceuticals, perfumes, and toothpastes.

What **plants** can be used to determine **blood type**?

Lectins—proteins that bind to carbohydrates on cell surfaces—found in lotus plants as well as jack and lima beans can be used to determine a person's blood type. Lectins bind to glycoproteins present on the plasma membrane of red blood cells. Because the cells of different blood types have distinct glycoproteins, cells of each blood type bind to a specific lectin.

What was one of the **most-famous criminal cases** involving **forensic botany**?

Forensic botany is the identification of plants or plant products; this form of study can be used to produce evidence for legal trials. One of the first criminal cases to use foren-

> ## Which war was fought over plants?
>
> The first Opium War (1839–1842) was fought between China and Britain over the trading rights for opium, an extract of the opium poppy (*Papaver somniferum*). Britain won the war and claimed ownership of Hong Kong as part of its victory.

sic botany was the famous 1935 trial of Bruno Hauptmann (1899–1936), who was accused, and later convicted, of kidnapping and murdering the son of Charles and Anne Morrow Lindbergh. The botanical evidence presented in the case centered on a homemade wooden ladder used during the kidnapping and left at the scene of the crime. After extensive investigation, the plant anatomist Arthur Koehler (1885–1967) showed that parts of the ladder were made from wooden planks taken from Hauptmann's attic floor.

How did the **search for cinnamon** lead to the **discovery of North and South America**?

Christopher Columbus (1451–1506) was one of many explorers trying to find a direct sea route to Asia, which during the fifteenth century was thought to be rich with spices. Cinnamon and other spices were so valued in Columbus's era that a new, direct route to Asia would have brought untold wealth to the discoverer and his country.

How did the introduction of the **potato** to Europe lead to a devastating **famine** in **Ireland**?

The white potato (*Solanum tuberosum*), native to South America, was first introduced to Spain in the middle of the sixteenth century. It was not widely accepted as a food crop since European relatives of the potato, such as nightshade, mandrake, and henbane, were known to be poisonous or hallucinogenic. In fact, all of the aboveground parts of a potato plant are poisonous and only the tuber is edible. The potato was established as a food crop in Ireland as early as 1625 and became a staple of the diet, especially among the poor, during the eighteenth and early nineteenth centuries. The widespread dependence on potatoes as a main source of food led to massive starvation when the plant pathogen *Phytophthora infestans* destroyed potato fields in the 1840s. Over one million Irish people died from starvation or subsequent disease; another 1.5 million emigrated from Ireland.

How has **dill** been used throughout **history**?

Dill has long been used for medicinal purposes. The Egyptians used dill (*Anethum graveolens*) as a soothing medicine. Greeks habitually used the herb to cure the hic-

Worker picking coffee beans. Coffee was widely used in the Arab world before its introduction to European society in the seventeenth century. North and South American coffee plantations were started in the eighteenth century.

cups. During the Middle Ages dill was prized for the protection it purportedly provided against witchcraft. Magicians and alchemists used dill to concoct spells, while a commonly known wives' tale stated that dill added to wine could enhance passion. Colonial settlers brought dill to North America, where it became known as "meetin' seed," because children were given dill seed to chew during long sermons in church.

How has **anise** been used throughout **history**?

The Romans brought the licorice-flavored herb anise (*Pimpinella anisum*) from Egypt to Europe, where they used it as payment for their taxes. It became a popular flavoring for cakes, cookies, bread, and candy.

Is **coffee native** to **Columbia** and **Brazil**?

Although premium coffee is today grown in the mountains of Central and South America, the coffee tree (*Coffea arabica*) is native to Ethiopia. Coffee was widely used in the Arab world before its introduction to European society in the seventeenth century. North and South American coffee plantations were started in the eighteenth century.

What **wildflower** was used by **Native Americans** to make **red dye**?

Native Americans painted their faces and dyed their clothes red with the root of the bloodroot wildflower (*Sanguinaria canadensis*), which is also called redroot, Indian

187

paint, and tetterwort. Bloodroot, found in shady, damp, and woodsy soils, blooms in May and has white flowers that are 2 in (5 cm) wide.

Which **plants** have been used to create **dyes**?

Natural materials, including many plants, were the source of all dyes until the late-nineteenth century. Blue dye was historically rare and was obtained from the indigo plant (*Indigofera tinctoria*). Another color difficult to obtain for dye was red. The madder plant (*Rubia tinctorum*) was an excellent source of red dye and was used for the famous "red coats" of the British Army. Other, more common, natural dyes derived from plant sources are summarized in the following chart:

Common Name	Scientific Name	Part of Plant Used	Color
Black walnut	*Juglans nigra*	Hulls	Dark brown, black
Coreopsis	*Coreopsis*	Flower heads	Orange
Lilac	*Syringa*	Purple flowers	Green
Red cabbage	*Brassica oleracea-capitata*	Outer leaves	Blue, lavender
Turmeric	*Curcuma longa*	Rhizome	Yellow
Yellow onion	*Allium cepa*	Brown, outer leaves	Burnt orange

What is **wormwood**?

Artemisia absinthium, known as wormwood, is a hardy, fragrant perennial that grows to heights of 2 to 4 ft (.6 to 1.2 m). Wormwood is native to Europe but has been widely naturalized in North America. Absinthe, a liquor, is distilled and flavored using this plant. Absinthe was banned in the United States in the early 1900s because it is considered habit forming and hazardous to one's health.

What is the **origin** of the name "**Jimson weed**"?

Jimson weed (*Datura stramonium*) is a corruption of the name "Jamestown weed." The colonists of Jamestown, Virginia, were familiar with this weed. It is also known as thorn apple, mad apple, stinkwort, angel's trumpet, devil's trumpet, stinkweed, dewtry, and white man's weed. Even when consumed in moderate amounts, every part of this plant is poisonous and potentially deadly. Even so, some of the alkaloids found in this plant are used by doctors as a pre-anesthetic.

What is the **first wildflower to bloom** each year in the northern portions of the **United States**?

The first flower to bloom during spring in the northern section of the United States is rarely seen because it blooms in swamps. *Spathyema foetidus*, commonly called skunk cabbage, appears in northern swamps during February. The first wildflower to bloom each year in New England and the Midwest is typically *Hepatica*, also known as liverleaf, which blooms in March or early April.

What **property** makes **skunk cabbage** unique?

Skunk cabbage (*Spathyema foetidus*) blooms while there is still snow on the ground. The root of the plant acts like a metabolic furnace, providing heat to the flowering shoot. It melts the surrounding snow as it pushes up through the frozen ground. Plant botanists have been unable to identify the cause of this phenomenon. Different theories have been proposed, with some experts believing it is a special adaptation for the cold weather climate and others speculating that it is an evolutionary remnant of a feature of a tropical plant.

Which **part** of the **wheat** plant (*Triticum aestivum*) is used to **make flour**?

Wheat is a monocotyledon grass whose fruit, the grain or kernel, contains one seed. The endosperm and embryo of the wheat plant are surrounded by the pericarp, or fruit wall, and the remains of the seed coat. More than 80 percent of the volume of the wheat kernel is made up of the starchy endosperm. White flour is made by milling the starchy endosperm. Wheat bran constitutes approximately 14 percent of the wheat kernel and is found in the covering layers and the outermost layer of the kernel, called the aleurone layer. Wheat germ is the embryo of the wheat plant and represents approximately 3 percent of the wheat kernel. Athough there are nearly two dozen species of wheat, the most important ones for commercial use are common wheat (*Triticum aestivum*, sometimes referred to as *Triticum vulgare*) and durum wheat (*Triticum durum*). The varieties of common wheat account for 90 percent of the wheat grown worldwide. Durum wheat accounts for 5 to 7 percent of the wheat grown, and all other species account for the remainder of the wheat grown.

How much do **plants contribute** to the **human diet**?

In the United States and western Europe approximately 65 percent of a human being's total caloric intake and 35 percent of consumed protein are obtained from plants or plant products. Soybeans are an example of a plant with high protein content. In developing nations almost 90 percent of calories and more than 80 percent of protein in a person's diet are from plants.

What plant **native to Central and South America** can be used as both a **poison** and a **healing remedy**?

Chondrodendron tomentosum, a plant that produces curare, has properties that are both healing and poisonous. In both Central and South America the plant has been used by many different Indian tribes to develop a poisonous mixture. The poisonous stems and roots of the plant are crushed and cooked until taking on a syrupy consistency. Indian tribes often dipped the tips of arrows and other weapons into the poisonous paste before battle. However, the root of the vine also has healing properties. In Brazil, especially, it is used as a diuretic and fever reducer and is commonly used to treat tissue inflammation, kidney stones, bruises, contusions, and edema.

What are **luffa sponges**?

Luffas are nonwoody vines of the cucumber family. A fibrous skeleton lies inside of the fruit, and this structure is often used as a sponge. The term "loofah" is commonly used when this material is used as a sponge. Dishcloth gourd, rag gourd, and vegetable sponge are other popular names for this sponge.

What **plants** are commonly used in the **perfume industry**?

Perfumes are made of a mixture of a large variety of scents. Although many perfumes are created synthetically, the expensive designer scents still use natural essential oils extracted from plants. The perfume industry uses all parts of the plant to create a unique blend of scents. Some commonly used plant materials for essential-oil extraction are:

Plant Organ	Source
Bark	Indonesia and Ceylon cinnamons, and cassia
Flowers	Rose, carnation, orange blossoms, ylang-ylang, violet, and lavender

Plant Organ	Source
Gums	Balsam and myrrh
Leaves and stems	Rosemary, geranium, citronella, lemon grass, and a variety of mints
Rhizomes	Ginger
Roots	Sassafras
Seeds and fruits	Orange, lemon, and nutmeg
Wood	Cedar, sandalwood, and pine

Why were **tomatoes** often called "**love apples**" and considered **aphrodisiacs**"?

Tomatoes belong to the nightshade family; they were cultivated in Peru and introduced to Europe by Spanish explorers. Tomatoes were introduced to Italy from Morocco, so the Italian name for the fruit was *pomi de Mori* (meaning "apples of the Moors"). The French called the tomato *pommes d'amore* (meaning "apples of love"). This latter name may have referred to the fact that tomatoes were thought to have aphrodisiac powers, or it may have been a corruption of the Italian name. When tomato plants were first introduced to Europe, many people viewed them with suspicion, since poisonous members of the nightshade family were commonly known. Although the tomato is neither poisonous nor an aphrodisiac, it took centuries for it to fully overcome its undeserved reputation.

What parts of **plants** are **sources** for **spices**?

Spices are aromatic seasonings derived from many different parts of plants including the bark, buds, fruit, roots, seeds, and stems. Some common spices and their sources are:

Spice	Scientific Name of Plant	Part Used
Allspice	*Pimenta dioica*	Fruit
Black pepper	*Piper nigrum*	Fruit
Capsicum peppers	*Capsicum annum*; *Capsicum baccatum*; *Capsicum chinense*; *Capsicum frutescens*	Fruit
Cassia	*Cinnamomum cassia*	Bark
Cinnamon	*Cinnamomum zeylanicum*	Inner bark
Cloves	*Eugenia caryophyllata*	Flower
Ginger	*Zingiber officinale*	Rhizome
Mace	*Myristica fragrans*	Seed
Nutmeg	*Myristica fragrans*	Seed
Saffron	*Crocus sativus*	Stigma
Turmeric	*Curcuma longa*	Rhizome
Vanilla	*Vanilla planifolia*	Fruit

What is the most **expensive spice** in the world?

The world's most expensive spice is saffron. The spice was highly sought after by the ancient civilizations of Egypt, Assyria, Phoenicia, Persia, Crete, Greece, and Rome. The term "saffron" comes from the Arabic word *za'faran*, meaning "yellow." The spice is obtained from the delicate stigmas of an autumn crocus, *Crocus sativus*, a species native to eastern Mediterranean countries and Asia Minor. *Crocus sativus* is propagated by corms. The blooming period for the crocus is approximately two weeks, after which the flowers must be picked while they are in full bloom and before any signs of wilting. Once picked, the three-part stigmas are removed from the petals before the petals wilt; this is a time-consuming process that can only be done by hand as the stigmas are very fragile. Then the stigmas are roasted

A bee on a saffron flower. Saffron is the world's most expensive spice.

and sold either as whole threads (whole stigmas) or powder. In order to harvest 1 lb (0.45 kg) of the spice, between 75,000 and 100,000 flowers must be picked. Approximately 4,000 stigmas yield 1 oz (28 gr) of the spice. In 1998 the retail price of saffron was approximately $240.00 per ounce ($8.50 per gram), confirming its place as the world's most costly spice.

Although **plants** were **used as medicines** in the past, has their **use decreased** in modern times?

Approximately 25 percent of all prescription drugs currently used in the Western world contain ingredients derived from plants. The United States National Cancer

Institute has identified 3,000 plants from which anti-cancer drugs can be or are made. Among this group are ginseng (*Panax quinquefolius*), Asian mayapple (*Podophyllum hexandrum*), western yew (*Taxus brevifolia*), and rosy periwinkle. Of the 3,000 plants used for anti-cancer drugs, 70 percent come from rainforests, which are also a source of plants used to treat countless other diseases and infections. However, 80 percent of the world's population does not use prescription drugs; rather, this group relies almost exclusively on herbal medicines. Therefore, the medicinal use of plants has not decreased.

What are some **common culinary herbs**?

Herbs are often used to enhance flavors in food. They are usually from the leaves of nonwoody plants.

Common Name	Scientific Name	Part Used
Basil	*Ocimum basilicum*	Leaves
Bay leaves	*Laurus nobilis*	Leaves
Cumin	*Cuminum cyminum*	Fruit
Dill	*Anethum graveolens*	Fruit, leaves
Garlic	*Allium satiavum*	Bulbets
Mustard	*Brassica alba; Brassica nigra*	Seed
Onion	*Allium cepa*	Bulb
Oregano	*Origanum vulgare*	Leaves
Parsley	*Petroselinum crispum*	Leaves
Peppermint	*Mentha piperita*	Leaves
Sage	*Salvia officinalis*	Leaves
Tarragon	*Artemesia dracunculus*	Leaves
Thyme	*Thymus vulgaris*	Leaves

What **significance** did *De Materia Medica*, written by the Greek physician **Dioscorides**, have?

De Materia Medica (About medicinal materials) was written by the Greek physician Dioscorides (*ca.* 40–90 C.E.) in the first century C.E. The manuscript included the names and uses of the 600 plants that, at the time, were known to have medicinal properties. The purpose of the publication was to improve medical service in the Roman Empire. In addition to its medical use, it became the book most often used for plant classification in the Western world for nearly 1,500 years. During the fifteenth and sixteenth centuries European botanists and physicians used *De Materia Medica* to formulate their "herbals", illustrated books on the presumed medicinal uses of plants.

What are some specific **medications** that have been **obtained** from **rainforest plants**?

Drug	Medicinal Use	Source
Cocaine	Analgesic	Coca bush
Cortisone	Anti-inflammatory	Mexican yam
Diosgenin	Birth control	Mexican yam
Morphine	Analgesic	Opium poppy
Quinine	Treatment for malaria	Chincona tree bark
Reserpine	Treatment for hypertension	Rauwolfia plant
Vinblastine	Treatment for Hodgkin's disease and leukemia	Rosy periwinkle plant

What is **herbal medicine**?

Herbal medicine treats disease and promotes health using plant materials. For centuries herbal medicine was the primary method of administering medically active compounds.

Common Herbal Remedies

Herb	Botanical name	Common use
Aloe	*Aloe vera*	Skin, gastritis
Black cohosh	*Cimicifuga racemosa*	Menstrual problems, menopause
Dong quai	*Angelica sinensis*	Menstrual problems, menopause
Echinacea	*Echinacea angustifolia*	Colds, immune health
Ephedra (ma huang)	*Ephedra sinica*	Asthma, energy, weight loss
Evening primrose oil	*Oenothera biennis*	Eczema, psoriasis, premenstrual syndrome, breast pain
Feverfew	*Tanacetum parthenium*	Migraine headaches
Garlic	*Allium sativum*	Cholesterol, hypertension
Ginkgo biloba	*Ginkgo biloba*	Cerebrovascular insufficiency, memory problems
Ginseng	*Panax ginseng, Panax quinquifolius, Panax pseudoginseng, Eleutherococcus senticosus*	Energy, immunity, libido
Goldenseal	*Hydrastis candensis*	Immune health, colds
Hawthorne	*Crateaegus laeviagata*	Cardiac function
Kava kava	*Piper methysticum*	Anxiety

Which common houseplants are poisonous?

Philodendron (*Philodendron* and *Monstera*) and dumbcane (*Dieffenbachia*) are some of the most-common poisonous houseplants. All parts of both of these plants are poisonous.

Herb	Botanical name	Common use
Milk thistle	*Silybum marianum*	Liver disease
Peppermint	*Mentha piperita*	Dyspepsia, irritable bowel syndrome
Saw palmetto	*Serona repens*	Prostate problems
St. John's Wort	*Hypericum perforatum*	Depression, anxiety, insomnia
Tea tree oil	*Malaleuca alternifolia*	Skin infections
Valerian	*Valeriana officinalis*	Anxiety, insomnia

Which important **pharmacological compound** was once obtained from the **gymnosperm** *Ephedra*?

In the past, the drug ephedrine, used in the treatment of respiratory problems, was extracted from species of the genus *Ephedra* (common name: Ma huang) found in China. This process has now been largely replaced with the preparation of synthetic ephedrines. Ephedra was a dietary supplement used to aid weight loss, enhance sports performance, and increase energy. In late 2003 the U.S. Food and Drug Administration banned the use of ephedra as a dietary supplement because it can pose an unreasonable health risk.

What is the **most dangerous poison** produced by a **plant**?

In North America the poisonous hemlock, *Conium maculatum*, is probably the most dangerous plant. The South American lana tree is another dangerous plant. Native Americans used its sap to make curare, a poisonous substance applied to their arrows and spears. It can cause death in a matter of minutes.

What **plants** produce **essential oils** that are commonly used in **aromatherapy**?

Aromatherapy is a holistic approach to healing using essential oils extracted from plants. Holistic medicine looks at the health of the whole individual, and treatments emphasize the connection of mind, body, and spirit. The term "aromatherapy" was first used by Rene Gattefosse, a French perfume chemist. He discovered the healing powers of lavender oil following a laboratory accident during which he burned his hand. Gattefosse began to investigate the properties of lavender oil and other essential oils and published a book on plant extracts. During aromatherapy treatments, essential oils are absorbed through breath or the pores of the skin; this process triggers certain physiological responses. Examples of essential oils and their uses are:

Essential Oil	Common Uses
Cypress	Antiseptic, asthma, coughing, relaxation
Eucalyptus	Anti-inflammatory, arthritis, relaxation
Frankincense	Coughing, bronchitis
Geranium	Dermatitis, relaxation, depression
Ginger	Bronchitis, arthritis, stimulant
Juniper	Antiseptic, aches, pains, relaxation
Lavender	Antiseptic, respiratory infections, relaxation
Marjoram	Respiratory infections, relaxation
Pine	Asthma, arthritis, depression
Roman chamomile	Toothaches, arthritis, tension
Bulgarian rose	Antiseptic, insomnia, relaxation
Rosemary	Bronchitis, depression, mental alertness
Sandalwood	Acne, bronchitis, depression
Tea tree	Respiratory infections, acne, depression

Who developed **plant breeding** into a **modern science**?

Luther Burbank (1849–1926) developed plant breeding as a modern science. His breeding techniques included crosses of plant strains native to North America and foreign strains. He obtained seedlings that were then grafted onto fully developed plants for an appraisal of hybrid characteristics. His keen sense of observation allowed him to recognize desirable characteristics, enabling him to select only varieties that would be useful. One of his earliest hybridization successes was the Burbank potato, from which more than 800 new strains and varieties of plants—including 113 varieties of plums and prunes—were developed. More than twenty of these plums and prunes are still commercially important today.

When was the **first plant patent** issued?

Henry F. Bosenberg, a landscape gardener, received U.S. Plant Patent no. 1 on August 18, 1931, for a climbing or trailing rose.

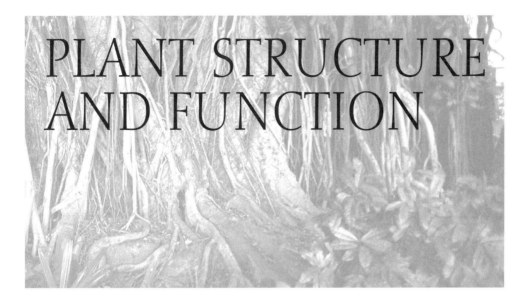

PLANT STRUCTURE AND FUNCTION

PLANT CELLS AND TISSUES

What are the **general characteristics** of a **plant**?

A plant is a multicellular, eukaryotic organism with cellulose-rich cell walls and chloroplasts that has starch as the primary carbohydrate food reserve. Plants are primarily terrestrial, autotrophic (capable of making their own food) organisms. Most plants contain chlorophylls *a* and *b* and xanthophylls (yellow pigments) as well as carotenes (orange pigments).

What are the **major parts** of **vascular plants**?

Vascular plants consist of roots, shoots, and leaves. The root system penetrates the soil and is below ground. The shoot system consists of the stem and the leaves.

What is the difference between the **root system** and the **shoot system** of **vascular plants**?

The root system is the part of the plant below ground level. It consists of the roots that absorb water and various ions necessary for plant nutrition. The root system anchors the plant in the ground. The shoot system is the part of the plant above ground level. It consists of the stem and leaves. The stem provides the framework for the positioning of the leaves. The leaves are the sites of photosynthesis.

Is there a **relationship** between the **size** of the **root system** and the size of the **shoot system**?

Growing plants maintain a balance between the size of the root system (the surface area available for the absorption of water and minerals) and the shoot system (the photosynthesizing surface). The total water- and mineral-absorbing surface area in young seedlings usually far exceeds the photosynthesizing surface area. As the plant ages, the root-to-shoot ratio decreases. Additionally, if the root system is damaged, reducing the water- and mineral-absorbing surface area, shoot growth is reduced by lack of water, minerals, and root-produced hormones. Similarly, reducing the size of the shoot system limits root growth by decreasing the availability of carbohydrates and shoot-produced hormones to the roots.

What are the **specialized cells** in plants?

All plant cells have several common features, such as chloroplasts, a cell wall, and a large vacuole. In addition, a number of specialized cells are found only in vascular plants. They include:

Parenchyma cells—Parenchyma (from the Greek *para*, meaning "beside," and *en + chein*, meaning "to pour in") cells are the most common cells found in leaves, stems, and roots. They are often spherical in shape with only primary cell walls. Parenchyma cells play a role in food storage, photosynthesis, and aerobic respiration. They are living cells at maturity. Most nutrients in plants such as corn and potatoes are contained in starch-laden parenchyma cells. These cells comprise the photosynthetic tissue of a leaf, the flesh of fruit, and the storage tissue of roots and seeds.

Collenchyma cells—Collenchyma (from the Greek term *kola*, meaning "glue") cells have thickened primary cell walls and lack secondary cell walls. They form strands or continuous cylinders just below the surfaces of stems or leaf stalks. The most common function of collenchyma cells is to provide support for parts of the plant that are still growing, such as the stem. Similar to parenchyma cells, collenchyma cells are living cells at maturity.

Sclerenchyma cells—Sclerenchyma (from the Greek term *skleros*, meaning "hard") cells have tough, rigid, thick secondary cell walls. These secondary cell walls are hardened with lignin, which is the main chemical component of wood. It makes the cell walls more rigid. Sclerenchyma cells provide rigid support for the plant. There are two types of sclerenchyma cells—fiber and sclereid. Fiber cells are long, slender cells that usually form strands or bundles. Sclereid cells, sometimes called stone cells, occur singly or in groups and have various forms. They have a thick, very hard secondary cell wall. Most sclerenchyma cells are dead cells at maturity.

Xylem—Xylem (from the Greek term *xylos*, meaning "wood") is the main water-conducting tissue of plants and consists of dead, hollow, tubular cells arranged end to end. The water transported in xylem replaces that lost via evaporation through stomata. The two types of water-conducting cells are tracheids and vessel elements. Water flows from the roots of a plant up through

> ### Which vegetable consists of a large number of collenchyma cells?
>
> Celery! The long "strings" in the leaf stalk or petiole of celery (the part we eat) consist mainly of collenchyma cells.

the shoot via pits in the secondary walls of the tracheids. Vessel elements have perforations in their end walls to allow the water to flow between cells.

Phloem—The two kinds of cells in the food-conducting tissue of plants, the phloem (from the Greek term *phloios*, meaning "bark"), are sieve cells and sieve-tube members. Sieve cells are found in seedless vascular plants and gymnosperms, while sieve-tube members are found in angiosperms. Both types of cells are elongated, slender, tube-like cells arranged end to end with clusters of pores at each cell junction. Sugars (especially sucrose), other compounds, and some mineral ions move between adjacent food-conducting cells. Sieve-tube members have thin primary cell walls but lack secondary cell walls. They are living cells at maturity.

Epidermis—Several types of specialized cells occur in the epidermis including guard cells, trichomes, and root hairs. Flattened epidermal cells, one layer thick and coated by a thick layer of cuticle, cover all parts of the primary plant body.

What types of **tissue systems** are found in plants?

Vascular plants have three tissue systems: the vascular tissue system, the ground tissue system, and the dermal or epidermis tissue system. The tissue systems are found in all parts of the plant—the roots, stem, and leaves.

What is the **function** of each of the **different tissue systems**?

The vascular tissue system consists of two kinds of conducting tissue: the xylem and the phloem. The xylem conducts water and dissolved minerals. The phloem conducts carbohydrates (mainly sucrose), hormones, amino acids, and other substances for the plant's growth and nutrition. The ground tissue system consists of three cell types—parenchyma, collenchyma, and sclerenchyma—that have relatively thin walls and living protoplasts for storage, photosynthesis, and secretion. The dermal tissue system is the outer protective covering of the plant including the cuticle.

What are the main **cell types** of each **tissue system**?

Tissue System	Tissue	Cell Type	Location	Functions
Dermal	Epidermis	Guard cells;	Outermost	Protection;

Tissue System	Tissue	Cell Type	Location	Functions
		trichomes	layer of cells of the primary plant body	minimizes water loss
Dermal	Periderm	Cork cells	Initial periderm beneath epidermis; subsequently formed periderm deeper in bark	Replaces epidermis as protective layer in plants with secondary growth
Ground	Parenchyma tissue	Parenchyma	Throughout the plant	Metabolic processes such as respiration, digestion, and photosynthesis; wound healing
Ground	Collenchyma tissue	Collenchyma	Beneath the epidermis in young elongating stems; in ribs along some leaves	Support in primary plant body
Ground	Sclerenchyma tissue	Fiber	Xylem and phloem; leaves of monocotyledons	Support and storage
Ground	Sclerenchyma tissue	Sclereid	Throughout the plant	Mechanical; protective
Vascular	Xylem	Tracheid	Throughout the plant	Main water-conducting element in gymnosperms and seedless vascular plants
Vascular	Xylem	Vessel element	Throughout the plant	Main water-conducting element in angiosperms
Vascular	Phloem	Sieve cell	Throughout the plant	Food-conducting element in gymnosperms
Vascular	Phloem	Albuminous cell	Throughout the plant	Believed to play a role in the delivery of substances to the sieve cell
Vascular	Phloem	Sieve-tube element	Throughout the plant	Food-conducting element in

Tissue System	Tissue	Cell Type	Location	Functions
Vascular	Phloem	Companion cell	Throughout the plant	angiosperms Believed to play a role in the delivery of substances to the sieve cell

How are **fibers classified**?

Fibers are classified in several ways. One way to classify fibers is based on their location—that is, whether or not they are in the xylem. Fibers in xylem are called xylary fibers, while those found in other tissues are called extraxylary fibers. Extraxylary fibers are usually longer than xylary ones. Fibers may also be classified according to their hardness. Hard fibers are from monocots and include xylem, thus making them lignified and stiff. An example of a hard fiber is rope made from sisal (*Agave sisalana*). Soft fibers, also called bast fibers, are harvested from dicots lacking lignin and are generally stronger and more durable than fibers from monocots. An example of a soft fiber is linen from flax (*Linum usitatissimum*).

What are **meristems**?

Meristems (from the Greek term *meristos*, meaning "divided") are unspecialized cells that divide and generate new cells and tissues. Apical meristems, found at the tips of all roots and stems, are responsible for a plant's primary growth. The vascular cambium and cork cambium are the meristems responsible for a plant's secondary growth.

How does the **apical meristem** of the **root grow**?

The apical meristem of the root divides and produces cells inwardly and outwardly. The cells that are produced inwardly grow backwards up the root, while the cells that are produced outwardly grow forward in the direction the root is growing. The outward cell division creates a root cap.

Do **plants** ever **stop growing**?

Unlike many organisms that stop growing when they reach maturity, plants continue to grow during their entire life span. Unlimited, prolonged plant growth is described as indeterminate. The apical meristem produces an unrestricted number of lateral organs indefinitely.

What are the **precursors** of the **vascular tissue, ground tissue, and epidermis**?

Soon after the plant embryo forms, the primary meristems develop. The protoderm, the precursor of the epidermis, forms in the outermost cells of the embryo. Vertical

201

divisions within the embryo result in the distinction between the procambium, the precursor of the vascular tissue, and the ground meristem, the precursor of the ground tissue.

What is the **difference** between **primary growth** and **secondary growth**?

Primary growth occurs in the tips of stems and roots in plants, thus increasing the length of the stems and roots. Secondary growth allows a plant to increase its diameter. The results of secondary growth form the division of a cylinder of cells around the plant's periphery.

SEEDS

What is a **seed**?

A seed is a mature, fertilized ovule. It consists of the seed embryo and the nutrient-rich tissue called the endosperm. The embryo consists of a miniature root and shoot. Once the seed is protected and enclosed in a seed coat, it ceases further development and becomes dormant.

What are the advantages of **seed dormancy**?

The time during which a seed is dormant (when growth and development do not occur) allows for the dispersal of seeds. The plant can send its seeds into new environments. Dormancy assures survival of the plant since germination does not occur until conditions are favorable for plant growth.

What conditions are necessary for **seed germination**?

Seeds remain dormant until the optimum conditions of temperature, oxygen, and moisture are available for germination and further development. In addition to these external factors, some seeds undergo a series of enzymatic and biochemical changes prior to germination.

What are the **best temperatures** for **seed germination**?

The best temperatures for seed germination in most plants are 77°F to 86°F (25°C to 30°C). Some varieties of seeds are able to germinate at temperatures ranging from 41°F to 86°F (5°C to 30°C). However, some seeds, such as those of the Rocky Mountain lodgepole pine tree (*Pinus contorta*) require extreme heat for germination. The cones of the Rocky Mountain lodgepole pine are covered with a resin that binds the scales of the cone. The high temperatures (113–120°F, or 45–50°C) associated with moderate to severe forest fires are required to melt the resin and release the seeds.

When a plant is said to be "**double dormant**," what does that mean?

Plants that are double dormant require a unique sort of layering or stratification in order for their seeds to germinate. The seeds of these plants must have a period of warmth and moisture followed by a cold spell. Both the seed coat and the seed embryo require this double dormancy if they are to germinate. In nature, this process usually takes two years. Some well-known plants that live the life of double dormancy include lilies, dogwoods, junipers, lilacs, tree peonies, and viburnums.

SHOOTS AND LEAVES

How does the **shoot develop** following germination?

Shoot development is classified based on whether the cotyledons (seed leaves) are carried above ground or remain below ground. Seed germination during which the cotyledons are carried above ground is called epigeous. The food stored in the cotyledons is digested and the products are transported to the growing parts of the young seedling. When the seedling becomes established and is no longer dependent upon the stored food in the seed for nutrition, the cotyledons gradually decrease in size, wither, and fall off. Seed germination during which the cotyledons remain underground is called hypogeous. The seedling uses the stored food from the cotyledons for growth, and then the cotyledons decompose. The cotyledons remain in the soil during the entire process.

What **protects** the **shoot tip** as it pushes through the soil?

Many seedlings form a bend or hook in their hypocotyls to protect the delicate shoot tip. The shoot tip is pulled rather than pushed through the soil.

What are the **parts of a stem**?

A stem has nodes and internodes. The nodes are the points where the leaves are attached to the stem. The internodes are the parts of the stem between the nodes.

What are the **functions** of **stems**?

The four main functions of stems are: 1) to support leaves; 2) produce carbohydrates; 3) store materials such as water and starch; and 4) transport water and solutes between roots and leaves. Stems provide the link between the water and dissolved nutrients of the soil and the leaves.

What types of **variations** are there in **plant stems**?

Plant stems show variation in size and shape. In addition, some plants have modified stems. For example, the strawberry plant has stolons or runners, which are horizontal stems that grow along the surface of the ground. Iris plants also have horizontal stems called rhizomes. Rhizomes are large, brownish, rootlike structures found just below the surface of the ground. The rhizomes store food and can spread to form new plants. White potato plants also have rhizomes. The rhizomes of the white potato plant end in large, round structures called tubers. The tubers are the part of the plant that we call potatoes and eat. Tendrils and twining shoots of the morning glory and sweet potato coil around objects and help support the plant.

What are **buds** on stems?

Buds may be terminal or axillary on a stem. The terminal bud is at the apex of the stem where the plant growth is concentrated. It contains developing leaves and a compact series of nodes and internodes. The axillary buds are found in the angles formed by a leaf and the stem. The axillary buds are usually dormant.

What is **apical dominance**?

Apical dominance is a phenomenon in which the terminal bud produces hormones inhibiting the growth of axillary buds. This allows the plant to grow taller, increasing its exposure to light. Under certain conditions, the axillary buds begin to grow, producing branches. When the terminal bud is pruned ("pinched back") on houseplants and fruit trees, axillary bud growth is stimulated, producing bushy, full-looking plants.

How does one distinguish between **thorns, spines and prickles**?

Thorns are modified branches or stems arising from the axils of leaves. One purpose of thorns is to protect the plant from grazing animals. Hawthorn trees are an example of a plant with true thorns. Spines are modified leaves such as those found on cactus plants. Prickles are sharp outgrowths from the epidermis of various plant structures including leaves and stems.

What are the **major functions** of **leaves**?

Leaves are the main photosynthetic organ. However, they are also important in gas exchange and water movement throughout the whole plant.

What are the **parts of a leaf**?

Leaves, outgrowths of the shoot apex, are found in a variety of shapes, sizes, and arrangements. Most leaves have a blade, a petiole, stipules, and veins. The blade is the flattened portion of the leaf. The petiole is the slender stalk of the leaf. The stipules, found on some but not all leaves, are located at the base of the petiole where it joins

How many leaves are on a mature tree?

Leaves are one of the most conspicuous parts of a tree. A maple tree (genus *Acer*) with a trunk one yard (one meter) wide has approximately 100,000 leaves. Oak (genus *Quercus*) trees have approximately 700,000 leaves. Mature American elm (*Ulmus americana*) trees can produce more than five million leaves per season.

the stem. Stipules may be leaflike and show considerable variation in size. Veins, xylem, and phloem run through the leaf.

What are some examples of **modified leaves**?

Some leaves are modified to perform functions other than photosynthesis. While the tendrils of some plants are modifications to the stems and provide support for the plant, in other species, such as pea plants (*Pisum sativum*), the tendrils are modified leaves. In carnivorous plants, such as the Venus flytrap (*Dionaea muscipula*) and the pitcher plant (*Sarracenia purpurea*), the leaves attract, capture, and digest the insects with enzymes. The leaves of many desert plants grow mainly underground with only a small transparent "window" tip protruding above the soil surface. The soil covering the leaf protects it from dehydration by the harsh desert winds. The "window" allows light to penetrate and reach the site of photosynthesis.

What **important organelle** is found in the **mesophyll layer** of the **leaf**?

The mesophyll (from the Greek terms *mesos*, meaning "middle," and *phyllum*, meaning "leaf") of a leaf consists of masses of parenchyma cells that are packed with chloroplasts important for photosynthesis. The palisade parenchyma comprises columnar layers of parenchyma cells found beneath the epidermis of many leaves. The spongy parenchyma is a mass of cells that are irregular in shape and often highly branched. There are large intercellular spaces in the spongy parenchyma that function in gas exchange and the passage of water vapor from the leaves. These spaces are connected to the stomata.

What is the **difference** between **simple leaves** and **compound leaves**?

The blades of simple leaves are undivided, although they may have deep lobes. By contrast, the blades of compound leaves consist of clearly separated leaflets. Each leaflet usually has its own petiole, which is called a petiolule. There are two types of compound leaves: pinnately compound leaves and palmately compound leaves. The leaflets in pinnately compound leaves arise from either side of an axis called the

205

rachis, which is an extension of the petiole. The leaflets in palmately compound leaves originate at the tip of the petiole and lack the rachis.

What characteristics **distinguish a leaf** from a **leaflet**?

The two criteria used to distinguish a leaf from a leaflet are: 1) buds are found in the axils of leaves but not in the axils of leaflets; and 2) leaves extend from the stem in various planes while the leaflets of a given leaf all lie in the same plane.

What are the **three common types** of **leaf arrangement** on a stem?

The three most common types of leaf arrangement are: 1) alternate (or spiral); 2) opposite; and 3) whorled. In many plants the leaves are arranged in a pattern alternating on either side of the stem. In other plants the leaves are opposite each other in pairs on the stem. In still other plants the leaves are whorled with three or more attached to one level of the stem.

In what ways are **leaves economically important**?

Leaves are used for food and beverages, dyes and fibers, and medicinal and other industrial uses. Certain plants, such as cabbage (*Brassica oleracea*), lettuce (*Lactuca sativa*), spinach (*Spinacia oleracea*), and most herbs—including parsley (*Petroselinum crispum*) and thyme (*Thymus vulgaris*)—are grown for their leaves. Bearberry leaves (*Arctostaphylos uva-ursi*) contain a natural yellow dye, while henna leaves (*Lawsonia inermis*) contain a natural red dye. The leaves of palm trees are used to make clothing, brooms, and thatched huts in tropical climates. Aloe leaves (*Aloe vera*) are well known for treating burns and are also used in manufacturing medicated soaps and creams.

How does the **cuticle protect plants**?

The cuticle contains a waxy substance, called cutin, that covers the parts of the plant exposed to the air—the stem and leaves. It is relatively impermeable and provides a barrier to water loss, thus protecting the plant from desiccation.

What is the **purpose** of **stomata**?

Stomata (singular "stoma" from the Greek term *stoma*, meaning "mouth") are specialized pores in the leaves and sometimes in the green portions of the stems as well as flowers and fruits. Carbon dioxide (CO_2) enters the plant through the stomata, while water vapor escapes through the same pores. The guard cells that border the stomata expand and contract to control the passage of water, carbon dioxide (CO_2), and oxygen (O_2).

What is the **effect** of **ozone** on **leaves**?

Ozone can enter the leaves of a plant through the stomata along with carbon dioxide (CO_2). Ozone causes the cells surrounding the stomata to decrease in turgidity, thus reducing the size of the opening. This protects the plant from further exposure to ozone. However, once inside the leaf, ozone is highly reactive and can destroy the leaf cells.

How does the **number of stomata** on the **epidermis of leaves** compare among **various species**?

There is great variation among species in the number of stomata on the upper and lower epidermis of leaves. Horizontally oriented leaves usually have more stomata on the protected lower side of the epidermis than on the exposed upper side. Conversely, in vertically oriented leaves there are usually similar numbers of stomata on the lower and upper sides of the epidermis. The following chart shows average number of stomata per square centimeter for several different species.

Common Name	Leaf Orientation	Upper Epidermis	Lower Epidermis
Apple, *Pyrus malus*	Horizontal	0	38,760
Black oak, *Quercus velutina*	Horizontal	0	58,140
Scarlet oak, *Quercus coccinea*	Horizontal	0	103,800
Mulberry, *Morus alba*	Horizontal	0	48,000
Pumpkin, *Cucurbita pepo*	Horizontal	2,791	27,132
Geranium, *Pelagonium domesticum*	Horizontal	1,900	5,900
Pea, *Pisum sativum*	Horizontal	10,100	21,600
Corn, *Zea mays*	Vertical	9,800	10,800
Onion, *Allium cepa*	Vertical	17,500	17,500
Pine, *Pinus sylvestris*	Vertical	12,000	12,000

Are the **guard cells** the same in all plants?

The guard cells in all plants have the same function: they regulate the exchange of gases and water by opening and closing the stomatal pore. However, structurally they are different. Guard cells in dicots are kidney-shaped, while those in monocots are shaped like dumbbells.

What are **trichomes**?

Trichomes are hairlike outgrowths of the plant epidermis. They are often found on stems, leaves ("wooly" or "fuzzy" leaves are covered by trichomes), and reproductive organs.

What are the **functions** of **trichomes**?

Trichomes have many functions. They help increase the rate of water absorption and reduce the rate of water loss due to evaporation in order to keep the leaf surface cool. Trichomes also provide a defense against insects, since the "hairiness" of the leaf impedes insect infestation. Glandular or secretory hairs provide a chemical defense against potential herbivores.

What is the **periderm**?

Periderm replaces the epidermis as the protective covering in stems and roots of plants with secondary growth. The periderm consists of three structures: 1) the cork or phellem; 2) the cork cambium, or phellogen; and 3) the phelloderm. The cork or phellem is nonliving and is the protective tissue formed to the outside by the cork cambium. The cork cambium or phellogen is the meristem that produces the periderm. The phelloderm is a living parenchyma tissue formed to the inside of the meristem.

How is **commercial cork cultivated**?

Commercial cork is the outer bark of the cork oak (*Quercus suber*) grown in the western Mediterranean. The first periderm is commercially useless. It is removed from the tree and discarded when the tree is approximately ten years old. When the tree is 20 to 25 years old and has a diameter of approximately 15.75 in (40 cm), a usable cork layer 1.2 to 3.9 in (3 to 10 cm) thick can be harvested. A similar layer can be harvested approximately once every ten years until the tree is approximately 150 years old. The cork of a cork oak breaks away at the cork cambium and can be peeled off without harming the tree.

What are the **properties** of **cork** that make it ideal for a variety of **commercial uses**?

Cork consists of densely packed cells (about one million cells per cubic centimeter) that contain the plant wax suberin, making cork impermeable to liquids and gases.

Dead bark on a tree. Barks may be classified as to whether they are firmly attached to the tree trunk or whether they peel away from the surface of the trunk easily.

Half of its volume is trapped air. Therefore, it is four times lighter than water. It is virtually indestructible, fire-resistant, and durable; resists friction; and absorbs vibration and sound. Among cork's uses are as stoppers for wine bottles, insulation for the space shuttle, and grips on symphony conductors' batons.

What is **replacing natural cork** for wine stoppers?

Plastic "corks" are replacing natural cork for wine stoppers. During the 1980s and early 1990s bad cork was traced to the fungal contaminant 2,4,6-trichloroanisole, TCA. TCA flattens the taste of the wine, removing the flavors the winemaker worked hard to produce. Additionally, the demand for wine in bottles has grown faster than the supply of cork from cork oaks. Although natural cork is still used for the best and most expensive wines—those that are aged for twenty or more years—plastic cork is becoming widely used at the lower end of the wine industry. Most plastic corks use a high-grade plastic that eliminates taste and odor problems of contaminated natural cork.

What is **wood**?

Wood is the accumulated secondary xylem of a plant. Generally, the wood used commercially is from plant stems rather than plant roots. Wood located near the center of a tree trunk is called heartwood. Its cells are infiltrated with gums and resins from the aging secondary xylem. Heartwood is often darker in color than wood nearer to the vascular cambium. Wood closer to the vascular cambium is called sapwood. It is actively involved in transporting water within the plant.

What is the function of **bark**?

Bark protects the interior tissues of a tree from external physical threats such as rain, hail, and snow, and biological threats such as animals, fungi, and bacteria. Bark also serves as a dumping ground in which the tree rids itself of waste products from its metabolism. Nutrients are transported through the phloem of the bark from the roots to the crown of the tree.

Can **bark** be used to **identify trees**?

Tree bark varies in texture and color between species. It is used as an aid in identifying a particular species, especially when there are no leaves on the tree. Barks may be classified as to whether they are firmly attached to the tree trunk or whether they peel away from the surface of the trunk easily. Barks that are firmly attached to the tree trunk may have vertical fissures or furrows that may be deep or shallow. In other varieties of tree, the bark may look as if it is divided into squares or rectangles or other irregular shapes resembling scales. An example of scale bark is found in various species of the pine (genus *Pinus*). Shag bark has long overlapping thin sheets, such as is found in eucalyptus (genus *Eucalyptus*). Paper birch (*Betula papyrifera*) is an example of a species with bark that peels away from the tree easily, also known as ring bark.

How **thick** is the **bark** of **common trees**?

Common Name	Scientific Name (Genus)	Thickness of Bark (mm)
Juniper	*Juniperus*	2–6
Spruce	*Picea*	5–30
Pine	*Pinus*	5–50
Maple	*Acer*	5–20
Poplar	*Populus*	5–80
Oak	*Quercus*	5–40
Beech	*Fagus*	2–10

Why is the white birch known as "**paper birch**"?

The outer layer of the birch tree grows in sheetlike layers. Hence it has been known as "paper birch." It is the same tree that Native Americans used to make paper birch canoes.

How long do the root cap cells live?

From their origin until they are sloughed, root cap cells live four to nine days depending on the length of the root cap and the species.

What is **pith**?

Pith is the ground tissue, consisting usually of parenchyma, in the center of the root or stem within the vascular cylinder.

What is the **structure** of **stems** of **seed plants**?

There are three basic types of organization in the primary structure of stems: 1) the vascular system may appear as a continuous hollow cylinder around the pith; 2) discrete vascular bundles may form a single ring around the pith; or 3) the vascular bundles may appear scattered throughout the ground tissue.

ROOTS

What is the **first structure** to emerge **following germination**?

The first structure to emerge following germination is the radicle or embryonic root. The radicle allows the developing seedling to become anchored in the soil and to absorb water. This first root, the primary root, develops branch roots called lateral roots. The lateral roots send out additional lateral roots, eventually creating the multibranched root system.

What is the **root cap**?

The root cap is a thimblelike mass of parenchyma cells that covers and protects the growing root tip as it penetrates the soil. The root cap is pushed forward as the root tip grows longer. The cells on the periphery of the root cap are sloughed as the root cap is pushed forward and new cells are added by the apical meristem. The root cap protects the apical meristem, aids the root as it penetrates the soil, and plays an important role in controlling the response of the root to gravity (gravitropism).

What are the **functions** of the **root system**?

The major functions of roots are: 1) anchorage in soil; 2) storage of energy resources such as the carrot and sugar beet; 3) absorption of water and minerals from the soil; and 4) conduction of water and minerals to and from the shoot. The roots store the

An uprooted pine tree. The food (energy resources) of each plant are stored in its roots.

food (energy resources) of the plant. The food is either used by the roots themselves or digested, and the products of digestion are transported back up through the phloem to the above-ground portions of the plant. The roots of some plants are harvested as food for human consumption. Plant hormones are synthesized in the meristematic regions of the roots and transported upward in the xylem to the aerial part of the plant to stimulate growth and development.

How **deep** does the **root system** penetrate the soil?

The depth to which the root system penetrates the soil is dependent on moisture, temperature, the composition of the soil, and specific plant. Most of the roots actively absorbing water and minerals, the "feeder roots," are found in the upper 3 ft (1 m) of the soil. The feeder roots of many trees are mainly in the upper 6 in (15 cm) of the soil—the part of the soil richest in organic matter.

Which plant has the **deepest root system**?

Roots of the desert shrub mesquite (*Prosopis juliflora*) have been found growing nearly 175 ft (53.5 m) deep near Tucson, Arizona.

How wide is the **lateral spread** of the **root system** of **trees**?

The lateral spread of a tree's root system is four to seven times greater than the spread of the crown of the tree.

The aerial roots of a banyan tree are called prop roots because they help support the structure of the tree.

What are some **causes** of **root damage**?

Roots may be damaged by temperature extremes, drought, nematodes, and other soil microfauna such as springtails that nibble the succulent roots. When there is known damage to the roots, cutting back the shoot system helps to reestablish the balance between the root system and the shoot system. Transplanting seedlings and other plants also damages roots, especially the root hairs.

What are **adventitious roots**?

Adventitious roots (from the Latin term *adventicius,* meaning "not belonging to") are roots that form on organs other than roots, such as a leaf or stem. In some plants, adventitious roots are a means of vegetative propagation such as for raspberries, apples, and cabbage.

What are **aerial roots**?

Aerial roots are adventitious roots formed on above-ground structures such as stems. Aerial roots serve different functions in different species. In some species, such as the banyan tree (*Ficus benghalensis*) and red mangrove (*Rhizophora mangle*), the aerial roots are called prop roots since they support the plant. The aerial roots of ivy (*Hedera helix*), ball moss (*Tillandsia recurvata*), and Spanish moss (*Tillandsia usneoides*) cling to the surface of an object providing support for the stem. The aerial roots of the vanilla orchid (*Vanilla planifolia*) and *Philodendron* are photosynthetic.

213

Garlic seed bulbs. Bulbs consist of fleshy scales containing a small basal plate (a modified stem from which the roots emerge) and a shoot.

How does the **root system** of a **monocot** differ from the root system of a **dicot**?

The root system of a monocot is a fibrous mass of roots providing broad exposure to soil water and minerals. The root system of a dicot consists of one large taproot with many small secondary lateral roots growing out of it.

What are **root hairs**?

Root hairs are tiny projections and outgrowths on the outermost layer of the root epidermis. They occur near the tips of roots, where they are abundant. Root hairs are short-lived, and new ones are produced at approximately the same rate as older ones die. Some plants have as many as 40,000 root hairs per square centimeter.

What is the **function** of **root hairs**?

Root hairs increase the surface area of the root system, allowing the roots to absorb water and minerals more efficiently. In a study on one rye plant, it was estimated that the plant had approximately 14 billion root hairs, with an absorbing surface area of 480 sq yds (401 sq m). If these root hairs were placed end to end, they would extend well over 6,214 mi (10,000 km)!

What is the **economic importance** of **roots**?

Carrots, beets, turnips, radishes, horseradish, sugar beets, and sweet potatoes are all taproots that have been used as food for human consumption for centuries. The spices licorice, sassafras, and sarsaparilla (the flavoring used to make root beer) are derived from roots. The drugs aconite, gentian, ipecac, ginseng, reserpine (a tranquilizer), and protoveratrine (a heart relaxant) are all extracted from the roots of plants.

How does a **bulb** differ from a **corm**, a **tuber**, and a **rhizome**?

Each of these structures is a modified stem that grows below ground. Many times the term *bulb* is applied to any underground storage organ in which a plant stores energy for its dormant period. Dormancy is a device a plant utilizes to get through difficult weather conditions (winter cold or summer drought). A true bulb consists of fleshy scales containing a small basal plate (a modified stem from which the roots emerge) and a shoot. The scales that surround the embryo are modified leaves that contain the nutrients for the bulb during dormancy and early growth. Some bulbs have a tunic (a paper-thin covering) around the scales. The basal plate can also hold the scales together. New bulbs form from the lateral buds on the basal plate. Tulips, daffodils, lilies, and hyacinths are examples of bulb flowers.

A corm is actually a stem that has been modified into a mass of storage tissue. The eye or eyes at the top of the corm are growing points. The corm is covered by dry leaf bases similar to the tunic covering of the bulb. Roots grow from the basal plate on the underside of the corm. New corms form on top of or beside the old one. Corm-type flowers include gladiolus, freesia, and crocus.

A tuber is a solid underground mass of stem like a corm, but it lacks both a basal plate and a tunic. Roots and shoots grow from eyes (growth buds) out of its sides, bottom, and sometimes its top. Some tubers are roundish. Others are flattened and lumpy. Some examples of tubers are gloxinia, caladium, ranunculuses, and anemone.

A tuberous root is a swollen root that has taken in moisture and nutrients. It resembles a tuber. New growth occurs on the base of the old stem, where it joins the root. A tuberous root can be divided by cutting off a section with an eye-bearing portion from where the old stem was attached. Dahlias have tuberous roots.

A rhizome or a rootstock is a thickened, branching storage stem that usually grows laterally along or slightly below the soil surface. Roots develop downward on the bottom surface, while buds and leaves sprout upwards from the top of the rhizome. A rhizome is propagated by cutting the parent plant into sections. Japanese, Siberian, and bearded irises; cannas; calla lilies; and trilliums are rhizomes.

FLOWERS

What are the **parts of a flower**?

In a generalized flower, there are four main parts:

215

Sepals—found on the outside of the bud or on the underside of the open flower. They serve to protect the flower bud from drying out. Some sepals ward off predators by their spines or chemicals. Collectively, the sepals form the calyx.

Petals—serve to attract pollinators and are usually dropped shortly after pollination occurs. Collectively, the petals form the corolla.

Stamen—the male part of a flower. Consists of a filament and anther where pollen is produced.

Pistil—the female part of a flower. Consists of the stigma, style, and ovary containing ovules. After fertilization, the ovules mature into seeds.

If a flower has all of these parts, it is called complete; if it lacks any of them, it is called incomplete. In terms of sexual reproduction in flowers, only stamens and pistils are necessary. Flowers with both structures are called perfect, but if they lack either one of the other they are called imperfect.

What are **effective** and **efficient types** of **pollination**?

Effective pollination occurs when viable pollen is transferred to a plant's stigmas, ovule-bearing organs, or ovules (seed precursors). Without pollination, there would be no fertilization. Since plants are immobile organisms, they usually need external agents to transport their pollen from where it is produced in the plant to where fertilization can occur. This situation produces cross-pollination, wherein one plant's pollen is moved by an agent to another plant's stigma. Some plants are able to self-pollinate—transfer their own pollen to their own stigmas. But of the two methods, cross-pollination seems more advantageous, for it allows new genetic material to be introduced.

Cross-pollination agents include insects, wind, birds, mammals, and water. Many times flowers offer one or more "rewards" to attract these agents—sugary nectar, oil, solid food bodies, perfume, a place to sleep, or sometimes the pollen itself. Other times the plant can "trap" the agent into transporting the pollen. Generally, plants use color and fragrances as attractants to lure these agents. For example, a few orchids use a combination of smell and color to mimic the female of certain species of bees and wasps so successfully that the corresponding males will attempt to mate with them. Through this process (pseudocopulation) the orchids achieve pollination. While some plants cater to a variety of agents, other plants are very selective and are pollinated by a single species of insect only. This extreme pollinator specificity tends to maintain the purity of a plant species.

Plant structure can accommodate the type of agent used. For example, plants such as grasses and conifers, whose pollen is carried by the wind, tend to have a simple structure lacking petals, with freely exposed and branched stigmas to catch airborne pollen and dangling anthers (pollen-producing parts) on long filaments. This type of anther allows the light round pollen to be easily caught by the wind. These plants are found in areas such as prairies and mountains, where insect agents are rare. In contrast, semi-enclosed, nonsymmetrical, long-lived flowers such as irises, roses, and snapdragons have a "landing platform" and nectar in the flower base to accommodate insect agents such as the bee. The sticky, abundant pollen can easily become attached to the insect to be borne away to another flower.

What are **nectaries**?

Plants secrete a variety of substances from specialized structures called secretory structures. Nectaries are structures that secrete nectar, a sugary compound that attracts insects, birds, or other animals. Most nectaries are associated with flowers and are called floral nectaries. Nectar is 10 to 50

Pollen-carrying insects such as bees are a crucial part of plant fertilization.

percent sugar, especially sucrose, glucose, and fructose. Plants usually produce small amounts of nectar, which forces foraging animals to visit several flowers before obtaining a full meal. A single insect or bird can, therefore, pollinate tens or hundreds of plants.

How does **water move up** a tree?

Water is carried up a tree through the xylem tissue in a process called transpiration. Constant evaporation from the leaf creates a flow of water from root to shoot. The roots of a tree absorb the vast majority of water that a tree needs. The properties of cohesion and adhesion allow the water to move up a tree regardless of its height. Cohesion allows the individual water molecules to stick together in one continuous stream. Adhesion permits the water molecules to adhere to the cellulose molecules in the walls of xylem cells. When the water reaches a leaf, water is evaporated, thus allowing additional water molecules to be drawn up through the tree.

What **compounds** are important in the **coloration** of **flowers**?

Color is one of the most conspicuous features of angiosperm flowers. All flower colors are produced by a small number of pigments. Many red, orange, or yellow flowers owe their color to the presence of carotenoid pigments similar to those that occur in leaves. The most important pigments in floral coloration are flavonoids. An important group of flavonoids, the anthocyanins, are major determinants of flower color. The three major anthocyanin pigments are: 1) pelargonidin (red); 2) cyanidin (violet); and 3) delphinidin (blue). Related compounds, known as flavanols, are yellow, ivory, or white. Mixtures of these different pigments, together with changes in cellular pH, produce the entire range of flower color in the angiosperms.

SOILS

What are the **different types** of **soil**?

Soil is the weathered outer layer of the earth's crust. It is a mixture of tiny rock fragments and organic matter. There are three broad categories of soils: clay, sandy, and loam. Clay soils are heavy with the particles sticking close together. Most plants have a hard time absorbing the nutrients in clay soil, and the soil tends to become waterlogged. Clay soils can be good for a few deep-rooted plants, such as mint, peas, and broad beans.

Sandy soils are light and have particles that do not stick together. Sandy soil is good for many alpine and arid plants, some herbs such as tarragon and thyme, and vegetables such as onions, carrots, and tomatoes. Loam soils are a well-balanced mix of smaller and larger particles. They provide nutrients to plant roots easily, they drain well, and they also retain water very well. Loams are considered ideal for plant growth.

What are the **essential nutrient elements** required for plant growth?

Essential nutrients are chemical elements that are necessary for plant growth. An element is essential for plant growth when: 1) it is required for a plant to complete its life cycle (to produce viable seeds); 2) it is part of a molecule or component of the plant that is itself essential to the plant, such as the magnesium in the chlorophyll molecule; and 3) the plant displays symptoms of deficiency in the absence of the element. Essential nutrients are also referred to as essential minerals and essential inorganic nutrients.

What are the **macronutrients** and **micronutrients** of plants?

The macronutrients of plants are carbon, hydrogen, oxygen, nitrogen, potassium, calcium, phosphorus, magnesium, and sulfur. These all are nearly or in some cases far greater than 1 percent of the dry weight of a plant. The micronutrients are iron, chlorine, copper, manganese, zinc, molybdenum, and boron. Each of the micronutrients constitutes less than one to several hundred parts per million in plants. Sodium, silicon, cobalt, and selenium are beneficial elements. Research has not shown that these elements are essential for plant growth and development.

What is the **function** of **plant nutrients**?

Element	Approximate Percent of Dry Weight	Important Functions
Macronutrients		
Carbon	44	Major component of organic molecules
Oxygen	44	Major component of organic molecules
Hydrogen	6	Major component of organic molecules
Nitrogen	1–4	Component of amino acids, proteins, nucleotides, nucleic acids, chlorophyll, coenzymes
Potassium	0.5–6	Component of enzymes, protein synthesis, operation of stomata
Calcium	0.2–3.5	Component of cell walls, maintenance of membrane structure and permeability, activates some enzymes
Magnesium	0.1–0.8	Component of chlorophyll molecule, activates many enzymes
Phosphorus	0.1–0.8	Component of ADP and ATP, nucleic acids, phospholipids, several coenzymes
Sulfur	0.05–1	Components of some amino acids and proteins, coenzyme A
Micronutrients (concentrations in parts per million)		
Chlorine	100–10,000	Osmosis and ionic balance
Iron	25–300	Chlorophyll synthesis, cytochromes, nitrogenase

Element	Approximate Percent of Dry Weight	Important Functions
Manganese	15–800	Activator of certain enzymes
Zinc	15–100	Activator of many enzymes, active in formation of chlorophyll
Boron	5–75	Possibly involved in carbohydrate transport and nucleic acid synthesis
Copper	4–30	Activator or component of certain enzymes
Molybdenum	0.1–5	Nitrogen fixation, nitrate reduction

What do the numbers on a bag of **fertilizer** indicate?

The three numbers, such as 15-20-15, refer to the percentages by weight of macronutrients found in the fertilizer. The first number stands for nitrogen, the second for phosphorus, and the third for potassium. In order to determine the actual amount of each element in the fertilizer, multiply the percentage by the fertilizer's total weight in pounds. For example, in a 50-pound bag of 15-20-15, there are 7.5 pounds of nitrogen, 10 pounds of phosphorus, and 7.5 pounds of potassium. The remaining pounds are filler.

What does "**pH**" mean when applied to **soil**?

Literally, pH stands for "potential of hydrogen" and is the term used by soil scientists to represent the hydrogen ion concentration in a soil sample. The relative alkalinity-acidity is commonly expressed in terms of the symbol pH. The neutral point in the scale is seven. Soil testing below seven is said to be acidic; soil testing above pH seven is alkaline. The pH values are based on logarithms with a base of ten. Thus, a soil testing pH five is ten times as acidic as soil testing pH six; while a soil testing pH four is one hundred times as acidic as soil testing pH six.

What is the **best soil pH** for growing plants?

Nutrients such as phosphorous, calcium, potassium, and magnesium are most available to plants when the soil pH is between 6.0 and 7.5. Under highly acidic (low pH) conditions, these nutrients become insoluble and relatively unavailable for uptake by plants. However, some plants such as rhododendrons grow better in acidic soils. High soil pH can also decrease the availability of nutrients. If the soil is more alkaline than pH 8, phosphorous, iron, and many trace elements become insoluble and unavailable for plant uptake.

What does the term **hydroponics** mean?

This term refers to growing plants in some medium other than soil; the inorganic plant nutrients (such as potassium, sulfur, magnesium, and nitrogen) are continuously supplied to the plants in solution. Hydroponics is mostly used in areas where there is little soil or unsuitable soil. Since it allows precise control of nutrient levels and oxygenation of the roots, it is often used to grow plants used for research purposes. Julius von Sachs (1832–1897), a researcher in plant nutrition, pioneered modern hydroponics. Research plants have been grown in solution culture since the mid-1800s. William Gericke, a scientist at the University of California, defined the word hydroponics in 1937. In the fifty years that hydroponics has been used on a commercial basis, it has been adapted to many situations. NASA will be using hydroponics in the space station for crop production and to recycle carbon dioxide into oxygen. Although successful for research, hydroponics has many limitations and may prove frustrating for the amateur gardener.

RESPONSE TO STIMULI

What is **tropism**?

Tropism is the movement of a plant in response to a stimulus. The categories include:

Chemotropism—a response to chemicals by plants in which incurling of leaves may occur.

Gravitropism—Formerly called geotropism, a response to gravity in which the plant moves in relation to gravity. Shoots of a plant are negatively geotropic (growing upward), while roots are positively geotropic (growing downward).

Hydrotropism—a response to water or moisture in which roots grow toward the water source.

Paraheliotropism—a response by the plant leaves to avoid exposure to the Sun.

Phototropism—a response to light in which the plant may be positively phototropic (moving toward the light source) or negatively phototropic (moving away from the light source). Main axes of shoots are usually positively phototropic, whereas roots are generally insensitive to light.

Thermotropism—a response to temperature by plants.

Thigmotropism or haptotropism—a response to touch by the climbing organs of a plant. For example, the plant's tendrils may curl around a support in a springlike manner.

What are **turgor movements**?

Turgor movements in plants are movements that are reversible. These movements are caused by changes in the turgor pressure of specific cells. For example, some plants exhibit different leaf positions during the day than at night.

221

How do plants **maximize exposure to light**?

The leaves of many plants move. Leaves often orient themselves perpendicularly to sunlight, thereby increasing the amount of light absorbed for photosynthesis. Leaves also form unusual patterns of layering, called mosaics, that minimize the shading of leaves by each other.

What are the **major classes** of **plant hormones**?

The five major classes of plant hormones are auxins, gibberellins, cytokinins, ethylene, and abscisic acid.

Hormone	Principal Action	Where Produced or Found in Plant
Auxins	Elongate cells in seedlings, shoot tips, embryos, leaves	Shoot apical meristem
Gibberellins	Elongate and divide cells in seeds, roots, shoots, young leaves	Apical portions of roots and shoots
Cytokinins	Stimulate cell division (cytokinesis) in seeds, roots, young leaves, fruits	Roots
Ethylene	Hastens fruit ripening	Leaves, stems, young fruits
Abscisic acid	Inhibits growth; closes stomata	Mature leaves, fruits, root caps

When were the major classes of **plant hormones identified** and who is associated with their identification?

Auxins—Charles Darwin (1809–1882) and his son, Francis (1848–1925), performed some of the first experiments on growth-regulating substances. They published their results in 1881 in *The Power of Movement in Plants*. In 1926 Frits W. Went (1903–1990) isolated the chemical substance responsible for elongating cells in the tips of oat (genus *Avena*) seedlings. He named this substance auxin, from the Greek term *auxein*, meaning "to increase."

Gibberellins—In 1926 the Japanese scientist Eiichi Kurosawa discovered a substance produced by a fungus, *Gibberella fujikuroi*, that caused a disease ("foolish seedling disease") in rice (*Oryza sativa*) seedlings in which the seedlings would grow rapidly but appear sickly and then fall over. The Japanese chemists Teijiro Yabuta and Yasuke Sumiki isolated the compound and named it gibberellin in 1938.

Cytokinins—Johannes van Overbeek discovered a potent growth factor in coconut (*Cocos nucifera*) milk in 1941. In the 1950s Folke Skoog (1908–2001) was able to produce a thousand-fold purification of the growth factor but was unable to isolate it. Carlos O. Miller (1923–), Skoog, and their colleagues succeeded in isolating and identifying the chemical nature of the growth factor. They named the substance kinetin and the group of growth regulators to which it belonged cytokinins because of their involvement in cytokinesis or cell division.

Ethylene—Even before the discovery of auxin in 1926, ethylene was known to have effects on plants. In ancient times the Egyptians would use ethylene gas to ripen fruit. During the 1800s shade trees along streets with lamps that burned ethylene, the illuminating gas, would become defoliated from leaking gas. In 1901 Dimitry Neljubov demonstrated that ethylene was the active component of illuminating gas.

Abscisic acid—Philip F. Wareing discovered large amounts of a growth inhibitor in the dormant buds of ash and potatoes that he called dormin. Several years later in the 1960s, Frederick T. Addicott (1912–) reported the discovery in leaves and fruits of a substance capable of accelerating abscisission that he called abscisin. It was soon discovered that dormin and abscisin were identical chemically.

What are more **recently discovered chemical regulators** of plant growth?

Scientists have recently discovered that plants also react to growth regulators other than the five major classes of plant hormones. Some of the more recently discovered plant growth regulators are summarized in the following chart:

Regulator	Chemical Nature	Effects
Brassinolides	Steroids	Stimulate cell division and elongation for normal plant growth
Salicylic acid	Phenolic compound	Activates pathogen defense genes
Jasmonates	Volatile fatty acid derivatives	Regulate seed germination, root growth, storage-protein accumulation, and synthesis of defense proteins
Systemin	Small peptide	Produced in wounded tissue systems; may induce defenses against herbivores in remote tissues

What are some **commercial uses** of **plant hormones**?

Plant hormones are used in a variety of applications to control some aspect of plant development. Auxins are used in commercial herbicides as weed killers. Another use of auxins is to stimulate root formation. It is often referred to as the "rooting hormone" and applied to cuttings prior to planting. Some hormones are used to increase fruit production and prevent preharvest fruit drop. Gibberellins are sprayed on Thompson seedless grapes during the flowering stage to thin the flowers on each cluster, thus allowing the remaining flowers to spread out and develop larger fruit. Gibberellins are also used to enhance germination and stimulate the early emergence of seedlings in grapes, citrus fruits, apples, peaches, and cherries. When used on cucumber plants, gibberellins promote the formation of male flowers, which is useful in the production of hybrid seeds.

What is the difference between **short-day plants** and **long-day plants**?

Short-day and long-day plants exhibit a response to photoperiodism, or the changes in light and dark in a twenty-four-hour cycle. Short-day plants form flowers when the days become shorter than a critical length, while long-day plants form flowers when the days become longer than a critical length. Short-day plants bloom in late summer or autumn in middle latitudes. Examples of short-day plants are chrysanthemums, goldenrods, poinsettias, soybeans, and ragweed. Long-day plants bloom in spring and early summer. Some examples of long-day plants are clover, irises, and hollyhocks. Florists and commercial plant growers can adjust the amount of light a plant receives to force it to bloom out of season.

What was the "**flower clock**" of Linnaeus?

Carolus Linnaeus (1707–1778), who was responsible for the binomial nomenclature classification system of living organisms, invented a floral clock to tell the time of day. He had observed over a number of years that certain plants constantly opened and closed their flowers at particular times of the day, these times varying from species to species. One could deduce the approximate time of day according to which species had opened or closed its flowers. Linnaeus planted a garden displaying local flowers, arranged in sequence of flowering throughout the day, that would flower even on cloudy or cold days. He called it a "horologium florae" or "flower clock."

Who was the **first scientist** to be associated with **plant tissue culture**?

Tissue culture is a technique for growing fragments of plants in an artificial medium. The basis for tissue culture was proposed in 1902 by the German botanist Gottlieb Haberlandt (1854–1945), who suggested that plant cells were totipotent. In this proposal, every cell has the same genes and the same genetic potential to make all cells other cell types. Botanists began testing Haberlandt's idea, and the proof of totipotency would be the regeneration of an entire plant from one or a few nonzygotic cells. The earliest experiments failed. Cultured cells remained alive for a short time but did not divide and soon died.

Who eventually showed that **plant cells** were **totipotent**?

In 1958 Frederick Campion Steward (1904–1993), a botanist at Cornell University, successfully regenerated an entire carrot plant from a tiny piece of phloem. Small pieces of tissue from carrots were grown in a nutrient broth. Cells that broke free from the fragments dedifferentiated, meaning that they reverted to unspecialized cells. However, as these unspecialized cells grew, they divided and redifferentiated back into specialized cell types. Eventually, cell division and redifferentiation produced entire new plants. Each unspecialized cell from the nutrient broth expressed its genetic potential to make all the other cell types in a plant. Why was Steward successful? Like previous investigators, he supplied the cultured cells with sugars, minerals, and vitamins. In addition, he also added a new ingredient: coconut milk. Coconut milk contains, among other things, a substance that induces cell division. Subsequent research identified this material as cytokinins, a group of plant hormones (growth regulators) that stimulate cell division. Once the cultured cells began dividing, they were transplanted on agar media, where they formed roots and shoots and developed into plants.

If **plants contain chlorophyll**, how can they be **parasitic**?

Parasitic plants harm other plants as they obtain nutrients from them. Cancer root (*Orobanche uniflora*) and mistletoe (genus *Phoradendron*) parasitize hardwood trees, and trees such as sandalwood (genus *Santalum*) obtain their nutrients from nearby grasses. Indian pipe (*Monotropa uniflora*) gets its nutrients from mycorrhizae of trees. Many parasitic plants lack chlorophyll and cannot carry out photosynthesis, and thus they depend entirely on their host for nutrients. In some cases the presence of chlorophyll does not guarantee an independent lifestyle. Mistletoe and witchweed (genus *Striga*) are green yet grow only as parasites. The green portions of these parasites contain only small amounts of certain enzymes required in photosynthesis. The link between parasites and their hosts is called a haustorium. In many plants the link is one or more xylem-to-xylem connections between the two plants. The parasitic plant depends largely on the evaporation of water from its leaves as a means of pulling nutrient-containing water from the xylem of its host. The stomata of many parasitic plants always remain at least partially open, even at night, ensuring a continuous supply of nutrients from the host.

APPLICATIONS

What **important fibers** are obtained from the sclerenchyma of plants?

Many different types of commonly used fibers are obtained the sclerenchyma of plants.

Scientific Name	Common Name	Use
Musa textilis	Manila hemp	Ropes and cords

Scientific Name	Common Name	Use
Agave sisalana	Sisal; century plant	Coarse ropes and twines
Furcraea gigantea	Mauritius hemp	Ropes, cords, and coarse fabrics
Cannabis sativa	Hemp; marijuana	Canvas, twine, and rope
Linum usitatissimum	Flax	Linen
Boehmeria nivea	Ramie	Oriental textiles and clothing
Corchorus capsularis	Jute	Coarse fabrics, including carpet backing, bags, burlap, and sacks

Which fiber has the **longest fiber cells**?

Ramie, *Boehmeria nivea*, has the longest fiber cells with lengths of more than 12 in (30 cm) long. Often used in fabric, these fibers are also stronger than cotton fibers.

What was the **historical significance** of **hemp**?

During the early years of colonial America, hemp (*Agave sisalana*) was as common as cotton is now. It was an easy crop to grow, requiring little water, no fertilizers, and no pesticides. The fabric looks and feels like linen. It was used for uniforms of soldiers, paper (the first two drafts of the Declaration of Independence were written on hemp paper), and an all-purpose fabric. Betsy Ross's flag was made of red, white, and blue hemp.

Where do **sclereids** occur in plants?

Sclereids occur in the roots, stems, and seed coats of plants. Groups of sclereids in the soft flesh give pears their characteristic gritty texture. Sclereids also give nutshells and seed coats their hardness.

What part of the **papyrus** plant (*Cyperus papyrus*) was used to make **paper**?

The central pith of the stalk or stem of the papyrus plant was cut into thin strips, pressed together, and dried to form a smooth writing surface. The thin strips were placed side-by-side longitudinally and crossed at right angles with another set of strips. The finished product was a pure-white sheet of paper.

How are **tree rings** used to date historical events?

The study of tree rings is known as dendrochronology. Every year, trees produce an annular ring composed of one wide, light ring and one narrow, dark ring. During spring and early summer, tree stem cells grow rapidly and larger, thus producing the wide, light ring. In winter, growth is greatly reduced and cells are much smaller, producing the narrow, dark ring. In the coldest part of winter or the dry heat of summer, no cells are produced. Comparing pieces of dead trees of unknown age with the rings of living trees allows scientists to establish the date when the fragment was part of a living tree. This technique has been used to date the ancient pueblos throughout the southwestern United States. A subfield of dendrochronology is dendro-

The flexibility and soft texture of cotton are due to the absence of lignin—in fact, cotton fibers are 95 percent cellulose.

climatology. Scientists study the tree rings of very old trees to determine climatic conditions of the past. The effects of droughts, pollution, insect infestations, fires, volcanoes, and earthquakes are all visible in tree rings.

How do **hardwoods** differ from **softwoods**?

"Hardwood" and "softwood" are terms used commercially to distinguish woods. Hardwoods are the woods of dicots, regardless of how hard or soft they are, while softwoods are the woods of conifers. Many hardwoods come from the tropics, while almost all softwoods come from the forests of the northern temperate zone.

227

Why do tree **leaves turn color** in the fall?

The carotenoids (pigments in the photosynthesizing cells), which are responsible for the fall colors, are present in the leaves during the growing season. However, the colors are eclipsed by the green chlorophyll. Toward the end of summer, when the chlorophyll production ceases, the other colors of the carotenoids (such as yellow, orange, red, or purple) become visible. Listed below are the autumn leaf colors of some common trees.

Tree	Color
Sugar maple, sumac	Flame red and orange
Red maple, dogwood, sassafras, scarlet oak	Dark red
Poplar, birch, tulip tree, willow	Yellow
Ash	Plum purple
Oak, beech, larch, elm, hickory, sycamore	Tan or brown
Locust	Stays green until leaves drop
Black walnut, butternut	Drops leaves before they turn color

The rings in the cross-section of a tree trunk indicate the tree's age.

Why are fall leaves **bright red in some years** and **dull in others**?

Two factors are necessary in the production of red autumn leaves. There must be warm, bright, sunny days during which the leaves manufacture sugar. Warm days must be followed by cool nights with temperatures below 45°F (7°C). This weather combination traps the sugar and other materials in the leaves, thus resulting in the manufacture of red anthocyanin. A warm cloudy day restricts the formation of bright colors. With decreased sunlight, sugar production is decreased, and this small amount of sugar is transported back to the trunk and roots, where it has no color effect.

How is **maple syrup harvested**?

Maple syrup is harvested from the trunks of sugar maple trees (*Acer saccharum*). Production of maple syrup requires daily temperatures to fluctuate between freezing and thawing. During cold nights

> ## Who first observed that the number of rings in the cross-section of a tree trunk indicates its age?
>
> The painter Leonardo da Vinci (1452–1519) noticed this phenomenon. He also saw that the year's dampness can be determined by the space between the tree's rings. The farther apart the rings, the more moisture there was in the ground around the tree.

(below freezing), starch made during the previous summer and stored in wood is converted to sugar. During the day when temperatures rise above freezing, a positive pressure is created in the xylem's sapwood. When tubes called spiles are driven into the sapwood, the positive pressure pushes the sugary sap out of the tree trunk at a rate of 100 to 400 drops per minute. The flow stops when temperatures drop below freezing.

What are **secondary metabolites** and what are some **specific examples** of such compounds **produced by plants**?

Secondary metabolites are compounds produced by plants that are important for the survival and propagation of the plants that produce them. Secondary metabolites serve as chemical signals enabling the plant to respond to environmental cues or functioning in the defense of the producer against herbivores, pathogens, or competitors. Other secondary metabolites provide protection from radiation from the Sun or assist in pollen and seed dispersal. They are produced at various sites within the cell and stored primarily within the vacuole. The three major classes of secondary plant compounds are alkaloids, terpenoids, and phenolics.

Compound	Source	Observation
Alkaloids		
Morphine	*Papaver somniferum* (opium poppy)	Analgesic (painkiller)
Cocaine	*Erythroxylum coca* (coca)	Anesthetic in eye surgery and dentistry; often abused as an illegal substance
Caffeine	*Coffea arabica* (coffee), *Cannellia sinensis* (tea), and *Theobroma cacao* (cocoa)	Stimulant found in many beverages
Nicotine	*Nicotiana tabacum* (tobacco plant leaves)	Stimulant; highly toxic, causing harmful effects from cigarette smoking
Conine	*Conium maculatum* (poison hemlock)	Nerve toxin; killer of Socrates

229

Varieties of maple trees produce some of the most colorful leaves in the fall.

Compound	Source	Observation
Strychnine	Strychnine tree	Potent nerve stimulant and convulsant
Tubocurarine	Curare tree	Used as a muscle relaxant during surgery; component of arrow poisons
Codeine	*Papaver somniferum* (opium poppy)	Cough suppressant
Atropine	*Hyoscyamus muticus* (Egyptian henbane) and *Atropa belladonna* (belladonna nightshade)	Used in eye exams to dilate pupils and as an antidote to nerve gas
Vincristine	*Catharanthus roseus* (Madagascar periwinkle)	Main treatment for certain kinds of leukemia
Quinine	*Cinchona officinalis* (quinine tree)	Bitter flavor of gin and tonic; used to prevent malaria
Terpenoids		
Menthol	Mints and eucalyptus tree	Strong aroma; used in cough medicines
Camphor	*Cinnamomum camphora* (camphor tree)	Component of disinfectants and plasticizers

Tobacco seeds. The nicotine from tobacco leaves is an alkaloid that is a powerful stimulant and highly toxic, causing harmful effects from cigarette smoking.

Compound	Source	Observation
Nepetalactone	*Nepeta catoria* (catnip)	Very attractive to cats
Digitalin	*Digitalis purpurea* (purple foxglove)	Cardiotonic used to stimulate heart action
Oleandrin	*Nerium oleander* (oleander)	Heart poison (similar to digitalin)
Hycopene	*Lycopersicon esculentum* (tomatoes)	Red/orange pigment
Rubber	*Hevea brasiliensis* (rubber tree)	Component of rubber tires
Taxol	*Taxus brevifolia* (Pacific yew)	Drug that inhibits cancerous tumors, especially of ovarian cancer
Phenolics		
Salican	*Salix* spp. (willow tree)	Folk medicine against headaches and fever; precursor of aspirin
Salic acid	*Juglans* (walnut husks)	Main component of some tannins
Myristicin	*Manodora myristica* (nutmeg)	Main flavor of the spice
Rutin	*Fagopyrum esculentum* (buckwheat)	Common "bioflavonoid" sold in nutrition stores

What is **allelopathy**?

Allelopathy is the release of chemicals by certain plants that inhibit the growth and development of competing plants. The chemicals are usually terpenes or phenols and may be found in roots, stems, leaves, fruits, or seeds. An example of this relationship among plants is the black walnut tree (*Juglans nigra*). A chemical compound in the leaves and green stems of the black walnut tree is leached by rainfall into the soil. The chemical compound from the black walnut tree leached by rainfall into the soil is hydrolyzed and oxidized into another compound called juglone. Juglone has been shown to be very toxic to many plants as well as an inhibitor of seed germination. Tomatoes and alfalfa wilt when grown near black walnuts, and their seedlings die if their roots contact walnut roots. Similarly, white pine (*Pinus strobus*) and black locust (*Robinia pseudoacacia*) are often killed by black walnuts growing in their vicinity. Another example of allelopathy is the production of camphor and cineole by the shrubs sage (*Salvia leucophylla*) and artemesia (*Artemesia californica*). Areas of 10 to 12 ft (3.0 to 3.6 m) in diameter around these plants are void of other plants.

What is **locoweed**?

The legumirous locoweeds or milk vetches (genera *Astragalus* and *Oxytropis*) have been a severe problem for ranchers in the western half of the United Sates and are considered to be some of the most toxic plants for horses, sheep, goats, and cattle. *Loco* is the Spanish word for "crazy" and refers to the staggering and trembling behavior of poisoned animals who walk into things and react to unseen objects. The poisonous compounds are an unusual group of alkaloids that affect certain cells of the central nervous system, explaining the behavioral changes observed.

What are some **products derived** from **secondary xylem (wood)** and **secondary phloem (bark)**?

More than 5,000 products are made from secondary xylem and phloem from the trees that grow in the United States. The greatest number of products use lumber, including fences, telephone poles, homes, furniture, flooring, sports equipment (oars, skis, bats, bowling pins, croquet balls, pool sticks), and musical instruments (violins). Lumber is also used to make veneers, plywood, and particleboard. Other products derived from wood are pulp and paper, fuel (wood-burning stoves), charcoal, fabrics, ropes, spices such as cinnamon, dyes, drugs (quinine, taxol), and tannins.

ANIMAL DIVERSITY

INTRODUCTION AND HISTORICAL BACKGROUND

What are the **main characteristics** of **animals**?

Animals are a very diverse groups of organisms, but all of them share a number of characteristics. Animals are multicellular eukaryotes that are heterotrophic, ingesting and digesting food inside the body. Animal cells lack the cell walls that provide support in the bodies of plants and fungi. Most animals have muscle systems and nervous systems, responsible for movement and rapid response to stimuli in their environment. In addition, most animals reproduce sexually with the diploid stage dominating the life cycle. In most species a large, nonmotile egg is fertilized by a small, flagellated sperm, thus forming a diploid zygote. The transformation of a zygote to an animal of specific form depends on the controlled expression in the developing embryo of special regulatory genes.

Who is considered the "**father of zoology**"?

Aristotle (384–322 B.C.E.) is considered the "father of zoology." His contributions to zoology include vast quantities of information about the variety, structure, and behavior of animals; the analysis of the parts of living organisms; and the beginnings of the science of taxonomy.

Who is considered the "**father of modern zoology**"?

Conrad Gessner (1516–1565), a Swiss naturalist, is often credited as the "father of modern zoology," based on his three-volume *Historia Animalium*, which served as a standard reference work throughout Europe in the sixteenth and seventeenth centuries.

Who is considered the **founder** of **experimental zoology**?

Abraham Trembley (1710–1784), a Swiss scientist, is considered the founder of experimental zoology. Much of his research involved studying the regeneration of hydras.

How are **animals classified**?

Animals belong to the kingdom Animalia. Most biologists divide the kingdom into two subkingdoms: 1) Parazoa (from the Greek *para*, meaning "alongside," and *zoa*, meaning "animal"); and 2) Eumetazoa (from the Greek *eu*, meaning "true"; *meta*, meaning "later"; and *zoa*, meaning "animal"). The only existing animals classified as parazoa are the sponges (phylum Porifera). Sponges are very different from other animals and function much like colonial, unicellular protoza even though they are multicellular. Cells of sponges can be versatile and change form and function and are not organized into tissues and organs. They also lack symmetry. All other animals have true tissues, are symmetrical, and are classified as eumetazoa.

How can animals be grouped according to **body symmetry**?

Symmetry refers to the arrangement of body structures in relation to the axis of the body. Most animals exhibit either radial or bilateral body symmetry. Animals such as jellyfishes, sea anemones, and starfishes have radial symmetry. In radial symmetry the body has the general form of a wheel or cylinder, and similar structures are arranged as spokes from a central axis. The bodies of all other animals are marked by bilateral symmetry, a design in which the body has right and left halves that are mirror images of each other. A bilaterally symmetrical body plan has a top and a bottom, also known respectively as the dorsal and ventral portions of the body. It also has a front (or anterior) end and a back (or posterior) end.

How can animals be grouped according to type of **body cavity**?

The structures of most animals develop from three embryonic tissue layers. The outer layer, the ectoderm, gives rise to the outer covering of the body and the nervous system. The inner layer, the endoderm, forms the lining of the digestive tube and other digestive organs. The middle layer, the mesoderm, gives rise to most other body structures, including muscles, skeletal structures, and the circulatory system. One system of grouping animals with three germ layers (triploblastic) is based on the presence and type of body cavity, or coelom, a fluid-filled space between the body wall and digestive tube. Acoelomates such a flatworms (phylum Platyhelminthes) have a solid body with no body cavity. Pseudocoelomates have a "false coelom," since the body cavity is not completely lined with mesoderm; pseudocoelomates include such animals as roundworms (phylum Nematoda) and rotifers (phylum Rotifera). Coelomates, which have a body cavity that is completely lined with mesoderm, include most other animals ranging from earthworms (phylum Annelida) to vertebrates (phylum Chordata).

What are the **noncoelomates**?

The noncoelomates are animals that lack a body cavity (a coelom) and include the sponges, jellyfishes, and simple worms.

How many different animals are there?

Biologists have described and named more than one million species of animals, and some biologists believe that there are several million to ten million more species that remain to be discovered, classified, and named.

What is the **difference between** an **invertebrate** and a **vertebrate**?

Invertebrates are animals that lack a backbone. Almost all animals (99 percent) are invertebrates. Of the more than one million identified animals, only 42,500 have a vertebral column; these are referred to as vertebrates. Most biologists believe the millions of species that have yet to be discovered are exclusively invertebrates.

What are the **largest** and **smallest vertebrates**?

	Name	Length and weight
Largest vertebrates		
Sea mammal	Blue or sulphur-bottom whale (*Balaenoptera musculus*)	100–110 ft (30.5–33.5 m) long; weighs 135–209 tons (122.4–189.6 tonnes)
Land mammal	African bush elephant (*Loxodonta africana*)	Bull is 10.5 ft (3.2 m) tall at shoulder; weighs 5.25–6.2 tons (4.8–5.6 tonnes)
Living bird	North African ostrich (*Struthio c. camelus*)	8–9 ft (2.4–2.7 m) tall; weighs 345 lb (156.5 kg)
Fish	Whale shark (*Rhincodon typus*)	41 ft (12.5 m) long; weighs 16.5 tons (15 tonnes)
Reptile	Saltwater crocodile (*Crocodylus porosus*)	14–16 ft (4.3–4.9 m) long; weighs 900–1,500 lb (408–680 kg)
Rodent	Capybara (*Hydrochoerus hydrochaeris*)	3.25–4.5 ft (1–1.4 m) long; weighs 250 lb (113.4 kg)
Smallest vertebrates		
Sea mammal	Commerson's dolphin (*Cephalorhynchus commersonii*)	Weighs 50–70 lb (236.7–31.8 kg)
Land mammal	Bumblebee or Kitti's hog-nosed bat (*Craseonycteris thong longyai*) or the pygmy shrew (*Suncus erruscus*)	Bat is 1 in (2.54 cm) long; weighs .062–.07 oz (1.6–2 g); shrew is 1.5–2 in (3.8–5 cm) long; weighs 0.052–.09 oz (1.5–2.6 g)

235

Name	Length and weight	
Bird	Bee hummingbird (*Mellisuga helenea*) long; weighs 0.056 oz (1.6 g)	2.25 in (5.7 cm)
Fish	Dwarf pygmy goby (*Trimmatam nanus*) long	0.35 in (8.9 mm)
Reptile	Gecko (*Spaerodactylus parthenopion*) long	0.67 in (1.7 cm)
Rodent	Pygmy mouse (*Baiomys taylori*) 4.3 in (10.9 cm) long; weighs 0.24–0.28 oz (6.8–7.9 g)	

What **names** are used for **male** and **female animals**?

Animal	Male name	Female name
Alligator	Bull	Cow
Ant	Drone	Queen
Ass	Jack, jackass	Jenny
Bear	Boar or he-bear	Sow or she-bear
Bee	Drone	Queen or queen bee
Camel	Bull	Cow
Caribou	Bull, stag, or hart	Cow or doe
Cat	Tom, tomcat, gib, gibeat, boarcat, or ramcat	Tabby, grimalkin, malkin, pussy, or queen
Chicken	Rooster, cock, stag, or chanticleer	Hen, partlet, or biddy
Cougar	Tom or Lion	Lioness, she-lion, or pantheress
Coyote	Dog	Bitch
Deer	Buck or stag	Doe
Dog	Dog	Bitch
Duck	Drake or stag	Duck
Fox	Fox, dog-fox, stag, Reynard, or renard	Vixen, bitch, or she-fox
Giraffe	Bull	Cow
Goat	Buck, billy, billie, or billie-goat	She-goat, nanny, nannie, or nannie-goat
Goose	Gander or stag	Goose or dame
Guinea pig	Boar	Sow
Horse	Stallion, stag, horse, stable horse, sire, or rig	Mare or dam
Impala	Ram	Ewe
Kangaroo	Buck	Doe
Leopard	Leopard	Leopardess

Animal	Male name	Female name
Lion	Lion or tom	Lioness or she-lion
Lobster	Cock	Hen
Manatee	Bull	Cow
Mink	Boar	Sow
Moose	Bull	Cow
Mule	Stallion or jackass	She-ass or mare
Ostrich	Cock	Hen
Otter	Dog	Bitch
Ox	Ox, beef, steer, or bullock	Cow or beef
Partridge	Cock	Hen
Peacock	Peacock	Peahen
Pigeon	Cock	Hen
Quail	Cock	Hen
Rabbit	Buck	Doe
Reindeer	Buck	Doe
Robin	Cock	Hen
Seal	Bull	Cow
Sheep	Buck, ram, male-sheep, or mutton	Ewe or dam
Swan	Cob	Pen
Termite	King	Queen
Tiger	Tiger	Tigress
Turkey	Gobbler or tom	Hen
Walrus	Bull	Cow
Whale	Bull	Cow
Woodchuck	He-chuck	She-chuck
Zebra	Stallion	Mare

What names are used for **juvenile animals**?

Animal	Name for young
Ant	Antling
Antelope	Calf, fawn, kid, or yearling
Bear	Cub
Beaver	Kit or kitten
Bird	Nestling
Bobcat	Kitten or cub
Buffalo	Calf, yearling, or spike-bull
Camel	Calf or colt
Canary	Chick

Animal	Name for young
Caribou	Calf or fawn
Cat	Kit, kitling, kitty, or pussy
Chicken	Chick, chicken, poult, cockerel, or pullet
Chimpanzee	Infant
Cicada	Nymph
Clam	Littleneck
Cod	Codling, scrod, or sprag
Condor	Chick
Cougar	Kitten or cub
Cow	Calf (m. bullcalf; f. heifer)
Coyote	Cub, pup, or puppy
Deer	Fawn
Dog	Whelp
Dove	Pigeon or squab
Duck	Duckling or flapper
Eagle	Eaglet
Eel	Fry or elver
Elephant	Calf
Elk	Calf
Fish	Fry, fingerling, minnow, or spawn
Fly	Grub or maggot
Frog	Polliwog or tadpole
Giraffe	Calf
Goat	Kid
Goose	Gosling
Grouse	Chick, poult, squealer, or cheeper
Horse	Colt, foal, stot, stag, filly, hog-colt, youngster, yearling, or hogget
Kangaroo	Joey
Leopard	Cub
Lion	Shelp, cub, or lionet
Louse	Nit
Mink	Kit or cub
Monkey	Suckling, yearling, or infant
Mosquito	Larva, flapper, wriggler, or wiggler
Muskrat	Kit
Ostrich	Chick
Otter	Pup, kitten, whelp, or cub
Owl	Owlet or howlet
Oyster	Set seed, spat, or brood
Partridge	Cheeper

Animal	Name for young
Pelican	Chick or nestling
Penguin	Fledgling or chick
Pheasant	Chick or poult
Pigeon	Squab, nestling, or squealer
Quail	Cheeper, chick, or squealer
Rabbit	Kitten or bunny
Raccoon	Kit or cub
Reindeer	Fawn
Rhinoceros	Calf
Sea Lion	Pup
Seal	Whelp, pup, cub, or bachelor
Shark	Cub
Sheep	Lamb, lambkin, shearling, or yearling
Skunk	Kitten
Squirrel	Dray
Swan	Cygnet
Swine	Shoat, trotter, pig, or piglet
Termite	Nymph
Tiger	Whelp or cub
Toad	Tadpole
Turkey	Chick or poult
Turtle	Chicken
Walrus	Cub
Weasel	Kit
Whale	Calf
Wolf	Cub or pup
Woodchuck	Kit or cub
Zebra	Colt or foal

SPONGES AND COELENTERATES

What is the **most primitive group** of animals?

Sponges (phylum Porifera, from the Latin terms *porus*, meaning "pore," and *fera*, meaning "bearing") represent the most primitive animals. These organisms are aggregates of specialized cells without true tissues or organs, with little differentiation and integration, and with no body symmetry. A sponge's body is perforated by holes that lead to an inner water chamber. Sponges pump water through those pores and expel it through a large opening at the top of the chamber. While water is passing through the body, nutrients are engulfed, oxygen is absorbed, and waste is eliminated.

Sponges are distinctive in possessing choanocytes, special flagellated cells whose beating drives water through the body cavity and that characterize them as suspension feeders (also known as filter feeders).

What is the **basic composition** of a **sponge**?

A sponge is supported by a skeleton made of hard crystals called spicules whose shape and composition are important features in taxonomy. Calcareous sponges have spicules of calcium carbonates, the material of marble and limestone. The silica spicules of the hexactinellid, or glass, sponges are formed into a delicate, glassy network. Demosponges have siliceous spicules and a network of fibrous protein, spongir, that is similar to collagen. The demosponges are the source of natural household sponges, which are made by soaking dead sponges in shallow water until all the cellular material has decayed, leaving the spongin network behind. However, most sponges sold now for household use are plastic and have nothing to do with real sponges.

How much **water** does an average **sponge circulate** during a day?

A sponge that is 4 in (10 cm) tall and 0.4 in (1 cm) in diameter pumps about 23 qts (22.5 l) of water through its body in one day. To obtain enough food to grow by 3 oz (100 g), a sponge must filter about 275 gal (1,000 kg) of seawater.

Are there more **marine** or **freshwater sponges**?

There are approximately 5,000 species of marine (saltwater) sponges and 150 species of freshwater sponges.

What accounts for the **various colors** of **sponges**?

Living sponges may be brightly colored—green, blue, yellow, orange, red, or purple— or they may be white or drab. The bright colors are due to bacteria or algae that live on or within the sponge.

What **animals** are members of the phylum **Cnidaria**?

Corals, jellyfishes, sea anemones, and hydras are members of the phylum Cnidaria. The name Cnidaria (from the Greek term *knide*, meaning "nettle," and Latin term *aria*, meaning "like" or "connected with") refers to the stinging structures that are characteristic of some of these animals. These organisms have a digestive cavity with only one opening to the outside; this opening is surrounded by a ring of tentacles used to capture food and defend against predators. Cells in the tentacles and outer body surface contain stinging, harpoonlike structures called nematocysts. Cnidarians are the first group in the animal hierarchy to have their cells organized into tissues.

Are the stings of jellyfishes and Portuguese man-of-war fatal to humans?

The stings of a jellyfish can be very painful and dangerous to humans, but they are generally not fatal. Most stings cause a painful, burning sensation that lasts for several hours. Welts and itchy skin rashes may also appear. Only the sting of the box jelly, or sea wasp (*Chironex fleckeri*), can result in death in humans. The box jelly is the only jellyfish for which a lifesaving, specific antidote exists. Stings from a Portuguese man-of-war produce immediate burning sensations and redness that may contain small white lesions. In severe cases, blisters and welts that look like a string of beads may appear, but they are not fatal. Detached tentacles of the Portuguese man-of-war found on beaches may remain dangerous for months.

What are the **two distinct body forms** of **cnidarians**?

The two forms are called the polyp stage and the medusa (plural, medusae), or jellyfish, stage. Polyps generally live attached to a hard surface and bud to produce more polyps and, in some cnidarians, to produce the medusa stage of the life cycle. These medusae, or jellyfish, drift with the ocean currents or swim by pulsating their umbrella-shaped bodies. Medusae release sperm and eggs into the water, where external fertilization occurs. After fertilization, the embryo develops into a larva that eventually settles to the bottom to become another polyp, thus completing the life cycle. Not all cnidarians go through both polyp and medusa stages. Some, such as corals and sea anemones, exist only as polyps.

What is the **largest jellyfish**?

The largest jellyfish is the genus *Cyanea capillata*. It may be more than 6.5 ft (2 m) in diameter and have tentacles of 98 ft (30 m) long and is among the largest invertebrates.

What are some **interesting features** of **jellyfishes**?

Jellyfishes live close to the shores of most oceans and spend most of their time floating near the surface. Jellyfishes have bell-shaped bodies that are between 95 percent and 96 percent water. They have a muscular ring around the margin of the bell that contracts rhythmically to propel them through the water. Jellyfishes are carnivores, subduing their prey with stinging tentacles and drawing the paralyzed animal into the digestive cavity. Jellyfishes are gelatinous—you can see through their bodies.

How does a **nematocyst** work?

A nematocyst is a specialized organelle found in all cnidarians. Each nematocyst features a coiled, threadlike tube lined with a series of barbed spines. The nematocyst is used to capture prey and may also be used for defense purposes. When it is triggered to discharge, the extremely high osmotic pressure within the nematocyst (140 atmospheres) causes water to rush into the capsule, increasing the hydrostatic pressure and expelling the thread with great force. The barb instantly penetrates the prey, stinging it with a highly toxic protein.

Which fishes **form symbiotic relationships** with the **Portuguese man-of-war**?

The Portuguese man-of-war (*Physalia physalis*), a member of the phylum Cnidaria, is a floating hydrozoan. It is a colony of four types of polyps—a pneumatophore, or float; dactylozooids, or tentacles; gastrozooids, or feeding zooids; and gonozooids, which produce gametes. A number of species of fishes from several genera form symbiotic relationships with the Portuguese man-of-war, including the genus *Nomeus* (a minnowlike fish), the clownfish (also called the Man of War fish), and the yellowjack. Most of these fishes live within the tentacles of the Portuguese man-of-war. Some of these fishes, in particular the clownfish, produce a slimy mucus that causes the man-of-war not to fire its nematocysts. The *Nomeus* fish do not produce this protective, slimy mucus but instead rely on a specialized swimming pattern—they swim near the surface in a large circular pattern in both clockwise and counterclockwise directions—to avoid the man-of-war's stings.

What members of **Cnidaria** are **economically important**?

Reef-building corals are among the most important members of Cnidaria. Coral reefs are among the most productive of all ecosystems. They are large formations of calcium carbonate (limestone) in tropical seas laid down by living organisms over thousands of years. Fishes and other animals associated with reefs provide an important source of food for humans, and reefs serve as tourist attractions. Many terrestrial organisms also benefit from coral reefs, which form and maintain the foundation of thousands of islands. By providing a barrier against waves, reefs also protect shorelines against storms and erosion.

Is **coral bleaching** related to changes in the **environment**?

Although corals can capture prey, many tropical species are dependent on photosynthetic algae (zooxanthellae) for nutrition. These algae live within the cells that line the digestive cavity of the coral. The symbiotic relationship between coral and zooxanthellae is mutually beneficial. The algae provide the coral with oxygen and carbon and nitrogen compounds. The coral supplies the algae with ammonia (waste product), from which the algae make nitrogenous compounds for both partners. Coral bleaching is the stress-induced loss of colorful algae (zooxanthellae) that live in coral cells. In coral bleaching the algae lose their pigmentation or are expelled from coral cells. Without the algae, coral become malnourished and die. The causes of coral bleaching are not completely understood, but it is believed that environmental factors are involved. Pollution, invasive bacteria such as *Vibrio*, salinity changes, temperature changes, and high concentrations of ultraviolet radiation (associated with the destruction of the ozone layer) all contribute to coral bleaching.

How are **coral reefs formed** and how fast are they built?

Coral reefs grow only in warm, shallow water. The calcium carbonate skeletons of dead corals serve as a framework upon which layers of successively younger animals attach themselves. Such accumulations, combined with rising water levels, slowly lead to the formation of reefs that can be hundreds of meters deep and long. The coral animal, or polyp, has a columnar form; its lower end is attached to the hard floor of the reef, while the upper end is free to extend into the water. A whole colony consists of thousands of individuals. There are two kinds of corals, hard and soft, depending on the type of skeleton secreted. The polyps of hard corals deposit around themselves a solid skeleton of calcium carbonate (chalk), so most swimmers see only the skeleton of the coral; the animal is in a cuplike formation into which it withdraws during the daytime. The major reef builder in Florida and Caribbean waters, *Montastrea annularis* (star coral) requires about 100 years to form a reef just 3 ft (1 m) high.

What is the origin of the name "**hydra**"?

Hydra, a well-known member of phylum Cnidaria is a tiny (0.4 in or 1 cm in length) organism found in freshwater ponds. It exists as a single polyp that sits on a basal disk that it uses to glide around. It can also move by somersaulting. It usually has six to ten tentacles, which it uses to capture food. Hydras reproduce both sexually and asexually (budding). Hydras are named after the multiheaded monster of Greek mythology that was able to grow two new heads for each head cut off. When a hydra is cut into several pieces, each piece is able to regrow all the missing parts and become a whole animal.

WORMS

What three groups are included in the **flatworms**?

Flatworms belong to the phylum Platyhelminthes. They are flat, elongated, acoelomate animals that exhibit bilateral symmetry and have primitive organs. The members of the flatworms are: 1) planarians, 2) flukes, and 3) tapeworms.

What are the **most common tapeworm infections** in humans?

Tapeworms, members of the class Cestoda, have long, flat bodies in which there is a linear series of sets of reproductive organs. Each set or segment is called a proglottid.

Tapeworm	Means of infection
Beef tapeworm (*Taenia saginata*)	Eating rare beef; most common of all tapeworms in humans
Pork tapeworm (*Taenia solium*)	Eating rare pork; less common than beef tapeworm
Fish tapeworm (*Diphyllobothrium latum*)	Eating rare or poorly cooked fish; fairly common in Great Lakes region of United States

What is unusual about the **fish tapeworm**?

The fish tapeworm is the largest cestode that infects humans. It can grow to a length of 66 ft (20 m). By comparison, the beef tapeworm may only reach a length of 33 ft (10 m).

How **numerous** are **roundworms**?

Roundworms, or nematodes, are members of the phylum Nematoda (from the Greek term *nematos*, meaning "thread") and are numerous in two respects: 1) number of known and potential species; and 2) the total number of these organisms in a habitat. Approximately 12,000 species of nematodes have been named, but it has been estimated that if all species were known, the number would be closer to 500,000. Nematodes live in a variety of habitats ranging from the sea to soil. Six cubic inches (100 cubic centimeters) of soil may contain several thousand nematodes. A square yard (.85 square meter) of woodland or agricultural soil may contain several million of them. Good topsoil may contain billions of nematodes per acre.

What is the most **famous roundworm**?

One soil nematode, *Caenorhabditis elegans*, is widely cultured and has become a model research organism in developmental biology. The study of this animal was begun in 1963 by Sydney Brenner (1927–) who received the Nobel Prize in Physiology or Medicine in 2002. The species normally lives in soil but is easily grown in the laboratory in petri dishes. It is only about .06 in (1.5 mm) long, has a simple, transparent body consisting of only 959 cells, and grows from zygote to mature adult in only three and a half days. The genome (genetic material) of *C. elegans*, consisting of 14,000 genes, was the first animal genome to be completely mapped and sequenced. The small transparent body of this nematode allows researchers to locate cells in which a specific, developmentally important gene is active. These cells show up as bright green spots in a photograph because they have been genetically engineered to produce a green fluorescent protein known as GFP. The complete "wiring diagram" of its nervous system is known, including all the neurons and all connections between them. Much of the knowledge of nematode genetics and development gained from the study of *C. elegans* is transferable to the study of other animals.

What are the **most common roundworm infections** in humans in the United States?

Roundworm	Means of infection
Hookworm (*Ancylostoma duodenale* and *Necator americanus*)	Contact with soil-based juveniles; common in southern states
Pinworm (*Enterobius vermicularis*)	Inhalation of dust that contains ova and by contamination with fingers; most common worm parasite in United States
Intestinal roundworm (*Ascaris lumbricoides*)	Ingestion of embryonated ova in contaminated food; common in rural Appalachia and southeastern states
Trichina worm (*Trichinella spiralis*)	Ingestion of infected meat; occurs occasionally in humans throughout North America

What are the major groups of **segmented worms**?

Members of the phylum Annelida, the segmented worms, have bilateral symmetry and a tubular body that may have 100 to 175 ringlike segments. The three classes of segmented worms are: 1) Polychaeta, the sandworms and tubeworms; 2) Oligochaeta, the earthworms; and 3) Hirudinea, the leeches.

In what ways are **earthworms beneficial**?

Earthworms help maintain fertile soil. An earthworm literally eats its way through soil and decaying vegetation. As it moves about, the soil is turned, aerated, and enriched by nitrogenous wastes. Charles Darwin (1809–1882) calculated that a single earthworm could eat its own weight in soil every day. Much of what is eaten is then excreted on the earth's surface in the form of "casts." The worms then re-bury these casts with their burrowing process. In addition, Darwin claimed that 2.5 acres (1 hectare) of soil might contain 155,000 earthworms, which in one year would bring 18 tons of soil to the surface and in 20 years might build a new layer 3 in (11 cm) thick.

What are **giant tube worms**?

These worms were discovered near the hydrothermal (hot water) ocean vents in 1977 when the submersible Alvin was exploring the ocean floor of the Galapagos Ridge (located 1.5 mi [2.4 km] below the Pacific Ocean surface and 200 mi [322 km] from the Galapagos Islands). Growing to lengths of 5 ft (1.5 m), *Riftia pachyptila Jones*, named after worm expert Meredith Jones of the Smithsonian Museum of Natural History, lack both mouth and gut and are topped with feathery plumes composed of over 200,000 tiny tentacles. The phenomenal growth of these worms is due to their internal food source—symbiotic bacteria, over 100 billion per ounce of tissue, that live within the worms' troposome tissues. To these troposome tissues, the tube worms transport absorbed oxygen from the water, together with carbon dioxide and hydrogen sulfide. Utilizing this supply, the bacteria in turn produce carbohydrates and proteins that the worms need to thrive.

What is the **largest leech**?

Most leeches are between .75 in and 2 in (2 and 6 cm) in length, but some "medicinal" leeches reach 8 in (20 cm). The giant of all leeches is the Amazonian *Haementeria ghilanii* (from the Greek term *haimateros*, meaning "bloody"), which reaches 12 in (30 cm) in length.

Why are **leeches** important in the field of **medicine**?

The medical leech, *Hirudo medicinalis*, is used to remove blood that has accumulated within tissues as a result of injury or disease. Leeches have also been applied to fingers or toes that have been surgically reattached to the body. The sucking by the leech unclogs small blood vessels, permitting blood to flow normally again through the body part. The leech releases hirudin, secreted by the salivary glands, which is an anticoagulant that prevents blood from clotting and dissolves preexisting clots. Other salivary ingredients dilate blood vessels and act as an anesthetic. A medicinal leech can absorb as much as five to ten times its body weight in blood. Complete digestion of this blood takes a long time, and these leeches feed only once or twice a year in this manner.

How long have **leeches** been used for **medicinal purposes**?

Leeches have been used in the practice of medicine since ancient times. During the 1800s leeches were widely used for bloodletting because of the mistaken idea that body disorders and fevers were caused by an excess of blood. Leech collecting and culture were practiced on a commercial scale during this time. William Wordsworth's (1770–1850) poem "The Leech-Gatherer" was based on this use of leeches.

MOLLUSKS

What are the major groups of **mollusks**?

There are four major groups of mollusks: 1) chitons; 2) gastropods, which include snails, slugs, and nudibranches; 3) bivalves, which include clams, oysters, and mussels; and 4) cephalopods, which include squids and octopods. Although mollusks vary widely in external appearance, most share the following body plan: 1) a muscular foot, usually used for movement; 2) a visceral mass containing most of the internal organs; and 3) a mantle—a fold of tissue that drapes over the visceral mass and secretes a shell (in organisms that have a shell).

What is the **largest group** of **mollusks**?

The gastropods (class Gastropoda), which include snails, slugs, and their relatives, is the largest and most diverse group of mollusks. It includes more than 40,000 different species and comprises the second largest group of related animals. Only the insects comprise a larger group. Most gastropods are marine animals, but there are many freshwater species. Garden snails and slugs have adapted to land.

How fast does a **snail move**?

Many snails move at a speed of less than 3 in (8 cm) per minute. This means that if a snail did not stop to rest or eat, it could travel 16 ft (4.8 m) per hour.

How many **tentacles** do the **cephalopods** have?

Octopods have eight tentacles or arms, squids have ten tentacles, and there are as many as ninety in the chambered nautilus.

What is the **largest invertebrate**?

The largest invertebrate is the giant squid, *Architeuthis dux*, which averages 30 to 53 ft (9 to 16 m) in length including its tentacles. It may reach a length of 69 ft (21 m). These animals have the largest eyes, up to 10 in (25 cm) in diameter, in the animal kingdom. It is believed that they generally live on or near the ocean bottom at a depth of 3281 ft (1000 m), or slightly more than a half mile below the surface of the sea.

What is the **heaviest invertebrate**?

The giant clam, *Tridacna maxima*, is the heaviest invertebrate—it may weigh as much as 122 lb (270 kg). The shells of this bivalve may be as long as 5 ft (1.5 m).

Are freshwater **clams** an **endangered group**?

Although freshwater clams are found on every continent except Antarctica, they are now considered one of the most jeopardized groups of animals in the world. Approxi- 247

Octopods are cephalopods that have eight tentacles.

mately 270 species belong to the family Unionidae, found in North America. A total of 72 percent of our 270 native mussel species are listed as recently extinct, endangered, threatened, or of special concern due to human impact on aquatic habitat, commercial harvesting, the introduction of carp, water pollution, and the invasion of zebra mussels.

What has been the impact of **zebra mussles** on North American waterways?

Zebra mussels (*Dreissena polymorpha*) are black-and-white-striped bivalve mollusks. They are hard-shelled species that adhere to hard surfaces with byssal threads. They were probably introduced to North America in 1985 or 1986 via discharge of a foreign ship's ballast water into Lake St. Clair. They have spread throughout the Great Lakes, the Mississippi River, and as far east as the Hudson River. High densities of zebra mussels have been found in the intakes, pipes, and heat exchangers of waterways throughout the world. They can clog the water intakes of power plants, industrial sites, and public drinking water systems; foul boat hulls and engine cooling water systems; and disrupt aquatic ecosystems. Water-processing facilities must be cleaned manually to rid the systems of the mussels. Zebra mussels are a threat to surface water resources because they reproduce quickly, have free-swimming larva and rapid growth, lack competitors for space or food, and have no predators.

How are **pearls** created?

Pearls are formed in saltwater oysters and freshwater clams. There is a curtainlike tissue called the mantle within the body of these mollusks. Certain cells on the side of

---

Which mollusks produce the most cultured pearls?

Cultured pearls are produced by both freshwater and marine mollusks. Most of the world's cultured pearls (known as freshwater pearls) are produced by freshwater mussels belonging to the family Unionoidae. Most saltwater pearls are produced by three species of oysters belonging to the genus *Pinctada*, including *Pinctada imbricata, Pinctada maxima,* and *Pinctada margaritifera.*

---

the mantle toward the shell secrete nacre, also known as mother-of-pearl, during a specific stage of the shell-building process. A pearl is the result of an oyster's reaction to a foreign body, such as a piece of sand or a parasite, within the oyster's shell. The oyster neutralizes the invader by secreting thin layers of nacre around the foreign body, eventually building it into a pearl. The thin layers are alternately composed of calcium carbonate, argonite, and conchiolin. Irritants intentionally placed within an oyster result in the production of what is called cultured pearls.

ARTHROPODS

Why are **arthropods** considered the **most biologically successful phylum** of animals?

Members of the phylum Arthropoda are characterized by jointed appendages and an exoskeleton of chitin. There are more than one million species of arthropods currently known to science, and many biologists believe there are millions more to be identified. Arthropods are the most biologically successful group of animals because they are the most diverse and live in a greater range of habitats than do the members of any other phylum of animals.

What are the **major groups** of arthropods?

Subphylum	Class	Examples
Chelicerata	Merostomata	Horseshoe crabs
Chelicerata	Arachnida	Spiders, scorpions, ticks, mites
Crustacea	Malacostraca	Lobsters, crabs, shrimps, isopods
Crustacea	Copepoda	Copepods
Crustacea	Cirripedia	Barnacles
Unirania	Insecta	Grasshoppers, roaches, ants, bees, butterflies, flies, beetles
Unirania	Chilopoda	Centipedes
Unirania	Diplopoda	Millipedes

How large is the **arthropod population**?

Zoologists estimate that the arthropod population of the world, including crustaceans, spiders, and insects, numbers about a billion million (10^{18}) individuals. More than one million arthropod species have been described, with insects making up the vast majority of them. In fact, two out of every three organisms known on Earth are arthropods, and the phylum is represented in nearly all habitats of the biosphere. About 90 percent of all arthropods are insects, and about half of the named species of insects are beetles.

What are the only **sessile crustaceans**?

Barnacles are the only sessile (permanently attached to one location) crustaceans. They were described by the nineteenth-century naturalist Louis Agassiz (1807–1873) as "nothing more than a little shrimplike animal standing on its head in a limestone house and licking food into its mouth." Accumulations of barnacles may become so great that the speed of a ship may be reduced by 30 percent to 40 percent, necessitating dry-docking the ship to remove the barnacles.

Do **centipedes** actually have one hundred **legs** and **millipedes** have more than one thousand legs?

Centipedes (class Chilopoda) always have an uneven number of pairs of walking legs, varying from 15 to more than 171. The true centipedes (order Scolopendromorpha) have 21 or 23 pairs of legs. Common house centipedes (*Scutigera coleoptrato*) have 15 pairs of legs. Centipedes are all carnivorous and feed mainly on insects. Millipedes (class Diplopoda) have thirty or more pairs of legs. They are herbivores, feeding mainly on decaying vegetation.

What **arthropods** are of **medical importance** in the United States?

Arthropod	Effect on human health
Black widow spider (*Latrodectus mactans*)	Venomous bite
Brown recluse or violin spider (*Loxosceles reclusa*)	Venomous bite
Scorpion *(Centruroides exilicauda)*	Venomous bite
Chiggers (*Trombiculid* mites)	Dermatitis

Which is stronger: steel or the silk from a spider's web?

Spider silk is stronger. Well known for its strength and elasticity, the strongest spider silk has tensile strength second only to fused quartz fibers and five times greater than that of steel of equivalent weight. Tensile strength is the longitudinal stress that a substance can bear without tearing apart.

Arthropod	Effect on human health
Itch mite (*Sarcoptes scabiei*)	Scabies
Deer tick (*Ixodes dammini*)	Bite transmits Lyme disease
Dog tick, wood tick (*Dermacentor* species)	Bite transmits Rocky Mountain spotted fever
Mosquitoes	Bite transmits disease (West Nile virus, encephalitis, filarial worms)
Horseflies, deerflies	Female has painful bite
Houseflies	Many transmit bacteria and viruses
Fleas	Dermatitis
Bees, wasps, ants	Venomous stings (single stings not dangerous unless person is allergic)

How long does it take the average **spider** to weave a **complete web**?

The average orb-weaver spider takes thirty to sixty minutes to completely spin its web. These species of spiders (order Araneae) use silk to capture their food in a variety of ways, ranging from the simple trip wires used by large bird-eating spiders to the complicated and beautiful webs spun by orb spiders. Some species produce funnel-shaped webs, and other communities of spiders build communal webs.

A completed web features several spokes leading from the initial structure. The number and nature of the spokes depend on the species. The spider replaces any damaged threads by gathering up the thread in front of it and producing a new one behind it. The orb web must be replaced every few days because it loses its stickiness (and its ability to entrap food).

What are the **largest** and **smallest aerial spider webs**?

The largest aerial webs are spun by the tropical orb weavers of the genus *Nephila*, which produce webs that measure up to 18.9 ft (6 m) in circumference. The smallest

251

The average orb-weaver spider takes thirty to sixty minutes to completely spin its web.

webs are produced by the species *Glyphesis cottonae*; their webs cover an area of about 0.75 sq in (4.84 sq cm).

Are **spiders really dangerous**?

Most spiders are harmless organisms that, rather than being dangerous to humans, are actually allies in the continuing battle to control insects. Most venom produced by spiders to kill prey is usually harmless to humans. However, there are two spiders in the United States that can produce severe or even fatal bites. They are the black widow spider (*Latrodectus mactans*) and the brown recluse spider (*Loxosceles reclusa*). Black widows are shiny black, with a bright red "hourglass" on the underside of the abdomen. The venom of the black widow is neurotoxic and affects the nervous system. About four or five of each 1,000 black widow bites have been reported as fatal. Brown recluse spiders have a violin-shaped strip on their back. The venom of the brown recluse is hemolytic and causes the death of tissues and skin surrounding the bite. Their bite can be mild to serious and sometimes fatal.

Which **first aid measures** can be used for a bite by a **black widow spider**?

The black widow spider (*Latrodectus mactans*) is common throughout the United States. Its bite is severely poisonous, but no first aid measures are of value. Age, body size, and degree of sensitivity determine the severity of symptoms, which include an initial pinprick with a dull numbing pain, followed by swelling. An ice cube may be placed over the bite to relieve pain. Between ten and forty minutes after the bite, severe abdominal pain and rigidity of stomach muscles develop. Muscle spasms in the extremities, ascending paralysis, and difficulty in swallowing and breathing may follow. The mortality rate is less than 1 percent, but anyone who has been bitten should see a doctor; the elderly, infants, and those with allergies are most at risk and may require hospitalization.

Do male mosquitoes bite humans?

No. Male mosquitoes live on plant juices, sugary saps, and liquids arising from decomposition. They do not have a biting mouth that can penetrate human skin as female mosquitoes do. In some species the females, who lay as many as 200 eggs, need blood to lay their eggs. These are the species that bite humans and other animals.

Why do some biologists consider the insects the most successful group of animals?

With more than one million described species (and perhaps millions more not yet identified), class Insecta is the most successful group of animals on Earth in terms of diversity, geographic distribution, number of species, and number of individuals. More species of insects have been identified than of all other groups of animals combined. What insects lack in size, they make up for in sheer numbers. If we could weigh all the insects in the world, their weight would exceed that of all the remaining terrestrial animals. About 200 million insects are alive at any one time for each human.

How has flight contributed to the success of insects?

Flight is one key to the great success of insects. An animal that can fly can escape many predators, find food and mates, and disperse to new habitats much faster than an animal that must crawl about on the ground.

What is a "bug," biologically speaking?

The biological meaning of the word "bug" is significantly more restrictive than in common usage. People often refer to all insects as "bugs," even using the word to include such organisms as bacteria and viruses as well as glitches in computer programs. In the strictest biological sense, a "bug" is a member of the order Hemiptera, also called true bugs. Members of Hemiptera include bedbugs, squash bugs, clinch bugs, stink bugs, and water striders.

What is the largest group of insects that have been identified and classified?

The largest group of insects that has been identified and classified is the order Coleoptera (beetles, weevils, and fireflies), with some 350,000 to 400,000 species. Beetles are the dominant form of life on Earth, as one of every five living species is a beetle.   253

What is the heaviest insect in the world?

The heaviest insect in the world is the goliath beetle (*Goliathus goliatus*) from Africa, which can weigh up to 3.5 oz (100 g).

What are the stages of **insect metamorphosis**?

There are two types of metamorphoses (marked structural changes in the growth processes): complete and incomplete. In complete metamorphosis the insect (such as an ant, moth, butterfly, termite, wasp, or beetle) goes through all the distinct stages of growth to reach adulthood. In incomplete metamorphosis the insect (such as a grasshopper, cricket, or louse) does not go through all the stages of complete metamorphoses.

Complete metamorphosis.

Egg: One egg is laid at a time or many (as much as 10,000).

Larva: What hatches from the eggs is called a larva. A larva can look like a worm.

Pupa: After reaching its full growth, the larva hibernates, developing a shell or pupal case for protection. A few insects (e.g., the moth) spin a hard covering called a cocoon. The resting insect is called a pupa (except the butterfly is called a chrysalis), and remains in the hibernation state for several weeks or months.

Adult: During hibernation, the insect develops its adult body parts. When it has matured physically, the fully grown insect emerges from its case or cocoon.

Incomplete metamorphosis.

Egg: One egg or many eggs are laid.

Early stage nymph: Hatched insect resembles an adult but is smaller in size. However, those insects that would normally have wings have not yet developed them.

Late-stage nymph: At this time the skin begins to molt (shed), and the wings begin to bud.

Adult: The insect is now fully grown.

What are some **beneficial insects**?

Beneficial insects include bees, wasps, flies, butterflies, moths, and others that pollinate plants. Many fruits and vegetables depend on insect pollinators for the production of seeds. Insects are an important source of food for birds, fishes, and many animals. In some countries such insects as termites, caterpillars, ants, and bees are eaten as food by people. Products derived from insects include honey and beeswax, shellac, and silk. Some predators such as mantises, ladybugs or lady beetles, and lacewings feed on other harmful insects. Other helpful insects are parasites that live on or in the body of harmful insects. For example, some wasps lay their eggs in caterpillars that damage tomato plants.

Why are insects often **found in amber**?

People have long been infatuated with amber, the fossilized form of ancient tree resin, a semiprecious stone used for jewelry and mosaics. Amber from the Dominican Republic contains an average of one insect in every hundred pieces. Some pieces of amber contain thousands of insects—both whole insects and insect fragments. These insects were probably crawling or lodged on the outside of a tree about 30 million years ago and became trapped by a glob of sticky tree resin, which continued to ooze around the animal matter and eventually fossilized. Scientists are able to study these insects, many of which are extinct but may turn out to be missing links to modern-day species.

How is the **light** in **fireflies produced**?

The light produced by fireflies (*Photinus pyroles*), or lightning bugs, is a kind of heatless light called bioluminescence. It is caused by a chemical reaction in which the substance luciferin undergoes oxidation when the enzyme luciferase is present. The flash is a photon of visible light that radiates when the oxidating chemicals produce a high-energy state, which then reverts back to the normal state. The flashing is controlled by the nervous system and takes place in special cells called photocytes. The nervous system, photocytes, and the tracheal end organs control the flashing rate. The air temperature also seems to be correlated with the flashing rate. The higher the temperature, the shorter the interval between flashes—eight seconds at 65°F (18.3°C) and four seconds at 82°F (27.7°C). Scientists are uncertain as to why this flashing occurs. The rhythmic flashes could be a means of attracting prey or enabling mating fireflies to signal in heliographic codes (that differ from one species to another), or they could serve as a warning signal.

What is the **most destructive insect** in the world?

The most destructive insect is the desert locust (*Schistocera gregaria*), the locust of the Bible, whose habitat ranges from the dry and semi-arid regions of Africa and the Middle East through Pakistan and northern India. This short-horn grasshopper can eat its own weight in food a day, and during long migratory flights a large swarm can consume 20,000 tons (18,144,000 kg) of grain and vegetation a day.

Who introduced the **gypsy moth** into the United States?

In 1869 Professor Leopold Trouvelot (1827–1895) brought gypsy moth egg masses from France to Medford, Massachusetts. His intention was to breed the gypsy moth with the silkworm to overcome a wilt disease of the silkworm. He placed the egg masses on a window ledge, and evidently the wind blew them away. About ten years later these caterpillars were numerous on trees in that vicinity, and in twenty years trees in eastern Massachusetts were being defoliated. In 1911 a contaminated plant shipment from Holland also introduced the gypsy moth to Massachusetts. These pests have now spread to twenty-five states, especially in the northeastern United States. Scattered locations in Michigan and Oregon have also reported occurrences of gypsy moth infestations.

The gypsy moth (*Porthetria dispar*) lays its eggs on the leaves of oaks, birches, maples, and other hardwood trees. When the yellow hairy caterpillars hatch from the eggs, they devour the leaves in such quantities that the tree becomes temporarily defoliated. Sometimes this causes the tree to die. The caterpillars grow from 0.5 in (3 mm) to about 2 in (5.1 cm) before they spin a pupa, in which they will metamorphose into adult moths.

Are there any **natural predators** of **gypsy moth caterpillars**?

About forty-five kinds of birds, squirrels, chipmunks, and white-footed mice eat this serious insect pest. Among the thirteen imported natural enemies of the moth, two flies—*Compislura concinnata* (a tachnid fly) and *Sturnia scutellata*—parasitize the caterpillar. Other parasites and various wasps have also been tried as controls, as well as spraying and male sterilization.

How does a **butterfly** differ from a **moth**?

Characteristic	Butterflies	Moths
Antennae	Knobbed	Unknobbed
Active time of day	Day	Night
Coloration	Bright	Dull
Resting wing position	Vertically above body	Horizontally beside body

While these guidelines generally hold true, there are exceptions. Moths have hairy bodies, and most have tiny hooks or bristles linking the forewing to the hind wing; butterflies do not have either characteristic.

What is the most popular **state insect**?

The honeybee is by far the most popular state insect, having been selected by sixteen states: Arkansas, California (nicknamed the Beehive State), Georgia, Kansas, Louisiana, Maine, Mississippi, Missouri, Nebraska, New Jersey, North Carolina, Oklahoma, South Dakota, Tennessee, Vermont, and Wisconsin.

Has the U.S. selected a national insect?

A group of citizens has petitioned the United States Congress to name the monarch butterfly as the national insect, but to date they have not been successful.

What are "**killer bees**"?

Africanized honeybees—the term entomologists prefer rather than killer bees—are a hybrid originating in Brazil, where African honeybees were imported in 1956. The breeders, hoping to produce a bee better suited to producing more honey in the tropics, instead found that African bees soon hybridized with and mostly displaced the familiar European honeybees. Although they produce more honey, Africanized honeybees (*Apis mellifera scutellata*) also are more dangerous than European bees because they attack intruders in greater numbers. Since their introduction they have been responsible for approximately 1,000 human deaths. In addition to such safety issues, concern is growing regarding the effect of possible hybridization on the U.S. beekeeping industry.

In October 1990 the bees crossed the Mexican border into the United States. They reached Arizona in 1993. In 1996, six years after their arrival in the United States, Africanized honey bees could be found in parts of Texas, Arizona, New Mexico, Nevada, and California. Their migration northward has slowed, partially because they are a tropical insect and cannot live in colder climates. Experts have suggested two possible ways of limiting the spread of the Africanized honeybees. The first is drone-flooding, a process by which large numbers of European drones are kept in areas where commercially reared European queen bees mate, thereby ensuring that only limited mating occurs between Africanized drones and European queens. The second method is frequent requeening, in which a beekeeper replaces a colony's queen with one of his or her own choosing. The beekeeper can then be assured that the queens are European and that they have already mated with European drones.

What are **migratory beekeepers**?

A migratory beekeeper is a person who transports his or her bee colonies to different areas to produce better honey or to collect fees for pollinating such crops as fruit trees, almonds, and alfalfa. The beekeepers frequently travel north in the spring and summer to pollinate crops and then back south in the fall and winter to maintain the colonies in the warmer southern weather. Approximately 1,000 migratory beekeepers operate in the United States, transporting approximately two million bee colonies a year.

> ### How many bees are in a bee colony?
>
> **O**n average, a bee colony contains from 50,000 to 70,000 bees, which produce a harvest of from 60 to 100 lb (27 to 45 kg) of honey per year. A little more than one third of the honey produced by the bees is retained in the hive to sustain the population.

How many flowers need to be tapped for **bees** to gather enough **nectar** to produce one pound of **honey**?

Bees must gather 4 lb (18 kg) of nectar, which requires the bees to tap about two million flowers, in order to produce 1 lb (454 g) of honey. The honey is gathered by worker bees, whose life span is three to six weeks, long enough to collect about a teaspoon of nectar.

Do **termites** have any natural predators?

Birds, ants, spiders, lizards, and dragonflies have been seen preying on young, winged termites when they emerge and fly from a home colony to establish new colonies. Termites are generally most vulnerable to predators when they emerge from their home colony. Chimpanzees are also known to use sticks as tools to forage for termites.

How are **ants** distinguished from **termites**?

Both insect orders—ants (order Hymenoptera) and termites (order Isoptera)—have segmented bodies with multijointed legs. Listed below are some differences.

Characteristic	Ant	Termite
Wings	Two pairs with the front pair being much longer than the back pair	Two pairs of equal length
Antenna	Bends at right angle	Straight
Abdomen	Wasp-waist (pinched in)	No wasp-waist

What is a "**daddy longlegs**"?

The name applies to two different kinds of invertebrates. The first is a harmless, nonbiting, long-legged arachnid. Also called a harvestman, it is often mistaken for a spider, but it lacks the segmented body shape that a spider has. Although it has the same number of legs (eight) as a spider, the harvestman's legs are far longer and thinner. These very long legs enable it to raise its body high enough to avoid ants or other small enemies. Harvestmen are largely carnivorous, feeding on a variety of small invertebrates such as insects, spiders, and mites. They never spin webs as spiders do. They also eat some juicy plants and in captivity can be fed almost anything edible, from bread and milk to meat. Harvestmen also need to drink frequently. The term "daddy longlegs" also is used for a cranefly—a thin-bodied insect with long thin legs that has a snoutlike proboscis with which it sucks water and nectar.

How long have **cockroaches** been on the earth?

The earliest cockroach fossils are about 280 million years old. Cockroaches (order Dictyoptera) are nocturnal scavenging insects that eat not only human food but book bindings, ink, and whitewash as well.

What **invertebrate lives** in both **marine** and **freshwaters** and is one of the most important of all animals?

Copepods, tiny crustaceans, are the link between the photosynthetic life in the ocean or pond and the rest of the aquatic food web. They are primary consumers grazing on algae in the waters of the oceans and ponds. These organisms, among the most abundant multicellular animals on Earth, are then consumed by a variety of small predators, which are eaten by larger predators, and so on. Virtually all animal life in the ocean depends on the copepods, either directly or indirectly. Although humans do not eat copepods directly, our sources of food from the ocean would disappear without the copepods.

ECHINODERMS

What are the **major groups** of **echinoderms**?

There are six principle groups of echinoderms (from the Greek terms *echina*, meaning "spiny," and *derma*, meaning "skin"): 1)class Crinoidea, sea lilies and feather stars; 2) class Asteroidea, sea stars; 3) class Ophiuroidea, basket stars and brittle stars; 4) class Eichinoidea, sea urchins and sand dollars; 5) class Holothuroidea, sea cucumbers; and 6) class Concentricycloidea, sea daisies, which live on water-logged wood in the deep sea and were first discovered in 1986.

Do all **starfishes**, or **sea stars**, have **five arms**?

Starfishes, or sea stars, are members of the class Asteroidea. Their bodies consist of a central disk from which radiate from five to more than twenty arms or rays.

Why are **sea urchins** used to study **embryonic development**?

Sea urchins are a useful model system for studying many problems in early animal development. Historically, sea urchins were a key system in elucidating a variety of classic developmental problems, including the mechanisms of fertilization, egg activation, cleavage, gastrulation, and the regulation of differentiation in the early embryo. In addition, early studies of the molecular basis of early embryology were carried out in this system. Gametes can be obtained easily, sterility is not required, and the eggs and early embryos of many species are beautifully transparent. The early development of sea urchin embryos is also highly synchronous—that is, when a batch of eggs is fertilized, all of the resulting embryos typically develop on the same schedule, making possible biochemical and molecular studies of early embryos.

CHORDATES

What are the major **characteristics** of all **chordates**?

All chordates share a notochord, dorsal nerve cord, and pharyngeal gill pouches. The notochord, a cartilaginous supporting rod, runs along the dorsal part of the body. It is always found in embryos, but in most vertebrates it is replaced during development by a backbone of bony or cartilaginous vertebrae. The tubular dorsal nerve cord, dorsal to the notochord, is formed during development by an infolding of the ectoderm. In vertebrates the nerve cord eventually becomes encased and thus protected by the backbone. The pharyngeal gill pouches appear during embryonic development on both sides of the throat region, the pharynx.

What are the three **major groups** of **chordates**?

The chordates are divided into three subphyla: Tunicata, Cephalochordata, and Vertebrata. Tunicates are like little leathery bags that are either free living or attach to pilings, rocks, and seaweeds. They are also called sea squirts because a disturbed animal may contract and shoot streams of water from both of its siphons.

The subphylum Cephalochordata contains the amphioxus or lancelet (*Branchiostoma*), which looks like a small fish and has the three chordate features as an adult. Amphioxus also shows clear serial segmentation or metamerism (from the Greek terms *meta*, meaning "between, among, after," and *meros*, meaning "part"). It is divided lengthwise into a series of muscle segments. Vertebrates, which comprise the third chordate subphylum, retain the same metamerism in internal structures.

What are the **major features** shown by all **vertebrates**?

Animals in the subphylum Vertebrata are distinguished from other chordates by several features. Most prominent is the endoskeleton of bone or cartilage, centering around the vertebral column (spine or backbone). Composed of separate vertebrae (showing internal metamerism), a vertebral column combines flexibility with enough strength to support even a large body. Other vertebrate features include: 1) complex dorsal kid-

> ## Which members of the agnathans are still living?
>
> The only living agnathans are cyclostomes, known commonly as lampreys and hagfishes.

neys; 2) a tail (lost via evolution in some groups) extending between the anus; 3) a closed circulatory systems with a single, well-developed heart; 4) a brain at the anterior end of the spinal cord, with ten or more pairs of cranial nerves; 5) a cranium (skull) protecting the brain; 6) paired sex organs in both males and females; and 7) two pairs of movable appendages—fins in the fishes, which evolved into legs in land vertebrates.

What was the **first group** of **vertebrates**?

The first vertebrates were fishes that appeared 500 million years ago. They were agnathans (from the Greek terms *a*, meaning "without," and *gnath*, meaning "jaw"), small, jawless fishes up to about 8 in (20 cm) long and also known as ostracoderms ("shell skin") because their bodies were covered with bony plates, most notably a head shield protecting the brain.

What is the **largest group of vertebrates**?

The largest group of vertebrates is fishes. They are a diverse group and include almost 21,000 species—more than all other kinds of vertebrates combined. Most members of this group are osteichythes, or "bony fishes," which include basses, trout, and salmon.

What **general characteristics** do all **fishes** have in common?

All fishes have the following characteristics: 1) gills that extract oxygen from water; 2) an internal skeleton with a skin that surrounds the dorsal nerve cord; 3) single-loop blood circulation in which the blood is pumped from the heart to the gills and then to the rest of the body before returning to the heart; 4) nutritional deficiencies, particularly some amino acids that must be consumed and cannot be synthesized.

What are **chondrichthyes**?

Chondrichthyes are fishes that have a cartilaginous skeleton rather than a bony skeleton; they include such organisms as sharks, skates, and rays.

How many kinds of **sharks** are there and how many are **dangerous**?

The United Nations' Food and Agricultural Organization lists 354 species of sharks, ranging in length from 6 in (15 cm) to 49 ft (15 m). While thirty-five species are

261

Is the whale shark a mammal or a fish?

The whale shark (*Rhincodon typus*) is a shark, not a whale. It is, therefore, a fish. This species' name merely indicates that it is the largest of all shark species (weighing 40,000 lb [18,144 kg] or more and growing to lengths of 49 ft [15 m] or more) and the largest fish species in the world. However, it is completely harmless to humans.

known to have attacked humans at least once, only a dozen do so on a regular basis. The relatively rare great white shark (*Carcharodan carcharias*) is the largest predatory fish. The largest specimen accurately measured was 20 ft, 4 in (6.2 m) long and weighed 5,000 lb (2,270 kg).

How far from shore do **shark attacks** occur?

In a study of 570 shark attacks, it was found that most shark attacks occur near shore. These data are not surprising since most people who enter the water stay close to the shore.

Distance from shore	Percentage of shark attacks	Percentage of people who swim at this distance
50 ft (15 m)	31	39
100 ft (30 m)	11	15
200 ft (60 m)	9	12
300 ft (90 m)	8	11
400 ft (120 m)	2	2
500 ft (150 m)	3	5
1000 ft (300 m)	6	9
1 mi (1.6 km)	8	6
>1 mi (>1.6 km)	22	1

What is unusual about the **teeth of sharks**?

Sharks were among the first vertebrates to develop teeth. The teeth are not set into the jaw but rather sit atop it. They are not firmly anchored and are easily lost. The teeth are arranged in six to twenty rows, with the ones in front doing the biting and cutting. Behind these teeth, others grow. When a tooth breaks or is worn down, a replacement moves forward. One shark may eventually develop and use more than 20,000 teeth in a lifetime.

What group of animals was the first to make a **partial transition** from **water to land**?

Amphibians have made a partial transition to terrestrial life. The living amphibians include newts, salamanders, frogs, and toads. Although lungfish made a partial transition to living out of the water, amphibians were the first to struggle onto land and become adapted to a life of breathing air while not constantly surrounded by water.

What does the word "**amphibian**" mean?

The word "amphibian," from the Greek term *amphibia*, means "both lives" and refers to the animals' double life on land and in water. The usual life cycle of amphibians begins with eggs laid in water, which develop into aquatic larvae with external gills; in a development that recapitulates its evolution, the fishlike larva develops lungs and limbs and becomes an adult.

What are the **major groups of amphibians**?

The following chart illustrates the three major groups of amphibians:

Examples	Order	Number of Living Species
Frogs and toads	Anura (Salientia)	3,450
Salamanders and newts	Caudata (Urodela)	360
Caecilians	Apoda (Gymnophiona)	160

What features of **reptiles** enabled them to become **true land vertebrates**?

Legs were arranged to support the body's weight more effectively than in amphibians, allowing reptile bodies to be larger and to run. Reptilian lungs were more developed with a greatly increased surface area for gas exchange than the saclike lungs of amphibians. The three-chambered heart of reptiles was more efficient than the amphibian heart. In addition, the skin was covered with hard, dry scales to minimize water loss. However, the most important evolutionary adaptation was the amniotic egg, in which an embryo could survive and develop on land. The eggs were surrounded by a protective shell that prevented the developing embryo from drying out.

What is the difference between a **reptile** and an **amphibian**?

Reptiles are clad in scales, shields, or plates, and their toes have claws; amphibians have moist, glandular skins, and their toes lack claws. Reptile eggs have a thick, hard, or parchmentlike shell that protects the developing embryo from moisture loss, even on dry land. The eggs of amphibians lack this protective outer covering and are always laid in water or in damp places. Young reptiles are miniature replicas of their parents in general appearance if not always in coloration and pattern. Juvenile amphibians pass through a larval, usually aquatic, stage before they metamorphose (change in form and structure) into the adult form. Reptiles include alligators, crocodiles, turtles, and snakes. Amphibians include salamanders, toads, and frogs.

What groups of **reptiles** are **living today**?

The three orders of reptiles that are alive today are: 1) Chelonia, which includes turtles, terrapins, and tortoises; 2) Squamata, which includes lizards and snakes; and 3) Crocodilia, which includes crocodiles and alligators.

Which **poisonous snakes** are native to the United States?

Snake	Average Length
Rattlesnakes	
Eastern diamondback (*Crotalus adamateus*)	33–65 in (84–165 cm)
Western diamondback (*Crotalus atrox*)	30–65 in (76–419 cm)
Timber rattlesnake (*Crotalus horridus horridus*)	32–54 in (81–137 cm)
Prairie rattlesnake (*Crotalus viridis viridis*)	32–46 in (81–117 cm)
Great Basin rattlesnake (*Crotalus viridis lutosus*)	32–46 in (81–117 cm)
Southern Pacific rattlesnake (*Crotalus viridis helleri*)	30–48 in (76–122 cm)
Red diamond rattlesnake (*Crotalus ruber ruber*)	30–52 in (76–132 cm)
Mojave rattlesnake (*Crotalus scutulatus*)	22–40 in (56–102 cm)
Sidewinder (*Crotalus cerastes*)	18–30 in (46–76 cm)
Moccasins	
Cottonmouth (*Agkistrodon piscivorus*)	30–50 in (76–127 cm)
Copperhead (*Agkistrodon contortrix*)	24–36 in (61–91 cm)
Cantil (*Agkistrodon bilineatus*)	30–42 in (76–107 cm)
Coral snakes	
Eastern coral snake (*Micrurus fulvius*)	16–28 in (41–71 cm)

When was the **term "dinosaur"** first used?

The term "dinosaur" was first used by Richard Owen (1804–1892) in 1841 in his report on British fossil reptiles. The term, meaning "fearful lizard," was used to describe the group of large extinct reptiles whose fossil remains had been found by many collectors.

What were the **smallest and largest dinosaurs**?

Compsognathus, a carnivore from the late Jurassic period (131 million years ago), was about the size of a chicken and measured, at most, 35 in (89 cm) from the tip of its snout to the tip of its tail. The average weight was about 6.8 lb (3 kg), but individuals could be as much as 15 lb (6.8 kg).

The largest species for which a whole skeleton is known is *Brachiosaurus*. A specimen in Humboldt Museum in Berlin measures 72.75 ft (22.2 m) long and 46 ft (14 m) high. It weighed an estimated 34.7 tons (31,480 kg). *Brachiosaurus* was a four-footed, plant-eating dinosaur with a long neck and tail and lived from about 155 to 131 million years ago.

What was the **typical life span** of **dinosaurs**?

The life span has been estimated at 75 to 300 years. Such estimates are educated guesses. From examination of the microstructure of dinosaur bones, scientists have inferred that dinosaurs matured slowly and probably had proportionately long life spans.

Did **dinosaurs** and **humans** ever coexist?

No. Dinosaurs first appeared in the Triassic period (about 220 million years ago) and disappeared at the end of the Cretaceous period (about 65 million years ago). Modern humans (*Homo sapiens*) appeared only about 25,000 years ago. Movies that show humans and dinosaurs existing together are only Hollywood fantasies.

Why did **dinosaurs** become **extinct**?

There are many theories as to why dinosaurs disappeared from Earth about 65 million years ago. Scientists debate whether dinosaurs became extinct gradually or all at once. The gradualists believe that the dinosaur population steadily declined at the end of the Cretaceous period. Numerous reasons have been proposed for this. Some claim the dinosaurs' extinction was caused by biological changes that made them less competitive with other organisms, especially the mammals that were just beginning to appear. Overpopulation has been argued, as has the theory that mammals ate so many dinosaur eggs that dinosaur reproduction was irrevocably harmed. Others believe that disease—everything from rickets to constipation—wiped the dinosaurs out. Changes in climate, continental drift, volcanic eruptions, and shifts in Earth's axis, orbit, and/or magnetic field have also been held responsible.

The catastrophists argue that a single disastrous event caused the extinction not only of the dinosaurs but also of a large number of other species that coexisted with them. In 1980 the American physicist Luis Alvarez (1911–1988) and his geologist son Walter Alvarez (1940–) proposed that a large comet or meteoroid struck Earth 65 million years ago. They pointed out that there is a high concentration of the element iridium in the sediments at the boundary between the Cretaceous and Tertiary periods. Iridium is rare on Earth, so the only source of such a large amount of it had to be outer space. This iridium anomaly has since been discovered at over fifty sites around the world. In 1990 tiny glass fragments, which could have been caused by the extreme heat of an impact, were identified in Haiti. A 110-mi (177-km) wide crater in the Yucatan Peninsula, long covered by sediments, has been dated to 64.98 million years ago, making it a leading candidate for the site of this impact.

A hit by a large extraterrestrial object, perhaps as much as 6 mi (9.3 km) wide, would have had a catastrophic effect upon the world's climate. Huge amounts of dust and debris would have been thrown into the atmosphere, reducing the amount of sunlight reaching the surface. Heat from the blast may also have caused large forest fires, which would have added smoke and ash to the air. Lack of sunlight would kill off plants and have a dominolike effect on other organisms in the food chain, including the dinosaurs.

It is possible that the reason for the dinosaurs' extinction may have been a combination of both theories. The dinosaurs may have been gradually declining, for whatever reason. The impact of a large object from space merely delivered the coup de grace.

The fact that dinosaurs became extinct has been cited as proof of their inferiority and that they were evolutionary failures. However, these animals flourished for 150 million years. By comparison, the earliest ancestors of humanity appeared only about three million years ago. Humans have a long way to go before they can claim the same sort of success as the dinosaurs.

Are **tortoises** and **terrapins** the same as **turtles**?

The terms "turtle," "tortoise," and "terrapin" are used for various members of the order Testudines (from the Latin term *testudo*, meaning "tortoise"). In North American usage they are all correctly called turtles. The term "tortoise" is often used for land turtles. In British usage the term "tortoise" is the inclusive term, and "turtle" is only applied to aquatic members of the order.

Are **turtles endangered**?

Worldwide turtle populations have declined due to several reasons, including habitat destruction; exploitation of species by humans for their eggs, leather, and meat; and their becoming accidentally caught in the nets of fishermen. In particular danger are sea turtles, such as Kemp's ridley sea turtle (*Lepidochelys kempii*), which is believed to have a population of only a few hundred. Other threatened species include the Central American river turtle (*Dermatemys mawii*), the green sea turtle (*Chelonia mydas)*, the leatherback sea turtle (*Geochelone yniphora*), the desert tortoise (*Gopherus agassizii*), and the Galapagos tortoise (*Geochelone elephantopus*).

Common name	Scientific name	Status
Turtle, Alabama red-belly	(*Pseudemys alabamensis*)	Endangered
Turtle, bog (=Muhlenberg) (southern)	(*Clemmys muhlenbergii*)	Threatened
Turtle, bog (=Muhlenberg) (northern)	(*Clemmys muhlenbergii*)	Threatened
Turtle, flattened musk (species range clarified)	(*Sternotherus depressus*)	Threatened
Turtle, ringed map	(*Graptemys oculifera*)	Threatened
Turtle, yellow-blotched map	(*Graptemys flavimaculata*)	Threatened
Sea turtle, green	(*Chelonia mydas*)	Endangered/Threatened
Sea turtle, hawksbill	(*Eretmochelys imbricata*)	Endangered
Sea turtle, Kemp's ridley	(*Lepidochelys kempii*)	Endangered
Sea turtle, leatherback	(*Dermochelys coriacea*)	Endangered
Sea turtle, loggerhead	(*Caretta caretta*)	Threatened
Sea turtle, olive ridley	(*Lepidochelys olivacea*)	Threatened

Source: U.S. Fish and Wildlife Service

What is the **most successful** and diverse group of **terrestrial vertebrates**?

Birds, members of the class Aves, are the most successful of all terrestrial vertebrates. There are twenty-eight orders of living birds with almost 10,000 species distributed over almost the entire earth. The success of birds is basically due to the development of the feather.

What accounts for the **different colors of bird feathers**?

The vivid color of feathers is of two kinds: 1) pigmentary, and 2) structural. Red, orange and yellow feathers are colored by pigments called lipochromes deposited in the feather barbules as they are formed. Black, brown, and gray colors are from another pigment, melanin. Blue feathers depend not on pigment but on scattering of shorter wavelengths of light by particles within the feather. These are structural feathers. Green colors are almost always a combination of yellow pigment and blue feather structure. Another kind of structural color is the beautiful iridescent color of many birds, which ranges from red, orange, copper, and gold to green, blue, and violet. Iridescent color is based on interference that causes light waves to reinforce, weaken, or eliminate each other. Iridescent colors may change with the angle of view.

The bald eagle was adopted as the national bird of the United States in 1782.

How are **birds related to dinosaurs**?

Birds are essentially modified dinosaurs with feathers. Robert T. Bakker (1945–) and John H. Ostrom (1928–) did extensive research on the relationship between birds and dinosaurs in the 1970s and concluded that the bony structure of small dinosaurs was very similar to *Archaeopteryx*, the first animal classified as a bird, but that dinosaur fossils showed no evidence of feathers. They proposed that birds and dinosaurs evolved from the same source.

Why is *Archaeopteryx* important?

Archaeopteryx is the first known bird. It had true feathers that provided insulation and allowed this animal to form scoops with its wings for catching prey.

What bird has the **biggest wing span**?

Three members of the albatross family, the wandering albatross (*Diomedea exculans*), the royal albatross (*Diomedea epomophora*), and the Amsterdam Island albatross (*Diomeda amsterdiamensis*), have the greatest wingspan of any bird species with a spread of 8 to 11 ft (2.5 to 3.3 m).

When was the **bald eagle** adopted as the national bird of the United States?

On June 20, 1782, the citizens of the newly independent United States of America adopted the bald or "American" eagle as their national emblem. At first the heraldic artists depicted a bird that could have been a member of any of the larger species, but by 1902 the bird portrayed on the seal of the United States of America had assumed its proper white plumage on head and tail. The choice of the bald eagle was not unanimous; Benjamin Franklin (1706–1790) preferred the wild turkey. Oftentimes a tongue-in-cheek humorist, Franklin thought the turkey a wily but brave, intelligent, and prudent bird. He viewed the eagle on the other hand as having "a bad moral character" and "not getting his living honestly," preferring instead to steal fish from hardworking fishhawks. He also found the eagle a coward who readily flees from the irritating attacks of the much smaller kingbird.

What is the name of the bird that perches on the black **rhinoceros's back**?

The bird, a relative of the starling, is called an oxpecker (a member of the Sturnidae family). Found only in Africa, the yellow-billed oxpecker (*Buphagus africanuswingspan*) is widespread over much of western and central Africa, while the red-billed oxpecker (*Buphagus erythrorhynchuswingspan*) lives in eastern Africa from the Red Sea to Natal.

Seven to 8 in (17 to 20 cm) long with a coffee-brown body, the oxpecker feeds on more than twenty species of ticks that live in the hide of the black rhinoceros (*Diceros bicornis*), also called the hook-lipped rhino. The bird spends most of its time on the rhinoceros or on other animals, such as the antelope, zebra, giraffe, or buffalo. The bird has even been known to roost on the body of its host.

The relationship between the oxpecker and the rhinoceros is called mutualism. The bird feeds on the rhinoceros's ticks, benefiting both the bird and the rhinoceros. In addition, the oxpecker, having much better eyesight than the nearsighted rhinoceros, alerts its host with its shrill cries and flight when danger approaches.

In what year was the **European starling** (*Sturnus vulgariswingspan*) imported into the United States?

Eugene Schieffelin (1826–1906) imported the European starling into the United States in 1890. Schieffelin wanted to establish in the United States every bird found in Shakespeare's works. He also imported English sparrows to New York City in 1860.

Will wild birds reject baby birds that have been **touched by humans**?

No. Contrary to popular belief, birds generally will not reject hatchlings touched by human hands. The best thing to do for newborn birds that have fallen or have been pushed out of the nest is to locate the nest as quickly as possible and gently put them back.

What names are used for **groups** or **companies** of **mammals**?

Mammal	Group name
Antelopes	Herd

Mammal	Group name
Apes	Shrewdness
Asses	Pace, drove, or herd
Baboons	Troop
Bears	Sloth
Beavers	Family or colony
Boars	Sounder
Buffaloes	Troop, herd, or gang
Camels	Flock, train, or caravan
Caribou	Herd
Cattle	Drove or herd
Deer	Herd or leash
Elephants	Herd
Elks	Gang or herd
Foxes	Cloud, skulk, or troop
Giraffes	Herd, corps, or troop
Goats	Flock, trip, herd, or tribe
Gorillas	Band
Horses	Haras, stable, remuda, stud, herd, string, field, set, team, or stable
Jackrabbits	Husk
Kangaroos	Troop, mob, or herd
Leopards	Leap
Lions	Pride, troop, flock, sawt, or souse
Mice	Nest
Monkeys	Troop or cartload
Moose	Herd
Mules	Barren or span
Oxen	Team, yoke, drove, or herd
Porpoises	School, crowd, herd, shoal, or gam
Reindeer	Herd
Rhinoceri	Crash
Seals	Pod, herd, trip, rookery, or harem
Sheep	Flock, hirsel, drove, trip, or pack
Squirrels	Dray
Swine	Sounder, drift, herd, or trip
Walruses	Pod or herd
Weasels	Pack, colony, gam, herd, pod, or school
Whales	School, gam, mob, pod, or herd
Wolves	Rout, route, or pack
Zebras	Herd

How fast does a humpback whale swim?

Humpback whales have a cruising speed of up to 9 mph (14.5 kph) but are capable of speeds up to 17 mph (27 kph).

What freshwater mammal is **venomous**?

The male duck-billed platypus (*Ornithorhynchus anatinus*) has venomous spurs located on its hind legs. When threatened, the animal will drive the spurs into the skin of a potential enemy, inflicting a painful sting. The venom this action releases is relatively mild and generally not harmful to humans.

What is the difference between **porpoises** and **dolphins**?

Marine dolphins (family Delphinidae) and porpoises (family Phocoenidae) together comprise about 40 species. The chief differences between dolphins and porpoises occur in the snout and teeth. True dolphins have a beaklike snout and cone-shaped teeth. True porpoises have a rounded snout and flat or spade-shaped teeth.

How do the **great whales** compare in weight and length?

Whale	Average weight	Greatest length
Sperm	35 tons (31,752 kg)	59 ft (18 m)
Blue	84 tons (76,204 kg)	98.4 ft (30 m)
Finback	50 tons (45,360 kg)	82 ft (25 m)
Humpback	33 tons (29,937 kg)	49.2 ft (15 m)
Right	50 tons (est.) (45,360 kg) (est.)	55.7 ft (17 m)
Sei	17 tons (15,422 kg)	49.2 ft (15 m)
Gray	20 tons (18,144 kg)	39.3 ft (12 m)
Bowhead	50 tons (45,360 kg)	59 ft (18 m)
Byrde's	17 tons (15,422 kg)	49.2 ft (15 m)
Minke	10 tons (9,072 kg)	29.5 ft (9 m)

What is the **fastest swimming whale**?

The orca or killer whale (*Orcinus orca*) is the fastest swimming whale. In fact, it is the fastest swimming marine mammal with speeds that reach 31 mph (50 kph).

What is the name of the **seal-like animal** in Florida?

The West Indian manatee (*Trichechus manatus*), in the winter, moves to more temperate parts of Florida, such as the warm headwaters of the Crystal and Homosassa rivers in central Florida or the tropical waters of southern Florida. When the air temperature rises to 50°F (10°C), it will wander back along the Gulf Coast and up the Atlantic Coast as far as Virginia. Long-range offshore migrations to the coast of Guyana and South America have been documented. In 1983, when the population of manatees in Florida was reduced to several thousand, the state gave it legal protection from being hunted or commercially exploited. However, many animals continue to be killed or injured by the encroachment of humans. Entrapment in locks and dams, collisions with barges and power boat propellers, and so on cause at least 30 percent of manatee deaths, which total 125 to 130 annually.

What is the only **four-horned animal** in the world?

The four-horned antelope (*Tetracerus quadricornis*) is a native of central India. The males have two short horns, usually 4 in (10 cm) in length, between their ears, and an even shorter pair, 1 to 2 in (2.5 to 5 cm) long, between the brow ridges over their eyes. Not all males have four horns, and in some the second pair eventually falls off. The females have no horns at all.

Is there a **cat** that lives in the **desert**?

The sand cat (*Felis margarita*) is the only member of the cat family tied directly to desert regions. Found in the deserts of North Africa, the Arabian Peninsula, Turkmenistan, Uzbekistan, and western Pakistan, the sand cat has adapted to extremely arid desert areas. The padding on the soles of its feet is well suited to the loose sandy soil, and it can live without drinking free-standing water. Having sandy or grayish-ochre dense fur, its body length is 17.5 to 22 in (45 to 57 cm). Mainly nocturnal (active at night), the cat feeds on rodents, hares, birds, and reptiles.

The Chinese desert cat (*Felis bieti*) does not live in the desert as its name implies, but inhabits the steppe country and mountains. Likewise, the Asiatic desert cat (*Felis silvestris ornata*) inhabits the open plains of India, Pakistan, Iran, and Asiatic Russia.

What is the only **American canine** that can **climb trees**?

The gray fox (*Urocyon cinereoargenteus*) is the only American canine that can climb trees.

Which **bear** lives in a **tropical rain forest**?

The Malayan sun bear (*Ursus malayanus*) is one of the rarest animals in the tropical forests of Sumatra, the Malay Peninsula, Borneo, Burma, Thailand, and southern China. The smallest bear species, with a length of 3.3 to 4.6 ft (1 to 1.4 m) and weighing 60 to 143 lb (27 to 65 kg), it has a strong, stocky body. Against its black, short fur it has a characteristic orange-yellow-colored crescent across its chest, which according to legend represents the rising sun. With powerful paws having long, curved claws to help it climb trees in the dense forests, it is an expert tree climber. The sun bear tears at tree bark to expose insects, larvae, and the nests of bees and termites. Fruit, coconut palms, and small rodents are also part of its diet. Sleeping and sunbathing during the day, it is active at night. Unusually shy and retiring, cautious and intelligent, the sun bear is declining in population as its native forests are being destroyed.

What is the **largest terrestrial mammal** in North America?

The bison (*Bison bison*) is the largest terrestrial mammal in North America. It weighs 3,100 lb (1406 kg) and is 6 ft (1.8 m) high.

Do **camels** store water in their humps?

The hump or humps do not store water, since they are fat reservoirs. The ability to go long periods without drinking water, up to ten months if there is plenty of green vegetation and dew to feed on, results from a number of physiological adaptations. One major factor is that camels can lose up to 40 percent of their body weight with no ill effects. A camel can also withstand a variation of its body temperature by as much as 14°F (8°C). A camel can drink 30 gallons (113.5 liters) of water in 10 minutes and up to 50 gallons (189 liters) over several hours. A one-humped camel is called a dromedary or Arabian camel; a Bactrian camel has two humps and lives in the wild on the Gobi Desert. Today, the Bactrian is confined to Asia, while most of the Arabian camels are on African soil.

How many **quills** does a **porcupine** have?

For its defensive weapon, the average North American porcupine has about 30,000 quills or specialized hairs, comparable in hardness and flexibility to slivers of celluloid and so sharply pointed they can penetrate any hide. The quills that do the most damage are the short ones that stud the porcupine's muscular tail. With a few lashes, the porcupine can send a rain of quills that have tiny scalelike barbs into the skin of its adversary. The quills work their way inward because of their barbs and the involuntary muscular action of the victim. Sometimes the quills can work themselves out, but other times the quills pierce vital organs, and the victim dies.

Slow-footed and stocky, porcupines spend much of their time in the trees, using their formidable incisors to strip off bark and foliage for their food, and supplement their diets with fruits and grasses. Porcupines have a ravenous appetite for salt; as herbivores (plant-eating animals), their diets have insufficient salt. So natural salt licks, animal bones left by carnivores (meat-eating animals), yellow pond lilies, and other items having a high salt content (including paints, plywood adhesives, and human clothing that bears traces of sweat) have a strong appeal to porcupines.

What is the difference between an **African elephant** and an **Indian elephant**?

The African elephant (*Loxodonta africana*) is the largest living land animal, weighing up to 8.25 tons (7,500 kg) and standing 10 to 13 ft (3 to 4 m) at the shoulder. The Indian elephant (*Elephas maximus*) weighs about 6 tons (5,500 kg) with a shoulder height of 10 ft (3 m). Other differences are:

African elephant	Indian elephant
Larger ears	Smaller ears
Gestation period of about 670 days	Gestation period of about 610 days
Ear tops turn backwards	Ear tops turn forwards
Concave back	Convex back
Three toenails on hind feet	Four toenails on hind feet
Larger tusks	Smaller tusks
Two fingerlike lips at tips of trunk	One lip at tip of trunk

How does a **mastodon** differ from a **mammoth**?

Although the words are sometimes used interchangeably, the mammoth and the mastodon were two different species. The mastodon seems to have appeared first, and a side branch may have led to the mammoth. The mastodon lived in Africa, Europe, Asia, and North and South America. It appears in the Oligocene era (25 to 38 million years ago) and survived until less than one million years ago. It stood a maximum of 10 ft (3 m) tall and was covered with dense, woolly hair. Its tusks were straight forward and nearly parallel to each other.

The mammoth evolved less than two million years ago and died out about 10,000 years ago. It lived in North America, Europe, and Asia. Like the mastodon, the mammoth was covered with dense, woolly hair, with a long, coarse layer of outer hair to protect it from the cold. It was somewhat larger than the mastodon, standing 9 to 15 ft (2.7 to 4.5 m). The mammoth's tusks tended to spiral outward, then up.

The gradual warming of the earth's climate and the change in environment were probably primary factors in the mammoth's extinction. But early man killed many mammoths as well, perhaps hastening the process.

What is the name of the **early Jurassic mammal** that is now extinct?

The fossil site of the mammal *Hadrocodium wui* was in Yunnan Province, China. This newly described mammal is at least 195 million years old. The estimated weight of the whole mammal is about .07 oz (2 g). Its tiny skull was smaller than a human thumb nail.

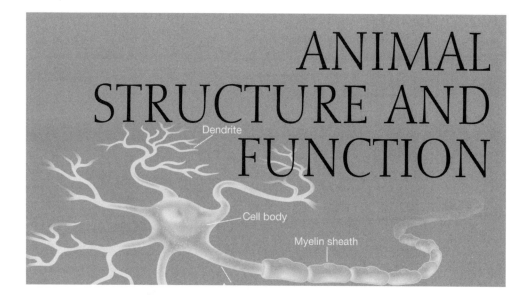

ANIMAL STRUCTURE AND FUNCTION

Dendrite

Cell body

Myelin sheath

INTRODUCTION AND HISTORICAL BACKGROUND

When was the term "**physiology**" first used?

The term "physiology" was first used by the Greeks as early as 600 B.C.E. to describe a philosophical inquiry into the nature of things. It was not until the sixteenth century that the term was used in reference to vital activities of healthy humans. During the nineteenth century its usage was expanded to include the study of all living organisms using chemical, physical, and anatomical experimental methods.

Who is considered the founder of **physiology**?

As an experimenter, Claude Bernard (1813–1878) enriched physiology by introducing numerous new concepts into the field. The most famous of these concepts is that of the French *milieu intérieur* or internal environment. The complex functions of the various organs are closely interrelated and are all directed to maintaining the constancy of internal conditions despite external changes. All cells exist in this aqueous (blood and lymph) internal environment, which bathes the cells and provides a medium for the simple exchange of nutrients and waste material.

When was the term "**homeostasis**" first used?

Walter Bradford Cannon (1871–1945), who elaborated on Claude Bernard's concept of the *milieu intérieur*, used the term "homeostasis" to describe the body's ability to maintain a relative constancy in its internal environment.

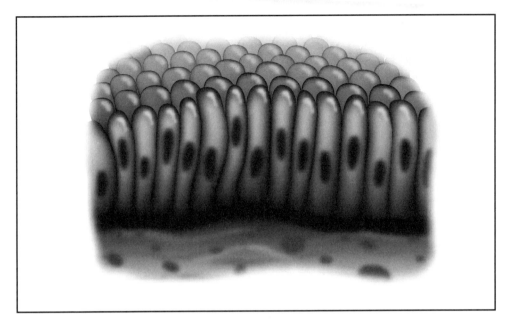

Columnar-shaped epithelial cells.

What was the **first professional organization** of physiologists?

The first organization of physiologists was the Physiological Society, founded in 1876 in England. In 1878 the *Journal of Physiology* began publication as the first journal dedicated to reporting results of research in physiology. The American counterpart, the American Physiological Society, was founded in 1887. The American Physiological Society's sponsored publication, the *American Journal of Physiology*, began in 1898.

What are the **four levels** of **structural organization** in animals?

Every animal has four levels of hierarchical organization: cell, tissue, organ, and organ system. Each level in the hierarchy is of increasing complexity, and all organ systems work together to form an organism.

TISSUE

What are the **four major types** of **tissue**?

A tissue (from the Latin *texere*, meaning "to weave") is a group of similar cells that perform a specific function. The four major types of tissue are epithelial, connective, muscle, and nerve.

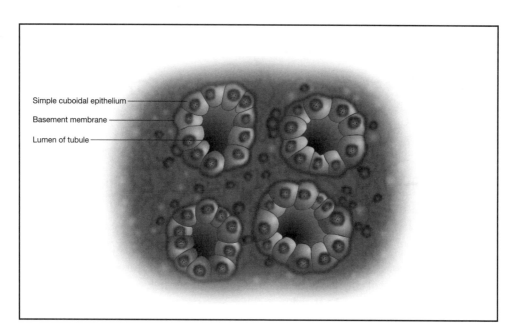

Simple cuboidal epithelium
Basement membrane
Lumen of tubule

Simple cuboidal epithelium. Epithelial tissue may have squamous-, cuboidal-, or columnar-shaped cells.

Where is **epithelial tissue found**?

Epithelial tissue, also called epithelium, (from the Greek *epi*, meaning "on," and *thele*, meaning "nipple") covers every surface, both external and internal, of the body. The outer layer of the skin, the epidermis, is one example of epithelial tissue. Other examples of epithelial tissue are the lining of the lungs, kidney tubules, and the inner surfaces of the digestive system, including the esophagus, stomach, and intestines. Epithelial tissue also includes the lining of parts of the respiratory system.

What are the **different shapes** and **functions** of the **epithelium**?

Epithelial tissue consists of densely packed cells. Epithelial tissues are either simple or stratified based on the number of cell layers. Simple epithelium has one layer of cells, while stratified epithelium has multiple layers. Epithelial tissue may have squamous-, cuboidal-, or columnar-shaped cells. Squamous cells are flat, square cells. Cuboidal cells form a tube. Columnar cells are stacked, forming a column taller than they are wide. There are two surfaces to epithelial tissue: one side is firmly attached to the underlying structure, while the other forms the lining. The epithelium forms a barrier, allowing the passage of certain substances while impeding the passage of other substances.

How often is the **epithelium replaced**?

Epithelial cells are constantly being replaced and regenerated during an animal's lifetime. The epidermis (outer layer of the skin) is renewed every two weeks, while the

279

epithelial lining of the stomach is replaced every two to three days. The liver, a gland consisting of epithelial tissue, easily regenerates after portions are removed surgically.

What is the **unique characteristic** of **connective tissue**?

The cells of connective tissue are spaced widely apart and are scattered through a nonliving extracellular material called a matrix. The matrix may be a liquid, jelly, or solid and varies in the different types of connective tissue.

What are the **major types** of **connective tissue** and their **function**?

The major types of connective tissue are: 1) loose connective tissue; 2) adipose tissue; 3) blood; 4) collagen, sometimes called fibrous or dense connective tissue; 5) cartilage; and 6) bone.

Loose connective tissue is a mass of widely scattered cells whose matrix is a loose weave of fibers. Many of the fibers are strong protein fibers called collagen. Loose connective tissue is found beneath the skin and between organs. It is a binding and packing material whose main purpose is to provide support to hold other tissues and organs in place.

Adipose tissue consists of adipose cells in loose connective tissue. Each adipose cell stores a large droplet of fat that swells when fat is stored and shrinks when fat is used to provide energy. Adipose tissue pads and insulates the animal body.

Blood is a loose connective tissue whose matrix is a liquid called plasma. Blood consists of red blood cells, erythrocytes, white blood cells, leukocytes, and thrombocytes or platelets, which are pieces of bone marrow cell. Plasma also contains water, salts, sugars, lipids, and amino acids. Blood is approximately 55 percent plasma and 45 percent formed elements. Blood transports substances from one part of the body to another and plays an important role in the immune system of the animal.

Collagen (from the Greek *kolla,* meaning "glue," and *genos,* meaning "descent") is a dense connective tissue, also known as fibrous connective tissue. It has a matrix of densely packed collagen fibers. There are two types of collagen: regular and irregular. The collagen fibers of regular dense connective tissue are lined up in parallel. Tendons, which bind muscle to bone, and ligaments, which join bones together, are examples of dense regular connective tissue. The strong covering of various organs, such as kidneys and muscle, is dense irregular connective tissue.

Cartilage (from the Latin *cartilago,* meaning "gristle") is a connective tissue with an abundant number of collagen fibers in a rubbery matrix. It is both strong and flexible. Cartilage provides support and cushioning. It is found between the discs of the vertebrae in the spine, surrounding the ends of joints such as knees, and in the nose and ears.

Bone is a rigid connective tissue that has a matrix of collagen fibers embedded in calcium salts. It is the hardest tissue in the body, although it is not brittle. Most of the skeletal system is comprised of bone, which provides support for muscle attachment and protects the internal organs.

How does **brown fat** differ from **white fat**?

Many mammals have both brown adipose tissue and white adipose tissue. They are both triglyceride lipids, but brown fat tissue has the ability to generate heat. Although called brown, it varies in color from dark red to tan, reflecting lipid content. It is most commonly found in newborn animals and in most species disappears by adulthood.

Which group of **animals** has exceptionally **well-developed brown fat**?

Mammalian hibernators have exceptionally well-developed brown fat. Some scientists even refer to it as "the hibernation gland." The supply of brown fat is used during the winter months when the animal hibernates and is built up during the spring and summer. Brown fat is important in the rewarming process that the animal undergoes after hibernation.

How many **types** of **muscle tissue** are found in **vertebrates**, and what is the **function** of **muscle tissue**?

Vertebrates have three types of muscle tissue: 1) smooth, 2) skeletal, and 3) cardiac muscle. Muscle tissue, consisting of bundles of long cells called muscle fibers, provides the capability of movement for the organism or for the movement of substances within the body of the organism.

What are the **differences** between **smooth, skeletal, and cardiac** muscle tissue?

Smooth muscle tissue is organized into sheets of long cells shaped like spindles. Each cell contains a single nucleus. Smooth muscle tissue lines the walls of the digestive tract (stomach and intestines), blood vessels, urinary bladder, and iris of the eye. Smooth muscle contraction is involuntary since it occurs without intervention of the animal.

Skeletal muscle tissue consists of numerous, very long muscle fibers that lie parallel to each other. Since the muscle fibers are formed by the fusion of several muscle cells, each long fiber has many nuclei. Muscle fibers have alternating light and dark bands, giving the appearance of a striped or striated fiber. Tendons attach skeletal muscles to the bone. When skeletal muscles contract, they cause the bone to move at the joint. Skeletal muscles are voluntary since the animal consciously contracts them. Skeletal muscles allow animals to move, lift, and utter sounds.

Cardiac muscle tissue is found in the hearts of vertebrates. It consists of striated muscle fibers like skeletal muscle tissue but is involuntary like smooth muscle tissue.

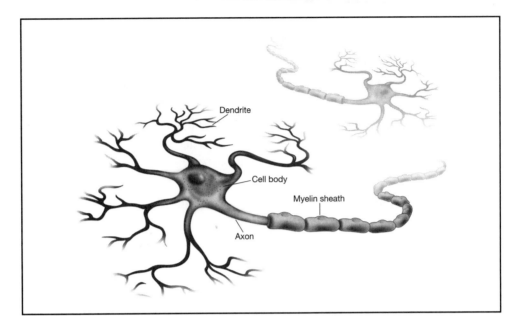

The structure of a neuron.

The cardiac muscle tissue consists of small, interconnected cells, each with a single nucleus. The ends of the cells form a tight latticework, allowing signals to spread from cell to cell and causing the heart to contract.

Does **exercise increase** the number of **muscle cells**?

Exercise does not increase the number of muscle cells. Adult animals have a fixed number of skeletal muscle cells. Exercise, however, does enlarge existing skeletal muscle cells.

What type of **cell** is found in **nerve tissue**?

Neurons are specialized cells that produce and conduct "impulses," or nerve signals. Neurons consist of a cell body containing a nucleus and two types of cytoplasmic extensions, dendrites and axons. Dendrites are thin, highly branched extensions that receive signals. Axons are tubular extensions that transmit nerve impulses away from the cell body, often to another neuron. Other cells in nerve tissue nourish the neurons, insulate the dendrites and axons, and promote quicker transmission of signals.

How many **different types** of **neurons** are found in nerve tissue?

There are three main types of neurons: sensory neurons, motor neurons, and interneurons (also called association neurons). Sensory neurons conduct impulses from sensory organs (eyes, ears, surface of the skin) into the central nervous system. Motor neurons conduct impulses from the central nervous system to muscles or glands. Interneurons are neither sensory neurons nor motor neurons. They permit elaborate processing of information to generate complex behaviors. Interneurons comprise the majority of neurons in the central nervous system.

What is the **function** of **nerve tissue**?

Nerve tissue serves as the communication system of an animal. Nerve tissue allows an animal to receive stimuli from its environment and to relay an appropriate response.

ORGANS AND ORGAN SYSTEMS

What is an **organ**?

An organ is a group of several different tissues working together as a unit to perform a specific function or functions. Each organ performs functions that none of the component tissues can perform alone. This cooperative interaction of different tissues is a basic feature of animals. The heart is an example of an organ. It consists of cardiac muscle wrapped in connective tissue. The heart chambers are lined with epithelium. Nerve tissue controls the rhythmic contractions of the cardiac muscles.

What is an **organ system**?

An organ system is a group of organs working together to perform a vital body function. Vertebrate animals have twelve major organ systems.

Organ System	Components	Functions
Cardiovascular and circulatory	Heart, blood, and blood vessels	Transports blood throughout the body, supplying nutrients and carrying oxygen to the lungs and wastes to kidneys
Digestive	Mouth, esophagus, stomach, intestines, liver, and pancreas	Ingests food and breaks it down into smaller chemical units
Endocrine	Pituitary, adrenal, thyroid, and other ductless glands	Coordinates and regulates the activities of the body
Excretory	Kidneys, bladder, and urethra	Removes wastes from the bloodstream
Immune	Lymphocytes, macrophages, and antibodies	Removes foreign substances
Integumentary	Skin, hair, nails, and sweat glands	Protects the body

Organ System	Components	Functions
Lymphatic	Lymph nodes, lymphatic capillaries, lymphatic vessels, spleen, and thymus	Captures fluid and returns it to the cardiovascular system
Muscular	Skeletal muscle, cardiac muscle, and smooth muscle	Allows body movements
Nervous	Nerves, sense organs, brain, and spinal cord	Receives external stimuli, processes information, and directs activities
Reproductive	Testes, ovaries, and related organs	Carries out reproduction
Respiratory	Lungs, trachea, and other air passageways	Exchanges gases—captures oxygen (O_2) and disposes of carbon dioxide (CO_2)
Skeletal	Bones, cartilage, and ligaments	Protects the body and provides support for locomotion and movement

DIGESTION

What are the steps of **food processing** for animals?

The first step is for animals to ingest food. The food is then broken down via the digestive process into molecules that the organism can absorb for energy. Once the food is digested, it is absorbed through the digestive tract to provide energy for the organism. The final step of food processing is elimination. During elimination, undigested material is passed out of the digestive tract.

How are **animals classified** based on what they eat?

Animals are classified based on whether they eat plants, other animals, or a combination of both. Animals that eat only plant matter are called herbivores (from the Latin *herba*, meaning "green crop," and *vorus*, meaning "devouring"). Examples of herbivores are cattle, deer, and many aquatic species that eat algae. Animals that eat other animals are called carnivores (from the Latin *carne*, meaning "flesh," and *vorus*, meaning "devouring"). Lions, sharks, snakes, and hawks are examples of carnivores. Animals that eat both plants and other animals are called omnivores (from the Latin *omnis*, meaning "all" and *vorus*, meaning "devouring"). Humans, crows, and raccoons are examples of omnivores.

How does the **dentition** of animals **reflect** their **diet**?

Herbivores have sharp incisors to bite off blades of grass and other plant matter. They also have a system of flat premolars and molars for grinding and crushing grasses and plant matter. Carnivores have pointed incisors and enlarged canine teeth to tear off pieces of meat. Their premolars and molars are jagged to aid in chewing flesh. Omnivores have nonspecialized teeth to accommodate a diet of both plant material and animals.

How do **continuous feeders** differ from **discontinuous feeders**?

Continuous feeders, also known as filter feeders, are aquatic animals that constantly feed by having water filled with food particles (e.g., small plankton or fishes) entering through the mouth. Continuous feeders do not need a storage area, such as a stomach, for food. Discontinuous feeders must hunt for food on a regular basis. They need a storage area to house food until it is digested.

What are the **main organs** of the **digestive system** and their function?

The digestive system includes the mouth, alimentary canal or gastrovascular cavity, esophagus, stomach, small intestine, large intestine, and anus. The mouth is the opening through which food is ingested. In animals with a second opening for elimination, the digestive system contains an alimentary canal, a tube allowing for the passage of food from the mouth to the anus. In contrast, animals with only one opening have a gastrovascular cavity that serves as the site of digestive activities. The esophagus is another channel through which food passes on the way to the stomach. The stomach (or "crop" in certain species, such as birds) stores food and is the primary site of chemical digestion. Following digestion in the stomach where food is broken down with acid and enzymes, the food passes into the small intestine. Nutrients are absorbed via the intestine. Much shorter than the small intestine, though greater in diameter, is the large intestine, also called the colon. Here, the solid material remaining after digestion is compacted and then eliminated via the anus.

Why do **cows** have **four stomachs**?

Cows have four stomachs in order to process their low-quality diet of grass. The four sections are: 1) the rumen, 2) the reticulum, 3) the omasum, and 4) the abomasum.

Ruminants, animals that chew their cud, include cattle, bison, buffalo, goats, antelopes, sheep, deer, giraffe, and okapis.

Cows are a type of mammal known as a ruminant. Ruminants eat rapidly and do not chew much of their food completely before they swallow it. The liquid part of their food enters the reticulum first, while the solid part of their food enters the rumen, where it softens. Bacteria in the rumen initially break the food material down as a first step in digestion. Ruminants later regurgitate the partially liquefied plant parts into the mouth, where they continue to munch it in a process known as "chewing their cud." Cows chew their cud about six to eight times per day, spending a total of five to seven hours in rumination. The chewed cud goes directly into the other chambers of the stomach, where various microorganisms assist in further digestion. Herbivores have a longer small intestine to allow maximum time for the absorption of nutrients.

How do **rodents, rabbits, and hares** digest **cellulose**?

Unlike cows, which have a rumen to digest cellulose, rodents, rabbits, and hares have a cecum, a large pouch to digest cellulose with the assistance of microorganisms. The cecum is located at the junction between the small and large intestines. It is impossible for these animals to regurgitate the contents of their stomachs (like ruminants), because the cecum is located beyond the stomach. Instead, these animals pass their food through their digestive tract a second time by ingesting their feces. When feces pass through the digestive tract, it is possible for these animals to absorb the nutrients produced by the microorganisms in the cecum.

RESPIRATION

What is **respiration**?

Respiration is the exchange of gases (oxygen and carbon dioxide) between an animal and its environment. There are three phases to the process of respiration (gas exchange): 1) breathing, when an animal inhales oxygen and exhales carbon dioxide; 2) transport of gases via the blood (circulatory system) to the body's tissues; and 3) at the cellular level, when the cells take in oxygen from the blood and in return add carbon dioxide to the blood.

Where does **respiration take place**?

Different types of animals have different respiratory organs for gas exchange. Four types of respiratory organs are: 1) skin, 2) gills, 3) tracheae, and 4) lungs. Many invertebrates and some vertebrate animals, including amphibians, breathe through their skin. Many of the animals that breathe through their skin (a process known as cutaneous respiration) are small, long, and flattened—for example, earthworms and flatworms. All animals that rely on their skin for respiration live in moist, damp places in order to keep their body surfaces moist. Capillaries, small blood vessels, bring blood rich in carbon dioxide and deficient in oxygen to the skin's surface, where gaseous exchange takes place via diffusion.

Gills may be external extensions of the body surface such as those found in aquatic insect larvae and some aquatic amphibians. Diffusion of oxygen occurs across the gill surface into capillaries, while carbon dioxide diffuses out of the capillaries into the environment. Fish and some other marine animals have internal gills. Water enters the animals through the mouth, then flows over the gills in a steady stream and out through gill slits. Although some animals with gills spend part of the time on land, they all must spend some time in moist, wet environments for the gills to function.

Lungs are internal structures found in most terrestrial animals where gas exchange occurs. The lungs are lined with moist epithelium to avoid their becoming desiccated. Some animals, including lungfish, amphibians, reptiles, birds, and mammals, have special muscles to help move air in and out of the lungs. Some animals have lungs connected to the outside surface with special openings and do not require special muscles to move air in and out of the lungs.

Insects have a system of internal tubes, called tracheae, that lead from the outside world to internal regions of the body via spiracles. Gaseous exchange takes place in the tracheae. Some insects rely on muscles to pump the air in and out of the tracheae, while in others the process is a passive exchange of gases.

Some spiders have book lungs in addition to trachae. Book lungs are hollow, leaflike structures through which the blood flows. These lungs hang in an open space that is connected to a tube. The other side of the tube is in open contact with the air.

What is the **respiration rate** of various **animals**?

The respiratory rate for various animals is:

Animal	Breaths per minute
Diamondback snake	4
Horse	10
Human	12
Dog	18
Pigeon	25–30
Cow	30
Giraffe	32
Shark	40

Animal	Breaths per minute
Trout	77
Mouse	163

How does the **breath-holding capability** of a **human** compare with **other mammals**?

Mammal	Average time in minutes
Human	1
Polar bear	1.5
Pearl diver (human)	2.5
Sea otter	5
Platypus	10
Muskrat	12
Hippopotamus	15
Sea cow	16
Beaver	20
Porpoise	15
Seal	15–28
Greenland whale	60
Sperm whale	90
Bottlenose whale	120

How do **air-breathing mammals** such as whales and seals **dive underwater** for extended periods of time?

Seals and whales are able to dive underwater for extended periods of time because they are able to store oxygen. While humans store 36 percent of their oxygen in their lungs and 51 percent in the blood, seals store only approximately 5 percent of their oxygen in their lungs and 70 percent in their blood. They also store more oxygen in the muscle tissue—25 percent compared with only 13 percent in human muscle tissue. While underwater, these mammals' heart rates and oxygen consumption rates decrease, allowing some species to remain underwater for up to twenty minutes at a time.

How long do **marine mammals remain underwater**?

Below are the maximum depths and the longest durations of time underwater by various aquatic mammals:

Mammal	Maximum depth	Maximum time underwater
Weddell seal	1,968 feet/600 meters	70 minutes

How many breaths does a human take each year?

Each year an adult human may take between four million and ten million breaths. The volume of air in each breath ranges from 500 milliliters for breaths at rest and 3500 to 4800 milliliters for each breath during strenuous exercise.

Mammal	Maximum depth	Maximum time underwater
Porpoise	984 feet/300 meters	6 minutes
Bottlenose whale	1,476 feet/450 meters	120 minutes
Fin whale	1,148 feet/350 meters	20 minutes
Sperm whale	>6,562 feet/>2,000 meters	75–90 minutes

CIRCULATION

What are the **functions** of the **circulatory system**?

The primary function of the circulatory system is to transport oxygen and nutrients to the all the cells of an organism. The circulatory system also transports wastes from the cells to waste-disposal organs such as the lungs for carbon dioxide and the kidneys for other metabolic wastes. In addition, the circulatory system plays a vital role in maintaining homeostasis.

What are the **differences** between an **open** and a **closed circulatory system**?

In an open circulatory system, found in many invertebrates (e.g., spiders, crayfish, and grasshoppers), the blood is not always contained within the blood vessels. Periodically, the blood leaves the blood vessels to bathe the tissues with blood and then returns to the heart. There is no interstitial body fluid separate from the blood. A closed circulatory system, also called a cardiovascular system, is found in all vertebrate animals and many invertebrates; in a closed system the blood never leaves the blood vessels.

What are the **components** of the **circulatory system**?

The components of the circulatory system are vessels, heart, and blood. The three types of vessels in a closed circulatory system are arteries, capillaries, and veins. Arteries transport blood away from the heart to the various organs in the body. Veins return blood to the heart after it circulates through the body. Capillaries form an elaborate network of tiny vessels that convey blood between arteries and veins.

Do all **animals** have **blood**?

Some invertebrates, such as flatworms and cnidarians, lack a circulatory system that contains blood. These animals possess a clear, watery tissue that contains some phagocytic cells, a little protein, and a mixture of salts similar to seawater. Invertebrates with an open circulatory system have a fluid that is more complex. It is usually referred to as hemolymph (from the Greek term *haimo*, meaning "blood," and the Latin term *lympha*, meaning "water"). Invertebrates with a closed circulatory system have blood that is contained within blood vessels. Vertebrates have blood composed of plasma and formed elements.

Why do **small animals** not have a **circulatory system**?

Smaller animals such as hydras do not have a separate circulatory system, since their cells are able to efficiently exchange materials (nutrients, gases, and wastes) through diffusion. The cells of these animals are close to the surface and thus can exchange nutrients effectively.

Who **first demonstrated** that **blood circulates**?

William Harvey (1578–1657) was the first person to demonstrate that blood circulates in the bodies of humans and other animals. Harvey's hypothesis was that the heart is a pump for the circulatory system, with blood flowing in a closed circuit. Harvey conducted his research on live organisms as well as dissection of dead organisms to demonstrate that when the heart pumps, blood flows into the aorta. He observed that when an artery is slit, all the blood in the system empties. Finally, Harvey demonstrated that the valves in the veins serve to return blood to the heart.

What are **blood groups**?

There are more than twenty genetically determined blood group systems among humans known today, but the AB0 and Rh systems are the most important ones used to type blood for human blood transfusions. Different species of animals have varying numbers of blood groups.

Species	Number of blood groups
Human	20+

Species	Number of blood groups
Pig	16
Cow	12
Chicken	11
Horse	9
Sheep	7
Dog	7
Rhesus monkey	6
Mink	5
Rabbit	5
Mouse	4
Rat	4
Cat	2

Do all animals have **red blood**?

The color of blood is related to the compounds that transport oxygen. Hemoglobin, containing iron, is red and is found in all vertebrates and a few invertebrates. Annelids (segmented worms) have either a green pigment, chlorocruorin, or a red pigment, hemerythrin. Some crustaceans (arthropods having divided bodies and generally having gills) have a blue pigment, hemocyanin, in their blood.

How does a **human's heartbeat** compare with those of **other mammals**?

Mammal	Resting Heart Rate (beats per minute)
Human	75
Horse	48
Cow	45–60

Mammal	Resting Heart Rate (beats per minute)
Dog	90–100
Cat	110–140
Rat	360
Mouse	498

EXCRETORY SYSTEM

What are the **functions** of the **excretory system**?

The excretory system is responsible for removing waste products. It also plays a vital role in regulating the water and salt balance in the organism.

How do the **excretory organs** vary in **different species**?

Many animals, such as sponges, jellyfish, tapeworms, and other small organisms, do not have distinct excretory organs. Rather, they rid their bodies of waste through diffusion. Larger, more complex animals require specialized, often tubular, organs to rid their bodies of waste. For example, flatworms such as planarians have tubules that collect wastes and expel them to the outside via pores. Segmented worms such as earthworms have nephridia (tubules with a ciliated opening) in each segment. Fluid from the body cavity is propelled through the nephridia. Wastes are expelled through a pore to the outside while certain substances are reabsorbed. Insects have a unique excretory system that consists of Malpighian tubules. Waste products enter the Malpighian tubes from the body cavity. Water and other useful substances are reabsorbed, while uric acid passes out of the body. Vertebrate animals have kidneys to dispose of metabolic wastes.

Which **type of nitrogenous** waste do **various species** of animals excrete?

Since it is highly toxic, excretion of pure ammonia is possible only for aquatic animals, because ammonia is very soluble in water. Urea and uric acid are excreted by terrestrial animals. Urea is approximately 100,000 times less toxic than ammonia, so it may be stored in the body and eliminated with relatively little water loss. Uric acid requires very little water for disposal and is often excreted as a paste or dry powder. An example is guano, the solid white droppings of seabirds and bats.

Type of Nitrogenous Waste	Animals	Animal Habitat
Ammonia	Aquatic invertebrates, Bony fishes, amphibian larvae	Water
Urea	Adult amphibians, mammals, sharks	Land; water (sharks)
Uric acid	Insects, birds, reptiles	Land

How are **nitrogenous wastes metabolized**?

Ammonia, urea, and uric acid are nitrogenous waste products. They are the result of the breakdown of various molecules including nucleic acids and amino acids. Some amino acids are used for new protein synthesis, while others are oxidized to generate energy or converted to fats or carbohydrates that can be stored. Once broken down, the amino group -NH_2 (containing one nitrogen and two hydrogens) must be removed or the animal will eventually be poisoned. Ammonia is the most toxic of the nitrogenous wastes. It is formed by the addition of a third hydrogen atom to the -NH_2 group. Both urea and uric acid are less toxic forms of nitrogenous waste but have greater energy requirements for their production.

Do **fish drink water**?

Marine bony fishes such as tuna, flounder, and halibut drink seawater almost constantly to replace water lost by osmosis and through their gills. It is estimated that they drink an amount equal to 1 percent of their body weight each hour, an amount comparable to a human drinking 1.5 pints or nearly three cups (700 milliliters) of water every hour around the clock. The gills eliminate most of the excess salts obtained by drinking large quantities of seawater. The fishes excrete small quantities of urine that is isotonic to their body fluids. By contrast, cartilaginous fishes (e.g., sharks and rays) do not need to drink water to maintain the balance of water (osmotic balance) in their bodies. They reabsorb the waste product urea, creating and maintaining a blood urea concentration that is 100 times higher than that of mammals. Their kidneys and gills thus do not have to remove large quantities of salts from their bodies.

Freshwater fishes never drink water separate from ingesting food. These fishes are prone to gain water since their body fluids are hypotonic (containing a lesser concentration of salts) to the surrounding water. They imbibe water through their gills to maintain the correct balance of salts in their bodies and excrete large quantities of diluted urine daily. It is estimated that freshwater fishes eliminate a quantity of urine equal to one-third of their body weight each day.

What are the **two types** of **euryhaline fishes**?

There are two types of euryhaline (from the Greek terms *eurys*, meaning "broad," and *hals*, meaning "salt") fishes. One type includes flounders, sculpins, and killifish that live in estuaries or intertidal areas where the salinity of the water fluctuates throughout the day. The second type includes salmon, shad, and eels that spend part of their life cycles in freshwater and the balance in seawater.

Are any other **animals able to drink seawater**?

Birds and reptiles that live near the sea are also able to drink seawater. These animals have nasal salt glands near their eyes through which they excrete the excess quantities of salt solution.

Salmon are euryhaline, spending part of their lives in freshwater and part of their lives in seawater. While living in freshwater, they behave like other freshwater fishes and do not drink water. However, when they move to the ocean, their gills and kidneys adapt to the different environment, and they compensate for water loss and salt balance by drinking water.

What is the **composition** of **urine**?

Urine is composed mostly of water containing organic wastes as well as some salts. The composition of urine can vary according to diet, time of day, and diseases. In one measure, the make-up of urine is 95 percent water and 5 percent solids. In terms of organic wastes (per 1,500 ml), urine contains 30 g of urea, 1–2 g each of creatinine and ammonia, and 1 g of uric acid. In terms of salts or ions, 25 g per 1,500 ml of urine contain the positive ions sodium, potassium, magnesium, and calcium, as well as the negative ions chlorides, sulfates, phosphates.

How does **kidney dialysis** remove waste products from the body?

Damaged or diseased kidneys are not capable of removing toxic waste substances from the body. Kidney dialysis removes the nitrogenous waste and regulates the pH of the blood when the kidneys do not work. Blood is pumped from an artery through a series of tubes made of a permeable membrane and a dialyzing solution. Urea and excess salts diffuse out of the blood as it circulates through the dialyzing machine and are then discarded. Necessary ions diffuse from the dialyzing solution back into the blood. The cleansed blood is then returned to the body.

ENDOCRINE SYSTEM

What is the **function** of the **endocrine system**?

The endocrine system is the main chemical-regulating system of an organism. Hormones, chemicals made and secreted by endocrine glands or neurosecretory cells, are the main messengers of the endocrine system. Hormones are transported in the blood to all parts of the body and interact with target cells (cells that contain hormone receptors), and they regulate metabolic rate, growth, maturation, and reproduction.

What are **neurosecretory cells**?

Neurosecretory cells are specialized nerve cells that produce and secrete hormones. Well-known examples of neurosecretory cells are oxytocin- and vasopressin-secreting neurons in the hypothalamus and cells in the adrenal medulla. These cells are found in vertebrates and invertebrates.

Who **discovered** the **first known hormone**?

The British physiologists William Bayliss (1860–1924) and Ernest Starling (1866–1927) discovered secretin in 1902. They used the term "hormone" (from the Greek word *horman*, meaning "to set in motion") to describe the chemical substance they had discovered that stimulated an organ at a distance from the chemical's site of origin. Their famous experiment using anesthetized dogs demonstrated that dilute hydrochloric acid, mixed with partially digested food, activated a chemical substance in the duodenum. This activated substance (secretin) was released into the blood-stream and came in contact with cells of the pancreas. In the pancreas it stimulated secretion of digestive juice into the intestine through the pancreatic duct.

What are some **vertebrate endocrine glands** and their **hormones**?

There are ten major endocrine glands in vertebrates.

Endocrine gland; hormone	Target tissue	Principal function
Posterior pituitary		
Antidiuretic hormone (ADH)	Kidneys	Stimulates water reabsorption by kidneys
Oxytocin	Uterus, mammary glands	Stimulates uterine contractions and milk ejection
Anterior pituitary		
Growth hormone (GH)	General	Stimulates growth, especially cell division and bone growth
Adrenocorticotropic hormone (ACTH)	Adrenal cortex	Stimulates adrenal cortex
Thyroid-stimulating hormone (TSH)	Thyroid gland	Stimulates thyroid
Luteinizing hormone (LH)	Gonads	Stimulates ovaries and testes
Follicle-stimulating hormone (FSH)	Gonads	Controls egg and sperm production
Prolactin (PRL)	Mammary glands	Stimulates milk production
Melanocyte-stimulating hormone (MSH)	Skin	Regulates skin color in reptiles and amphibians; unknown function in humans

Endocrine gland hormone	Target tissue	Principal function
Thyroid		
Calcitonin	Bone	Lowers blood calcium level
Parathyroid		
Parathyroid hormone (PTH)	Bone, kidneys, digestive tract	Raises blood calcium level
Adrenal medulla		
Epinephrine (adrenaline) and norepinephrine (noradrenaline)	Skeletal muscle, cardiac muscle, blood vessels	Initiates stress responses; raises heart rate, blood pressure, metabolic rates; constricts certain blood vessels
Adrenal cortex		
Aldosterone	Kidney tubules	Stimulates kidneys to reabsorb sodium and excrete potassium
Cortisol	General	Increases blood glucose
Pancreas		
Insulin	Liver	Lowers blood glucose level; stimulates formation and storage of glycogen
Glucagon	Liver, adipose tissue	Raises blood glucose level
Ovary		
Estrogens	General; female reproductive structures	Stimulates development of secondary sex characteristics in females and uterine lining
Progesterone	Uterus, breasts	Promotes growth of uterine lining; stimulates breast development
Testes		
Androgens (testosterone)	General; male reproductive structures	Stimulates development of male sex organs and spermatogenesis
Pineal gland		
Melatonin	Gonads, pigment cells	Involved in daily and seasonal rhythmic activities (circadian cycles); influences pigmentation in some species

What are the "**fight-or-flight**" hormones?

Epinephrine and norephinephrine are released by the adrenal glands in times of stress. The familiar feelings of a pounding, racing heart, increased respiration, elevated blood pressure, and goose bumps on the skin are responses to stressful circumstances.

How do **steroid hormones** differ from **nonsteroid hormones**?

Steroid hormones such as estrogen and testosterone enter target cells and directly interact with the DNA in the nucleus. Nonsteroid hormones such as adrenaline generally do not enter the target cell but instead bind to a receptor protein found on external cell membranes. This then causes a sequence of metabolic effects.

How are **anabolic steroids** harmful to those who use them?

Anabolic (protein-building) steroids are drugs that mimic the effects of testosterone and other male sex hormones. They can build muscle tissue, strengthen bone, and speed muscle recovery following exercise or injury. They are sometimes prescribed to treat some types of anemia as well as osteoporosis in postmenopausal women. Anabolic steroids have become a lightning rod of controversy in competitive sports. The drugs are banned from most organized competitions because of the dangers they pose to health and to prevent athletes from gaining an unfair advantage.

Adverse effects of anabolic steroids include hypertension, acne, edema, and damage to liver, heart, and adrenal glands. Psychiatric symptoms can include hallucinations, paranoid delusions, and manic episodes. In men anabolic steroids can cause infertility, impotence, and premature balding. Women can develop masculine characteristics such as excessive hair growth, male-pattern balding, disruption of menstruation, and deepening of the voice. Children and adolescents can develop problems in growing bones, leading to short stature.

What is a **goiter**?

A goiter is an enlargement of the thyroid gland caused by hypothyroidism (too little thyroxin). An insufficient dietary intake of iodine is a common cause of goiter.

What is the difference between **Type I and Type II diabetes**?

Diabetes mellitus is a hormonal disease that occurs when the body's cells are unable to absorb glucose form the blood. Type I is insulin-dependent diabetes mellitus (IDDM), and Type II is non-insulin-dependent diabetes mellitus (NIDDM). In Type I diabetes there is an absolute deficiency of insulin. In Type II diabetes insulin secretion may be normal, but the target cells for insulin are less responsive than normal.

NERVOUS SYSTEM

What is the **nervous system**?

The nervous system is an intricately organized, interconnected system of nerve cells that relays messages to and from the brain and spinal cord of an organism in verte-

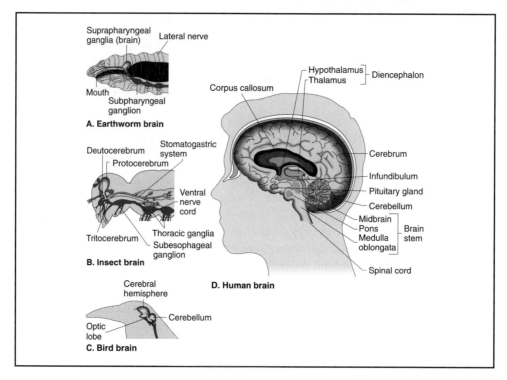

A. Earthworm brain
- Suprapharyngeal ganglia (brain)
- Lateral nerve
- Mouth
- Subpharyngeal ganglion

B. Insect brain
- Deutocerebrum
- Stomatogastric system
- Protocerebrum
- Ventral nerve cord
- Tritocerebrum
- Thoracic ganglia
- Subesophageal ganglion

C. Bird brain
- Cerebral hemisphere
- Cerebellum
- Optic lobe

D. Human brain
- Corpus callosum
- Hypothalamus
- Thalamus
- Diencephalon
- Cerebrum
- Infundibulum
- Pituitary gland
- Cerebellum
- Midbrain
- Pons
- Medulla oblongata
- Brain stem
- Spinal cord

A comparison of the brains of an earthworm, an insect, a bird, and a human.

brates. It receives sensory input, processes the input, and then sends messages to the tissues and organs for an appropriate response. In vertebrates there are two parts to the nervous system: 1) the central nervous system, consisting of the brain and spinal cord; and 2) the peripheral system, consisting of nerves that carry signals to and from the central nervous system.

How does the **nervous system** of **invertebrates** differ from that of **vertebrates**?

The least complex nervous system is the nerve net of cnidarians such as hydras. The nerve net is a network of neurons located throughout the radially symmetric body. The neurons are in contact with one another and with muscle fibers within epidermal cells. These animals lack a head and brain. Invertebrates that display bilateral symmetry such as planarians, annelids, and arthropods have a brain (a concentration of neurons at the anterior or head end) and one or more nerve cords and the presence of a central nervous system. Vertebrates have a central nervous system and a peripheral nervous system.

What is **myelin**?

Myelin forms an insulating wrapping around large nerve axons. In the peripheral nervous system myelin is formed by Schwann cells (a type of supporting cell) that wrap repeatedly around the axon. In the central nervous system myelin is formed by repeated wrappings of processes of oligodendrocytes (a different type of supporting cell). The process of each cell forms part of the myelin sheath. The space between the myelin from individual Schwann cells or oligodendrocyte processes is a bare region of the axon called the node of Ranvier. Nerve conduction is faster in myelinated fibers because it jumps from one node of Ranvier to the next. For this reason it is called saltatory (jumping) conduction.

What are **demyelinating diseases**?

Demyelinating diseases involve damage to the myelin sheath of neurons in either the peripheral or central nervous system. Multiple sclerosis (MS) is a chronic, potentially debilitating disease that affects the myelin sheath of the central nervous system. The illness is probably an autoimmune disease. In MS the body directs antibodies and white blood cells against proteins in the myelin sheath surrounding nerves in the brain and spinal cord. This causes inflammation and injury to the myelin sheath. Demyelination is the term used for a loss of myelin, a substance in the white matter that insulates nerve endings. Myelin helps the nerves receive and interpret messages from the brain at maximum speed. When nerve endings lose this substance, they cannot function properly, leading to patches of scarring, or "sclerosis." The result may be multiple areas of sclerosis. The damage slows or blocks muscle coordination, visual sensation, and other functions that rely on nerve signals.

In the autoimmune disorder known as Guillain-Barrè syndrome, the body's immune system attacks part of the peripheral nervous system. The immune system starts to destroy the myelin sheath that surrounds the axons of many peripheral nerves, or even the axons themselves. The myelin sheath surrounding the axons speeds up the transmission of nerve signals and allows the transmission of signals over long distances. In diseases such as Guillain-Barrè in which the peripheral nerves' myelin sheaths are injured or degraded, the nerves cannot transmit signals efficiently. Consequently, muscles begin to lose their ability to respond to the brain's commands, commands that must be carried through the nerve network. The brain also receives fewer sensory signals from the rest of the body, resulting in an inability to feel textures, heat, pain, and other sensations. Alternately, the brain may receive inappropriate signals that result in tingling, "crawling-skin," or painful sensations. Because the signals to and from the arms and legs must travel the longest distances, these extremities are most vulnerable to interruption. The first symptoms of this disorder include varying degrees of weakness or tingling sensations in the legs. In many instances the weakness and abnormal sensations spread to the arms and upper body. In severe cases the patient may be almost totally paralyzed since the muscles cannot be used at all. In these cases the disorder is life threatening—potentially interfering with breathing and, at times, with blood pressure or heart rate—and is considered a medical emergency. Such a patient is often put on a respirator to assist with breathing and is watched closely for problems such as an abnormal heart beat, infections, blood clots, and high or low blood pressure. Most patients, however, recover from even the most severe cases of Guillain-Barrè syndrome, although some continue to have a certain degree of weakness.

How is the **peripheral nervous system** organized in vertebrates?

There are two divisions to the peripheral nervous system: the sensory division and the motor division. The sensory division has two sets of neurons. One set (from the eyes, ears, and other external sense organs) brings in information about the outside environment, while the other set supplies the central nervous system with information about the body itself, such as the acidity of the blood. The motor division includes the somatic nervous system and the autonomic nervous system. The somatic nervous system carries signals to skeletal muscles and skin, mostly in response to external stimuli. It controls voluntary actions. The neurons of the autonomic nervous system are involuntary. This latter system is further divided into the sympathetic and parasympathetic divisions. The sympathetic division prepares the body for intense activities. It is responsible for the "fight or flight" response. The parasympathetic division, or "housekeeper system," is involved in all responses associated with a relaxed state such as digestion.

How is the **vertebrate brain organized**?

The vertebrate brain is divided into three regions: the hindbrain, the midbrain, and the forebrain. The size of each region of the brain varies from species to species. The hindbrain may be considered an extension of the spinal cord. Hence, it is often described as the most primitive portion of the brain. The primary function of the hindbrain is to coordinate motor reflexes. The midbrain is responsible for processing visual information. The forebrain is the center for processing sensory information in fish, amphibians, reptiles, birds, and mammals.

What is a **reflex**?

A reflex is an involuntary response formulated in the spinal cord to a specific stimulus.

Who proposed that the **left side of the brain** has different functions than the **right side of the brain**?

Roger Sperry (1913–1994) conducted the pioneering research on the different functions of the left side and right side of the brain. The left side of the brain controls lan-

When was Parkinson's disease first described?

Parkinson's disease was first formally described by James Parkinson (1755–1824), a London physician, in "An Essay on the Shaking Palsy," published in 1817.

guage, logic, and mathematical abilities. In contrast, the right side of the brain is associated with imagination, spatial perception, artistic and musical abilities, and emotions. Sperry received the Nobel Prize in Physiology or Medicine in 1981 for his work.

What are some **diseases** that affect the **nervous system**?

Epilepsy, multiple sclerosis, and Parkinson's disease are all diseases of the nervous system. Epilepsy is a nervous system disorder in which clusters of neurons in the brain sometimes signal abnormally. In epilepsy the normal pattern of neuronal activity becomes disturbed, causing strange sensations, emotions, and behavior or sometimes convulsions, muscle spasms, and loss of consciousness. Epilepsy is a disorder with many possible causes. Anything that disturbs the normal pattern of neuron activity—from illness to brain damage to abnormal brain development—can lead to seizures. Epilepsy may develop because of an abnormality in brain wiring, an imbalance of nerve signaling chemicals called neurotransmitters, or some combination of these factors.

Parkinson's disease is a progressive neurological disorder that results from degeneration of neurons in a region of the brain that controls movement. This degeneration creates a shortage of the brain signaling chemical (neurotransmitter) known as dopamine, causing the movement impairments that characterize the disease.

What are two of the **most common forms** of **dementia**?

The term "dementia" describes a group of symptoms that are caused by changes in brain function. The two most common forms of dementia in older people are Alzheimer's disease and multi-infarct dementia (sometimes called vascular dementia). These types of dementia are irreversible, which means they cannot be cured. In Alzheimer's disease nerve-cell changes in certain parts of the brain result in the death of a large number of cells. The symptoms of Alzheimer's disease range from mild forgetfulness to serious impairments in thinking, judgment, and the ability to perform daily activities.

In multi-infarct dementia a series of small strokes or changes in the brain's blood supply may result in the death of brain tissue. The location in the brain where the small strokes occur determines the seriousness of the problem and the symptoms that arise. Symptoms that begin suddenly may be a sign of this kind of dementia. People with multi-infarct dementia are likely to show signs of improvement or remain stable for long periods of time, then quickly develop new symptoms if more strokes occur. In many people with multi-infarct dementia, high blood pressure is to blame.

How is **sensory information transmitted** to the central nervous system?

Sensory information is transmitted to the central nervous system through a process that includes stimulation, transduction, and transmission. A physical stimulus (e.g., light or sound pressure) is converted into nerve cell electrical activity in a process called transduction. The electrical activity is then transmitted as action potentials to the central nervous system.

What are the **main types of receptors**?

Receptor cells are cells that receive stimuli. Each type of receptor responds to a particular stimulus. The five main types of receptors are pain receptors, thermoreceptors, mechanoreceptors, chemoreceptors, and electromagnetic receptors.

Pain receptors are probably found in all animals. However, it is difficult to understand nonhuman perception of pain. Pain often indicates danger, and the animal or individual retreats to safety.

Thermoreceptors in the skin are sensitive to changes in temperature. Thermoreceptors in the brain monitor the temperature of the blood to maintain proper body temperature.

Mechanoreceptors are sensitive to touch and pressure, sound waves and gravity. The sense of hearing relies on mechanoreceptors.

Chemoreceptors are responsible for taste and smell.

Electromagnetic receptors are sensitive to energy of various wavelengths including electricity, magnetism, and light. The most common types of electromagnetic receptors are photoreceptors that detect light and control vision.

How do animals and people **identify smells**?

The sense of smell allows animals and humans as well as other organisms to identify food, mates, and predators. This sense also provides sensory pleasure (e.g., flowers) and warnings of danger (e.g., chemical dangers). There are specialized receptor cells in the nose that have proteins that bind chemical odorants and cause the receptor cells to send electrical signals to the olfactory bulb of the brain. Cells in the olfactory bulb relay this information to olfactory areas of the forebrain to generate perception of smells.

Which **insect** has the **best sense of smell**?

Giant male silk moths (*Bombyx mori*) may have the best sense of smell in the world. Their antennae are covered with about 65,000 tiny bristles. Most of the bristles are chemoreceptors. The moths can smell a female's perfume nearly 7 mi (11 km) away.

How are **taste and smell** related?

By convention, air-breathing vertebrates, including humans, associate taste with materials that come in direct contact with the animal, usually through the mouth. By contrast, smell is associated with substances that reach the animal from a distance, usually through the nose. However, the distinction between the two becomes blurred when considering animals that live in water. Although fishes have well-developed chemoreceptors, scientists generally do not refer to the senses of "taste" and "smell" in fishes.

What are the three types of **photoreceptors** among **invertebrates**?

There are three different types of eyes, represented by different types of photoreceptors, among invertebrates. They are: 1) eye cup, 2) compound eye, and 3) single-lens eye. The eye cup is a cluster of photoreceptors cells that partially shield adjacent photoreceptor cells. The compound eye consists of many tiny light detectors (photoreceptors). Crayfish, crabs, and nearly all insects have compound eyes. Single-lens eyes, found in cephalopods such as squids and octopi, are similar to cameras. They have a small opening, the pupil, through which light enters.

Do animals have **color vision**?

Most reptiles, fishes, insects, and birds appear to have a well-developed color sense. Most mammals are color-blind. Apes and monkeys have the ability to tell colors apart. Dogs and cats seem to be color-blind and only see shades of black, white, and gray.

Can **animals** hear different **sound frequencies** than humans?

The frequency of a sound is the pitch. Frequency is expressed in Hertz (Hz). Sounds are classified as infrasounds (below the human range of hearing), sonic range (within the range of human hearing), and ultrasound (above the range of human hearing).

Animal	Frequency range heard (Hz)
Dog	15 to 50,000
Human	20 to 20,000
Cat	60 to 65,000
Dolphin	150 to 150,000
Bat	1,000 to 120,000

How sensitive is the **hearing of birds**?

In most species of birds the most important sense after sight is hearing. Birds' ears are close to their bodies and covered with feathers. However, the feathers covering the ears do not have barbules, which would obstruct sound. Ears of different heights allow the bird to locate of sound. Nocturnal raptors such as the great horned owl have a very well-developed sense of hearing in order to be able to capture their prey in total darkness.

How do **fish swimming** in a school **change their direction** simultaneously?

The movements of a school of fish, which confuse predators, happen because the fish detect pressure changes in the water. The detection system, called the lateral line, is found along each side of the fish's body. Along the line are clusters of tiny hairs inside cups filled with a jellylike substance. If a fish becomes alarmed and turns sharply, it causes a pressure wave in the water around it. This wave pressure deforms the "jelly" in the lateral line of nearby fish. The deformation moves the hairs that trigger nerves, and a signal is sent to the brain telling the fish to turn.

How much **electricity** does an **electric eel** generate?

An electric eel (*Electrophorus electricus*) has current-producing organs made up of electric plates (modified muscle cells) on both sides of its vertebral column running almost its entire body length. The charge—on the average of 350 volts, but as great as 550 volts—is released by the central nervous system. The shock consists of four to eight separate charges, each of which lasts only two- to three-thousandths of a second.

These shocks, used as a defense mechanism, can be repeated up to 150 times per hour without any visible fatigue to the eel. The most powerful electric eel, found in the rivers of Brazil, Colombia, Venezuela, and Peru, produces a shock of 400 to 650 volts.

What is the difference between **ectotherms** and **endotherms**?

Ectotherms, also known as cold-blooded animals, warm their bodies by absorbing heat form their surroundings. These animals have large variations in normal body temperature due to their changing environment. Most invertebrates, fishes, reptiles, and amphibians are ectotherms. The body temperature of endotherms, also known as warm-blooded animals, depends on the heat produced by the animal's metabolism. Mammals, birds, some fishes, and some insects are endotherms. Their normal body temperature is fairly constant even when there are vast differences in the temperature of their environment.

What is **normal body temperature**?

Normal body temperature is the acceptable temperature for an animal. The following chart identifies normal body temperature for a variety of ectotherms and endotherms.

Animal	Normal Temperature °F	Normal Temperature °C
Human (endotherm)	98.6	37
Cat (endotherm)	101.5	38.5
Dog (endotherm)	102	38.9
Cow (endotherm)	101	38.3
Mare (endotherm)	100	37.8
Pig (endotherm)	102.5	39.2
Goat (endotherm)	102.3	39.1
Rabbit (endotherm)	103.1	39.5
Sheep (endotherm)	102.3	39.1
Pigeon (endotherm)	106.6	41
Lizard (ectotherm)	87.8–95	31–35
Salmon (ectotherm)	41–62.6	5–17
Rainbow trout (ectotherm)	53.6–64.4	12–18
Rattlesnake(ectotherm)	59–98.6	15–37
Grasshopper(ectotherm)	101.5–108	38.6–42.2

Do animals other than humans have **fingerprints**?

It is known that gorillas and other primates have fingerprints. Of special interest, however, is that our closest relative, the chimpanzee does not. Koala bears also have fingerprints. Researchers in Australia have determined that the fingerprints of koala bears closely resemble those of human fingerprints in size, shape, and pattern.

IMMUNE SYSTEM

How does the **immune system** work?

The immune system has two main components: white blood cells and antibodies circulating in the blood. The antigen-antibody reaction forms the basis for this immunity. When an antigen (antibody generator)—a harmful bacterium, virus, fungus, parasite, or other foreign substance—invades the body, a specific antibody is generated to attack the antigen. The antibody is produced by B lymphocytes (B cells) in the spleen or lymph nodes. An antibody may either destroy the antigen directly or it may "label" it so that a white blood cell (called a macrophage, or scavenger cell) can engulf the foreign intruder. After a human has been exposed to an antigen, a later exposure to the same antigen will produce a faster immune system reaction. The necessary antibodies will be produced more rapidly and in larger amounts. Artificial immunization uses this antigen-antibody reaction to protect the human body from certain diseases, by exposing the body to a safe dose of antigen to produce effective antibodies as well as a "readiness" for any future attacks of the harmful antigen.

What are **nonspecific defenses**?

Nonspecific defenses do not differentiate between various invaders. Barriers such as skin, hide, and the mucous membrane lining the respiratory and digestive tracts; phagocytic white blood cells; and chemicals are nonspecific defenses. The nonspecific defenses are the first to respond to a foreign substance in the body.

How do **T cells** differ from **B lymphocytes**?

Lymphocytes are one variety of white blood cells and are part of the body's immune system. The immune system fights invading organisms that have penetrated the body's general defenses. T cells, responsible for dealing with most viruses, for handling some bacteria and fungi, and for cancer surveillance, are one of the two main classes of lymphocytes. T lymphocytes, or T cells, compose about 60 to 80 percent of the lymphocytes circulating in the blood. They have been "educated" in the thymus to perform particular functions. Killer T cells are sensitized to multiply when they come into contact with antigens (foreign proteins) on abnormal body cells (cells that have been invaded by viruses, cells in transplanted tissue, or tumor cells). These killer T cells attach themselves to the abnormal cells and release chemicals (lymphokines) to destroy them. Helper T cells assist killer cells in their activities and control other aspects of the immune response. When B lymphocytes, which compose approximately 10 to 15 percent of total lymphocytes, contact the antigens on abnormal cells, the lymphocytes enlarge and divide to become plasma cells. Then the plasma cells secrete vast numbers of immunoglobulins or antibodies into the blood, which attach themselves to the surfaces of the abnormal cells, to begin a process that will lead to the destruction of the invaders.

What are some typical **disorders** of the **immune system?**

Allergies, autoimmune diseases, and immunodeficiency diseases are different kinds of disorders of the immune system. Allergies are abnormal sensitivities to a substance that is harmless to many other people. Common allergens include pollen, certain foods, cosmetics, medications, fungal spores, and insect venom. The antibody immunoglobulin E (IgE) is responsible for most allergic reactions. When exposed to an allergen, IgE antibodies attach themselves to mast cells or basophils. Mast cells are normal body cells that produce histamines and other chemicals. When exposed to the same allergen at a later time, the individual may experience an allergic response when the allergen binds to the antibodies attached to mast cells, causing the cells to release histamine and other inflammatory chemicals. While most allergic reactions are expressed as a runny nose, difficulty in breathing, skin rashes and eruptions, or intestinal discomfort, a severe allergic reaction results in anaphylactic shock.

Autoimmune diseases are diseases in which the immune system rejects the body's own molecules. Insulin-dependent diabetes, rheumatoid arthritis, systemic lupus erythematosus, and rheumatic fever are autoimmune diseases. In contrast, in immunodeficiency diseases such as AIDS, the immune system is too weak to fight disease.

Do **animals suffer** from **allergies?**

Veterinarians report that dogs and cats suffer from allergies. They may be allergic to food, insect bites, dust, household chemicals, or pollen. Instead of having runny noses and watery eyes, animals experience itchy skin conditions, difficulty in breathing, or disruptions in the digestive tract.

REPRODUCTION

How does **asexual reproduction** differ from **sexual reproduction?**

Asexual reproduction produces offspring with the exact genetic material of the parent. Only one individual is needed to produce offspring via asexual reproduction. Sexual reproduction produces offspring by the fusion of two gametes (haploid cells) to form one zygote (diploid cell). The male gamete is the sperm, and the female gamete is the egg.

What are some **methods** of **asexual reproduction?**

Budding, fission, and fragmentation are methods of asexual reproduction. In budding a new individual begins as an outgrowth, or bud, of the parent. Eventually, the bud detaches from the parent and develops into a new individual. Budding is common among sponges and coelenterates such as hydras and anemones. Fission is the division of one individual into two or more individuals of almost equal size. Each new individual develops into a mature adult. Some corals reproduce by dividing longitudinally into two smaller but complete individuals. Fragmentation is the breaking of the parent into several pieces. It is accompanied by regeneration when each piece develops into a mature individual. Sea stars are well known for reproducing by fragmentation and regeneration.

Can animals **regenerate** parts of their bodies?

Regeneration occurs in a wide variety of animals; however, it progressively declines as the animal species becomes more complex. Regeneration frequently occurs among primitive invertebrates. For example, a planarian (flatworm) can split symmetrically, with the two sides turning into clones of one other. In higher invertebrates regeneration occurs in echinoderms such as starfish and arthropods such as insects and crustaceans. Starfish are known for their ability to develop into complete individuals from one cut off arm. Regeneration of appendages (limbs, wings, and antennae) occurs in insects such as cockroaches, fruit flies, and locusts and in crustaceans such as lobsters, crabs, and crayfish. For example, regeneration of a crayfish's missing claw occurs at its next molt (shedding of its hard cuticle exterior shell/skin in order to grow and the subsequent hardening of a new cuticle exterior). Sometimes the regenerated claw does not achieve the same size of the missing claw. However, after every molt (a process that occurs two to three times a year) the regenerated claw grows and will eventually become nearly as large as the original claw. On a very limited basis some amphibians and reptiles can replace a lost leg or tail.

What is **hermaphroditism**?

Hermaphroditic animals have both male and female reproductive systems. Hermaphroditism provides a means for animals to reproduce sexually without finding mates. For example, individuals in many species of tapeworms fertilize their own eggs. In other species such as earthworms, each individual serves as a male and female during mating, both donating and receiving sperm.

How does **external fertilization** differ from **internal fertilization**?

External fertilization is common among aquatic animals including fishes, amphibians, and aquatic invertebrates. Following an elaborate ritual of mating behavior to synchronize the release of eggs and sperm, both males and females deposit their gametes in the water at approximately the same time in close proximity to each other. The water protects the sperm and eggs from drying out. Fertilization occurs when the sperm reach the eggs. Internal fertilization requires that sperm be deposited in or close to the female reproductive tract. It is most common among terrestrial animals that either lay a shelled egg, such as reptiles and birds, or when the embryo develops for a period of time within the female body.

Which **aquatic animals** have **internal fertilization**?

Certain sharks, skates, and rays have internal fertilization. The pelvic fins are specialized to pass sperm to the female. In most of these species the embryos develop internally and are born alive.

Which animal has the **longest gestation** period?

The animal with the longest gestation period is not a mammal; it is the viviparous amphibian, the alpine black salamander, which can have a gestation period of up to thirty-eight months at altitudes above 4,600 ft (1,402 m) in the Swiss alps; it bears two fully metamorphosed young.

Which **mammal** has the shortest **gestation period**? Which one has the **longest period**?

Gestation is the period of time between fertilization and birth in oviparous animals. The shortest gestation period known among mammals is twelve to thirteen days, shared by three marsupials: the American or Virginian opossum (*Didelphis marsupialis*); the rare water opossum, or yapok (*Chironectes minimus*) of central and northern South America; and the eastern native cat (*Dasyurus viverrinus*) of Australia. The young of each of these marsupials are born while still immature and complete their development in the ventral pouch of their mother. While twelve to thirteen days is the average, the gestation period is sometimes as short as eight days. The longest gestation period for a mammal is that of the African elephant (*Loxodonta africana*), with an average of 660 days and a maximum of 760 days.

What is **delayed implantation**?

Delayed implantation is a phenomenon that lengthens the gestation period of many mammals. The blastocyst remains dormant while its implantation in the uterine wall is postponed for a period of time lasting from a few weeks to several months. Many mammals (including bears, seals, weasels, badgers, bats, and some deer) use this phenomenon to extend their gestation period through delayed implantation so they give birth at the time of year that offers the best chance of survival for their young.

How many **eggs** does a **spider** lay?

The number of eggs varies according to the species. Some larger spiders lay more than 2,000 eggs, but many tiny spiders lay one or two and perhaps no more than a dozen during their lifetime. Spiders of average size probably lay a hundred or so. Most spiders lay all their eggs at one time and enclose them in a single egg sac; others lay eggs over a period of time and enclose them in a number of egg sacs.

How many **eggs** are produced by **sea urchins**?

The number of eggs produced by sea urchins is enormous. It has been estimated that a female of the genus *Arbacia* contains about eight million eggs. In the much larger genus *Echinus* the number reaches twenty million.

How **long** do animals, in particular mammals, **live**?

Of the mammals, humans and fin whales live the longest. Below is the maximum lifespan for various animal species.

Animal	Latin name	Maximum life span in years
Marion's tortoise	*Testudo sumeirii*	152+
Quahog	*Venus mercenaria*	ca. 150
Common box tortoise	*Terrapene carolina*	138
European pond tortoise	*Emys orbicularis*	120+
Spur-thighed tortoise	*Testudo graeca*	116+
Fin whale	*Balaenoptera physalus*	116
Human	*Homo sapiens*	116
Deep-sea clam	*Tindaria callistiformis*	ca. 100
Killer whale	*Orcinus orca*	ca. 90
European eel	*Anguilla anguilla*	88
Lake sturgeon	*Acipenser fulvescens*	82
Freshwater mussel	*Margaritana margaritifera*	80 to 70
Asiatic elephant	*Elephas maximus*	78
Andean condor	*Vultur gryphus*	72+
Whale shark	*Rhiniodon typus*	ca. 70
African elephant	*Loxodonta africana*	ca. 70
Great eagle-owl	*Bubo bubo*	68+
American alligator	*Alligator mississipiensis*	66
Blue macaw	*Ara macao*	64
Ostrich	*Struthio camelus*	62.5
Horse	*Equus caballus*	62
Orangutan	*Pongo pygmaeus*	ca. 59
Bateleur eagle	*Terathopius ecaudatus*	55
Hippopotamus	*Hippopotamus amphibius*	54.5
Chimpanzee	*Pan troglodytes*	51

Animal	Latin name	Maximum life span in years
White pelican	*Pelecanus onocrotalus*	51
Gorilla	*Gorilla gorilla*	50+
Domestic goose	*Anser a. domesticus*	49.75
Grey parrot	*Psittacus erythacus*	49
Indian rhinoceros	*Rhinoceros unicornis*	49
European brown bear	*Ursus arctos arctos*	47
Grey seal	*Halichoerus gryphus*	46+
Blue whale	*Balaenoptera musculus*	ca. 45
Goldfish	*Carassius auratus*	41
Common toad	*Bufo bufo*	40
Roundworm	*Tylenchus polyhyprus*	39
Giraffe	*Giraffa camelopardalis*	36.25
Bactrian camel	*Camelus ferus*	35+
Brazilian tapir	*Tapirus terrestris*	35
Domestic cat	*Felis catus*	34
Canary	*Serinus caneria*	34
American bison	*Bison bison*	33
Bobcat	*Felis rufus*	32.3
Sperm whale	*Physeter macrocephalus*	32+
American manatee	*Trichechus manatus*	30
Red kangaroo	*Macropus rufus*	ca. 30
African buffalo	*Syncerus caffer*	29.5
Domestic dog	*Canis familiaris*	29.5
Lion	*Panthera leo*	ca. 29
African civet	*Viverra civetta*	28
Theraphosid spider	*Mygalomorphae*	ca. 28
Red deer	*Cervus elaphus*	26.75
Tiger	*Panthera tigris*	26.25
Giant panda	*Ailuropoda melanoleuca*	26
American badger	*Taxidea taxus*	26
Common wombat	*Vombatus ursinus*	26
Bottle-nosed dolphin	*Tursiops truncatus*	25
Domestic chicken	*Gallus g. domesticus*	25
Grey squirrel	*Sciurus carolinensis*	23.5
Aardvark	*Orycteropus afer*	23
Domestic duck	*Anas platyrhynchos domesticus*	23
Coyote	*Canis latrans*	21+
Canadian otter	*Lutra canadensis*	21

Animal	Latin name	Maximum life span in years
Domestic goat	*Capra hircus domesticus*	20.75
Queen ant	*Myrmecina graminicola*	18+
Common rabbit	*Oryctolagus cuniculus*	18+
White or beluga whale	*Delphinapterus leucuas*	17.25
Platypus	*Ornithorhynchus anatinus*	17
Walrus	*Odobenus rosmarus*	16.75
Domestic turkey	*Melagris gallapave domesticus*	16
American beaver	*Castor canadensis*	15+
Land snail	*Helix spiriplana*	15
Guinea pig	*Cavia porcellus*	14.8
Hedgehog	*Erinaceus europaeus*	14
Burmeister's armadillo	*Calyptophractus retusus*	12
Capybara	*Hydrochoerus hydrochaeris*	12
Chinchilla	*Chinchilla laniger*	11.3
Giant centipede	*Scolopendra gigantea*	10
Golden hamster	*Mesocricetus auratus*	10
Segmented worm	*Allolobophora longa*	10
Purse-web spider	*Atypus affinis*	9+
Greater Egyptian gerbil	*Gerbillus pyramidum*	8+
Spiny starfish	*Marthasterias glacialis*	7+
Millipede	*Cylindroiulus landinensis*	7
Coypu	*Myocastor coypus*	6+
House mouse	*Mus musculus*	6
Malagasy brown-tailed mongoose	*Salanoia concolor*	4.75
Cane rat	*Thryonomys swinderianus*	4.3
Siberian flying squirrel	*Pteromys volans*	3.75
Common octopus	*Octopus vulgaris*	2 to 3
Pygmy white-toothed shrew	*Suncus etruscus*	2
Pocket gopher	*Thomomys talpoides*	1.6
Monarch butterfly	*Danaus plexippus*	1.13
Bedbug	*Cimex lectularius*	0.5 or 182 days
Black widow spider	*Latrodectus mactans*	0.27 or 100 days
Common housefly	*Musca domesticus*	0.04 or 17 days

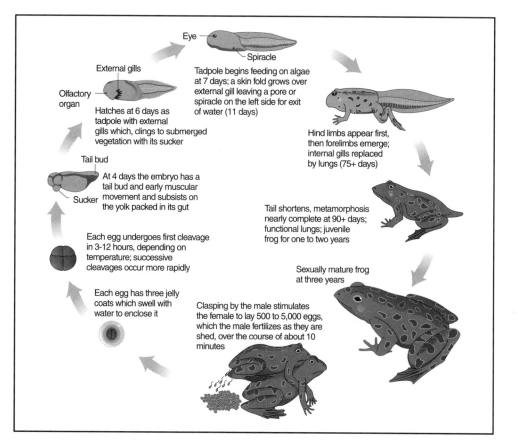

The life cycle of a frog.

What is the life-span of a **fruit fly**?

The length of adult life can vary considerably. Under ideal conditions an adult *Drosophila melanogaster* can live as long as forty days. In crowded conditions life-span may drop to twelve days. Under normal laboratory conditions, however, adults generally die after only six or seven days.

How can you tell **male** and **female lobsters** apart?

The differences between male and female lobsters can only be seen when they are turned on their backs. In the male lobster the two swimmerets (forked appendages used for swimming) nearest the carapace (the solid shell) are hard, sharp, and bony; in the female the same swimmerets are soft and feathery. The female also has a receptacle that appears as a shield wedged between the third pair of walking legs. During mating the male deposits sperm into this receptacle, where it remains for as long as several months until the female uses it to fertilize her eggs as they are laid.

What is a **mermaid's purse**?

Mermaid's purses are the protective cases in which the eggs of dogfish, skates, and rays are released into the environment. The rectangular purse is leathery and has long tendrils streaming from each corner. The tendrils anchor the case to seaweed or rocks, where the case is protected during the six to nine months it takes for the embryos to hatch. Empty cases often wash up on beaches.

How is the **gender of alligator embryos** determined?

The gender of an alligator is determined by the temperature at which the eggs are incubated. High temperatures of 90° to 93°F (32° to 34°C) result in males; low temperatures of 82° to 86°F (28° to 30°C) yield females. This determination takes place during the second and third week of the two-month incubation. Further temperature fluctuations before or after this time do not alter the gender of the young. The heat from the decaying matter on top of the nest incubates the eggs.

What is **unique** about **egg incubation** in some **amphibians**?

Unlike most toads and frogs the female Surinam toad (*Pipa pipa*) carries her eggs in special pockets in the skin on her back. Each egg develops in its own pocket in the female's skin. The tadpoles' tails are "plugged in" to the mother's system, similar to the placenta of mammals, exchanging nutrients and gases. The tadpoles develop quickly, undergoing metamorphosis while still in the pockets. Upon transformation into miniature frogs, they break free of their pocket walls to begin independent lives.

What is the **importance** of an **external egg** in reproduction?

Species that have an external egg usually produce a greater number of zygotes, because mating between males and females is not required for successful reproduction. The external egg of most species has a leathery outer covering to prevent desiccation.

Which **birds** lay the **largest and smallest eggs**?

The elephant bird (*Aepyornis maximus*), an extinct flightless bird of Madagascar, also known as the giant bird or roc, laid the largest known bird eggs. Some of these eggs measured as much as 13.5 in (34 cm) in length and 9.5 in (24 cm) in diameter. The largest egg produced by any living bird is that of the North African ostrich (*Struthio camelus*). The average size is 6 to 8 in (15 to 20.5 cm) in length and 4 to 6 in (5 to 15 cm) in diameter. The smallest mature egg, measuring less than 0.39 in (1 cm) in length, is that of the vervain hummingbird (*Mellisuga minima*) of Jamaica.

Generally speaking, the larger the bird, the larger the egg. However, when compared with the bird's body size, the ostrich egg is one of the smallest eggs, while the hummingbird's egg is one of the largest. The Kiwi bird of New Zealand lays the largest egg relative to body size of any living bird. Its egg weighs up to 1 lb (0.5 kg).

What is unusual about the way the **emperor penguin's eggs** are incubated?

Each female emperor penguin (*Aptenodytes forsteri*) lays one large egg. Initially, both sexes share in incubating the egg by carrying it on his or her feet covered with a fold of skin. After a few days of passing the egg back and forth, the female leaves to feed in the open water of the Arctic Ocean. Balancing their eggs on their feet, the male penguins shuffle about the rookery, periodically huddling together for warmth during blizzards and frigid weather. If an egg is inadvertently orphaned, a male with no egg will quickly adopt it. Two months after the female's departure, the chick hatches. The male feeds it with a milky substance he regurgitates until the female returns. Now padded with blubber, the females take over feeding the chicks with fish they have stored in their crops. The females do not return to their mate (and own offspring), however, but wander from male to male until one allows her to take his chick. It is then the males' turn to feed in open water and restore the fat layer they lost while incubating.

What are some **animals** that have **pouches**?

Marsupials (meaning "pouched" animals) differ from all other living mammals in their anatomical and physiological features of reproduction. Most female marsupials—kangaroos, bandicoots, wombats, banded anteaters, koalas, opossums, wallabies, Tasmanian devils, etc.—have an abdominal pouch called a marsupium, in which their young are carried. In some small terrestrial marsupials, however, the marsupium is not a true pouch but merely a fold of skin around the mammae (milk nipples).

The short gestation period in marsupials (in comparison to other similarly sized mammals) allows their young to be born in an "undeveloped" state. Consequently, these animals have been viewed as "primitive" or second-class mammals. However, some scientists now see that the reproductive process of marsupials has an advantage over that of placental mammals. A female marsupial invests relatively few resources during the brief gestation period, more so during the lactation (nursing period) when the young are in the marsupium. If the female marsupial loses its young, it can conceive again sooner than a placental mammal in a comparable situation.

Which **mammals lay eggs** and **suckle their young**?

The duck-billed platypus (*Ornithorhynchus anatinus*), the short-nosed echidna or spiny anteater (*Tachyglossus aculeatus*), and the long-nosed echidna (*Zaglossus bruijni*)—indigenous to Australia, Tasmania, and New Guinea, respectively—are the only three species of mammals that lay eggs (a nonmammalian feature) but suckle their young (a mammalian feature). These mammals (order Monotremata) resemble reptiles in that they lay rubbery shell-covered eggs that are incubated and hatched outside the mother's body. In addition, they resemble reptiles in their digestive, reproductive, and excretory systems and in a number of anatomical details (eye structure, presence of certain skull bones, pectoral [shoulder] girdle, and rib and vertebral structures). They are, however, classed as mammals because they have fur and a four-chambered heart, nurse their young from gland milk, are warm-blooded, and have some mammalian skeletal features.

SKELETAL SYSTEM

What is the **function** of the **skeletal system**?

The skeletal system is a multifunctional system. The skeletal system provides support, allows an animal to move, and protects the internal organs and soft parts of an animal's body.

What are the **three main types** of **skeletal systems**?

The three main types of skeletal systems are hydrostatic skeleton, exoskeleton, and endoskeleton. A hydrostatic skeleton consists of fluid under pressure. This type of skeletal system is most common in soft, flexible animals such as hydras, planarians, and earthworms and other segmented worms. Hydras and planarians have a fluid-filled gastrovascular cavity. The body cavity, or coelom, of an earthworm is also fluid-filled.

Many aquatic and certain terrestrial animals have an exoskeleton. The exoskeleton is rigid and hard. Mollusks have an exoskeleton made of calcium carbonate. It grows with the animal during its entire lifetime. Another type of exoskeleton common among insects and arthropods is made from chitin. Chitin is a strong flexible nitrogenous polysaccharide. While it provides excellent protection and allows for a large variety of movements, it does not grow with the animal. When an animal outgrows its skeleton, it must shed its skeleton and replace it with a larger one in a process known as molting.

An endoskeleton consists of bone and cartilage and grows with the animal throughout its life. It stores calcium salts and blood cells and consists of hard or leathery supporting elements situated among the soft tissues of an animal. Although most common among vertebrates, certain invertebrates such as sponges, sea stars, sea urchins, and other echinoderms have an endoskeleton of hard plates beneath their skin. This type of skeletal system allows for a wider range of movement than do the other two.

What is the chemical composition of **chitin**?

Chitin, found in the exoskeletons of insects and other arthropods, is a glucosamine polysaccharide with the formula of $C_{30}H_{50}O_{19}N_4$ and a molecular weight of 770.42. The basic units of this substance are linked together by condensation reactions to make up long chains. Hydrogen bonds link the chains together and help make chitin rigid and strong. It is a white, amorphous, semitransparent mass that is insoluble in common solvents like water and alcohol.

What are the upper and lower **shell** of a **turtle** called?

The turtle (order Testudines) uses its shell as a protective device. The upper shell is called the dorsal carapace and the lower shell is called the ventral plastron. The shell's sections are referred to as the scutes. The carapace and the plastron are joined at the sides.

How much **weight** can an **ant carry**?

Ants are incredibly strong in relation to their size. Most ants can carry objects ten to twenty times their own weight, and some ants can carry objects up to fifty times their own weight. Ants are able to carry these objects great distances and even climb trees while carrying them. This is comparable to a 100-pound person picking up a small car, carrying it seven to eight miles on his back, and then climbing the tallest mountain while still carrying the car!

How many **vertebrae** are in the **neck** of a **giraffe?**

A giraffe neck has seven vertebrae, the same as other mammals, but the vertebrae are greatly elongated.

LOCOMOTION

What are the **problems** an animal must **overcome** to **move**?

In contrast to other organisms, animals are able to move. The two forces an animal overcomes to move are gravity and friction. Aquatic animals do not have much difficulty overcoming gravity, since they are buoyant in water. However, because water is dense, the problem of resistance (friction) is greater for these animals. Many of them have sleek shapes to help them swim. Terrestrial animals tend to have fewer problems with friction since air poses fewer problems of resistance than does water. However, terrestrial animals must work harder to overcome gravity.

Which animals can run **faster than a human**?

The cheetah, the fastest mammal, can accelerate from 0 to 45 mph (64 kph) in 2 seconds; it has been timed at speeds of 70 mph (112 kph) over short distances. In most chases cheetahs average around 40 mph (63 kph). Humans can run very short distances at almost 28 mph (45 kph) maximum. Most of the speeds given in the table below are for distances of .25 mi (0.4 km).

Animal	Maximum speed (mph)	Maximum speed (kph)
Cheetah	70	112.6
Pronghorn antelope	61	98.1
Wildebeest	50	80.5
Lion	50	80.5
Thomson's gazelle	50	80.5
Quarter horse	47.5	76.4
Elk	45	72.4
Cape hunting dog	45	72.4
Coyote	43	69.2
Gray fox	42	67.6
Hyena	40	64.4
Zebra	40	64.4
Mongolian wild ass	40	64.4
Greyhound	39.4	63.3
Whippet	35.5	57.1
Rabbit (domestic)	35	56.3
Mule deer	35	56.3
Jackal	35	56.3
Reindeer	32	51.3
Giraffe	32	51.3
White-tailed deer	30	48.3
Wart hog	30	48.3
Grizzly bear	30	48.3
Cat (domestic)	30	48.3
Human	27.9	44.9

How do **fleas** jump so far?

The jumping power of fleas comes both from strong leg muscles and from pads of a rubber-like protein called resilin. The resilin is located above the flea's hind legs. To jump, the flea crouches, squeezing the resilin, and then it relaxes certain muscles. Stored energy from the resilin works like a spring, launching the flea. A flea can jump well both vertical-

How fast can a crocodile run on land?

In smaller crocodiles the running gait can change into a bounding gallop that can achieve speeds of 2 to 10 mph (3 to 17 kph).

ly and horizontally. Some species can jump 150 times their own length. To match that record, a human would have to spring over the length of two and a quarter football fields—or the height of a 100-story building—in a single bound. The common flea (*Pulex irritans*) has been known to jump 13 in (33 cm) in length and 7.25 in (18.4 cm) in height.

What causes the **Mexican jumping bean** to move?

The bean moth (*Carpocapa saltitans*) lays its eggs in the flower or in the seed pod of the spurge, a bush known as *Euphorbia sebastiana*. The egg hatches inside the seed pod, producing a larva or caterpillar. The jumping of the bean is caused by the active shifting of weight inside the shell as the caterpillar moves. The jumps of the bean are stimulated by sunshine or by heat from the palm of the hand.

At what **speeds** do fishes **swim**?

The maximum swimming speed of a fish is somewhat determined by the shape of its body and tail and by its internal temperature. The cosmopolitan sailfish (*Istiophorus platypterus*) is considered to be the fastest fish species, at least for short distances, swimming at greater than 60 mph (95 kph). Some American fishermen, however, believe that the bluefin tuna (*Thunnus thynnus*) is the fastest, but the fastest speed recorded for this species so far is 43.4 mph (69.8 kph). Data is extremely difficult to secure because of the practical difficulties in measuring the speeds. The yellowfin tuna (*Thunnus albacares*) and the wahoe (*Acanthocybium solandri*) are also fast, timed at 46.35 mph (74.5 kph) and 47.88 mph (77 kph) respectively during 10 to 20 second sprints. Flying fish swim at 40+ mph (64+ kph), dolphins at 37 mph (60 kph), trout at 15 mph (24 kph), and blenny at 5 mph (8 kph). Humans can swim 5.19 mph (8.3 kph).

What is the **fastest snake** on land?

The black mamba (*Dendroaspis polylepis*), a deadly poisonous African snake that can grow up to 13 ft (4 m) in length, has been recorded reaching a speed of 7 mph (11 kph). A particularly aggressive snake, it chases animals at high speeds, holding the front of its body above the ground.

Do all **birds fly**?

No. Among the flightless birds, the penguins and the ratites are the best known. Ratites include emus, kiwis, ostriches, rheas, and cassowaries. They are called ratites because they lack a keel on the breastbone. All of these birds have wings but lost their power to fly millions of years ago. Many birds that live isolated on oceanic islands (for example, the great auk) apparently became flightless in the absence of predators and the consequent gradual disuse of their wings for escape.

How fast does a **hummingbird** fly, and how far does the hummingbird **migrate**?

Hummingbirds fly at speeds up to 71 mph (114 kph). Small species beat their wings 50 to 80 times per second, higher in courtship displays. For comparison, the following table lists the flight speeds of some other birds:

Bird	Speed in miles per hour	Speed in kilometers per hour
Peregrine falcon	168–217	270.3–349.1
Swift	105.6	169.9
Merganser	65	104.6
Golden plover	50–70	80.5–112.6
Mallard	40.6	65.3
Wandering albatross	33.6	54.1
Carrion crow	31.3	50.4
Herring gull	22.3–24.6	35.9–39.6
House sparrow	17.9–31.3	28.8–50.4
Woodcock	5	8

The longest migratory flight of a hummingbird documented to date is the flight of a rufous hummingbird from Ramsey Canyon, Arizona, to near Mt. Saint Helens, Washington, a distance of 1,414 mi (2,277 km). Bird-banding studies are now in progress to verify that a few rufous hummingbirds do make a 11,000 to 11,500-mi (17,699 to 18,503-km) journey along a super Great Basin High route, a circuit that could take a year to complete. Hummingbird studies, however, are difficult to complete because so few banded birds are recovered.

How **fast** does a **hummingbird's wing** move?

Hummingbirds are the only family of birds that can truly hover in still air for any length of time. They need to do so in order to hang in front of a flower while they perform the delicate task of inserting their slim sharp bills into the flower's depths to drink nectar. Their thin wings are not contoured into the shape of aerofoils and do not generate lift in this way. Their paddle-shaped wings are, in effect, hands that swivel at the shoulder. They beat them in such a way that the tip of each wing follows the line of a figure-eight lying on its side. The wing moves forward and downward into the front loop of the eight, creating lift. As it begins to come up and goes back, the wing twists through 180 degrees so that once again it creates a downward thrust. The humming-

Can any bird fly upside down?

Hummingbirds are the only bird species that can fly upside down. They can do so because of their angled wing structure, but they can accomplish the maneuver for only a short period of time. Hummingbirds can also fly backwards in order to remove their bills from tube flowers.

bird's method of flying does have a major limitation. The smaller a wing, the faster it has to beat in order to produce sufficient downward thrust. An average-sized hummingbird beats its wings 25 times a second. The bee hummingbird, native to Cuba, is only 2 in (5 cm) long and beats its wings at an astonishing 200 times a second.

Do any **mammals fly**?

Bats (order Chiroptera with 986 species) are the only truly flying mammals, although several gliding mammals are referred to as "flying" (such as the flying squirrel and flying lemur). The "wings" of bats consist of double membranes of skin stretching from the sides of the body to the hind legs and tail, and are actually skin extensions of the back and belly. The wing membranes are supported by the elongated fingers of the forelimbs (or arms).

ANIMAL BEHAVIOR

INTRODUCTION AND HISTORICAL BACKGROUND

What is the definition of **behavior**?

In its broadest sense, behavior covers all kinds of movement and responses to environmental changes. In other words, behavior is the term used to describe what an animal does.

Who was the **first individual** to study **animal behavior**?

Aristotle (384–322 B.C.E.) wrote ten volumes on the *Natural History of Animals*. The Roman naturalist Pliny (23–79 C.E.) also extensively observed and recorded observations of organisms in his book *Natural History*. In more recent times Charles Darwin (1809–1882) recorded (in his journal) the behavior of the marine iguanas of the Galapagos Islands. Darwin also published a book, *The Expression of the Emotions of Man and Animals* (1872), in which he showed how natural selection would favor specialized behavioral patterns for survival. However, it was not until 1953, when Niko Tinbergen (1907–1988) documented and published his studies of gulls (*The Herring Gull's World*) and their begging techniques, that the field of ethology—the study of animal behavior—was established.

What is **ethology**?

Ethology emerged in the mid-1930s and was first recognized as a subdiscipline of biology in Europe. Ethology differed from traditional biological studies of animals in that scientific principles were applied to the study of animal behavior, with practition-

The term was coined from the French word *éthologie* by the zoologist Isodore Geoffroy Saint-Hilaire (1805–1861). It was first used in English by the American entomologist William Morton Wheeler (1865–1937), who employed it in 1902.

ers using both field observations and laboratory experiments. The field was developed and first recognized as a science in Europe, where experimental conditions were kept as natural as possible.

Who was **Niko Tinbergen**?

Niko Tinbergen (1907–1988), along with Konrad Lorenz (1903–1989) and Karl von Frisch (1886–1982), laid the four cornerstones of ethological study: causation, development, evolution, and function of behavior. Tinbergen's earliest studies were of the aggressive threat displays of adult male three-spined stickleback fish (*Gasterosteus aculeatus*). Tinbergen found that by presenting the male with stickleback models, he was able to elicit an aggressive response.

He began his career in animal behavior as a child by studying the mating and nesting behavior of sticklebacks in his backyard pond. He attended the University of Leiden, where he did his dissertation research on digger wasps; his thesis was only thirty-six pages long and was barely accepted by the faculty. In 1936 he met Lorenz at a symposium, and they became lifelong friends and colleagues. During World War II Tinbergen was imprisoned in a German hostage camp and Lorenz in a Russian camp; when they were reunited after the war, both resumed their studies on animal behavior.

Who was **Konrad Lorenz**?

Konrad Lorenz (1903–1989) became famous for his work in the field of avian ethology, particularly in his studies of imprinting. By raising goslings from the time they were hatched, Lorenz was able to make the goslings follow him rather than their own mother. This work led to the theory that the goslings were genetically programmed to exhibit a certain behavior with regard to any large organism that was near them during a critical early period of their life. In his publications he applied comparative methods to the study of behavior and the psychology of perception.

Who first won the **Nobel Prize** for work on animal **behavior**?

Although there is no specific Nobel Prize for the study of animal behavior, the first prize to be awarded for the scientific study of animal behavior was awarded jointly in

1973 in the category "Physiology or Medicine" to Konrad Lorenz (1903–1989), Karl von Frisch (1886–1982), and Niko Tinbergen (1907–1988). Each ethologist had a main research area for which he was well known: Lorenz for imprinting behavior of birds; von Frisch for the "dance" of the honeybees; and Tinbergen for aggressive behavior of stickleback fish.

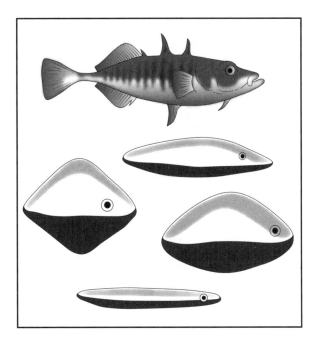

Niko Tinbergen made important studies of innate stickleback behavior.

What does the term *umwelt* refer to with regard to **animal behavior**?

Umwelt was first used by Jacob von Uexkuell (1864–1944) to describe the part of its environment that an animal can actually perceive with its sense organs and nervous system. Animals display this concept in the wild, but the same behavior can be seen with domestic animals. For example, if a stranger comes into your house and tries to pet your dog, this may be perceived as a threat within the dog's *umwelt*. The practical side of this is that you should be aware of your pet's *umwelt* when introducing it to new experiences and people.

What is **behavioral ecology**?

Behavioral ecology investigates the relationship between the environment and animal behavior. It emphasizes the evolutionary roots of the behavior, in contrast to the classical studies involving animals in laboratory settings. George C. Williams (1926–), in his book *Adaptation and Natural Selection* (1966), first posed the question as to how behavior affects evolutionary fitness. By showing that behavior is responsive to the environmental forces that drive natural selection (and evolutionary fitness), researchers have demonstrated that the environment plays a crucial role in determining which behaviors are exhibited in natural settings.

What is **anthropomorphism**?

Anthropomorphism is the attribution of human characteristics and feelings to nonhumans. An example of this is *Bambi*, a story written by Felix Salten (1869–1945) in 1923; **325**

Konrad Lorenz was a pioneer in the field of avian ethology. He was particularly known for his study of imprinting.

the inspiration for the story came from the wildlife he saw while vacationing in the Alps. When the story was eventually made into a Disney movie, Bambi had become a talking animal, complete with human feelings and emotions. Anthropomorphism can obscure the true motivation for an animal's behavior.

What is **sociobiology**?

Sociobiology, which is considered by some as a subdiscipline of behavioral ecology, is the study of the social organization of a species. Sociobiology attempts to develop rules that explain the evolution of certain social systems.

Who is **Jane Goodall**?

Jane Goodall (1934–) is a primate ethologist who is famous for her studies of chimpanzees in Tanzania. She began her career as a secretary in Nairobi, Kenya, for Louis B. Leakey (1903–1972). After more than forty years of research, Goodall showed that chimpanzees could make and use tools (a behavior previously attributed only to humans). She was also able to distinguish the individual personalities among the chimpanzees she studied. She currently continues her work through the Jane Goodall Institute.

Who was **Dian Fossey**?

Dian Fossey (1932–1985) was an occupational therapist who, inspired by the writings of the naturalist George Schaller (1933–), decided to study the endangered mountain gorilla of Africa. She was trained in field work by Jane Goodall and went on to watch and record the behavior of mountain gorillas in Zaire and Rwanda. She eventually obtained a Ph.D. in zoology from Cambridge University and in 1983 published a book

Who is the "father" of sociobiology?

Edward O. Wilson (1929–), with the publication of *Sociobiology: The New Synthesis* (1975), formulated the general biological principles that govern both social behavior and organization of all kinds of animals. He found that social insects, such as bees and wasps, have a complex hierarchy, with strict rules governing who reproduces, who gets food, and who defends the colony. Wilson became interested in animal behavior as a child when he watched ant colonies. He is one of the most prolific writers in ecology, having written more than twenty books, won two Pulitzer Prizes, and discovered hundreds of new species. The application of sociobiology to humans initially generated controversy as to whether the same forces drive highly organized animal and human soci-

on her studies, *Gorillas in the Mist*. In 1985 she was found murdered in her cabin in Rwanda; her death is still unsolved.

What movie was made about **Dian Fossey's** work with **lowland gorillas?**

The movie *Gorillas in the Mist* was released in 1988, with actress Sigourney Weaver playing the role of Dian Fossey. The movie was filmed in Rwanda and Kenya and galvanized support for the plight of the gorillas.

How is **animal behavior** studied in the **field**?

Animal behavior is studied by construction of an ethogram, which is a listing and description of all naturally observed behaviors. Behavior can also be studied through the use of manipulative investigations, both in the field and in the laboratory. These behaviors are then categorized. In order to be objective, all observers must record behavior patterns in exactly the same way. Observations can then be statistically analyzed.

How can **animal behavior** be **categorized**?

Animal behavior can be sorted into broad categories (e.g., courtship, feeding) or into more specific patterns (e.g., attack, chase, aggressiveness).

How does an **animal's energy budget** affect its **behavior**?

Every animal has a finite amount of energy available for use in a unit of time. This energy usage is its metabolic rate (measured in calories or kilocalories) plus the energy required for life activities. The energy budget places limitations on an animal's behavior. For example, a lizard, which is an ectotherm (cold-blooded), uses less ener-

Dian Fossey made important studies of mountain gorillas in Zaire and Rwanda.

gy because it does not maintain a constant body temperature; ectotherms such as amphibians and reptiles control body temperature by behavior. Endothermic (warm-blooded) animals (e.g., birds, mammals) have a higher energy budget, most of which is used to maintain internal body temperature. Thus, endotherms require more energy than ectotherms of a similar size and so may spend a greater portion of their day searching for food. Other factors that influence energy requirements are age, sex, size, type of diet, activity level, hormonal balance, and time of day.

Energy expenditures in kcal/kg per day

Animal	Energy expenditure Kcal/kg per day
Deer mouse	438
Penguin	233
Human	36.5
Python	5.5

LEARNING

What kinds of **behavior** do **protozoans** exhibit?

Protozoans react to changes in their immediate environment, but there is no evidence for any type of learning. For example, paramecia will avoid a strong chemical or physical stimulus by turning to locate an escape route. An example of a negative stimulus is cool water, which paramecia will swim away from since they prefer warmer temperatures.

What are the **different categories** of **simple behavior**?

Category	Description	Example
Kinesis	Change speed of random movement in response to stimulus	Wood lice: tendency to keep moving in dry environments and to stop moving in moist ones
Taxis	Directed movement toward/away from stimulus	Planaria: negative phototaxis away from light
Reflex	Movement of body part in response to stimuli	Human infant: finger in palm causes infant to close hand
Fixed action pattern	Stereotyped series of movement in response to specific stimulus	Grey goose: will roll a round white object (stimulus) back into its nest in response to specific releaser

What is a **fixed action pattern**?

A fixed action pattern (FAP) is an innate behavior pattern that is genetic and thus independent of individual learning. It consists of a series of stereotypic behaviors that are dependent on an external signal (sign stimulus) that causes the behavior. A well documented FAP is the response of the male three-spined stickleback fish to aggressive stimuli. When a male stickleback is presented with a fish model displaying a red belly characteristic of males, it will display a series of standard threat and aggressive attack behaviors.

Are **humans** the **only animals** who can **think**?

Before one can answer this question, one must define what is meant by thought. Thought can be defined in several ways: is it philosophical rumination or the processing of perceptions of the natural world? Because we are still trying to translate animal communication into human language, it is difficult to provide definitive proof of philosophical thought processes. Current studies by animal behaviorists indicate that animals that have a varied social life (such as chimpanzees) perceive the world in ways similar to humans. However, since we do not share a common verbal language with animals, it is impossible to know what they are thinking.

How can **animal thought processes** be studied?

Determining how an animal thinks will involve scanning the brain of the animal while performing a cognitive task. Data would then be compared with humans.

Although human functional MRI experiments are becoming common, such experiments on other animals are still quite rare.

Can **animals** recognize different **languages**?

Scientists have compared language discrimination in human newborns and cotton-top tamarin monkeys. Each group was presented with twenty sentences in Japanese and twenty sentences in Dutch. Infant reactions were gauged by their interest in sucking on a pacifier; when infants first heard sentences in Dutch, they sucked rapidly on their pacifiers, but after a while they grew bored with the Dutch sentences, and the rate of sucking slowed. When someone started speaking in Japanese, they showed increased interest by increasing their rate of sucking. Language discrimination was studied in tamarin monkeys by changes in the facial orientation toward or away from the loudspeaker. Similar to the infant reactions, the monkeys looked at a loudspeaker broadcasting Dutch sentences and looked away when bored. When someone started speaking Japanese sentences, they looked back at the loudspeaker. Results indicate shared sensitivities between monkeys and humans in the ability to discriminate between languages.

Besides humans, which **vertebrates** are the **most intelligent**?

According to the behavioral biologist Edward O. Wilson (1929–), the ten most intelligent animals are the following: 1) chimpanzee (two species); 2) gorilla; 3) orangutan; 4) baboon (seven species, including drill and mandrill); 5) gibbon (seven species); 6) monkey (many species, especially macaques, the patas, and the Celebes black ape); 7) smaller toothed whale (several species, especially killer whale); 8) dolphin (many of the approximately eighty species); 9) elephant (two species); 10) pig.

Among the **invertebrates**, which are the most **intelligent**?

Most specialists agree that the cephalopods—the octopi, squids, and nautilus—are the most intelligent invertebrates. These animals can make associations among stimuli and have been used as models for studying learning and memory. Octopi can be trained to perform many tasks, including distinguishing between objects and opening jars to obtain food.

Who was **Ivan Pavlov**?

Ivan Pavlov (1849–1936) was a Russian physiologist who became famous for his experiments with dogs in which the animals performed a specific behavior upon being confronted with a certain stimulus. In these well-known investigations, minor surgery was performed on a dog so that its saliva could be measured. The dog was deprived of food, a bell was sounded, and meat powder was placed in the dog's mouth. The meat powder

How does Pavlovian conditioning work with humans?

Pavlovian conditioning is used every day on humans. Advertising campaigns that link an unrelated stimulus (a beautiful woman) with a desired behavior (buying beer or cigarettes) are well-designed examples of classical conditioning.

caused the hungry dog to salivate; this is an example of an unconditioned reflex. However, eventually, after many trials, the dog would salivate at the sound of the bell, without meat powder being offered. This is an example of a conditioned reflex or classical Pavlovian conditioning. Although he never thought much of the then-fledgling science of psychology, Pavlov's work on conditioned reflexes has been far reaching, from elementary education to adult training programs. Pavlov was awarded the Nobel Prize in Physiology or Medicine (1904) for his study of the physiology of digestion.

What is **cognition**?

Cognition is the highest form of learning, and consists of the perception, storage, and processing of information gathered by sensory receptors.

Why do **animals play**?

Many animals (mammals and some birds, particularly) have been observed at play during different stages of development. Although play would seem to be random, ethologists have described three patterns: 1) social, for establishing relationships to other animals; 2) exercise, for development of muscles; 3) exploration of an object. All of these occur during the wrestling, chasing, and tumbling activities of the young of many species. Some juvenile play, such as lion cubs capturing mice, may be practice for adult activities such as hunting.

What animals have been studied for **play behavior**?

Coyotes, wolves, and dogs (all canids) have been studied extensively for comparisons of play behavior patterns. Neither dogs (specifically beagles) nor wolves showed agonistic behavior during play. In comparison, coyotes showed high levels of agonistic behavior throughout a comparable developmental period. Coyotes usually established dominance in a pack by fighting at an early age.

How is **B. F. Skinner** related to **operant conditioning**?

B. F. Skinner (1904–1990) was an American psychologist who extensively studied trial and error learning in animals (later known as operant conditioning). A standard setup

Who first studied animal emotions?

In 1872 Charles Darwin published *The Expression of the Emotions in Man and Animals*, in which he raised such questions as: Why do dogs wag their tails? Why do cats purr?

for his research involved the following: an animal is placed in a cage (known as a Skinner box) that has a bar or pedal that yields a food reward when pressed. Once the animal has practiced the behavior, it will continue to press the bar repeatedly, having learned to associate this activity with food. By releasing food only when the animal completes some task, the observer can train the subject to perform complex behaviors on demand. These operant conditioning techniques have been used to teach behaviors such as training pigeons to play table tennis with their beaks.

Do **animal studies** predict **human behavior**?

George Romanes (1848–1894) was one of the first scientists to investigate the comparative psychology of intelligence. He believed that by studying animal behavior, one could gain insights into human behavior. However, his theory was based on inferences rather than direct observations of comparable behavior.

What **emotions** do **animals** have?

Many pet owners say that they know when their animal is happy or sad, and now there is evidence to show that animals do exhibit emotion. Researchers have found that emotions are accompanied by biochemical changes in the brain that can be measured. When scientists examine the physiological changes found in humans that correlate with certain emotional states (e.g., anger, fear, lust), they find that these changes can also be observed in certain animal species. A study of stress among African baboons showed that social behavior, personality, and rank within the troop can influence the levels of stress hormones. There is increasing evidence that birds, reptiles, and fishes also experience some form of emotions.

Although the idea of animals feeling emotions raises skepticism, students of animal behavior agree that many creatures experience fear, which is largely instinctive and, in effect, is prewired into the brain. Field observations have recorded expressions that correlate with pleasure, play, grief, and depression. Jane Goodall, watching the reaction of a young chimp after the death of his mother, maintains that the animal "died of grief." Even with this evidence, it is impossible to truly know how another organism "feels."

What is the relationship between **blue jays, monarch butterflies, and milkweed**?

As part of their life cycle, female monarchs lay their eggs exclusively on milkweed plants. After a few days the eggs hatch and a yellow, black, and white-striped caterpillar emerges from each egg. These caterpillars are totally dependent on milkweed plants. Although the plants contain toxic substances (cardenolides) that are poisonous to other animals, the toxin is harmless to the monarch. Blue jays, who spend much of their day searching for food, will often eat insects such as adult monarch butterflies to supplement their otherwise vegetarian diet. However, if the food tastes bad, the blue jays will vomit up the food and will then learn individually to avoid the food in the future. Thus, wild monarch butterflies with high levels of cardenolide concentrations are less susceptible to natural predation by birds. This is an example of operant conditioning in the wild.

What is **habituation**?

Habituation is the decreased response to a stimulus that is repeated without reinforcement. Habituation can be very important to an animal in its natural surroundings. As an example, young ducklings run for cover when a shadow (a possible predator) passes overhead; gradually, however, the ducklings learn which types of shadows are dangerous and which are harmless.

How do **songbirds** learn to **sing**?

Through analysis of many bird species, ethologists have found two major types of song development: 1) imitating the songs of others, particularly adults of the same species; and 2) invention or improvisation of unique songs. Observations of male song sparrows, particularly during their first month of life, show that when the birds arrive at a new habitat, they memorize the songs of the males in that neighborhood.

When do **birds learn to sing**?

Male song sparrows generally learn to sing during a critical period between ten and fifty days after hatching. In some birds such as the mouse wren, the learning period for song development is influenced by photoperiod (amount of daylight) and social interactions with other adult birds.

What does **chest-beating in gorillas** signify?

Chest-beating in gorillas is part of an aggressive behavior display. It is usually presented by a silverback (male) against unrelated silverbacks. A chest-beating display, accompanied by hoots and barks, may also be used to impress females.

What are the **most primitive animals** that have been studied for their **behavior**?

Sponges are the most primitive animals studied for their behavior. Investigations have shown that sponges in their natural environment may move away from physical contact with competitors, areas with limited food, or sites that have excessive exposure to silt abrasion by water flows.

Can **animals learn** from other animals?

Yes, animals can learn from other animals. Researchers observing the behavior of Japanese macaques would leave pieces of potato on the beach of the island where the study occurred. Every day the macaques would spend their time carefully cleaning the sand off their treats. Then one day a young female carried her potato to the sea, where she rinsed it off. Soon her mother was following her example, and then other females as well until finally the entire troop had learned the behavior.

Who was the **first primate** taught to use **sign language**?

Although it was long known that primates use a number of methods of communication in the wild, early attempts (from 1900 to the 1930s) to teach primates simple words were failures. A 1925 scientific article suggested sign language as an alternative to verbal language in communicating with primates. In the 1960s researchers tried to teach chimps and gorillas a modified form of sign language. It began with Washoe the chimpanzee, followed by the gorillas Michael (now dead) and Koko. Washoe learned a little over 100 signs, but Koko has a working vocabulary of over 1,000 signs and understands about 2,000 words of spoken English.

Who first suggested that **apes** could use **language**?

Samuel Pepys (1633–1703), famous for his seventeenth-century diary, wrote about what he called a "baboone" and suggested that it might be taught to speak or make signs.

If **primates** are so close to **humans**, both genetically and evolutionarily, then why can't they **speak**?

Scientists used to think that apes were not intelligent enough to speak; however, it is now thought that an ape's vocal cords are not "built" for speech. After many years of

What is a historical example of imprinting?

The children's nursery rhyme "Mary Had A Little Lamb" is an excellent example of imprinting: "Mary had a little lamb/Whose fleece was white as snow./And everywhere that Mary went,/The lamb was sure to go!/It followed her to school one day,/which was against the rules./It made the children laugh and play,/To see a lamb at school.

observations, it is well known that apes do use vocal communication, but it is usually in the form of hoots and grunts, with accompanying gestures.

What does **genetics** have to do with **behavior**?

Some ethologists feel that all behavior is genetically programmed. If a behavior is under genetic control, then it is a sequence of events: sign stimulus, releasing mechanism, and fixed action pattern. However, the mainstream view is that most or all behavior is a result of a combination of genetic programming and environmental learning.

What is **imprinting**?

Imprinting occurs when an animal learns to make a response to a particular animal or object. Usually the behavior is learned by a young animal through exposure to a stimulus early in its development. There are two types of imprinting: filial (social attachment) and sexual. Perhaps the most famous example of imprinting is that of Konrad Lorenz (1903–1989) and the goslings who imprinted him as their mother and followed him in a classic example of filial imprinting.

Do all animals use **imprinting**?

Imprinting is a prime example of the relative influence of nature (genes) and nurture (environment) on behavior. The organism is born with a sketchy outline of an object (parent, reproductive partner) drawn from its genetic component, and this is then filled in by its experiences in the environment. Therefore, while not all species have been found to meet the scientific definition of imprinting, it is highly likely that all animals will exhibit at least some behaviors that combine genetic and environmental influences.

Do animals have **friends**?

Animals do form social attachments (friends) among their peers. For example, among savanna baboons, bonds between males and females are a central feature of a society. Behaviors like allogrooming (grooming another animal) and reciprocal altruism (helping another animal) allow unrelated animals to form enduring bonds.

BEHAVIORAL ECOLOGY

Why do animals **migrate**?

Animals migrate for a variety of reasons, including climate (too hot or too cold during part of the year), food availability (seasonal), and breeding (some animals need a specific environment to lay eggs or give birth). Animals move between locations to take advantage of optimal environments.

Why do the **swallows** return to **Capistrano**?

As part of their annual migration, thousands of swallows arrive each year about March 19 at the San Juan Capistrano mission in San Juan, California. They depart around October 23 for their 6,000 mi (9,656 km) flight to Goya, Argentina. March 19 and October 23 are the feast days for Saints Joseph and San Juan, respectively.

How does a **homing pigeon** find its way home?

Scientists currently have two hypotheses to explain the homing flight of pigeons. Neither has been proved to the satisfaction of all the experts. The first hypothesis involves an "odor map." This theory proposes that young pigeons learn how to return to their original point of departure by smelling different odors that reach their home in winds from varying directions. They would, for example, learn that a certain odor is carried on winds blowing from the east. If a pigeon were transported eastward, the odor would tell it to fly westward to return home. The second hypothesis proposes that a bird may be able to extract its home's latitude and longitude from the earth's magnetic field. It may be proven in the future that neither theory explains the pigeon's navigational abilities or that some synthesis of the two theories is the actual mechanism.

What animal makes the **longest migration**?

The arctic tern (*Sterna paradisaea*) flies a round trip from the Arctic of North America and Eurasia to the Antarctic, a round trip that can total as much as 20,000 mi (32,000 km)!

How do **animals** know when it is time to **migrate**?

Although there are several cues that animals could use to stimulate migration, the most consistent year after year would be the change in duration of daylight, also known as photoperiodism. Although it has been suggested that temperature changes might also play a role, light is still the best predictor of migratory onset.

How do animals know which **direction to travel**?

The ability to navigate from one place to another is found among a diverse group of species. While it is well known that birds can navigate over long distances, many peo-

Why do the buzzards return to Hinckley, Ohio?

After a massive game hunt in Hinckley in December 1818, most of the remains from the hunt were buried by snow. When the snow melted the following spring, the buzzards were drawn by the stench of the dead animals. Since that time buzzards return annually looking for another free lunch. The official date of the return is March 15, proclaimed "Buzzard Sunday" in Hinckley.

ple are unaware that bats, salmon, locusts, and frogs are also capable of this behavior. Animals use a variety of cues to find their way including the position of the Sun, Moon, and stars. Topographic features of the landscape, meteorological cues (e.g., prevailing winds), and magnetic fields are also used by certain animals.

Who decides which way a troop of **monkeys** will **travel**?

Among Savannah and Gelada baboons, females form the stable social structure of the troops while males transfer between troops. Observations of Gelada troops have shown that it is the females who determine the direction the group will travel during the daily foraging trips.

How do **salmon** find the way to their **spawning grounds**?

Scientists do not know exactly how a salmon remembers the way back to its native stream after an ocean journey possibly lasting several years and covering several thousand miles. They agree, however, that salmon, like homing pigeons, appear to have an innate compass or "search recognition" mechanism that operates independently of astronomical or physical signs. Some scientists theorize that this internal compass uses the infinitely small electrical voltages generated by the ocean currents as they travel through the earth's magnetic field. Others believe that the salmon's homing mechanism may take its cues from the varying salinities of the water or specific smells encountered along the journey.

Can animals behave **altruistically**?

Altruism is the performance of a behavior that will benefit the recipient at a cost to the donor, such as risking one's life to save another. There are numerous examples of animals exhibiting altruistic behavior. For instance, adult crows may act as "nannies" for other crows, instead of increasing their fitness by producing their own offspring. Another example is a ground squirrel that will warn others of the presence of a predator, even though making such a call may draw the attention of the predator to itself. In studying social insects, Edward O. Wilson (1929–) found that in many species of social insects, workers forego reproduction entirely (they are sterile) in order to help raise their sisters. There are two possible explanations for this behavior. The donor is performing the act either in hope that the recipient will someday return the favor (reciprocal altruism) or because the recipient is a family member. In the game of evolution, winners are those who leave the greatest number of copies of their genes in the subsequent generations, so this "kin selection" form of altruism may not be so altruistic after all.

What predicts whether one **animal** will **help another**?

Usually animals that help one another are of the same species (conspecific) and are most likely to be genetically related. The closer the relationship, the more likely helping will occur. This is demonstrated among vampire bats, where bats returning to the roost may share blood with those who have not fed. Most sharing occurs between close relatives (e.g., mother and offspring); however, unrelated but closely associated individuals may exhibit reciprocal altruism where help is given to those individuals most likely to return the favor.

Do all **females** take care of their **offspring**?

While females are the most common primary caregivers, in some species (e.g., seahorses) males are the primary caregivers for their offspring. Male parental care may be as simple as defending the nest against potential predators, or as time-consuming as providing food and the shelter of their bodies for young hatchlings. In other species such as guppies, no parental care by either sex is provided.

What is meant by a **dominant animal**?

When two animals attempt to acquire the same resource simultaneously (e.g., a food item) and one consistently wins the prize, that animal is considered dominant. Evolutionary theory suggests that dominance may be a way for animals to avoid actual fighting, since aggressive interaction may be physically costly even for the winner.

How do animals know who is the **boss in the herd**?

Determining dominance can be overt or quite cryptic. It may involve a physical challenge and some kind of stylized fighting, or it may rely on body language, like a change in the way a dog wags its tail. Sometimes animals determine dominance by cues too subtle for human observers to recognize.

What does it mean to be the **alpha male**?

In animal groups with multiple individuals, those at the top of the dominance hierarchy are designated as the alpha male (and alpha female). Those next in line would be the beta individuals, using the nomenclature of the Greek alphabet. Alpha individuals control the behavior of the other animals and may be the only individuals that mate within the group. They may also be the decision-makers, as in determining which direction the group travels, where the group sleeps, and so forth.

What is meant by the phrase "**pecking order**"?

Pecking order refers to the dominance hierarchy (i.e., relative ranking) of animals within the same species. It has been adapted in business to refer to one's position within the company.

Can **animals** commit **suicide**?

There is no evidence of animals committing suicide. In strict Darwinian terms this would be of no advantage to an individual's fitness.

Can **animals** commit **murder**?

If murder is defined as the killing of members of the same species, then some species do indeed commit murder. This could be the result of an altercation to determine dominance within the group or a battle over a resource like food or a mate. Animals, including lions and langur monkeys, have also been known to commit infanticide—the killing of infants. In these cases infanticide has been linked to the arrival of a new alpha male in the group. Scientists surmise that by killing the infants in the group (who were fathered by some other male), the new alpha can bring their mothers into sexual receptivity faster and thus ensure a chance for reproductive success.

How do **elephants** find each other across the **savannah**?

Although elephants are well known for the trumpeting calls that they make when angry or disturbed, they are also capable of using ultrasound (sounds above the range of human hearing) and infrasound (sounds below the range of human hearing) to communicate with one another. Researchers have concluded that elephants may be able to hear ultrasonic calls from as far away as 2.5 mi (4 km). In contrast, it is estimated that an infrasonic call by a male elephant could in fact cover an area of 11.6 sq mi (30 sq km).

339

Geese flying in V-formation.

Do **whales** really **talk** to one another?

Whales produce low-frequency sounds that allow them to communicate across long distances. Recent research has found that among the fin whales, only males produce these calls. The long low frequency sounds of male fin whales attract females to patches of food, where mating can then occur. This means that the increasing amounts of sonar activity from ships in the ocean may interfere with the ability of these males to find mates, and therefore threatens the species' survival.

Do **animals** ever **take advantage** of each other?

Any interaction where one actor benefits while the other is hurt in some way could be described as one animal taking advantage of another. Common examples of this type of behavior include predation and parasitism.

What is **parasitism**?

Parasitism is an interaction in which one organism (the parasite) co-opts the resources of another organism (the host). By definition, the host is hurt by the association while the parasite benefits. Parasitism can be physical, like the parasitic worms found in the internal organs of animals, or social, like the brood or nest parasitism found in some birds. In these species, resident birds are tricked into incubating the eggs (and raising the chicks) of interlopers.

Why do **birds flock**?

The old adage about "safety in numbers" certainly holds true for flocking birds and other social groups. Flocks, herds, schools, and others provide a refuge for the young and an anti-predator advantage for all. The simultaneous movements of a large number of prey are apt to confuse predators and decrease the likelihood that they will capture any one individual.

Why do **birds fly** in **formation**?

Some birds, like geese and cranes, often fly in V-formation because of the phenomenon of wingtip vortices. The lead bird in the V-formation "breaks up" the wall of air that the flock flies into. The swirling air then helps push along the birds behind. However, being the lead bird is hard work, and the leader may then drop back to allow another bird to take on the work of the lead bird. The V-formation also gives geese the ability to watch each other and communicate (via honking) about likely landing spots.

What names are used for **groups of birds**?

A group of birds in general is called a congregation, flight, flock, volery, or volley.

Bird	Group name
Bitterns	Siege or sedge
Budgerigars	Chatter
Chickens	Flock, run, brood, or clutch
Coots	Fleet or pod
Cormorants	Flight
Cranes	Herd or siege
Crows	Murder, clan, or hover
Curlews	Herd
Doves	Flight, flock, or dole
Ducks	Paddling, bed, brace, flock, flight, or raft
Eagles	Convocation
Geese	Gaggle or plump (on water), flock (on land), skein (in flight), or covert
Goldfinches	Charm, chattering, chirp, or drum
Grouses	Pack or brood
Gulls	Colony
Hawks	Cast
Hens	Brood or flock
Herons	Siege, sege, scattering, or sedge
Jays	Band
Larks	Exaltation, flight, or ascension

341

Bird	Group name
Magpies	Tiding or tittering
Mallards	Flush, sord, or sute
Nightingales	Watch
Partridges	Covey
Peacocks	Muster, ostentation, or pride
Penguins	Colony
Pheasants	Nye, brood, or nide
Pigeons	Flock or flight
Plovers	Stand, congregation, flock, or flight
Quails	Covey or bevy
Sparrows	Host
Starlings	Chattering or murmuration
Stork	Mustering
Swallows	Flight
Swans	Herd, team, bank, wedge, or bevy
Teals	Spring
Turkeys	Rafter
Turtle doves	Dule
Woodpeckers	Descent
Wrens	Herd

Why do **fishes** travel in **schools**?

About 80 percent of the approximately 20,000 fish species travel in schools. Fishes travel in schools for both protection and for efficiency. Safety in numbers (in a school) is a form of predator avoidance, because trying to catch one fish in a large, moving school can be a difficult for a predator. Secondly, fishes that travel in schools have less drag (friction) and therefore use less energy for swimming. Also, when fishes spawn, a school ensures that some eggs will evade predators and live to form another school.

How do **schooling fishes** swim so close without **colliding**?

A fish school travels in a smooth movement; each fish responds to movements of other fish, such that if one fish changes direction, all of the others will move accordingly. With eyes placed on the side of the head, fishes can see what is next to them and use this to move. A complex combination of hearing, lateral line system, sight, and smell allows fishes to establish position and direction in a school.

What is the difference between a **territory** and a **home range**?

A territory is a defended area. It can be as small as the space around a female red-winged blackbird's nest or as large as the backyard that your dog defends. A home

> ## Why do cats always rub against their owners and furniture?
>
> The cat may be marking you as part of her territory. By rubbing against objects (and people), the cat will be distributing her scent, and the scent will then serve as a warning to any potential interlopers.

range, in contrast, is simply the area where an animal spends its time. Home ranges may be shared with conspecifics (members of the same species) and may overlap those of other species.

What animal has the **largest home range**?

The home range of an adult polar bear may cover an area of 20,000 sq mi (50,000 sq km). This is an area about the size of Nova Scotia in Canada. The home range of each polar bear varies due to food availability and condition of the ice.

Why do animals **defend territories**?

Animals defend territories to protect their assets such as food, water, or mates. Animals will also defend the area where their offspring live, because in evolutionary terms offspring are the ultimate asset.

How do animals **mark** their **territory**?

Animals use pheromones to mark their territory. A pheromone is a type of airborne chemical. This can be accomplished by leaving bits of fur and dander on a visual object or by scent-marking with small amounts of urine.

Why do animals **pretend** to be **hurt**?

Among bird watchers, the female killdeer is well known for pretending to be hurt. When a potential predator appears within the vicinity of her nest, the killdeer will adopt a posture of wing-dragging, making it appear that she is injured and an easy catch. The female will gradually lead the predator away from her nest, eventually flying off when the predator is at a safe distance from the nest.

Do **opossums** play **dead**?

The common or Virginia opossum (*Didelphis virginiana*) is the only marsupial in the southern and eastern portions of the United States. When threatened or frightened, the opossum will lie quite still with stiffened limbs and a fixed gaze, appearing as if

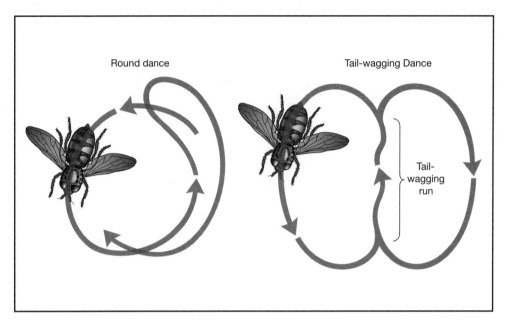

Round dance

Tail-wagging Dance

Tail-wagging run

The bee dance, also known as the waggle dance, is the most complex of all nonhuman communication systems.

dead. When the perceived threat is over, the opossum will return to its normal, nocturnal activities.

Why do **pufferfishes puff**?

Pufferfishes are any of a number of species found in warm seas that use a special adaptation of the gullet to inflate their bodies to nearly twice the normal size. Pufferfishes, blowfishes, and their like do this in response to a perceived threat. The increased size and unpalatable-looking spines make the potential prey look quite unappetizing to predators.

What is the most famous and **deadliest of the pufferfishes**?

Fugu rubripes, the Japanese pufferfish, is a specialty in Sushi restaurants. However, only specially trained chefs can safely prepare the fish for consumption by minimizing the presence of tetrodotoxin (1000 times deadlier than cyanide).

Can **cobras** really **hypnotize their prey**?

While hypnotism would be a fortunate side effect, it is thought that the semi-erect posture and swaying motion are probably used by the cobra to estimate the distance to its prey. African cobras don't strike at their target; rather, they spit venom at threats up to 6 ft (1.9 m) away. To do this, they need to take visual aim.

Who discovered **bee dancing**?

Bee dancing was studied by Karl von Frisch (1886-1982) and colleagues in the 1940s. Details and interpretation of the dance language were published in the 1967 book *The Dance Language and Orientation of Bees*. Von Frisch was able to study a group of bees by replacing one of the walls of a hive with glass.

What is the **significance of bee dancing**?

How is it that within a short time of a bee finding a food source, a great number of fellow bees will converge to gather pollen and nectar? Communication about the food location occurs through the bee dance. This symbolic language of bees, also known as the waggle dance, is the most complex of all nonhuman communication systems. By placing sources of food at varying distances and angles from the hives, Karl von Frisch found that forager bees perform either a round dance for food at short distances (66–656 ft or 20–200 m) or a waggle dance for food at longer distances. While "dancing," the forager also emits sound bursts. The specifics of the dance (e.g., area the dance occupies on the comb, duration of each cycle, and length of sound bursts) indicate the distance and direction to the food source.

Why are some animals so **brightly colored** while others are so **dull**?

There are two general functions of bright coloration. The individual with such coloration is trying to advertise either to members of its own species (its conspecifics) or to those of other species. Within species, communication revolves around reproductive behavior. Male redwinged blackbirds, for example, use their brightly colored shoulder patches to advertise their territory ownership to potential mates and rivals. An experiment with zebra finches found that females were more likely to choose males as mates if the identification bands on their legs were red rather than some other color. Between species, communication is usually threatening. Warning (or aposematic) coloration is a method used by animals with stings or poison to circumvent attacks. For example, stinging insects like wasps and bees use similar color patterns of yellow and black to advertise their arsenal. Poison dart frogs (family Dendrobatidae) also use bright coloration in this way. Dull or camouflage coloration (as in a flounder) is an alternative strategy. By hiding, these species hope to avoid predation.

What are examples of **courting behavior**?

Male stickleback fish swim in a stereotypical manner as they court potential mates. Male bowerbirds build elaborate towers of vegetation to entice females. Female moths release pheromones that attract males from up to a mile away. Male African elephants use low-frequency sounds to find females who are sexually receptive. And, of course, male birds and frogs use vocalizations to attract mates.

> ## Can elephants paint?
>
> The answer is yes! In Thailand 3,000 unemployed, domesticated elephants were on the verge of starvation due to the banning of logging. Two Russian-American artists, after hearing of the elephants' plight, started the Asian Elephant Art and Conservation Project. The elephants paint modern and impressionist pictures in minutes, and the paintings are immediately bought by tourists. The elephants also have been taught to play traditional Thai instruments and are known as the Thai Elephant orchestra.

Who extensively studied **fly behavior**?

Vincent Dethier (1914–1993) spent most of his life researching chemical responses in insects. He wrote numerous articles and books, both for the general public and specifically for children. *To Know A Fly*, perhaps his most famous book, is considered one of the classic books in entomology. The book details physiological research on flies, particularly with regard to chemoreception, with a touch of humor. Aside from his career as a distinguished research scientist, he was able to popularize what scientists do and what draws people to science.

Do animal **societies** have a **culture**?

Culture can be defined as the set of societal rules that are passed from one generation to the next. Parents or caregivers teach juveniles what they need to know in order to participate in their society. This phenomenon has been noted among nonhuman primates and elephants as well as other species.

Elephant families rely on memory and the knowledge of the matriarch (oldest female) to respond to social cues. The matriarch controls the direction in which the herd moves, and where and for how long feeding occurs; when there is danger, the other herd members cluster around the matriarch. Perhaps most striking is the tender loving care lavished on young elephants. The bond between mother and daughter elephants lasts up to fifty years.

How does an **animal's expression** reflect **behavior**?

Although only primates have the facial musculature to be truly expressive, many species use their appearance to convey information. For example, pufferfishes expand their bodies to look more threatening, as do dogs and cats when they raise their hackles and puff up their body fur. Cats, horses, and others also use their ears to relay information about their intentions. And certainly, many a pet owner has learned to read his dog's "mood" by the movement of the tail.

How do **animals** show **sexual readiness**?

Unlike female humans, most female mammals have an estrous cycle instead of a menstrual one. Around the time of ovulation, these females experience estrus, a period of sexual receptivity. Sexual readiness can be advertised in a variety of ways, including physical and behavioral changes. Among females in species with estrus, it is common to see a swelling of the external genitalia. Female chimpanzees and other primates demonstrate this phenomenon. Male mandrills exhibit sexual maturity and dominance by the vibrancy of the coloration on the facial ridges and posterior flesh. When male elephants reach sexual maturity, they experience musth, a condition typified by leakage of fluid from the penis as well as oozing of "tears" from the face. Of course, members of both sexes are likely to also exhibit changes in behavior, actively seeking out the opposite sex, for example. Cats and dogs in heat will go to extraordinary lengths to meet a potential mate.

Why do some animals **hunt** only at **night**?

In the arms race between predator and prey, a change in the behavior of one species can drive adaptation in another (coevolution). Many rodent species for example, are adapted to life at night, but their predators have gained adaptations to night work as well. Owls and foxes have special adaptations that allow them to hunt at night for nocturnally active rodents (e.g., field mice, moles, voles).

Do animals **laugh**?

So far, no example of animal laughter comparable to that of humans has been documented. However, researchers have reported that under certain conditions some species emit special vocalizations that could be considered akin to laughter. In one specific case a researcher has been able to identify a huffing sound unique to dogs at play. Scientists studying the behavior of nonhuman primates and rats have reported similar observations.

Do animals **cry**?

Crying in response to emotional distress has been documented only in humans. However, many animals, particularly the young, demonstrate their response to distress by changes in vocalizations and movement. While tear production (the lacrimal response) is found in a number of animals (but not crocodiles!), the tears are used to maintain the cleanliness and moisture of the eyes and not to display emotion.

What is **nest parasitism**?

From an evolutionary perspective, if breaking the rules is to your advantage, you should do so, as long as you don't get caught. Taking advantage of another's hard work has been documented in a number of species, most notably the brown-headed cowbird. Cowbirds find food by following large mammals (e.g., cows) across open country and eating the bugs disturbed by their hooves. Additionally, when it comes time to reproduce, female cowbirds simply fly off into nearby woods and lay their eggs in the nests of other species, which then raise the cowbird offspring with (or instead of) their own. This nest parasitism is thought to have contributed to the population decrease of species like the Eastern bluebird. Several other species are well known for this behavior, most notably the cuckoo, giving rise to the expression "a cuckoo in the nest," describing someone with a hidden agenda or who is operating under false pretenses.

Why do some animals only have a **few offspring**?

Females have a finite amount of energy that they can allocate to the growth and maintenance of offspring. Therefore, two basic strategies have evolved for successful reproduction: 1) produce many small offspring, only some of whom may survive; or 2) produce just a few larger offspring, each with a higher likelihood of survival.

What is the difference between **altricial** and **precocial** offspring?

"Altricial" and "precocial" are terms used to describe the development of young birds at hatching. Hatchlings that are able to find and consume their own food and have at least minimal antipredator defenses are described as precocial. Offspring requiring extensive parental care, like humans, are altricial. Examples of precocial nonmammalian species include chickens and the young of game birds. Examples of altricial nonmammalian species include songbirds, woodpeckers, and hummingbirds.

What is **camouflage behavior**?

Camouflage behavior allows an animal to avoid predation by appearing to be something it is not. Examples of this include the behavior known as freezing. In addition to the opossum, hog-nosed snakes also exhibit this behavior. Freezing may at first seem counterproductive as an antipredator strategy, but it turns out that some predators (e.g., frogs and wolves) will not attack a nonmoving target.

What is the role of the **queen bee**?

A honeybee society is made up of a single queen, a few male drones, and up to 80,000 female workers. The queen is the only female in the hive capable of reproduction, and she can lay as many as 1,000 eggs in a single day. While the queen's job is solely to lay eggs, the workers' jobs change over time. Young workers serve as nurses to the larvae that develop from the eggs. Older workers serve first as hive builders and then as foragers. And the males? Their job is to fertilize the new queens.

Is polygyny only applicable to nonhuman animals?

A survey of human societies found that polygyny is actually our most common type of mating system. This is reinforced by genetic data that suggests that there is little variation among Y chromosome sequences within populations. Thus, relatively few men have contributed a larger fraction of their chromosome pool in every generation.

What is **polygyny**?

Polygyny is a type of mating system that involves pair bonds between one male and several females. While this may at first seem unfair to females, in a polygynous system every female can usually have a mate, while only the most desirable males will find a mate.

How do **animals** know which **prey to eat**?

Once an animal has located a potential food, it must decide whether to consume or ignore it. In field studies, behavioral ecologists have observed that animals usually select food that will yield the higher rate of energy return for the energy spent "capturing" the food. Very few animals actually eat all of the food they are capable of consuming. This is known as optimal foraging strategy. As an example, crows that live in the Pacific Northwest often find littleneck clams, which they drop on rocks to crack the clams and then eat them. However, the crows do not eat all the clams they locate; they only eat those clams that are larger and thus contain more energy.

How do **bats hunt**?

Bats hunt a variety of prey. Some bats are insectivores, capable of catching their prey on the wing. Vampire bats hunt for large, slow-moving organisms to use for sustenance. Frugivorous or fruit-eating bats are adapted to finding ripe fruit distributed throughout the forest. Depending on the food type, bats may rely on their sight and/or echolocation to find their prey. In echolocation, bats emit ultrasonic sounds that bounce off the objects around them. These signals are fielded by the specialized folds of flesh on the face and around the ears, allowing the bat to judge the direction and distance of objects.

How is an **ant colony** organized?

Ants are the dominant social insects and, numerically, the most abundant. At any one time there are least 10^{15} (one quadrillion) ants alive on the planet! In general, ant colonies contain a variety of castes (groups of individuals with a common job) like workers or soldiers, and a queen who is in charge of egg-laying. For example, the largest and most aggressive workers make up the soldier caste, whose job it is to protect the colony against dangerous invaders. Because of the way that sex is determined among ants, males are haploid (contain one set of chromosomes) while females are diploid (two sets of chromosomes). This causes a close interrelatedness among members of a colony that is theorized as one of the reasons for the evolution of social colonies.

How does an animal's **niche** affect its **behavior**?

A niche can be defined as the natural history of a species, where it lives, what it eats, and so forth. Those constraints in turn have an effect on behavior. For example, lizard species found in deserts from different parts of the world use a very similar behavior to harvest water from the moist air in the early morning. They use the bumps or ridges on their heads as condensing spots and then orient their bodies so that the water rolls toward the snout, where it can be licked off by the tongue.

Can an **octopus learn**?

Cephalopods (squid and octopi) are unique among the invertebrates in the extent of their intelligence. Octopi can be taught, for example, to associate neutral geometric shapes with either punishment (a mild electric shock) or reward (food). This can then be used to train them to avoid one type of food and reach for another. Research has indicated that octopi are also tool users; with their flexible arms and suckers, octopi are able to manipulate their environment, as in building a simple home. After an octopus has selected a home site, it will narrow the entrance size by moving small rocks.

Why do **termites** march around a **ring made of ink**?

Termite soldiers and workers are blind and so use pheromones to navigate. Two chemical compounds have already been identified as termite pheromones. It seems likely that the chemical formulation of certain inks contains compounds that mimic these naturally produced signals.

How does **social deprivation** affect animals?

The effect of a lack of parental care on the social development of young monkeys was studied by Harry Harlow (1905–1981) and his colleagues beginning in the 1950s. In a now-classic experiment, Harlow was able to show that the mother-infant bond was so important to young rhesus monkeys that the infants preferred a soft cuddly fake mother to a fake mother built from wire even if it had a nursing bottle attached. Depending on the age of the monkey and the duration of the treatment (total isolation, isolation with fake mother, and so on), monkeys in these studies later exhibited a range of behavioral deficits including rocking and swaying, poor maternal behavior, and a failure in understanding communication signals from other monkeys.

Do **migrating** animals actually have **magnetic crystals** in their **brains**?

Scientists have long speculated that birds may have magnetic compasses that allow them to detect the earth's magnetic field and thus determine directions. Experiments in which magnets were strapped to the heads of pigeons caused the birds to become disoriented. Although magnetic iron has been found in bacteria and a variety of animal tissues, none has been clearly linked with a magnetic sense.

Why do animals that are **round** tend to show only very **simple behaviors**?

Examples of "round" animals include members of the phylum Cnidaria (hydras, jellyfishes, corals) and phylum Echinodermata (starfishes, sea urchins, sand dollars). An animal with radial symmetry usually has a nerve net that only allows very simple types of behavior. Animals that are round are usually sessile (i.e., nonmoving). This is in contrast to animals that display bilateral symmetry, in which there is a distinct head/tail and in which the animals can be divided into different planes. Bilaterally symmetrical animals usually move in a specific direction.

What is a **biological clock**?

A biological clock controls a biological rhythm; it involves an internal pacemaker with external (usually environmental) cues. An environmental signal that cues the clock for animals is called an *zeitgeber*, a German term meaning "time giver." Examples of *zeitgebers* include light and dark cycles; high and low tides; temperature; and food availability.

How are **biological rhythms** correlated with animal behavior?

A biological rhythm is a biological event or function that is repeated over time in the same order and with a specific interval. Biological rhythms are evident when an animal's behavior can be directly correlated to certain environmental features that occur at a distinct frequency. Biological clocks control animal behaviors such as when migration, mating, sleep, hibernation, and eating occur.

Examples of Biological Rhythms

Type of cycle	Organism/behavior
Tidal: 12.4 hrs	Oyster (feeding); fiddler crab (mating/egg laying)
Circadian: 24 hours	Fruit fly (adults emerge from pupa); deer mouse (general activity)
Circannual: 12 months	Woodchuck (hibernation); robin (migration/mating)
Intermittent: from several days to several years	Lion (feeding triggered by hunger); shiner (river fish; reproduction triggered by flooding)

How do animals **recognize** each other?

We know that animals can use scent, color, and sound to recognize individuals, and they may also be able to recognize other attributes as well. A recent study on sheep intelligence indicates that easily herded animals may be smarter than originally thought. The sheep were shown pictures of other sheep and were subsequently rewarded if they moved toward a selected picture. The sheep learned which face produced a reward. Ultimately, it was shown that sheep were able to pick a selected picture 80 percent of the time and could remember up to fifty images for two years.

Can animals use **tools**?

A tool can be defined as any object used by an animal to perform a specific task. Chimpanzees carefully select twigs that they then prepare as probes to fish out termites from mounds. Sea otters use rocks to crack open clam shells. Birds will drop clams onto rocks to crack their shells. Japanese macaques use the sea to wash sand off food items.

What different **types** of **aggressive behavior** do animals display?

Animals can show aggression through sound (e.g, growls, barks, trumpeting), sight (e.g., changing coloration, inflating body structures), and even scent. They can change the way they move, where they perch, or how much tooth enamel they display. For example, yawning among male mandrills is often not an expression of boredom but rather an opportunity to display their well-honed canine teeth.

How can some animals **communicate via electric fields**?

Sharks, skates, and rays (all cartilaginous fish) have specialized structures for sensing weak electric fields. These structures are used for finding prey and navigation and are also thought to be useful in finding mates as well.

How do animals **communicate by scent**?

A variety of animals use scent-marking to identify their territory. Some use scent-marking for possessions. In Asian muntjac deer, males have scent glands on their faces that they use to mark their mates.

What is **hibernation**?

Hibernation (from the Latin term *hiberna*, meaning "winter")is a period of dormancy practiced by animals to overcome wintry environmental conditions. Hibernation involves a decrease in metabolic rate (the rate of burning calories), heart rate, respiration, and other functions (e.g., urine production, rate of digestion). These rates dive so low that the animal's body temperature approaches that of its surroundings. Small animals whose increased metabolic rate forces them to find an alternative to starving during the winter months are more likely to hibernate than larger animals. Many rodents and bats hibernate, as do some Australian marsupials. Hummingbirds and some other species of birds hibernate as well. As for bears, while they are certainly less active during the winter, they do not truly hibernate. Instead, they take very long naps known as "winter sleep."

Do **bears** in a zoo **hibernate**?

Bears do not technically hibernate. In winter conditions they find protected areas like a cave or hollow log and conserve energy by taking extended naps (winter sleep). This is why it is always dangerous to disturb a wintering bear, because it is merely napping, not hibernating. In zoos, because temperatures in cages and enclosures remain warm throughout the year, and the bears have access to a food supply continually replenished by keepers, they remain active year-round.

What is **torpor**?

Torpor is a short-term decrease in body temperature and metabolic rate. Animals such as hummingbirds and bats go through daily periods of torpor that allow them to reduce their energy requirements at night or when hunting is poor. Torpor can be considered as a type of brief sleep, but it is distinct from a state from hibernation.

What is **estivation**?

Estivation (from the Latin term *aestas*, meaning "summer") is a process by which animals become dormant during the summer rather than the winter. Estivation may be used as a survival strategy against intense heat (ground squirrels), drought (snails), or both. The Columbian ground squirrel both hibernates and estivates, beginning its dormant period in the late summer and continuing it until the next May.

APPLICATIONS

What is unique about the **marine iguanas of the Galapagos**?

The marine iguanas of the Galapagos show evidence of unique behavioral adaptations to their environment: they are strictly vegetarian and are able to swim. In effect, they are the "cows" of the Galapagos, taking on the role of primary consumers.

Why do animals in a **zoo** pace?

Pacing is an indication of lack of stimulation. A recent doctoral study found that larger animals, which have a larger home range in the wild, are particularly prone to pacing in captivity, a behavior known as cage stereotypy. Most reputable zoos attempt to assuage pacing by providing what is known as "animal enrichment." Enrichment may include constantly changing objects, odors, and sounds to mimic the types of stimulations that would be found in the animal's natural habitat. As an example, in order to keep life interesting for poison dart frogs from South America, keepers will hide their crickets inside a coconut with holes drilled in it. The frogs then have to seek out the food inside the chirping coconut.

Do **humans** show **grooming behavior** similar to that of **other primates**?

Grooming is an important tool in cementing social relationships among primates. In studies of nonhuman primates, grooming, which is used ostensibly to remove external or ectoparasites, most often occurs between close relatives. In humans, grooming is also used as a relationship builder, although humans will groom nonrelatives as well.

Can animals suffer from **psychological disorders**?

Although animals are used as models for research into disorders like anxiety, depression, and even schizophrenia, these conditions have been induced in them either through surgery or behavioral treatments. Scientists are currently developing strains of mice and rats that are genetically predisposed toward these diseases in hopes of gaining a better understanding of symptoms and more effective treatments. As for animals in the wild, the answer lies in the still-unresolved issues of whether animals experience emotions and whether they have consciousness or the awareness of self.

Why do **cats swish their tails**?

A cat's tail is a means of communication and can represent a range of moods. Every cat owner is familiar with the slow sweep of a cat's tail from side to side. A happy cat will greet you with a tail raised high, while a slight movement indicates pleasure and/or anticipation. As the cat becomes more hunter-like, the flick of the tip

> ## What is meant by the expression "wag the dog"?
>
> Tail wagging is a good indication of a dog's intent. For example, exaggerated wagging is a greeting of pleasure and/or friendliness, although it is also used by subordinate animals to greet a more dominant individual. This behavior may be so extreme that it appears as if the tail is wagging the dog. A slow movement of the tail, especially when the tail is held stiffly in line with the back, indicates displeasure and warning. A tail clamped tightly between the hind legs is an indication of fear.

becomes first a twitch and then a thrashing. At this point the cat is highly engaged and likely to pounce.

Can you teach an **old dog new tricks**?

Effective dog training is based on using the dog's normal behavior to teach it new applications. For example, rewarding Fido when he randomly performs the desired behavior will allow the trainer to eventually generate that response on demand. Theoretically, positive reinforcement can be used at any stage of a dog's life.

Why will my **dog** play **fetch** but not my **cat**?

Domestication occurred differently for cats and dogs. While dogs came to work closely with humans, relying on them for food and shelter, cats maintained a more distant relationship. A dog plays fetch because it is an extension of the cooperative working-hunting relationship she has with humans, while the cat has no such relationship.

Can **horses really do math**?

At the end of the nineteenth century, a performing horse in Germany known as Clever (or Kluge) Hans was able to tap out the answers to mathematical problems written on a chalk board. Hans would use his right forefoot to indicate the single digits (0–9) and his left forefoot for the tens place (10, 20, 30, etc.). His amazing performances continued for a number of years until the psychologist Oskar Pfungst (1874–1932) was able to show that Hans was simply counting until his questioner indicated (subconsciously) that Hans had reached the correct sum. Even though the horse was not actually performing calculations, his ability to observe and respond to subtle changes in human behavior is still quite noteworthy.

Who was **Mr. Ed**?

Mr. Ed was the "the talking horse" and a star of a television show in the 1960s. When he appeared to talk, the horse was actually responding to cues from his trainer. Move-

Can human behavior change dog behavior?

Researchers at the Max Planck Institute for Evolutionary Anthropology have demonstrated that dogs will modify their behavior if they know that they are being observed by humans. In the study, dogs were less likely to approach a forbidden treat if the human sitting quietly nearby had open rather than closed eyes. Distracted humans with their backs turned or playing video games were also more likely to have food snatched.

ment of a small rope running from his halter through his mouth and held by the trainer off camera would cause Ed to move his lips as if he were speaking. In the real world, so far only birds have been able to mimic human speech.

How do you teach a **parrot** to **talk**?

Teaching a parrot or cockatoo to talk involves a combination of repetition and reward. The owner or trainer reinforces proper vocalization by the bird with treats or laudatory language.

Which animals are **most demanding as pets**?

There are several ways to answer this question. Physically, animals with very precise feeding requirements could be considered most demanding. An example would be the type of live food that a snake requires. On the other hand, animals with high learning capacity and communication skills can be quite exhausting for the owner. For many people, the physical and/or mental challenges of keeping a high maintenance pet like a chimpanzee or tiger can become overwhelming, which is why so many exotic pets wind up being abandoned or given up for adoption.

Do humans use **pheromones**?

Yes, human eggs release a chemical signal that allows them to communicate with sperm. Human females also respond to pheromones. In this case the signals actually coordinate the menstrual cycle. In fact, two human pheromones have been identified: the first increases the likelihood of ovulation, while the second suppresses ovulation. For years scientists debated the importance of pheromones to humans, but the close examination of the olfactory receptors and the vomeronasal organ in the nose has given researchers a mechanism by which humans may be able to recognize these compounds.

Do animals "**marry**"?

It is estimated that 90 percent of bird species are monogamous—that is, one male mates with one female to produce offspring. Some of these pair bonds may actually

What bird gives new meaning to the term "bird brain"?

Alex, an African gray parrot, and other African grays have been studied at the University of Arizona. These parrots are remarkable mimics: Alex can name over fifty different objects and can understand concepts of "same/different," "absence," "quantity," and "size."

extend beyond a single mating season and so could be considered a form of "marriage." The type of pair bonds a species will form is dependent on their ecological niche and is heavily influenced by the needs of their offspring. Altricial offspring, which require large amounts of parental care for survival (like humans), demand the efforts of two parents and therefore are more likely to be found in monogamous species.

DNA, RNA, AND CHROMOSOMES

INTRODUCTION AND HISTORICAL BACKGROUND

What term was originally used for **DNA**?

DNA was originally called nuclein because it was first isolated from the nuclei of cells. In the 1860s Friedrich Miescher (1844–1895), a Swiss biochemist working in Germany at the University of Tübingen lab of Felix Hoppe-Seyler (1825–1895), was given the task of researching the composition of white blood cells. He found a good source of white blood cells from the used bandages that he obtained at a nearby hospital. He washed off the pus and isolated a new molecule from the cell nucleus; white blood cells have very large nuclei. He called the substance nuclein (which we now call DNA). The substance was rich in nitrogen and phosphorus and also contained carbon, hydrogen, and oxygen. Hoppe-Seyler checked and verified the important work of his student Miescher.

Whose experiments showed that **nonlethal bacteria** could be **transformed** into **lethal bacteria**?

In 1928 an army medical officer, Frederick Griffith (1878–1939), was trying to find a vaccine against *Streptococcus pneumoniae*. In the course of his work, he found that there were two strains of the bacteria, one that had a smooth coat *S* but was lethal; the other form had a rough coat *R*, which was nonlethal when injected into mice. He decided to investigate what would happen if he injected both heat-killed *S* bacteria and live *R* bacteria into mice. To his surprise the mice injected with this combination died. Upon closer examination of the blood, living *S* bacteria were found. Something had occurred that transformed the nonlethal *R* bacteria into *S* bacteria. Subsequent

359

experiments throughout the 1940s attempted to find the identity of the transforming factor, which was eventually found to be DNA.

How did scientists decide that **DNA** was the **genetic material** for all cellular organisms?

The proof that the material basis for a gene is DNA came from the work of Oswald T. Avery (1877–1955), Colin M. MacLeod (1909–1972), and Maclyn McCarty (1911–) in a paper published in 1944. This group of scientists followed the work of Griffith in order to discover what causes nonlethal bacteria to transform to a lethal strain. Using specific enzymes, all parts of the *S* (lethal) bacteria were degraded, including the sugarlike coat, the proteins, and the RNA. The degradation of these substances by enzymes did not affect the transformation process. Finally, when the lethal bacteria were exposed to DNase, an enzyme that destroys DNA, all transformation activity ceased. The transforming factor was DNA.

Why were the **1950s** such an important era of **DNA research**?

In the wake of World War II, when research was primarily driven by wartime need, researchers again returned to basic rather than applied research. The 1950s was a period of intense study as scientists attempted to understand the nature of DNA and the gene. The problem was attacked from both biochemical and structural aspects.

What is the "**Race for the Double Helix**?"

Great minds on both sides of the Atlantic Ocean were intent on finding the structure of DNA. "Race for the Double Helix" is a BBC-TV production (1986) based on the research and personal interactions of those who worked on the structure of DNA during the 1950s. The main characters included James Watson (American, 1928–), Francis Crick (British, 1916–2004), Rosalind Franklin (British, 1920–1958), Maurice Wilkins (British, 1916–), Peter Pauling (Linus Pauling's son,

In this photo of Nobel Prize winners from 1962, Francis Crick is at the far left, Maurice Wilkins is second from left, and James Watson is fourth from left.

American, 1931–2003), John Randall (British, 1905–1984), and Edwin Chargaff (French, 1905–2002).

Why is **Rosalind Franklin** called the "Dark Lady of DNA"?

Rosalind Franklin (1920–1958) was a chemist by training; she worked at King's College, London, in 1951 in the lab of John Randall (1905–1984). Both Franklin and Randall were working on the structure of DNA using the relatively new field of X-ray crystallography. Through meticulous research on the DNA molecule, Franklin took photographs that indicated a helical structure. Randall presented Franklin's work at a seminar where it was then (without Franklin's knowledge) provided to the competitors (Watson and Crick) at Cambridge University. This research was crucial to the detailed description of DNA that was published in 1953. Because the Nobel Prize is only awarded to the living, Franklin, who died of cancer in 1958, did not share the award when it was given to Watson, Crick, and Wilkins in 1962.

What was the role of **Maurice Wilkins** in early DNA research?

Maurice Wilkins (1916–) was trained as a physicist and worked briefly on the Manhattan Project, where he became disillusioned and then turned to the field of biophysics. He worked at Kings College, London, with John Randall, where together they began to use X-ray crystallography to study DNA. Both Wilkins and Rosalind Franklin

Although trained as a physicist, Maurice Wilkins made pioneering studies of DNA using X-ray crystallography.

worked in the same laboratory, but their relationship was not one of cooperation. This ultimately slowed the progress of their work.

What is **Chargaff's rule**?

Chargaff's rule is based on the data generated by Edwin Chargaff (1905–2002) in his study of the base composition of DNA from various organisms. By comparing the nitrogen base composition of various organisms, he found that in all double-stranded DNA, the amount of adenine equals the amount thymine, and amount of guanine equals the amount cytosine (A=T, G=C).

DNA

What is **DNA**?

Deoxyribonucleic acid (DNA) is the genetic material for all cellular organisms. The discovery of DNA is considered the most important molecular discovery of the twentieth century.

What are the **component molecules of DNA**?

The full name of DNA is deoxyribonucleic acid, with the "nucleic" part coming from the location of DNA in the nuclei of eukaryotic cells. DNA is actually a polymer (long strand) of nucleotides. A nucleotide has three component parts: a phosphate group, a five-carbon sugar (deoxyribose), and a nitrogen base. If you visualize DNA as a ladder, the sides of the ladder are made of the phosphate and deoxyribose molecules, and the rungs are made of two different nitrogen bases. The nitrogen bases are the crucial part of the molecule with regard to genes. Specific sequences of nitrogen bases make up a gene.

How is a **DNA molecule** held together?

Although DNA is held together by several different kinds of chemical interactions, it is still a rather fragile molecule.

The nitrogen bases that constitute the "rungs" of the ladder are held together by hydrogen bonds. The "sides" of the ladder (the phosphate and deoxyribose molecules) are held together by a type of covalent bond called a phosphodiester bond. Because part of the DNA molecule is polar (the outside of the ladder), and the rungs (nitrogen bases) are nonpolar, there are other interactions—called hydrostatic interactions— that occur between the hydrogen and oxygen atoms of DNA and water. The internal part of the DNA tends to repel water, while the external sugar-phosphate molecules tend to attract water. This creates a kind of molecular pressure that glues the helix together.

What are the **nitrogenous bases of DNA**?

The nitrogenous bases have a ring of nitrogen and carbon atoms with various functional groups attached. There are two types of nitrogenous bases. They differ in their structure: thymine and cytosine have a single-ring structure, while adenine and guanine have double-ring structures. When Watson and Crick were imagining how the bases would join together, they knew that the pairing had to be such that there was always a uniform diameter for the molecule. It therefore became apparent that a double-ring base must always be paired with a single-ring base on the opposite strand.

What is the **law of complementary base pairing**?

The law of complementary base pairing refers to the pairing of nitrogenous bases in a specific manner: purines pair with pyrimidines. More specifically, adenine must

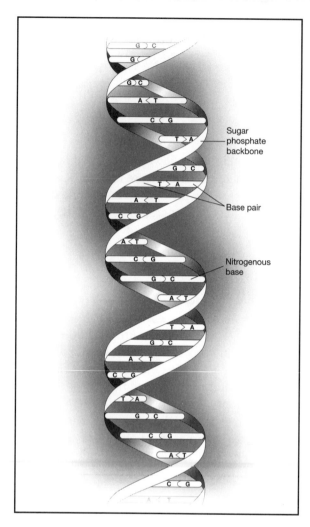

A depiction of the double helical structure of DNA.

Sugar phosphate backbone

Base pair

Nitrogenous base

always pair with thymine, and guanine must always pair with cytosine. The basis of this law came from the data obtained by Edwin Chargaff (1905–2002) and is known as Chargaff's law or rule.

The **DNA strands** are called **antiparallel.** What does this mean?

This means that structurally, the two strands of DNA (the two sides of the ladder) run in opposite directions. This property of antiparallelity allows for formation of the hydrogen bonds between the nitrogenous bases. These bonds are critical to the integrity of the molecule.

How do we know that **DNA** is **antiparallel**?

The central structure of a DNA molecule is the deoxyribose molecule, which has 5 carbons, each of which is numbered. At one end, phosphate can only be attached at the number 5 carbon (5'). At the other end, there is no phosphate attached, and this is called the number 3 carbon (3') end; here, a hydroxyl group is attached to the sugar instead of a phosphate.

What is **DNA supercoiling**?

When DNA is not being replicated or specific genes are not transcribed, its normal form is two strands twisted around a helical axis, much like a spiral staircase turning clockwise. However, during DNA replication or transcription, enzymes alter the structure of DNA such that additional twists are added (positive supercoiling) or subtracted (negative sueprcoiling). Either type of supercoiling makes DNA even more compact. A group of enzymes, the topoisomerases, are able to disentangle DNA strands. They are called topoisomerases because they control the topology of DNA.

How is **DNA unzipped**?

DNA is unzipped during its replication process; the two strands of the double helix are separated and a new complementary DNA strand is synthesized from the parent strands. Also, during DNA transcription, one DNA strand, known as the template strand, is transcribed (copied) into an mRNA strand. In order for the two strands of DNA to separate, the hydrogen bonds between the nitrogen bases must be broken. DNA helicase breaks the bonds. However, the enzyme does not actually unwind the

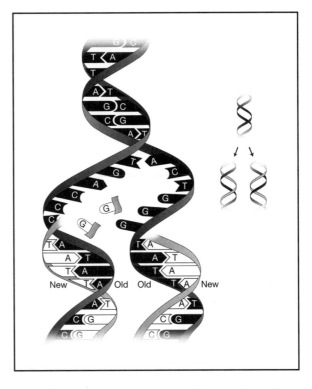

DNA replication is a complex process requiring more than a dozen enzymes, nucleotides, and energy.

DNA; there are special proteins that first separate the DNA strands at a specific site on the chromosome. These are called initiator proteins.

What is needed for **DNA replication**?

DNA replication is a complex process requiring more than a dozen enzymes, nucleotides, and energy. In eukaryotic cells there are multiple sites called origins of replication; at these sites, enzymes unwind the helix by breaking the double bonds between the nitrogen bases. Once the molecule is opened, separate strands are kept from rejoining by DNA stabilizing proteins. DNA polymerase molecules read the sequences in the strands being copied and catalyze the addition of complementary bases to form new strands.

What is a **mutation**?

A mutation is an alteration in the DNA sequence of a gene. Mutations are a source of variation to a population, but they can have detrimental effects in that they may cause diseases and disorders. One example of a disease caused by a mutation is sickle cell disease, in which there is a change in the amino acid sequence (valine is substituted for glutamic acid) of two of the four polypeptide chains that make up the oxygen-carrying protein known as hemoglobin.

Is DNA always copied exactly?

Considering how many cells there are in the human body and how often it occurs, DNA replication is fairly accurate. Spontaneous damage to DNA is low, occurring at the rate of 1–100 mutations per 10 billion cells in bacteria. The rate for eukaryotic genes is higher, about 1–10 mutations per million gametes. The rate of mutation can vary according to different genes in different organisms.

Mutation Rates of Different Genes in Different Organisms

Organism	Mutation	Rate	Unit
E. coli	Histidine requirement	2×10^{-8}	Per cell division
Corn	Kernel color	2.2×10^{-6}	Per gamete
Fruit fly	Eye color	4×10^{-5}	Per gamete
Mouse	Albino coat color	4.5×10^{-5}	Per gamete
Human	Huntington disease	1×10^{-6}	Per gamete

How fast is DNA copied?

In prokaryotes, about 1,000 nucleotides can be copied per second, so all of the 4.7Mb of *Escherichia coli* can be copied in about forty minutes. Since the eukaryotic genome is immense compared to the prokaryotic genome, one might think that the eukaryotic DNA replication would take a very long time. However, actual measurements show that the chromosomes in eukaryotes have multiple replication sites per chromosome. Eukaryotic cells can replicate about 500–5000 bases per minute; the actual time to copy the entire genome would depend on the size of their genome.

What is B-DNA?

There are three different structural forms of DNA. The secondary structure of DNA varies according to the amount of water surrounding the molecule. The structure that Watson and Crick described is B-DNA, and evidence supports that this is the predominate form of most cells. B-DNA is a beta helix with a clockwise spiral. Another structural form, A-DNA, exists when there is less water immediately surrounding the molecule; A-DNA is an alpha (right-handed) helix, but it is shorter and wider than B-DNA. The third form, Z-DNA, is a left-handed helix in which the sugar phosphate groups zigzag. This form is found when certain specific base sequences are present.

How does **DNA correct** its own **errors**?

Spontaneous damage to DNA occurs at a rate of one event per billion nucleotide pairs per minute. Assuming this rate in a human cell, there are 10,000 different sites in the body at which DNA is damaged every twenty-four hours. DNA has a number of quality control mechanisms. DNA polymerase (the enzyme that catalyzes DNA replication) has a proofreading function that immediately corrects 99 percent of these errors during replication. Those errors that pass through are corrected by a mismatch repair system. When a mismatch base (such as A-G) is detected, the incorrect strand is cut and the mismatch is removed. The gap is then filled in with the correct base, and the DNA is resealed.

Is all **DNA** found in the **nucleus**?

In addition to the nuclear DNA of eukaryotic cells, mitochondria (an organelle found in both plant and animal cells) and chloroplasts (found in plant and algal cells) both contain DNA. Mitochondrial DNA contains genes essential to cellular metabolism. Chloroplast DNA contains genetic information essential to photosynthesis.

Why is **DNA** such a **stable** molecule?

DNA is a stable molecule due to the stacking interaction between the adjacent base pairs in the helix and the hydrogen bonding between bases. This stability has allowed researchers to harvest DNA from extinct species.

Is **DNA** always **double-stranded**?

At temperatures greater than 176°F (80°C), eukaryotic DNA will become single-stranded. Single-stranded eukaryotic DNA does not always have the characteristic secondary helical structure. Single-stranded DNA can form a hairpin, stem, or a cross. Certain viruses have only single-stranded DNA.

What is **DNA acetylation**?

DNA acetylation refers to the addition of an acetyl (CH_3CO) group to one of the histone proteins that help hold DNA in its tightly wound configuration. When histones

367

are altered by this change, the binding between histones and DNA is relaxed. This promotes transcription in eukaryotic cells.

What is **DNA methylation**?

DNA methylation is a modification of DNA in which methyl (-CH$_3$) groups are added to certain positions on the nitrogen bases. In bacteria, adenine and cytosine are commonly methylated, while in eukaryotic cells, cytosine is most commonly methylated. In eukaryotic cells, methylation is a method of inhibiting gene expression.

How does **smoking** affect the **DNA** of **lung cells**?

Smoking appears to alter gene expression in lung cells. The proteins produced by the bronchial cells of smokers show increased synthesis of genes, which can eventually be carcinogenic. It appears that the expression of these important genes that control cancer development, cancer supression, or airway inflammation varies according to the number of years spent smoking. The good news is that two years after a person stops smoking, his/her gene expression levels begin to resemble those of people who have never smoked.

What is **cDNA**?

Complementary DNA (cDNA) is single-stranded DNA that is complementary to a certain sequence of messenger RNA. It is usually formed in a laboratory by the action of the enzyme reverse transcriptase on a messenger RNA template. Complementary DNA is a popular tool for molecular hybridization or cloning studies.

Other than **DNA**, what else can **nucleotides** be used for?

Nucleotides can also act as messengers or modulators. For example, the nucleotide adenosine (adenine plus ribose without the phosphate) may be the most important type of immunomodulator (a compound that increases or decreases neuronal activity). Adenosine is used clinically to stop the heart when it is beating erratically; the natural heart pacemaker cells then return the heart to its normal rhythm. Adenosine is also thought to play a role in feelings of fatigue or tiredness; neuroscientists currently theorize that caffeine works to maintain alertness by interfering with the reception of adenosine on the cell surface.

RNA

When was **RNA discovered**?

By the 1940s it was known that there was another kind of nucleic acid other than DNA, this one called RNA. Phoebus Levene (1869–1940), a Russian-born chemist, further

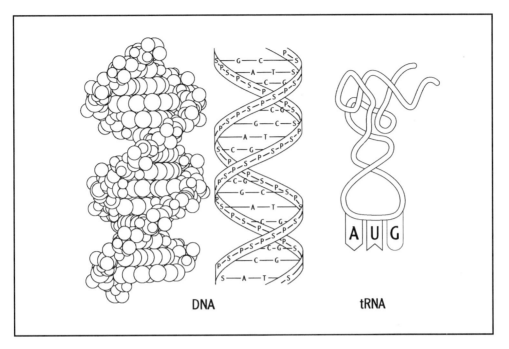

A comparison of DNA and RNA.

refined the work of Albrecht Kossel (1853–1927). Kossel was awarded a Nobel Prize in 1910 for determining the composition of nuclein. At the time of Kossel's work, it was not clear that DNA and RNA were different substances. In 1909 Levene isolated the carbohydrate portion of nucleic acid from yeast and identified it as the pentose sugar ribose. In 1929 he succeeded in identifying the carbohydrate portion of the nucleic acid isolated from the thymus of an animal. It was also a pentose sugar, but it differed from ribose in that it lacked one oxygen atom. Levene called the new substance deoxyribose. These studies defined the chemical differences between DNA and RNA by their sugar molecules.

Which came first—**DNA or RNA**?

The first information molecule had to be able to reproduce itself and carry out tasks similar to those done by proteins. However, proteins, even though bigger and more complicated than DNA, can't make copies of themselves without the help of DNA and RNA. Therefore, RNA was the likely candidate for first information molecule because scientists have found that RNA, unlike DNA, can replicate and then self edit.

What is the **difference** between **DNA and RNA**?

DNA and RNA are both nucleic acids formed from a repetition of simple building blocks called nucleotides. A nucleotide consists of a phosphate (PO_4), sugar, and a nitrogen base, of which there are five types: adenine (A), thymine (T), guanine (G), cytosine (C), and

369

James Watson was a key player in the race to uncover the double helical structure of DNA.

uracil (U). In a DNA molecule, this basic unit is repeated in a double helix structure made from two chains of nucleotides linked between the bases. The links are either between A and T or between G and C. The structure of the bases does not allow other kinds of links. RNA is also a nucleic acid, but it consists of a single chain, and the sugar is ribose rather than deoxyribose. The bases are the same except that the thymine (T) in DNA is replaced by another base called uracil (U), which also links to adenine (A).

Where is **RNA formed**?

All RNA is formed in the nucleus (eukaryotic cells) or in the nucleoid region (prokaryotic cells). The principal enzyme responsible for RNA synthesis is RNA polymerase.

What is the "**RNA Tie Club**"?

In 1953, shortly after the publication of James Watson and Francis Crick's paper on the structure of DNA, George Gamow (1904–1968), a Russian physicist, wrote to Watson and Crick and suggested a mathematical link between DNA structure and the structure of twenty amino acids. The "RNA Tie Club" was an outgrowth of this proposal; members included Watson, Crick, and seventeen others, each taking a nickname from one of the twenty amino acids. Each member of the club was presented with a specially designed tie corresponding to the appropriate amino acid.

What are **ribozymes**?

Ribozymes are often referred to as "molecular scissors" that cut RNA. They were discovered in the early 1980s by Sidney Altman (1939–) and Thomas Cech (1947–), who

What is meant by the term "RNA world"?

"**R**NA world" refers to a hypothetical stage in the origin of life on Earth in which RNA carried out two tasks: 1) storing information, and 2) acting as an enzyme. The term was first used in 1986 by Walter Gilbert (1932–). It is thought that RNA molecules assembled themselves randomly in the primordial soup and carried out simple metabolic activities.

won the Nobel Prize in Chemistry for their work in 1989. The ability of ribozymes to recognize and cut specific sequences of RNA allows certain genes to be turned off. Ribozymes are now being used in human genetic studies.

Why is **RNA** an **unstable molecule**?

Not all forms of RNA are unstable. Messenger RNA (mRNA) molecules, however, can vary in their stability, depending on their rate of degradation and synthesis in cells and the amount of a particular protein needed.

GENES AND CHROMOSOMES

How are **chromosomes assembled**?

Chromosomes are assembled on a scaffold of proteins (histones) that allow DNA to be tightly packed. There are five major types of histones, all of which have a positive charge; the positive charges of the histones attract the negative charges on the phosphates of DNA, thus holding the DNA in contact with the histones. These thicker strands of DNA and proteins are called chromatin. Chromatin is then packed to form the familiar structure of a chromosome. During mitosis, chromosomes acquire characteristic shapes that allow them to be counted and identified.

What is a **trinucleotide** repeat?

Dispersed throughout the human genome are sequences of repetitive DNA. These repeats will have one to six nucleotides (base pairs). One type in particular, involving three repeating nucleotide pairs, is called a trinucleotide repeat. As DNA is copied and then transferred to a new generation, the number of repeated sets can increase and cause diseases.

Can people have **missing or extra chromosomes**?

Yes, people can live with this chromosomal abnormality, depending on which chromosomes are copied or missing. Examples of these conditions include Down syndrome (an extra copy of chromosome 21) and Turner syndrome (a female with only one X chromosome), both of which can result in healthy babies. Conversely, almost 1 percent of all conceptions are triploid (three copies of each chromosome), but over 99 percent of these die before birth.

How has **DNA** become **commercialized**?

DNA has been extensively commercialized in its applications to plant biotechnology, genetically modified organisms, gene therapy, gene patents, and applications to forensic science. DNA jewelry, artwork, and apparel can be purchased. Music CDs have been created based on DNA sequences.

What is the **average size** of a **gene**?

The average size of a vertebrate gene is about 30,000 base pairs. Bacteria, because their sequences contain only coding material, have smaller genes of about 1,000 base pairs each. Human genes are in the 20,000–50,000 base pair range, although sizes greater than 100,000 base pairs have been suggested as well.

What is **aneuploidy**?

Cell division that results in an unequal separation of genetic material creates aneuploid (from the Greek terms *an*, meaning "not"; *eu*, meaning "true"; and *ploid*, meaning "number") cells. Examples would be human cells with 47 or 45 chromosomes. For a diploid organism, aneuploidy is usually written algebraically as 2n+1 or 2n-1, indicating an odd number of genes.

What is **polyploidy**?

Polyploidy is a condition where complete, extra sets of chromosomes are contained within cells. Polyploid individuals are described as being 3n (triploid), 4n (tetraploid), and so on. Humans cannot survive as polyploids and rarely as aneuploids.

How many **human RNA genes** are there?

There are about 250 genes that code for short RNA sequences instead of protein. These RNA strands appear to regulate the activity of other genes, particularly those involved in embryonic development.

> ## What most common type of genetic disorder linked to mental retardation has a trinucleotide repeat?
>
> There are several known trinucleotide repeats, but one that is well characterized is a CGG sequence found on the X chromosome. Multiple repeats of this sequence are responsible for Fragile X syndrome, the most common type of inherited mental retardation. While normal X chromosomes contain between 6 and 50 repeats, mutant X chromosomes of this type may contain up to 1,000 copies.

How are **genes controlled**?

Genes are controlled by regulatory mechanisms that vary by whether the organism is a prokaryote or a eukaryote. Bacteria (prokaryote) genes can be regulated by DNA binding proteins that influence the rate of transcription, or by global regulatory mechanisms that refer to an organism's response to specific environmental stimuli such as heat shock. This is particularly important in bacteria. Gene control in eukaryotes depends on a complex set of regulatory elements that turn genes off and on at specific times. Among these regulatory elements are DNA binding proteins as well as proteins that, in turn, control the activity of the DNA binding proteins.

What are the **shortest** and **longest chromosomes**?

Among human chromosomes, the length ranges from 50 to 250 million base pairs. The longest chromosome is chromosome 1, with 300 million bases (approximately 10 percent of the human genome), and the shortest is chromosome 21, with 50 million bases.

How can organisms with just **one set of chromosomes** reproduce?

An organism that is haploid (one set of chromosomes) can reproduce by mitosis to produce more haploid cells or a multicellular haploid organism. This is typical of some algae and fungi.

Can organisms with **more than two sets of chromosomes** reproduce?

Organisms with an odd number of chromosome sets cannot produce gametes with a balanced assortment of chromosomes. Polyploidy is very common in plants, with 30 to 70 percent of today's angiosperms thought to be polyploid. When a tetraploid (4N) plant tries to breed with a diploid (2N) plant, triploid offspring are formed. An example of this is the seedless grape.

Polyploidy in plants

Plant	Ancestral haploid number	Chromosome number	-ploidy
Peanut	10	40	4N

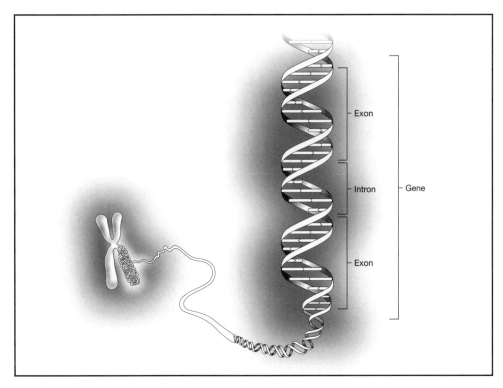

The structural components of a gene.

Plant	Ancestral haploid number	Chromosome number	-ploidy
Sugar cane	10	80	8N
Banana	11	22,33	2N,3N
Cotton	13	52	4N
Apple	17	34,51	2N,3N

What is the **p53 gene**?

The p53 protein interferes with tumor growth and is 53 kilodaltons in length. It is found in increased amounts in cells that have undergone treatment (such as radiation) that damages DNA. The gene that codes for this tumor suppressor is known as the p53 gene. Mutations that inactivate this gene are found in the majority of all human tumors.

What are the **components of a gene**?

The term "gene" describes a section of DNA that will be used as a template to build a strand of RNA or protein. In addition to this information, each gene also contains a promoter region, which indicates where the coding information actually begins, and a terminator, which delineates the end of the gene.

What is a **base pair**?

The phrase "Base pair" is a way of describing the length of DNA. Since most DNA is double-stranded, every nitrogen base containing nucleotide on one strand of the molecule is paired with a complementary base on the other strand. Thus, "10 base pairs" refers to a segment of DNA that is 10 nucleotides in length and 2 strands in width.

What is a **polytene chromosome**?

When DNA replication takes place without subsequent cell division, polytene chromosomes may result. These are giant chromosomes comprised of multiple chromatids (identical chromosome copies) arranged together like strands of a large cable. They are commonly found in the salivary glands of the fruit fly (*Drosophila melanogaster*) and are regions of active transcription.

Do **both sides of DNA** contain **genes**?

One strand of DNA contains the information that codes for genes, and it is called the "antisense strand" or "noncoding strand." It is this strand that is transcribed into mRNA and is designated as the template strand. The other, complementary strand is called the "coding strand" (because it contains codons) or "sense strand." Its sequence is identical to the mRNA strand, except for the substitution of U (uracil) for T (thymine).

What is **gene redundancy**?

Gene redundancy refers to having multiple copies of the same gene. This assures that essential genes that are required in large amounts, like those for ribosomal RNA (rRNA), can be transcribed at multiple sites. In addition, gene redundancy provides for a source of gene sequences that can be modified by natural selection without immediately destroying the organism.

What is a **homeobox gene**?

The homeobox is specific subset of nucleotides in homeotic genes that have important roles in embryonic development. Homeotic genes code for proteins that act as transcription factor, enhancing the rate at which certain DNA sequences are transcribed. The similarity of this sequence in different groups of species is another indication of the shared evolutionary history of species.

What is a **centimorgan**?

A centimorgan (cM) is a way of measuring the distances between genes on a chromosome. This unit was named in honor of Thomas H. Morgan (1866–1945), who was one of the first to map genes onto chromosomes. Typically, a distance of a cM corresponds to a distance of about one million base pairs.

What are **chiasmata**?

The site of "crossing over," where homologous chromosomes actually exchange material during the first round of meiosis, is known as a chiasma. The plural version is chiasmata.

What is a **chromosome map**?

A chromosome map lists the sequence of genes found on a given chromosome. Chromosome maps are usually determined by breeding experiments in which the ratio of the offspring with certain combinations of traits indicates how far apart those traits are on the chromosome.

What is a **pseudogene**?

As genes are copied and transposed, mutations may occur that make the copies nonfunctional. This is thought to be the origin of pseudogenes, which resemble actual genes but which, while stable, do not produce a polypeptide.

How can **genes** be **silenced**?

Genes can be silenced either by preventing their transcription or by interference with their transcript. Gene silencing is a method by which researchers can study gene expression.

What is a **telomere**?

At the end of eukaryotic chromosomes there lies a unique structure known as a telomere. Experiments have determined that without telomeres, the chromosome structure may be compromised; the DNA of the chromosome tends to stick to other pieces of DNA, and enzymes (deoxyribonucleases) are more likely to degrade or digest the ends of the chromosomes. Human telomeres have specific repetitive DNA sequences (TTAGGG) that may be repeated from 250 to 1500 times.

What is the relationship between **telomeres** and **aging**?

In many cells there appears to be an optimal number of repeats at the telomeres. Normal telomere length is restored by the enzyme telomerase. Studies have shown that when the telomerase gene is silenced, there is progressive shortening of telomeres as cells divide. Germ line and some stem cells express the telomerase gene, but somatic cells do not express the gene and therefore show telomere shortening with each cell division. Thus, it may be possible to slow body aging by slowing cellular senescence. This can be achieved by maintaining the activity of the telomerase gene.

What is the cellular "**fountain of youth**"?

Research on telomeres and telomerase shows that cells can be kept alive far beyond their normal life span. Specific studies have been carried out on the roundworm, *C. elegans*, in which longer life spans were observed in worms with longer telomeres.

What is the relationship between **telomeres** and **cancer**?

Increased telomerase activity can increase longevity of cells, but it is also implicated in cancer formation. Almost 90 percent of cancer cells have been found to have enhanced telomerase activity, and a cancer cell usually divides about eighty times before a tumor mass becomes large enough to be detected. In contrast, normal human cells usually divide thirty to fifty times before telomeres become too short and doubling stops.

TRANSCRIPTION AND TRANSLATION

What is **transcription**?

Transcription is the synthesis of an mRNA strand from a DNA template sequence, commonly known as a gene. The mRNA is then used as a pattern for building a polypeptide.

What is **translation**?

Translation is the process that produces polypeptides from an mRNA transcript. This process involves several different types of molecules; tRNA, rRNA, ribosomal proteins, and the energy molecule GTP.

Where do **transcription** and **translation** take place?

In bacteria, the process of protein synthesis takes place in the cytoplasm, as bacteria lack a nucleus or other membrane-bound organelles. In eukaryotes, transcription occurs in the nucleus. From there the mRNA moves into the cytoplasm, where it is translated into a polypeptide at a ribosome.

What is the **one gene–one enzyme hypothesis**?

In the 1930s George Beadle (1903–1989) and Boris Ephrussi (1901–1979) theorized that the variety of fruit fly mutations might be due to mutations of individual genes that code for each of the enzymes involved in a given pathway. Subsequently, Beadle and Edward Tatum (1909–1975) performed a series of experiments with the orange bread mold *Neurospora* that elucidated the enzymatic pathway required by the fungus to produce a specific nutritional requirement, arginine. The researchers were able to create a series of mutants, each lacking in a different enzyme in the pathway. In this way they were able to piece together the sequence of events required for the production of arginine and thereby show where each mutant fit. The work of Beadle and Tatum provided important support for the one gene–one enzyme hypothesis, which holds that the function of a gene is to produce a specific enzyme. Their work garnered a Nobel Prize in Physiology or Medicine in 1958.

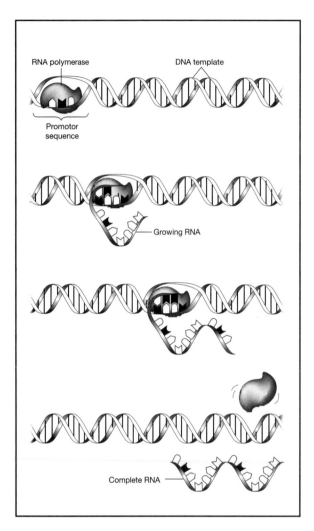

Transcription is the synthesis of an mRNA strand from a DNA template sequence, commonly known as a gene.

378

Has the one gene–one enzyme hypothesis been changed?

Some of the products of gene translation do not become enzymes. They may instead be used in a variety of roles, for example as transporters, receptors, or structural elements of cells. Because of this the Beadle-Tatum hypothesis has been renamed the one gene–one polypeptide hypothesis, although that may change again as we acknowledge the variety of uses that any given polypeptide may have in cells.

What is a **codon**?

A codon is the three-unit sequence (AUG, AGC, etc.) of mRNA nucleotides that codes for a specific amino acid. Since there are only twenty commonly used amino acids and sixty-four (4x4x4) possible codon sequences, the genetic code is described as both degenerate and unambiguous. Each codon codes for only one amino acid, but each amino acid may have more than one matching codon.

What is the **genetic code**?

The genetic code is a chart depicting the relationship between each of the possible mRNA codons and their associated amino acids. The codons are grouped according to the amino acid they code for. Present in the code as well are the "start" and "stop" codons. The start codon actually codes for methionine, which is always the first amino acid in the polypeptide sequence. Methionine may appear elsewhere in the polypeptide as well. Methionine is removed during post-translational processing.

What is a **stop codon**?

A stop codon signals the end of coding. Instead of a tRNA, with its amino acid in tow, a release factor matches the stop codon during translation, causing the polypeptide to be released from the ribosome. It is interesting to note that the genetic code contains only one start codon but three stop codons. This means that a random DNA mutation is much more likely to prevent successful protein synthesis than encourage it.

How does a **ribosome** participate in **protein synthesis**?

The ribosome serves as the site of translation. This combination of RNA and protein is a meeting place for mRNA and tRNA.

What is the **structure** of a **ribosome**?

A ribosome is composed of two parts known as the large and small subunits. Each of these is a combination of protein and a type of RNA known as rRNA. At the beginning of translation, the two subunits form a structure around the mRNA molecule as the first tRNA (the one matching the first methionine or fMet) arrives. The completed ribosome has niches that hold up to three tRNAs at a time. Because a cell has so many ribosomes at any one time, rRNA is the most common type of RNA found in cells.

Can an **RNA sequence** be made into **DNA**?

Yes, it can. A process known as reverse transcription can convert an RNA sequence into DNA by using an enzyme known as reverse transcriptase. First observed in retroviruses like HIV, reverse transcriptase has also been identified as playing a role in the copying of DNA segments from one site to another in the genome.

How is **RNA** made from **DNA**?

In eukaryotes, first the DNA of the specific gene unwinds. Then enzymes, known as RNA polymerases, use the DNA sequence, the pairing rules (U-A, G-C, C-G, A-T), and available RNA nucleotides to efficiently copy the DNA sequence into RNA.

DNA nucleotide	Matching RNA nucleotide
Adenine	Uracil
Thymine	Adenine
Cytosine	Guanine
Guanine	Cytosine

What is the role of **transfer RNA**?

Transfer RNA, or tRNA, is the translation molecule. It recognizes the nucleic acid message and converts it into polypeptide language. On one part of the molecule is a section of three nucleotides known as the "anticodon" that matches the pair rules for a specific codon. The other end of the molecule, which looks something like an inverted three-leaf clover, is the amino acid attachment site. A tRNA with a specific anticodon will bind to only one kind of amino acid, ensuring the accuracy of translation from mRNA to polypeptide.

RNA nucleotide	Matching RNA nucleotide
Adenine	Uracil
Uracil	Adenine
Cytosine	Guanine
Guanine	Cytosine

What is the **wobble hypothesis**?

It turns out that only the first two nucleotides of the codon and anticodon bond together securely during translation. The flexibility of bonding at the third site (the "wobble") allows cells to build only forty-five tRNA molecules to cover the sixty-four possible mRNA codons. By converting an adenine base into a molecule called iosine, the tRNA increases the number of mRNA nucleotides it can match in the third position. This hypothesis was first proposed by James Watson (1928–) and has since been borne out by a number of experiments.

Third base in the tRNA anticodon	Matching base in the mRNA codon
C	G
G	U or C
A	U
U	A or G
I	A, U, or C

What is a **reading frame**?

The reading frame is the three-nucleotide section of the mRNA molecule that is "read" by tRNA during translation. Normally, the reading frame matches the codon, but mutations that cause the gain or loss of nucleotides can result in the offsetting of the reading frame, leading to the production of a very different protein.

For example, a sentence comprised of three-letter words changes its meaning with the deletion of one letter: "The big fat cat ate one red rat." Delete the first letter E and this sentence becomes:

"Thb igf atc ata tet her edr at." Which effectively is nonsense!

What is **RNA splicing**?

In eukaryotes, the RNA transcript may be modified before it is sent to the cytoplasm to be translated into a protein. This process is known as RNA splicing or post-transcriptional processing. From an evolutionary standpoint, this ability to edit the message allows greater variety of expression from a finite set of genes. Current research indicates that most eukaryotic genes are interrupted or discontinuous—that is, the mRNA produced from the DNA sequence has had intervening nucleotides removed.

What is a **spliceosome**?

A spliceosome is a structure, comprised of RNA and protein, that is responsible for RNA splicing in the eukaryotic nucleus.

What is an **intron**?

An intron is the part of the RNA transcript that is removed by the spliceosome and therefore remains in the nucleus.

What is an **exon**?

An exon is a segment of the transcript that is tied together with other exons by the spliceosome to make the mRNA molecule. After the exons are spliced together, they exit the nucleus for the cytoplasm, where the mRNA is translated into a polypeptide.

What is **junk DNA**?

The noncoding portions of the genome were formerly referred to as "junk" DNA. Some of this material turns out to play regulatory roles in DNA replication and transcription, and the rest may be an evolutionary holdover.

What is an **operon**?

An operon is a segment of DNA containing all genes used to produce proteins in a specific metabolic pathway. A functional operon also contains the RNA polymerase–binding site known as a promoter as well as an on-off switch known as the operator. So far, operons have been identified only in bacteria, as this highly efficient grouping of related genes maximizes the information encoded on their single chromosome.

What is the difference between **inducible** and **repressible operons**?

Inducible operons are used to produce proteins only under specific conditions. Most bacteria use glucose as their main energy source. Therefore, the unusual arrival of the milk sugar lactose in the cell turns the operator switch to the "on" position, which allows transcription to occur. Repressible operons have operators that are always on (transcription occurring) unless a signal (for example, too much product) turns them off.

How long does **protein synthesis** take in bacteria, once **mRNA** is formed?

The bacterium *E.coli* can add an amino acid in only 0.05 second. That means that synthesizing a protein of 300 amino acids takes only 15 seconds.

How does an **antibiotic** affect **protein synthesis**?

Antibiotics can slow or shut down protein synthesis in bacteria. There are about 160 known antibiotics, and each acts differently and selectively, depending on the type of bacterial infection. As an example, streptomycin inhibits initiation of protein synthesis by interfering with final formation of the two subunits of a bacterial ribosome.

What is **post-translational processing**?

The process of translation uses an mRNA pattern to produce a string of amino acids technically known as a polypeptide. To turn a polypeptide into a protein with the shape that is crucial to its function requires several more steps. Sugars, lipids, phosphate groups, or other molecules may be added to the amino acids in the chain. The polypeptide may be "trimmed" as well; the first few amino acids may be removed. Some proteins, like insulin, only become functional after cleavage by an enzyme. Because of this ability to create variations from the basic polypeptide chain, a given gene may code for several different functional structures. This helps to explain the unexpectedly small size of the human genome. Our 35,000 or so genes actually code for several hundred thousand different proteins.

What are **chaperone proteins**?

While most proteins fold spontaneously into their specific conformation, some require the help of other molecules to achieve this. Chaperone proteins guide newly synthesized polypeptides into their functional three-dimensional shape.

GENETICS

MENDELIAN GENETICS

Who was **Mendel**?

Gregor Mendel (1822–1884) is the founding father of experimental genetics. His work with the garden pea, *Pisum sativuum* , was not consistent with the nineteenth-century ideas of inheritance. Mendel was the first to demonstrate transmission of distinct physical characteristics from generation to generation.

What is the **blending theory**?

Blending theory was the commonly held belief that characteristics were mixed in each generation. For example, breeding two horses, one with a light-colored coat, the other dark, would result in offspring that were all intermediate in coat color. If this held true, then eventually all organisms would become more alike in each generation. Although this theory persisted for many years, it was eventually supplanted by the work of Mendel and the modern geneticists.

What is the **homunculus theory?**

Illustrations from the late seventeenth century depict sperm containing a miniaturized adult. This preformed human was called a homunculus. It was thought at that time that the sperm (and its homunculus) contained all of the traits for building a baby and that the egg and womb served simply as incubators.

What **traits** of **peas** did Mendel study?

Mendel studied peas of distinct and recognizable plant varieties. Mendel's experiments included the characteristics of height, flower color, pea color, pea shape, pod color, and the position of flowers on the stem.

What is an **allele**?

An allele is an alternative form of a gene; there are usually two alleles for each gene, although the number may vary from one trait to another. Each individual inherits one allele from the mother and one from the father. Alleles for a trait are located on corresponding loci on each homologous chromosome.

What is meant by the terms **homozygous** and **heterozygous**?

Since each diploid organism has two alleles for every gene, the alleles can be alike (homozygous) or different (heterozygous). These terms are derived from the Greek terms *homos*, meaning "same," and *heteros*, meaning "different," plus the Greek suffix *zygotos*, meaning "yoked together."

How does a **dominant trait** differ from a **recessive trait**?

A dominant trait is one that is expressed whenever present, either as a homozygous genotype (DD) or a heterozygous one (Dd). In other words, dominance means that a heterozygote, carrying only one copy of the dominant allele, will display the dominant phenotype. The term "dominant" does not infer that this is more normal or common than a recessive trait. In contrast, recessive phenotypes always result from a genotype in which alleles are alike. Recessive traits seem to be more common in a population than dominant traits. Sometimes an allele is a "no signal" message, meaning that no functional polypeptide will be made from its expression.

What is meant by the term "**true breeding**"?

Individuals, who, when bred to others of the same genotype, produce only offspring of that genotype, are called true breeding. In other words, homozygous individuals (AA, aa) are true breeding when bred among themselves, while heterozygotes (Aa) are not.

What is a **hybrid**?

A hybrid is produced as the offspring of two true breeding organisms of different strains (AA x aa). If the hybrids (Aa) are then mated, the result is a hybrid cross.

What is the **law of segregation**?

This principle, which Mendel discovered, predicts the transmission of a single trait. When a gamete is produced, the two copies of a particular gene (alleles) separate during its formation. For example, if a pea plant has the genetic makeup of Rr, where R is round and r is wrinkled, then half the gametes will contain R and the other half r. The law of segregation comes into play during meiosis when an organism decreases its chromosome number by half when forming egg or sperm.

What is the **law of independent assortment**?

This Mendelian principle deals with the prediction of the outcome of dihybrid (two trait) crosses, such as the traits for seed shape (R=round and r=wrinkled) and seed color (G=green and g=yellow). If two true breeding plants are crossed, RRGG x rrgg, then all offspring would be heterozygous for both genes (RrGg) and would be round and green. By crossing these first-generation offspring (referred to as the F_1 generation) among themselves, four phenotypic and nine genotypic classes are generated. From this, Mendel concluded that traits are transmitted to offspring independently of one another. In other words, the separation of alleles for one trait does not influence or control the distribution of alleles from the second trait. This law holds true as long as the two traits in question are located on separate chromosomes or are so distant from each other on the same chromosome that they sort independently.

Did **Darwin** and **Mendel** know each other?

Although both men lived during the same time period in the nineteenth century, they did not know each other. Charles Darwin's (1809–1882) publication *On the Origin of*

Species (1859) popularized his theory of natural selection but raised many questions as to how organisms could display modified or new traits. In 1865 Mendel published his landmark paper "Experiments in Plant Hybridization." Researchers have been unable to document whether either man used the other's work in the development of their respective theories.

Why is **Mendel's work disputed**?

There are some scientists who feel that Mendel's work is too perfect to be accurate, even though the validity of his conclusions are not in doubt. A closer look at Mendel's work suggests that he may not have reported the inheritance of traits that did not show independent assortment, and thus he may have "skewed" some of his data. A British statistician and population geneticist, R. A. Fisher (1890–1962), pointed out in 1936 that Mendel's data fit the expected ratios much closer than chance alone would indicate. However, Mendel's published data is comparable to the work of others in his field, and his conclusions are still accepted as part of the core of genetics.

The sex-linked pattern of the inheritance of hemophilia within the royal family of Queen Victoria and Prince Albert is perhaps the most famous human pedigree

What is the relationship between **probability** and **genetics**?

Probability is a branch of mathematics used to predict the likelihood of an event occurring. It is also an important tool for understanding inheritance patterns of specific traits. There are two rules that are applied to genetic inheritance: 1) The rule of multiplication is used when determining the probability of any two events happening

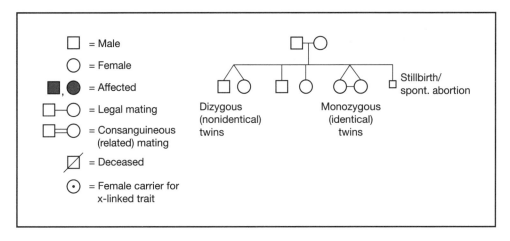

Common pedigree symbols and configurations.

simultaneously. For example, if the probability of having dimples is 1/4, and the probability of having a male child is 1/2, then the probability of having a boy with dimples is 1/4 x 1/2, or 1/8. 2) The rule of addition is used when determining the outcome of an event that can occur in two or more ways. For example, if we were to consider the probability of having a male child or of having a child with dimples, then the probability would be 1/2 + 1/4, or 3/4. In other words, chances are 3 out of 4 that a given birth will produce a male child or a child with dimples.

How does **phenotype** relate to **genotype**?

The phenotype is the physical manifestation of the genotype and the polypeptides it codes for. Consider the rose. The color and shape of the petals (the phenotype) are the result of chemical reactions within the cells of each petal. Polypeptides synthesized from the directions encoded within the cell's genes (the genotype) are part of those reactions. Different versions of a gene will produce different polypeptides, which in turn will cause different molecular interactions and ultimately a different phenotype.

What are the **predictive ratios** of common **Mendelian crosses**?

Common Mendelian Crosses and Ratios

Type of cross:	Phenotypic Ratios:	Genotypic Ratios:
Monohybrid: Aa x Aa	3 A_ : 1 aa	1 AA:2 Aa: 1 aa
Dihybrid: AaBb x AaBb	9 A_B_	1 AABB
	3 A_bb	2 AABb
	3 aaB_	1 AAbb

389

Type of cross:	Phenotypic Ratios:	Genotypic Ratios:
	1 aabb	2 AaBB
		4 AaBb
		2 Aabb
		1 aaBB
		2 aaBb
		1 aabb

A= dominant, a= recessive, B = dominant, b = recessive

How can the results of a genetic cross be **predicted**?

The results of a genetic cross can be predicted using a diagram called a Punnett square, which graphically represents Mendel's laws of segregation and independent assortment. On each side of the square, one places the possible alleles donated by each parent. Dominant alleles are represented by capital letters (G in this case) and recessive alleles by lowercase letters (g). Inside the boxes are the allelic combinations; in this case, each box would contain Gg (heterozygous combination).

	G	G
g	Gg	Gg
G	Gg	Gg

How does **meiosis** relate to **Mendel's laws**?

Mendel's law of segregation directly refers to meiosis. He had suggested that gametes contain "characters" that control specific traits, and these characters separate when gametes form. Gametes are formed by meiosis, in which the chromosome number is reduced by one half (from diploid to haploid).

What is a **pedigree**?

A pedigree is a genetic history of a family, which shows the inheritance of traits through several generations. Information that can be obtained from a pedigree includes birth order, sex of children, twins, marriages, deaths, stillbirths, and pattern of inheritance of specific genetic traits.

Find your generation!

Generation

Great-Great-Great-Grandparents	P
Great-Great-Grandparents	F_1
Great-Grandparents	F_2
Grandparents	F_3
Parents	F_4
Children	F_5

What is meant by **genetic determinism**?

Genetic determinism is a theory proposing that behavior and character are entirely shaped by genes. Since only genetic material passes from one generation to the next, it seemed logical at one time to assume that all of our traits (physical, behavioral, psychological) were the result of gene sequences. More recent and extensive research has demonstrated the importance of environmental factors in determining how or which genes are actually involved in the development of the individual.

How are **generations** described in **genetic crosses**?

Usually, genetic crosses begin with true breeding parents of different strains that are designated by letter *P*. The hybrid offspring are called the F_1 generation; each subsequent generation is called F_2, F_3, and so on.

What is meant by a **diploid** organism?

A diploid organism is one that has two sets of chromosomes, one from each parent.

What are **homologous chromosomes**?

In diploid organisms, chromosomes occur in matching pairs, with the exception of the sex chromosomes. Not all organisms are diploid; for example, bacteria have only one circular chromosome, and some insects may have an odd number of chromosomes. Every human somatic cell (excluding egg or sperm cells) has forty-six chromosomes—twenty-two somatic chromosome pairs and one sex chromosome pair. The somatic chromosome pairs, called homologous chromosomes, carry the same sequence of genes for the same inherited characteristics.

What are **autosomes**?

Autosomes are chromosomes that contain information (gene sequences) available to both sexes. These are the chromosomes commonly used in genetic examples.

What are **sex chromosomes**?

Sex (X and Y in mammals) chromosomes are the nonmatching pair. The Y chromosome, found only in males, has the sex-determining gene, SRY. The SRY determines whether or not an individual will develop testis and produce appreciable quantities of testosterone (the hormone that generates male characteristics). There is no known correlate gene on the X chromosome for the formation of ovaries. Therefore, it is the absence of the Y chromosome that determines a female.

NEW GENETICS

What is meant by the **modern era** of genetics?

Mendel's work was really not appreciated until advances in cytology enabled scientists to better study cells. In 1900, Hugo deVries (1848–1935) of Holland, Carl Correns (1864–1933) of Germany, and Erich von Tschermak (1871–1962) of Austria examined Mendel's original 1866 paper and repeated the experiments. In the following years chromosomes were discovered as discrete structures within the nucleus of a cell. In 1917 Thomas Hunt Morgan (1866–1945), a fruit fly geneticist at Columbia University, extended Mendel's findings to the structure and function of chromosomes. This and subsequent findings in the 1950s were the beginning of the modern era of genetics.

What is a **gene**?

A gene is a specific sequence of DNA that contains the molecular recipe for a polypeptide. A polypeptide is a subunit of a protein.

Can I see a **gene**?

No. A gene cannot be seen because it is submicroscopic. We can see a chromosome, which contains genes, and geneticists can pinpoint the location of a gene on that chromosome, but the actual gene cannot be seen.

What were the first uses of the terms *gene* and *genotype*?

The terms *gene*, from the Greek term *genos*, meaning "to give birth to," and *genotype* were first used in 1909 by a Danish botanist, Wilhelm Johannsen (1857–1927). Johannsen is considered to be one of the architects of modern genetics.

What is the **Hardy-Weinberg theorem**?

The Hardy-Weinberg theorem, formulated in 1909, extends Mendel's laws to natural populations. It explains in a formula what happens to the allelic and genotypic frequencies as alleles are passed between generations. The Hardy-Weinberg equilibrium is based on the following assumptions: 1) the population size is large; 2) individuals mate randomly within the population; 3) there are no mutations; 4) there is no input of new alleles from outside sources; and 5) there is no natural selection, such that each new allelic combination has equal chances of survival. The Hardy-Weinberg theorem can be expressed in two ways: 1) $p + q = 1$; and 2) $p^2 + 2pq + q^2 = 1$.

Variables of the Hardy-Weinberg Theorem

Variable	p	q	p^2	2pq	q^2
meaning	Frequency of dominant allele	Frequency of recessive allele	Frequency of homozygous dominant genotype	Frequency of heterozygous genotype	Frequency of recessive genotype

What is a **Barr body**?

In 1949 Dr. Murray Barr (1908–1995) noticed a dark body in the neurons of female cats. It was later identified as a structure found only in the nucleus of females. It was named a Barr body in honor of its discoverer.

What is the **Lyon hypothesis**?

The Lyon hypothesis refers directly to a Barr body. It was proposed by Mary Lyon (1925–) in 1961 that a Barr body is actually an inactivated X chromosome. According to this hypothesis, female mammals sequester one X chromosome in each of their cells during the early stages of development. This folded chromosome becomes the dark body of Barr's observation. This means that both males and females rely on the information from only a single X chromosome. Therefore, it is only one X chromosome that provides genetic information in both males and females.

Who first discovered the link between **genetics** and **metabolic disorders**?

Archibald Garrod (1857–1936) was a British physician who became interested in a rare but harmless disease called alkaptonuria. A unique characteristic of this condition is that the patient's urine turns black when exposed to air. Previously, physicians thought the disorder was due to a bacterial infection; however, Garrod observed that the disorder was more common in children of first-cousin marriages and followed the Mendelian description of an autosomal recessive disease. Garrod, a follower of the fledgling science of biochemistry, suspected that the urine turned dark because patients lacked the enzyme to break down the protein that caused the urine to darken. He referred to alkaptonuria as an inborn error of metabolism. Unfortunately, very few people understood metabolic pathways at that time, so his contributions remained largely unnoticed. By the 1950s his work was verified, earning Garrod the title of "father of chemical genetics."

What is **Thomas Hunt Morgan's** contribution to genetics?

Thomas Hunt Morgan (1866–1945) won the Nobel Prize in Physiology or Medicine in 1933 for his discovery of the role played by chromosomes in heredity. He is perhaps most noted for his "fly lab" at Columbia University in New York, where he collected *Drosophila* (fruit fly) mutants. Morgan studied fruit flies much in the same way that Mendel studied peas. He found that the inheritance of certain characteristics, such as eye color, was affected by the sex of the offspring.

What is **gene expression**?

Gene expression refers to the molecular product of a particular gene and is another way of describing the phenotype. The phenotype of an organism is a result of the chemical interactions that are guided by the DNA (gene blueprint) for a specific polypeptide.

What were the contributions of **Beadle and Tatum** to genetics?

The work of George Beadle (1903–1989) and Edward Tatum (1909–1975) in the 1930s and 1940s demonstrated the relationship between genes and proteins. They grew the

Who discovered jumping genes?

Barbara McClintock (1902–1992), who worked on the cytogenetics of maize during the 1950s at Cold Spring Harbor Laboratory in New York, discovered that certain mutable genes were transferred from cell to cell during development of the corn kernel. McClintock made this inference based on observations of changing patterns of coloration in maize kernels over many generations of controlled crosses. She was awarded the Nobel Prize in Physiology or Medicine in 1983 for her work.

orange bread mold *Neurospora* on a specific growth medium. After exposing the mold to ultraviolet (UV) radiation, the mold was unable to grow on a medium unless it was supplemented with specific amino acids. The UV radiation caused a single gene mutation that led to the production of a mutant enzyme. This enzyme, in turn, caused the mold to exhibit a mutant phenotype. Beadle's and Tatum's work was important in demonstrating that genes control phenotypes through the action of proteins in metabolic pathways. In other words, one gene affects one enzyme. Since Beadle and Tatum were awarded their Nobel Prize in 1958, scientists have discovered that the relationship between genes and their proteins is much more complex.

What is meant by a **jumping gene**?

A jumping gene is a gene that can move from one location to another in a chromosome or can even "jump" from one chromosome to another. Another name for a jumping gene is a transposon.

Are all **genes** in a **genome** used by an organism?

No. All genes are not used by an organism all the time. Since protein synthesis is an energy intensive cellular process, proteins are not produced unless they are needed for a specific cell function. For example, before a human has reached adult height, cells are continually producing human growth hormone, a protein that encourages bone and muscle growth. However, at a certain age (which varies by individual), the gene will become dormant and will no longer produce growth hormone.

What is the difference between a **gene** and a **chromosome**?

The human genome contains twenty-four distinct, physically separate units called chromosomes. Arranged linearly along the chromosomes are tens of thousands of genes. The term "gene" refers to a particular part of a DNA molecule defined by a specific sequence of nucleotides. It is the specific sequence of the nitrogen bases that encodes a gene. The human genome contains about 3 billion base pairs, and the length of genes varies widely.

What is a **karyotype**?

A karyotype is a snapshot of the genome. A karyotype can be used to detect extra or missing chromosomes, chromosomal rearrangements, or chromosomal breaks. Any cell that contains a nucleus can be used to make a karyotype. However, white blood cells seem to work best for human karyotypes. After the cells are cultured, they are killed by using a drug that halts mitosis, and the chromosomes are then stained, observed, and a size order chart is produced.

Why are some **species** more commonly used for **genetic studies** than others?

Species with a relatively small genome, with a short generation time from seed to seed, and that are adaptable to living in captivity are appealing as experimental organisms. Even though many of these species bear little physical resemblance to humans, they do share part of our genome and so can answer some of the questions we have about genetic inheritance and gene expression.

Species Commonly Used in Genetics Research

Species	Kingdom	Genome size (in million base pairs)
Arabidopsis thaliana (plant)	Plant	120
Neurospora (orange bread mold)	Fungi	40
Escherichia coli (bacteria)	Monera	4.64
Drosophila melanogaster (fruit fly)	Animal	170
Caenorhabditis elegans (roundworm)	Animal	97

How can one **gene control** another?

One gene cannot actually control another gene, but one gene can mask the effect of another gene. This is called epistasis. For example, a gene in Labrador dogs controls deposition of melanin. The dominant allele B causes deposition of large amounts of melanin. A recessive allele b causes less deposition. So a BB or Bb dog is black, while a bb dog is brown. Another gene controls whether or not melanin is deposited at all. This gene E allows deposition of melanin while the recessive form e does not. Therefore, an ee dog is yellow while an Ee or EE dog is not yellow and has melanin. The interaction of two genes, the B and E genes, thus controls coloration in Labrador dogs.

Correlation between genotype and coat color in Labrador dogs

Genotype	Coat color
BBEE, BbEE, BBEe, BbEe	Black

Genotype	Coat color
Bbee, BBee	Yellow
bbEE, bbEe	Brown
Bbee	Yellow

How can a given **trait** have more than **two alleles**?

An allele is simply an alternate form of a gene. It is possible to have a number of variations for the same gene sequence, hence the term multiple alleles. In cases of multiple alleles, there is usually a dominance hierarchy or relational schematic that demonstrates which of every possible pair of alleles will be preferentially expressed.

What is meant by a **codominant allele**?

Codominance is an example of non-Mendelian genetics. When alleles are codominant, all versions are expressed in the phenotype. An example of this can be found in chinchilla rabbits, where there are four genes that affect coat color.

Gene	Gene effect
C	Allows development of black and yellow pigments
C (chd) Chinchilla	Turns yellow pigments to white or pearl white
C (chl) Shading	Eliminates all yellow pigments; turns black into sepia brown
C Albino gene	Prevents development of any pigment

In rabbits the dominant allele C allows coat color, while cc is an albino. There are different versions of the allele c, allowing for a range of expression.

Genotype	Phenotype
C _	full color
c c(chl)	shaded
cc	albino

What would be an example of a cross involving **multiple alleles**?

Blood type in mammals is due to particular marker molecules (called glycoproteins) located on the red blood cell membrane. In humans the ABO blood type is determined by a gene with three alleles: A, B, and *i* (no membrane marker). A and B code for different types of marker molecules. While A and B are equally dominant, both are dominant to *i*. So a cross of A*i* x B*i* (blood type A x B) would generate the following ratios: 1/4 AB (Type AB), 1/4 A*i* (Type A), 1/4 B*i* (Type B), 1/4 *ii* (Type O).

How can the **environment** affect **genes**?

Some genes are sensitive to temperature. An enzyme that controls the synthesis of black pigment melanin in Siamese cats is only active at cool temperatures. The cooler the body part, the darker the pigment. Since the body is warmer than the extremities, it remains lighter in color. During the winter if your Siamese cat goes outdoors, the fur may become darker. Siamese kittens are usually white due to the warmth of the mother's body.

How can **chromosomes** become **damaged**?

Chromosomes can become damaged physically (in their appearance) and molecularly (in the specific DNA sequences they contain). Chromosomes can break randomly or due to exposure to ionizing radiation, which acts like a miniature cannonball, blasting the strands of DNA. Other factors such as physical trauma or chemical insult can also cause breaks. When a chromosome breaks, the broken part may then reattach to another chromosome, a process called translocation. Sometimes the broken chromosomes may attach to each other, forming a ring. Any type of chemical change is known as a mutation.

What are the different kinds of **mutations**?

Mutations may be germinal (occurring within the gametic cells) or somatic (occurring within any nonsex cell). The difference is that a germinal mutation will affect all cells of an individual, while a somatic mutation will only affect those cells produced by mitosis from the original mutated cell. There are various categories of mutations:

- A point mutation is a change in a single DNA base

- A deletion occurs when information is removed from a gene

- An insertion occurs when extra DNA is added to a gene

- Frameshift mutations occur when one or two bases are either added or deleted

Types of Mutations

Type of mutation	Original Sequence	Mutated Sequence
Substitution	ATC**C**TTAGGA	ATC**G**TTAGGA
Deletion	ATCC**TTA**GGA	ATCCGGA
Insertion	ATCCTTAGGA	ATCCTT**CCG**AGGA
Inversion	AT**CCTT**AGGA	AT**TTCC**AGGA
Duplication	AT**CCTT**AGGA	AT**CCTTCCTT**AGGA

What is a lethal mutation?

A mutation that prevents the organism (or embryo) from performing vital functions is known as a lethal mutation. Research indicates that many genes have the potential of forming lethal mutations. The average human is heterozygous for three to five lethal alleles and thus does not express the lethal condition. A human example of such a lethal allele is brachydactyly, which is expressed only in heterozygotes as abnormally short fingers; however, the genetic defect is fatal during infancy in homozygous recessive individuals due to major skeletal defects.

How can genes become **rearranged**?

Any rearrangement within a gene is a mutation, and such rearrangement is usually due to a random genetic accident. The mutation can be harmless and simply add to the variation within an organism's genome, or it can be harmful with dire consequences.

Are all **mutations** bad?

No, all mutations are not bad. Although people may use the term "mutant" in a disparaging manner, mutations are important to a population's gene pool because of the variation they contribute. Without mutations, there would be no variations and, following along Darwin's thinking, no natural selection.

What is **pleiotropy**?

Pleiotropy refers to a case where one gene may influence several other characteristics. An example of pleiotropy is sickle cell anemia, a disorder in which a single point mutation in the amino acid sequence for hemoglobin results in a spectrum of effects. Red blood cells produce abnormal hemoglobin molecules, which, because of their odd shape, tend to stick together and crystallize. Therefore, the normal disk shape of red blood cells changes to a sickle shape (hence the name of the disorder). Sickle-shaped red blood cells will clog small vessels, causing pain and the possibility of brain damage and heart failure. Since some of the hemoglobin in these cells is abnormal, there is less oxygen available, leading to physical weakness and anemia. If left untreated, the anemia can impair mental function.

What is meant by **heterozygote advantage**?

Heterozygote advantage, also known as balancing selection, arises when there is a survival (selective) advantage for hybrids (heterozygotes) as compared to that of either homozygous recessive genotype or homozygous dominant genotype. Some diseases are lethal only in the homozygous state. If an allele is lethal, then why would it not disappear from a population totally? In the example of cystic fibrosis, heterozygotes may be more resistant to life-threatening dehydration caused by diseases such as cholera, which is accompanied by massive diarrhea. In cholera, caused by the lethal bacterium *Vibrio cholera*, the bacteria produce a toxin that causes considerable loss of water, dehydration, and death within a short time. Another example of heterozygote advantage is the high frequency of heterozygotes for sickle cell anemia in Africa (and increased resistance to malaria).

How does **sex determination** differ in some animals?

Sex determination among various animal groups

Group	Sex determining factor
Humans and other mammals	Males XY, Females XX; presence of Y determines sex
Birds	Males WW, Females WZ
Bees	Females are diploid, males are haploid
Fruit flies	Ratio of sex chromosomes to autosomes. When this ratio is greater than or equal to 1.0, the fly is female; when it is 0.5 or less, the fly is male

What are some examples of **polygenic inheritance**?

Phenotypes of a given trait that are controlled by more than one gene are described as having polygenic inheritance. Examples of polygenic inheritance in humans include height, weight, skin color, and intelligence. Some congenital malformations (birth defects) like clubfoot, cleft palate, or neural tube defects are also the result of multiple gene interactions.

What are **multifactorial traits**?

Phenotypes that are the result of one or more environmental factors and at least two genes are called multifactorial traits. This designation reflects the many factors that may have a role in the physical manifestation of a gene sequence. There are many illnesses that are multifactorial; an example is favism, an inborn metabolic disorder with a predisposition to anemia. However, the anemia only develops when fava beans are eaten or the pollen is inhaled (the environmental factor).

Can an organism have more than **two sets of genes**?

Having more than two complete sets of chromosomes is known as polyploidy. In humans, it is virtually impossible to survive to adulthood as a polyploid. However, plants can often survive and even thrive as polyploids. In fact, this is a common method by which new plant species arise.

What is **crossing over**?

When sister (homologous) chromosomes line up during the early stages of meiosis I, segments of DNA may be moved from one chromosome to another when the two strands overlap. This process is known as crossing over, since gene sequences cross from one chromosome to its sister chromosome.

What are **linked genes**?

Linked genes are genes that tend to be inherited together. The human genome has twenty-four chromosome types and approximately 35,000 genes; therefore, at least 1,000 genes would, on average, be found on a single chromosome and travel through meiosis as a discrete unit. However, linked genes tend to be close enough that they are separated by crossing over less often than unlinked genes.

What are examples of **sex-linked traits**?

Sex-linked traits are most commonly found on the X chromosome and include color blindness (both red and green types), hemophilia (types A and B), icthyosis (a skin disorder causing large dark scales), and Duchenne muscular dystrophy. There are very few Y-linked traits. Hairy ears, a relatively rare trait, is a Y-linked trait.

What is a **gene family**?

Genes are organized into groups called gene families. Many genes have overlapping sequences with each other. Those that share between 30 and 90 percent of their sequences are grouped together in families. Gene families may range in size from just a few genes to several thousand. An example of a gene family is the group of genes that code for histones. Histones are proteins important for maintaining DNA in a particular shape.

What is **epigenetics**?

Just as the term "epidermis" refers to the layer of the skin above (or beyond) the dermis, the term "epigenetics" describes nongenetic causes of a phenotype. An example is genetic imprinting.

What is **genetic imprinting**?

Genetic imprinting is a form of epigenetic inheritance. It is demonstrated where there is a different expression of an allele depending on the parent that transmits it. An example in humans is that of two medical syndromes, both of which result from a deletion on chromosome 15. In Prader-Willi syndrome, originating from a deletion on the paternal chromosome 15, affected children have small hands and feet, short stature, mental retardation, and are obese. However, in Angelman syndrome, originating from a deletion on the maternal chromosome 15, affected children have a large mouth and tongue, severe mental and motor retardation, and a happy disposition, accompanied by excessive laughter.

HUMAN GENOME PROJECT

What is the **Human Genome Project (HGP)**, and what are its **goals**?

The HGP was begun in 1990 as a thirteen-year effort and was slated to be completed in 2003. According to the official HGP web site (http://www.doegenomes.org/), the goals are as follows:

- Identify all of the approximately 30,00–40,000 genes in human DNA
- Determine the sequence of the 3 billion chemical pairs of human DNA
- Store this information in public databases
- Improve tools for data analysis
- Transfer related technologies to the private sector
- Address the ethical, legal, and social issues that may stem from the project.

How has the **Human Genome Project** changed scientists' estimates of the number of genes in humans?

The Human Genome Project has caused a drastic re-evaluation of the probable number of genes in the genome. As a result of the project, estimates have gone from 100,000 genes to somewhere between 30,000 and 40,000.

According to the most recent estimate, **how many genes** do humans have?

As of April 2003 researchers estimated that building a human would require only about 24,500 genes (the mustard seed plant has about 25,000 and fruit flies about 13,000). When analysis of the draft human genome sequence was published in February 2001, there was estimated to be only about 30,000 to 40,000 protein-coding genes. This number will probably fluctuate as applications of the HGP are implemented.

What is the approximate cost of the HGP?

The HGP is reported to have a cost of $3 billion; however, this figure includes project funding over the thirteen-year period (1990–2003). The actual human genome sequencing accounts for only a small portion of the budget.

What is **ELSI** as it pertains to the **Human Genome Project**?

ELSI is a specific aspect of the Human Genome Project that deals with the ethical (E), legal (L), and social (S) issues (I) arising from increasing availability of genetic information. About 3 to 5 percent of the budget of the HGP is devoted to ELSI; it represents the largest bioethics study in the world.

What is **HUGO**?

HUGO, also known as the Human Genome Organization, is an international consortium created to coordinate the work of human geneticists around the world.

What government departments fund the **Human Genome Project**?

In the United States the project is funded primarily by the Department of Energy (DOE) and the National Institutes of Health (NIH). Funding from the DOE grew from a need to research the effects of both radiation and the chemical by-products of energy production on genes. In Europe, funding was provided by the European Commission, charities, and national research councils.

What is meant by a **genetic fingerprint**?

Just as a real fingerprint is used to identify people individually, a genetic (or DNA) fingerprint is a unique pattern of DNA sequences for each individual.

What is a **genome**?

A genome is the complete set of genes inherited from one's parents. Genome sizes vary from one species to another. The final number for humans is yet to be determined.

Genome Sizes by Species

Type	Species	Genome size (bases)	Estimated genes
Bacterium	*Escherichia coli*	4.6 million	3,200
Yeast	*Saccharomyces cerevisiae*	12.1 million	6,000
Worm	*Caenorhabiditis elegans*	97 million	19,099

Type	Species	Genome size (bases)	Estimated genes
Fruit fly	*Drosophila melanogaster*	137 million	13,000
Mustard weed	*Arabidopsis thaliana*	100 million	25,000
Pufferfish	*Fugu rubripes*	400 million	38,000
Mouse	*Mus musculus*	2.6 billion	30,000
Human	*Homo sapiens*	3 billion	30,000
Human immuno-deficiency virus	*HIV*	9,700	9

How much of the **human genome** actually codes for genetic information?

Only about 5 percent of the human genome appears to code for genetic information.

What is the **latest genome** to be made public?

In January 2004 the first draft of the honeybee genome was made available. Why would the honeybee genome be important? The honeybee (*Apis mellifera*) is a major pollinator of crops and is a social insect, making it significant for basic research on insects.

Who made the first **gene map**?

Although gene maps are a relatively recent (past thirty years) means of locating genes, geneticists of the early twentieth century had prototype gene maps. Edmund Beecher Wilson (1856–1939) and his colleagues were the first to demonstrate that the genetic differences between males and females were due to a special pair of chromosomes in the cell. Thomas Hunt Morgan (1866–1945), Calvin Bridges (1889–1938), and their colleagues were able to place a gene known to be inherited differently by males and females onto one of the sex chromosomes. This was the beginning of the first gene map.

APPLICATIONS

What are the largest **human genetic databases**?

Genomics research on entire populations is an area that is rapidly expanding. In order to control access to genetic information and provide ethical guidelines for its use, several countries have developed human genetic databases. Iceland and the United Kingdom have national genetic databases, while Canada and Sweden have regional databases.

What is the most common **gene deficiency**?

Glucose-6 phosphate dehydrogenase (G6PD) is the most common human enzyme deficiency in the world. It is estimated that about 400 million people worldwide are affected by this disorder. The enzyme G6PD is carried on the X allele and is therefore referred to as an X-linked trait; it is more likely to affect males than females. G6PD is a critical enzyme that catalyzes the oxidation/reduction reactions resulting in the production of NADPH; NADPH is a required cofactor in many cellular biosynthetic pathways. The stability of red blood cells is affected by G6PD deficiency since red blood cells are especially sensitive to oxidative stress. Clinical problems experienced by people with G6PD deficiency include neonatal jaundice and hemolytic anemia.

Is **lactose intolerance** a genetic disorder?

Lactose intolerance is due to reduced activity of the enzyme lactase. Lactase hydrolyzes the lactose of mammalian milk into the two monosaccharides glucose and galactose. Lactose intolerance is very common in adults, as expression of the lactase enzyme begins to decrease

Thomas Hunt Morgan made important discoveries in the role played by chromosomes in heredity.

in humans at about two years of age. Estimates of lactose intolerance in the U.S. population vary. Congenital lactase deficiency is a rare autosomal recessive disorder and results in no lactase production. This is in contrast to childhood and adult onset lactase deficiency, which is common and is inherited in an autosomal dominant pattern.

Lactose intolerance in the U.S. population

Population group	Incidence rate
American Caucasians (Northwest European ancestry)	6-25 percent
American Blacks	45-81 percent

Population group	Incidence rate
Mexican Americans	47-74 percent
American Indians, Aleuts, Eskimos	75 percent
American Asians	95 percent

Why are people from certain **geographic areas** more likely to have different **genetic traits**?

Human activities can influence the distribution of alleles. Some examples include social constraints that limit choice of a mate to language, religion, and economic status; the tendency of some isolated populations (such as native island peoples) to stay isolated; and worldwide movement of populations due to technological advances. Studies have been done on the distribution of the B allele of the ABO red blood cell system. Initially, the highest occurrence of the B allele was in the indigenous populations of Central Asia. However, as people migrated from Asia into Europe, the frequency of the B allele changed such that the highest frequency is now in eastern Europe and lowest in southwestern Europe.

Why do some genetic conditions **skip a generation**?

Conditions that are the result of homozygous recessive inheritance can often skip a generation. If a homozygous recessive individual produces offspring with a homozygous dominant individual, then all of their children will be heterozygous and not exhibit the condition. However, should two normal-appearing, heterozygous individuals marry, some of their children may have the condition. This also depends on whether the condition is the result of single trait inheritance or the combination of genes and environmental factors. Environmental factors absent in one generation but present in the next (drought, heat wave, etc.) might cause a trait to suddenly reappear.

Are **mental disorders** inherited?

Most mental disorders have a genetic component. A study on a large family in 1993 identified a link between a particular type of mental retardation, which included frequent aggressive and violent outbursts, with a region on the short arm of the X chromosome in some of the males. This is one of the few cases in which there appears to be a direct correlation between a single gene defect and a particular type of mental disorder.

Do **genes** affect your **mood**?

Mood disorders are conditions that go beyond an occasional "bad day" into the realm of severe emotional disturbance. Mood disorders include depression (the most common mood disorder), bipolar disorder, and schizophrenia. By examining family medical histories and adoption studies, researchers have concluded that there is a link between bipolar illness and genetics. However, the presence of certain gene sequences is not a mandate, as only about 60 percent of monozygotic ("identical") twins with the bipolar sequence actually become ill. It is obvious that environmental and societal factors also play roles in determining one's mood.

Is **aggressive behavior** inherited?

Because it is difficult to characterize a phenotype for aggression, more research is needed before specific genes are linked to aggressive behavior.

Is **alcoholism** an **inherited** disease?

If all of the adults in the U.S. were to be tested, an estimated 10 percent of the adult population would be classified as alcoholic. Of that number, the ratio of males to females is approximately 4 to 1. This disproportion could be attributed to both environment and genetics. Research involving male relatives, both biological and adoptive, points to a multifactorial source for alcoholism. In other words, while sons and brothers of male alcoholics are likely to be alcoholic as well (25 percent and 50 percent, respectively), genetic inheritance is not solely responsible for all of the cases of alcoholism. Alcoholism has been linked to depression; as more is learned about the genetic basis of mood disorders, the genetic basis of alcoholism may be unraveled.

Is **obesity** inherited?

Genetic clues to obesity have been studied in mice, where at least two genes, obese (ob) and diabetes (db), have been identified and analyzed. The ob gene codes for the weight controlling hormone leptin (from the Greek term *leptos*, meaning "slender"), which is produced by fat cells. Once leptin is released by fat cells, it travels to the hypothalamus (a specific area of the brain), where it binds to receptors. The db gene codes for these hypothalamic receptors. Once leptin is bound, it regulates the rate of energy consumption. However, body weight is a complex phenotype, and it is likely that there are more genes to be found that control body weight.

Is there a **gay gene**?

Homosexuality has a strong genetic component. One study found that among fifty-six pairs of monozygotic male twins, concordance (simultaneous incidence) for homosexuality was 52 percent, while it was only 22 percent for fraternal (dizygotic) twins and 11 percent for unrelated adopted brothers. This indicates that the closer two males are genetically, the more likely they are to have the same sexual orientation. Several studies have pointed to a likely location for male homosexuality traits at the tip of the X chromosome. It should be noted that not all males with this sequence identified themselves as homosexual, and so it seems likely that a variety of factors (environmental and genetic) may have roles in determining sexual orientation. So far, there are no correlate statistics for female homosexuality.

What is meant by **genetic discrimination**?

Genetic discrimination refers to the use of genetic information by health insurers and employers to determine eligibility, set premiums (for health insurance), or hire and fire people. The U.S. Senate passed the Genetic Information Nondiscrimination Act of 2003 on October 14, 2003. The act establishes basic legal protections that will prohibit discrimination in health insurance and employment on the basis of predictive genetic information. It will fully protect the privacy of genetic information.

What are some **famous figures** in history with **genetic disorders**?

One of the most common genetic connective tissue disorders is Marfan syndrome. People with Marfan syndrome are usually very tall, with skeletal malformation and loose joints. Abraham Lincoln, Paganini, and Rachmaninoff were believed to have Marfan syndrome. Because of their above-average height and loose joints, people with Marfan syndrome may excel in sports such as basketball or volleyball. Olympic volleyball star Flo Hyman died of a ruptured aorta as a result of Marfan's. King George III (1738–1820) of Great Britain suffered from porphyria, a disorder inherited as a dominant trait that causes occasional attacks of pain and dementia.

What are some human **characteristics** controlled by **one gene**?

Human One Gene Traits

Name	Phenotype
Hitchhiker's thumb	Recessive
Tongue rolling	*Dominant
Thumb crossing (right over left)	Recessive
Widow's peak	Dominant
Woolly hair	Dominant
Attached ear lobes	Recessive

* There is some disagreement about whether this trait is strictly autosomal dominant.

How many **Nobel Prizes** have been awarded in genetics?

Of the 162 Nobel Prizes awarded in the category for Physiology or Medicine, forty-two prizes (26 percent) have been awarded in genetics. This far exceeds other categories; the closest in number is disease treatment, with twenty-four prizes (15 percent).

It is believed that some of the early English colonists that settled in New England may have had Huntington's disease. Huntington's disease is an autosomal dominant disorder characterized by late onset symptoms (age forty to fifty) such as mild behavioral and neurological changes; as the disease progresses, psychiatric problems develop that frequently lead to insanity. Early descriptions of the odd behavior included names such as "that disorder" and "Saint Vitus's dance" to describe involuntary muscle jerks and twitches. Many of the witches who were on trial for possession may have had Huntington's disease, which causes uncontrollable movements and odd behavior.

What genetic diseases come from mutations of **mitochondrial DNA**?

Because mitochondria are primarily responsible for energy conversion, mutations in their DNA affect the ability of the cell to generate ATP, the immediate energy source for cells. Those cells in tissues with the highest energy demands demonstrate the most severe effects of mitochondrial mutations.

Examples of Mitochondrial Diseases and Their Phenotype

Syndrome	Phenotype
Kearns-Sayre	Short stature, degeneration of the retina
Leber optic atrophy (LHON)	Loss of vision in center of visual field
MELAS (mitochondrial encephalopathy lactic acidosis stroke)	Vomiting, seizures, stroke-like episodes
MERRF (myoclonic epilepsy ragged red fibers	Deficiencies in enzyme complexes associated with energy transfer
Oncocytoma	Benign tumors of the kidney

What is the historical evidence of a **genetic disease**?

One example of historical evidence of a genetic disease is found in Egyptian art of 3,500 years ago that shows Akhenaten, Pharaoh of Egypt during the eighteenth dynasty. Akhenaten is depicted as a man with a sharp chin, almond-shaped eyes, full lips, long arms and fingers, and enlarged breasts. One early statue shows the king in the nude and without genitalia. These features have intrigued archaeologists, and two theories have been proposed for his odd appearance. One is that Akhenaten suffered from either Froehlich syndrome or Marfan syndrome. Froehlich syndrome is an endocrine disorder found most commonly in men; this disorder affects development of secondary sex characteristics, which fits some of the characteristics associated with Akhenaten's appearance. However, Akhenaten, although he had some feminine characteristics, was known to have fathered at least six children, and he was not retarded (another characteristic of Froehlich syndrome). The more likely possibility is Marfan syndrome, which results in patients with long fingers, skeletal abnormalities, but normal intelligence and fertility. Also, evidence suggests that the rest of his family apparently showed some of these traits, suggesting that he had a genetic disorder.

How did Lysenko affect **genetics** in the **Soviet Union**?

Trofim Lysenko (1889–1976) was a dominant force in biology and agriculture in the Soviet Union from 1937 to 1964. Lysenko was a follower of Jean-Baptiste Lamarck (1744–1829), who believed that environmental experiences could change genetic inheritance. Lysenko sought to manipulate the environment in the hopes that this would increase agricultural productivity. By the time Lysenko had fallen out of favor, Soviet agricultural productivity had, in fact, actually decreased significantly, failing to keep pace with the advances of other developed countries.

Is **IQ** genetically controlled?

This is actually two questions. First, can intelligence be measured quantitatively? And second, is there a correlation between intelligence and certain genotypes? Both questions can be answered in the negative. As our definition of intelligence has evolved, it has become more difficult to assign a single number to the trait of intelligence. Therefore, it is almost impossible to demonstrate which gene sequences correlate to high IQ values; even though our understanding of the genome is broadening, so too is our definition of intelligence. It becomes less and less likely that we will be able to find one or two genes that determine all of the facets of intelligence, considering the importance of environment in determining phenotype. IQ is most likely a multifactorial trait.

What is **eugenics**?

Sir Francis Galton (1822–1911), a cousin of Charles Darwin, founded eugenics. After reading Darwin's work on natural selection, Galton thought that the human species could be improved by artificial selection—the selective breeding for desirable traits. This method is often used in domesticated animals. In Galton's plan, those with desirable traits would be encouraged to have large families, while those with undesirable traits would be kept from breeding. However, Galton's theory overlooked two important points: the importance of environmental factors on phenotypic expression, and the difficulty of removing recessive traits from the gene pool. Recessive alleles can be passed from one individual to the next as part of a heterozygous genotype, thereby escaping detection for generations. Galton's work was enthusiastically adopted in both the United States and Europe. In the United States between 1900 and 1930, eugenics gave rise to changes in federal immigration laws and the passage of state laws requiring the sterilization of "genetic defectives" and certain types of criminals. In Europe, eugenics became a cornerstone of the Nazi movement.

Will two people with the same **genes** for a particular trait have the same exact manifestation of the **disease**?

Penetrance is the term used to describe the probability that an individual with a given genotype will always demonstrate the matching phenotype. For example, a condition with 100 percent penetrance will be one where genotype always determines phenotype; 50 percent penetrance means that there is only a 50-50 chance of the genotype producing the characteristic phenotype. Myotonic dystrophy, an autosomal dominant genetic disorder, is an example in which there is reduced penetrance within a pedigree; in other words, some family members may have the trait while others may not.

Why is there such a range of **skin color** in humans?

Studies on the genetic components of skin color indicate that at least three and perhaps as many as four gene pairs interact to produce a given shade of skin color. Four gene pairs would generate 4x4x4x4 (256) possible genotypes, which is ample for the number of color variations found among humans.

BIOTECHNOLOGY AND GENETIC ENGINEERING

INTRODUCTION AND HISTORICAL BACKGROUND

What is **biotechnology**?

Biotechnology is the use of a living organism to produce a specific product. It includes any technology associated with the manipulation of living systems for industrial purposes. In its broadest sense, biotechnology includes the fields of chemical, pharmaceutical, and environmental technology as well as engineering and agriculture.

Who was awarded two **Nobel Prizes** for **DNA sequencing**?

Frederick Sanger (1918–) won the Nobel Prize in Chemistry in 1958 for his work on the structure of proteins (especially insulin) and shared the Nobel Prize in Chemistry in 1980, along with Walter Gilbert (1932–), for contributions in determining the nitrogen base sequences of DNA. This method was later referred to as the Sanger sequencing method for reading DNA.

What is **TIGR**?

TIGR is the Institute for Genomic Research, a nonprofit private research institute founded in 1992 by Craig Venter (1948–) and headquartered in Rockville, Maryland. Its specific interests are in the structural, functional, and comparative analysis of genomes and gene products. TIGR collaborates with institutions around the world. For example, it worked with the International Livestock Research Institute in Nairobi, Kenya, to determine the genome of the parasite that causes East Coast fever, a fatal disease of cattle in sub-Saharan Africa.

When was the term "biotechnology" first used?

The Hungarian agricultural economist Karl Ereky (1878–?) used the term in 1919 in reference to the production of products from raw materials with the aid of living organisms. Ereky was specifically referring to the large scale farming of pigs using sugar beets as a food source.

What are the **major achievements** of **biotechnology** in the twentieth century?

Date	Discovery
1968	Stanley Cohen (1922–) uses plasmids to transfer antibiotic resistance to bacterial cells
1970	Herb Boyer (1936–) discovers that certain bacteria can "restrict" some bacteriophages by producing enzymes (restriction enzymes): isolates EcoRI
1972	Paul Berg (1926–) splices together DNA from the SV 40 virus and *E. coli*, making recombinant DNA; shares 1980 Nobel Prize with Walter Gilbert (1932–) and Fred Sanger (1918–)
1974	Stanley Cohen (1922–), Annie Chang, and Herb Boyer (1936–) splice frog DNA into *E. coli*, producing the first recombinant organism
1975	DNA sequencing developed by Walter Gilbert (1932–), Allan Maxam (1952–), and Fred Sanger (1918–)
1978	Human insulin cloned in *E. coli* by biotech company Genentech
1986	Kary Mullis (1944–) develops the polymerase chain reaction (PCR), in which DNA polymerase can copy a DNA segment many times in a short period of time
1989	Human Genome Project (HGP) begins
1990	Researchers at National Institutes of Health (NIH) use gene therapy to treat a human patient
1994	Introduction of first transgenic food, the Flavr Savr tomato
1996	Dolly is cloned by Ian Wilmut (1944–)
1997	First human artificial chromosome is developed
2000	Completion of the first working draft (90 percent complete) of the HGP
2003	Glofish, the first genetically modified pets, are marketed and sold in the U.S.

Who was the first individual to find the **gene** for **breast cancer**?

Mary Claire King (1946–) determined that in 5 to 10 percent of those women with breast cancer, the cancer is the result of a mutation of a gene on chromosome 17, the BRCA1 (Breast Cancer 1) . The BRCA1 gene is a tumor suppressor gene and is also linked to ovarian cancer. Subsequently, other researchers were able to clone the gene and pinpoint its exact location on chromosome 17.

Who was the first person to receive **gene therapy**?

In 1990 Ashanti DeSilva, a four-year-old with severe combined immune deficiency (SCID), had her white blood cells removed and replaced with a normal copy of the defective gene. SCID is a life-threatening disease in which patients lack a healthy immune system due to the inability to produce an important enzyme (adenosine deaminase). Although the treatment proved safe and her immune system was strengthened, the treated cells failed to give rise to additional healthy cells. In order to maintain normal levels of adenosine deaminase, DeSilva must take doses of the enzyme itself.

What is **genetic engineering**?

Genetic engineering, also popularly known as molecular cloning or gene cloning, is the artificial recombination of nucleic acid molecules in a test tube; their insertion into a virus, bacterial plasmid, or other vector system; and the subsequent incorporation of the chimeric molecules into a host organism in which they are capable of continued propagation. The construction of such molecules has also been termed gene manipulation because it usually involves the production of novel genetic combinations by biochemical means.

Genetic engineering techniques include cell fusion and the use of recombinant DNA or gene-splicing. In cell fusion the tough outer membranes of sperm and egg cells are removed by enzymes, and then the fragile cells are mixed and combined with the aid of chemicals or viruses. The result may be the creation of a new life form from two species (a chimera). Recombinant DNA techniques transfer a specific genetic activity from one organism to the next through the use of bacterial plasmids (small circular pieces of DNA lying outside the main bacterial chromosome) and enzymes, such as restriction endonucleases (which cut the DNA strands); reverse transcriptase (which makes a DNA strand from an RNA strand); DNA ligase (which joins DNA strands together); and Taq polymerase (which can make a double-strand DNA molecule from a single-strand "primer" molecule). The recombinant DNA process begins with the isolation and fragmentation of suitable DNA strands. After these fragments are combined with vectors, they are carried into bacterial cells, where the DNA fragments are "spliced" on to plasmid DNA that has been opened up. These hybrid plasmids are then mixed with host cells to form transformed cells. Since only some of the transformed cells will exhibit the desired characteristic or gene activity, the transformed cells are separated and grown individually in cultures. This methodology has been successful in producing large quantities of hormones (such as insulin) for the biotechnology industry. However, it is more difficult to transform animal and plant cells. Yet the technique exists to make plants resistant to diseases and to make animals grow larger. Because genetic engineering interferes with the processes of heredity and can alter the genetic structure of our own species, there is much concern over the ethical ramifications of such power, as well as the possible health and ecological consequences of the creation of these bacterial forms. Some applications of genetic engineering in various fields are:

- Agriculture: Crops having larger yields, disease- and drought-resistancy; bacterial sprays to prevent crop damage from freezing temperatures; and livestock improvement through changes in animal traits.

- Industry: Use of bacteria to convert old newspaper and wood chips into sugar; oil- and toxin-absorbing bacteria for oil spill or toxic waste clean-ups; and yeasts to accelerate wine fermentation.

- Medicine: Alteration of human genes to eliminate disease (experimental stage); faster and more economical production of vital human substances to alleviate deficiency and disease symptoms (but not to cure them); substances include insulin, interferon (cancer therapy), vitamins, human growth hormone ADA, antibodies, vaccines, and antibiotics.

- Research: Modification of gene structure in medical research, especially cancer research.

- Food processing: Rennin (enzyme) in cheese aging.

What was the **first commercial use** of **genetic engineering**?

Commercial recombinant DNA technology was first used to produce human insulin in bacteria. In 1982 genetically engineered insulin was approved by the FDA for use by diabetics. Insulin is normally produced by the pancreas, and for more than fifty years the pancreas of slaughtered animals such as swine or sheep was used as an insulin source. To provide a reliable source of human insulin, researchers harvested the insulin gene from cellular DNA. Researchers made a copy of DNA carrying this insulin gene and spliced it into a bacterium. When the bacterium was cultured, the microbe split from one cell into two cells, and both cells got a copy of the insulin gene. Those two microbes grew, then divided into four, those four into eight, the eight into sixteen, and so forth. With each cell division, the two new cells each had a copy of the gene for human insulin. And because the cells had a copy of the genetic "recipe card" for insulin, they could make the insulin protein. Using genetic engineering to produce insulin was both cheaper and safer for patients, as some patients were allergic to insulin from other animals.

METHODS

What is a **vector**?

A vector is an agent used to carry genes into another organism. Specific examples of natural vectors include plasmids or viruses. In human gene therapy, vector viruses must be able to withstand the challenge of the patient's immune system. Once the vector manages to invade the immune system, it must be able to penetrate the cell membrane and, finally, must be able to combine its genome into that of the host cell. Vectors are also crucial to plant and animal genetic engineering.

What is **recombinant DNA**?

Recombinant DNA is hybrid DNA that has been created from more than one source. An example is the splicing of human DNA into bacterial DNA so that a human gene product is produced by a bacterial cell.

What is a **restriction endonuclease**?

A restriction endonuclease is an enzyme that cleaves DNA at specific sites. Restriction enzymes are made by bacteria as a means of bacterial warfare against invading bacteriophages (viruses that infect bacteria). These enzymes are now used extensively in biotechnology to cleave DNA into shorter fragments for analysis or to selectively cut plasmids so that foreign DNA can be inserted.

Characteristics of Some Restriction Enzymes

Enzyme Name	Source
BamHI	*Bacillus amyloliquefaciens H*
EcoRI	*E. Coli RY13*

417

Enzyme Name	Source
HindIII	*Haemophilus influenzae R_d*
SmaI	*Serratia marcescens*
HaeIII	*Haemophilus egyptius*

What are **RFLPs**?

RFLPs are restriction fragment length polymorphisms, which are variations in the short base sequences where restriction enzymes can cut DNA. By cutting two different DNA molecules with the same restriction enzyme, scientists can compare the lengths of the fragments; two identical molecules will have identical fragments, while two similar molecules may be largely alike, with perhaps a few differences in fragment size. These differences in restriction fragment lengths are called polymorphisms and are used in all types of DNA typing.

What is **polymerase chain reaction**?

Polymerase chain reaction, or PCR, is a laboratory technique that amplifies or copies any piece of DNA very quickly without using cells. The DNA is incubated in a test tube with a special kind of DNA polymerase, a supply of nucleotides, and short pieces of synthetic, single-strand DNA that serve as primers for DNA synthesis. With automation, PCR can make billions of copies of a particular segment of DNA in a few hours. Each cycle of the PCR procedure takes only about five minutes. At the end of the cycle the DNA segment—even one with hundreds of base pairs—has been doubled. A PCR machine repeats the cycle over and over. PCR is much faster than the days it takes to clone a piece of DNA by making a recombinant plasmid and letting it replicate within bacteria.

PCR was developed by the biochemist Kary Mullis (1944–) in 1983 at Cetus Corporation, a California biotechnology firm. In 1993 Mullis, along with Michael Smith (1932–), won the Nobel Prize in Chemistry for development of PCR.

What is **Taq polymerase**?

Taq polymerase is a DNA polymerase (enzyme) isolated from the heat-tolerant bacterium *Thermus aquaticus*. It is an integral enzyme to polymerase chain reaction (PCR), which allows for multiple copies of a DNA segment to be made in several hours using a thermocycler.

What is a **thermocycler**?

A thermocycler is a special machine used for PCR. Thermocyclers are used to vary the temperature of the DNA over a preset cycle. Since the DNA helix unravels at higher temperatures, this thermocylcing causes the original sample to separate and be used as a template for building multiple copies of the original sequence.

What is **DNA amplification**?

DNA amplification is a method by which a small piece of DNA is copied thousands of times using PCR. DNA amplification is used in cloning, to detect small amounts of DNA in a sample, and to distinguish different DNA samples (as in DNA fingerprinting).

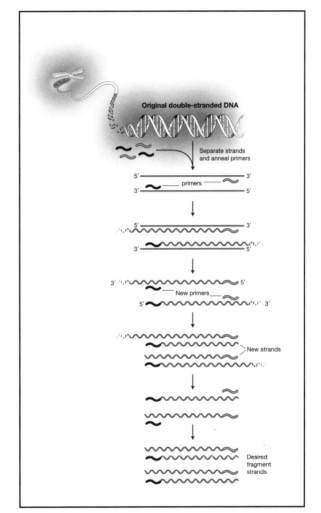

Polymerase chain reaction is a laboratory technique that amplifies or copies any piece of DNA very quickly without using cells.

What is **transformation**?

Transformation is the process by which a cell or an organism incorporates foreign DNA. Transformation usually occurs between a plasmid and a bacterium. The transformed cell or organism then produces the protein encoded by the foreign DNA. In order to determine whether cells have been transformed, the foreign DNA usually contains a marker, such as the gene for penicillin resistance. Only cells with the resistance gene will be able to grow in a culture medium containing penicillin.

What is **transduction**?

Transduction is the process by which a vector (usually a bacteriophage) carries DNA from one bacterium to another bacterium. It can be used experimentally to map bacterial genes.

What is a **gene library**?

A gene library is a collection of cloned DNA, usually from a specific organism. Just as a conventional library stores information in books and computer files, a gene library stores genetic information either for an entire genome, a single chromosome, or specific genes in a cell. For example, one can find the gene library of a specific disease such as cystic fibrosis, the chromosome where most cystic fibrosis mutations occur, or the entire genome of those individuals affected by the disease.

How is a **gene library** made?

To create a gene library, scientists extract the DNA of a specific organism and use restriction endonucleases to cut it. The scientists then insert the resulting fragments into vectors and make multiple copies of the fragments. The number of clones needed for a genomic library depends on the size of the genome and the size of the DNA fragments. Specific clones in the library are located using a DNA probe.

How are genes physically found in a specific **genome**?

Finding one gene out of a possible 30,000 to 40,000 genes in the human genome is a difficult task. However, the process is made easier if the protein product of the gene is known. As an example, if a researcher is looking to find the gene for mouse hemoglobin, he or she would isolate the hemoglobin from mouse blood and determine the amino acid sequence. The amino acid sequence could then be used as a template to generate the nucleotide sequence. Working backward again, a complimentary DNA probe to the sequence would be used to identify DNA molecules with the same sequence from the entire mouse genomic library.

However, if the protein product is not known, the task is more difficult. An example of this would be that of finding the susceptibility gene for late-onset Alzheimer's disease. DNA samples would be collected from family members of a patient with late-onset Alzheimer's disease. The DNA would be cut with restriction endonucleases, and restriction fragment length polymorphisms (RFLPs) would be compared among the family. If certain RFLPs are only found when the disease gene is present, then it is assumed that the distinctive fragments are markers for the gene. Geneticists then sequence the DNA in the same area of the chromosome where the marker was found, looking for potential gene candidates.

What is an **artificial chromosome**?

An artificial chromosome is a new type of vector that allows cloning of larger pieces of DNA. It consists of a telomere at each end, a centromere sequence, and specific sites

> ## What are the most frequently used artificial chromosomes?
>
> Artificial chromosomes include YACs (yeast artificial chromosome), BACs (bacterial artificial chromosomes), and MACs (mammalian artificial chromosomes).

at which foreign DNA can be inserted. Once the DNA fragment is spliced in, the engineered chromosome is reinserted into a yeast cell. The yeast then reproduces the chromosome as if it were part of the normal yeast genome. As a result, a colony of yeast cells would then all contain a specific fragment of DNA.

What is **antisense technology**?

Antisense technology involves targeting a protein at the RNA level as a means of regulating gene expression. The technology involves segments of DNA that are complementary to specific RNA targets. When the modified DNA binds to the RNA, the RNA can no longer produce a protein. Antisense technology is a new technique in cancer therapy. Human trials are ongoing on small-cell lung cancer and leukemia. Since cancer is a disease that can be characterized by excessive production of a particular protein, antisense technology works by interfering with the RNA that produces the protein.

What is a **gene probe**?

A gene probe is a specific segment of single-strand DNA that is complementary to a desired gene. For example, if the gene of interest contains the sequence AATGGCACA, then the probe will contain the complementary sequence TTACCGTGT. When added to the appropriate solution, the probe will match and then bind to the gene of interest. To facilitate locating the probe, scientists usually label it with a radioisotope or a fluorescent dye so that it can be visualized and identified.

What is a **gene gun**?

The gene gun, developed by Cornell University plant scientists, is a method of direct gene transfer that is used in plant biotechnology. In order to transfer genes into plants, gold or tungsten microspheres (1 micrometer in diameter) are coated with DNA from a specific gene. The microspheres are then accelerated toward target cells (contained in a petri dish) at high speed. Once inside the target cells, the DNA on the outside of the microsphere is released and can be incorporated into the plant's genome. This method is also known as microprojectile bombardment or biolistics. The survival rate of bombarded cells varies with the rate of penetration. For example, if the particle penetration rate reaches 21 per cell, approximately 80 percent of bombarded cells may die.

Currently, drug companies spend $23 to $47 billion dollars a year on drug research and development. Obviously, capturing a portion of that money would be helpful to any country's economy. The organisms involved in bio-prospecting are renewable resources, and tropical rainforests and coral reefs, in particular, are likely to benefit.

What is **bioprospecting**?

Bioprospecting involves the search for possible new plant or microbial strains, particularly from the world's largest rainforests and coral reefs. These organisms are then used to develop new phytopharmaceuticals. There is some controversy as to who owns the resources of these countries: the countries in which the resources reside or the company that turns them into valuable products.

What is a **bioreactor**?

A bioreactor is a large vessel in which a biological reaction or transformation occurs. Bioreactors are used in bioprocessing technology to carry out large scale mammalian cell culture and microbial fermentation.

Examples of Products and Organisms Made in Bioreactors

Product	Organisms
Single cell protein	*Candida utilis* (yeast)
Penicillin (antibiotic)	*Penicillium chrysogenum* (fungus)
Alpha amylase (enzyme)	*Bacillus amyloliquefaciens* (bacteria)
Riboflavin (vitamin)	*Eremothecium ashbyii* (bacteria)
Poliomyelitis vaccine	Monkey kidney or human cells
Insulin (hormone)	Recombinant *Escherichia coli* (bacteria)

What is a **chimera**?

The chimera from Greek mythology is a fire-breathing monster with a lion's head, goat's body, and a serpent's tail. The chimera of biotechnology is an animal formed from two different species or strains—that is, a mixture of cells from two very early embryos. Most chimeras used in research are made from different mouse strains. Chimeras cannot reproduce.

What is **cell culture**?

Cell culture is the cultivation of cells (outside the body) from a multicellular
organism. This technique is very important to biotechnology processes because

Can a human cell line be patented?

In 1995 an indigenous man of the Hagahai people from Papua New Guinea had cells that became Patent No. 5,397,696. The National Institutes of Health obtained the patent on the cells, and the foreign citizen essentially ceased to own his genetic material. Public outcry against the patent was great, particularly because there were only 260 Hagahai persons remaining at the time, and they had only come into contact with the outside world in 1984. Also, there is a fundamental difference between native belief and Western science at the philosophical level: native people hold that all life is sacred and believe that removing any part of it—even a blood sample—may restrict the individual's ability to pass into the next stage of life. In 1996 NIH relinquished all rights to the patent.

most research programs depend on the ability to grow cells outside the parent animal. Cells grown in culture usually require very special conditions (e.g., specific pH, temperature, nutrients, and growth factors). Cells can be grown in a variety of containers, ranging from a simple petri dish to large-scale cultures in roller bottles, which are bottles that are rolled gently to keep culture medium flowing over the cells.

What is **terminator gene technology**?

This is a method of biotechnology in which crops that are bioengineered for a specific desirable trait (such as drought resistance) would contain a lethal gene that would cause any seeds produced by the plant to be nonviable. The lethal gene could be activated by spraying with a solution sold by the same company that originally marketed the bioengineered plant. Thus, the plants would still provide seeds with nutritional value, but these seeds could not be used to produce new plants. This technique would allow companies to control the product's genes so that they would not spread into the general plant population. Seeds would have to be purchased by the grower for each season.

What is a **microarray**?

A microarray is a technique in which PCR-amplified DNA fragments are placed on a thin glass or silicon plate by cross-linking the DNA to the glass or silicon. Fluorescent dye–labeled mRNA or complementary DNA is then hybridized to the sample. When hybridization occurs, a specific fluorescent color is produced. For example, if two samples, one labeled with a red dye and one with a green dye, are both hybridized to the same DNA sequence on the microarray, a yellow color is produced. The amount of color produced also allows scientists to detect the level of gene expression.

423

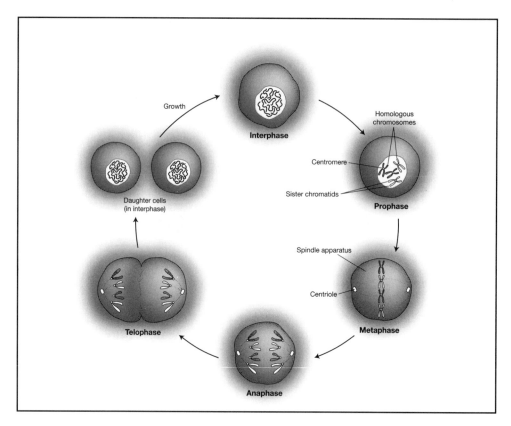

Mitosis, a fundamental process in cloning and cell regeneration, involves cell nucleus division in which each chromosome splits into two, such that both daughter cells have identical chromosomal make-up.

What is a **gene chip**?

A gene chip is part of the process of microarray profiling; it is also known as a biochip or a DNA chip. It is about the size of a postage stamp and is based on a glass wafer, holding as many as 400,000 tiny cells. Each tiny cell can hold DNA from a different human gene and can perform thousands of biological reactions in a few seconds. These chips can be used by pharmaceutical companies to discover what genes are involved in various disease processes. They can also be used to type single nucleotide polymorphisms (SNPs), which are base pair differences that are found approximately every 500 to 1,000 base pairs in DNA. There are more than 3 million SNPs in the human genome. They are very important in DNA typing because they represent about 98 percent of all DNA polymorphisms.

What is **DNA sequence analysis**?

DNA sequence analysis is part of a subfield of genomics known as structural genomics. DNA sequence analysis results in a complete description of a DNA molecule in terms of its order of nitrogen bases (adenine, thymine, guanine, cytosine). All sequencing projects are based on a similar technology: DNA fragmentation with restriction endonucleases, cloning of the DNA fragments, and overlapping analysis of the fragments to eventually provide a map.

What is the **Sanger technique**?

The Sanger technique was developed by Fred Sanger (1918–) in the 1970s. It is also referred to as shotgun sequencing and is a common method used for DNA sequencing. The basics of this method involve using dideoxynucleotides instead of the deoxynucleotides normally found in DNA. Dideoxynucleotides are nucleotides missing a hydroxyl group at the site where one nucleotide usually joins to another nucleotide. If the hydroxyl group is missing, chain elongation stops. The process can be done manually or by automation (the cost of sequencer is approximately $100,000). In the automated process, a short DNA primer is synthesized that is complementary to the template to be analyzed. Electrophoresis is used to separate the bands produced. A different fluorescent label is used on each of the dideoxynucleotides (adenine, guanine, thymine, cytosine), and these are added to the template. Once hybridization occurs, the chain elongation stops and large numbers of fluorescent bands are produced and "read" using a laser. The laser then transmits the sequence into nitrogen bases.

What is the **ELISA test**?

ELISA stands for enzyme-linked immunosorbent assay. It is a test used to determine if an individual has generated antibodies for a certain pathogen. In the test, inactivated antigens (for the pathogen) are coated onto a plate. In addition to antibodies from the patient serum, there is a second antihuman antibody linked to an enzyme. If the patient's antibodies bind to both the second antibody and the antigen, a color change occurs.

What is **green fluorescent protein**?

Green fluorescent protein is a protein found in a luminescent jellyfish (*Aquorea victoria*) that lives in the cold waters of the northern Pacific. Bioluminescence is the production of light by living organisms. These jellyfish contain two proteins: a bioluminescent protein called aequorin that emits blue light, and an accessory green fluorescent protein (GFP). However, what we actually see when the jellyfish fluoresces is the conversion of the blue light emitted by aequorin to a green light—a metabolic reaction facilitated by the GFP. Since GFP is simply a protein, it is often used both as a marker for gene transfer and for localization of proteins. There are a variety of green fluorescent proteins that can glow different colors.

What is a **knockout mouse**?

A knockout mouse is a mouse that has specific genes mutated ("knocked out") so that the lack of production of the gene product can be studied in an animal model. Knockout

> ## What is the GloFish?
>
> The fluorescent red zebrafish (*Danio rerio*) called GloFish are produced by bio-engineering the fish with the gene for red fluorescent protein from sea anemones and coral. The fish were originally bred by scientists at the National University of Singapore to detect water pollution, but today they are produced primarily to satisfy the demands of aquarium owners who like brightly colored creatures. Research continues on developing zebrafish that will selectively fluoresce when exposed to estrogen or heavy metals in the water. In the United States, the FDA, which has jurisdiction over genetically modified organisms, maintains that the fish should pose no threat to native fish populations, as they are tropical aquarium fish used exclusively for that purpose. Concerns about the GloFish include the ethics of the use of genetic engineering for production of uniquely colored fish for such a trivial purpose, the potential use of ornamental fish (such as the GloFish) as a food fish, or the chance that the GloFish might proliferate without safety monitoring. GloFish are not intended for human consumption.

mice are frequently used in pharmaceutical studies to test the potential for a particular human enzyme as a therapy target. Mice have identical or nearly identical copies of human proteins, making them useful in studies that model effects in humans.

What is **FISH**?

FISH (fluorescent in situ hybridization) is a method in which a clone gene is "painted" with a fluorescent dye and mapped to a chromosome. In FISH, cells are arrested in the metaphase stage of mitosis and placed on a slide where they burst open, spreading chromosomes over the surface. A fluorescent-labeled DNA probe of specific interest is placed on the slide and incubated long enough for hybridization to occur. The slide is then viewed under a fluorescence microscope that focuses ultraviolet light on the chromosomes. Hybridized regions of the chromosome fluoresce. This is a type of karyotypic analysis that can generate a physical map matching clones and gene markers to specific parts of a chromosome.

What is **flow cytometry**?

Flow cytometry is an exciting new biotechnology method in which single cells or subcellular fractions flowing past a light source are scanned based on measurement of visible or fluorescent light emission. The cells can be sorted into different fractions depending on fluorescence emitted by each droplet. Flow cytometry can be used to identify particular cell types such as malignant cells, or immune cells such as T or B cells.

What is **high throughput screening**?

High throughput screening is a large-scale automated search of a genome in order to determine how a cell (or larger biological system) uses its genetic information to create a specific product or total organism. The term "platform" is used to describe all components needed for high throughput screening.

What is **PAGE**?

PAGE (polyacrylamide gel electrophoresis) is a type of separation method in which polyacrylamide (a polymer of acrylamide) is used instead of agarose in the gel bed of the electrophoresis chamber. Polyacrylamide is preferable to agarose because it forms a tighter gel bed that is then capable of separating smaller molecules, especially those with molecular weights that are very similar. PAGE is used in DNA sequencing, DNA fingerprinting, and for protein separation.

What is a **transposon**?

A transposon is a nucleotide sequence in a gene that literally "jumps" around from one chromosome to another. Transposons can be problematic, as they may disrupt the normal function of an important gene by their random insertion into the middle of that gene. However, transposons are found in almost all organisms, and their presence in a genome indicates that genetic information is not fixed within the genome. By studying results of the Human Genome Project, scientists predict that about 45 percent of all human genes may be derived from transposons.

What is a **VNTR**?

VNTR refers to variable number of tandem repeats, which is a kind of molecular marker that contains sequences of DNA that have end-to-end repeats of different short DNA sequences. When the repeating unit is 2 to 4 base pairs long, it is called an STR (short tandem repeat), and when the repeating unit is 2 to 30 base pairs long, it is called a VNTR. An example of a VNTR is GGATGGATGGATGGATGGAT, which is four tandem repeats of the sequence GGAT. In addition, VNTRs are highly variable regions, and thus there may be many alleles (gene variations) at many loci (physical location of a gene on a chromosome). VNTRs were discovered by Alec Jeffreys (1950–) and colleagues as they were working on the basics of DNA fingerprinting.

What is the **biological basis** for **DNA fingerprinting**?

DNA fingerprinting, also known as DNA typing or DNA profiling, is based on the unique genetic differences that exist between individuals. Most DNA sequences are identical, but out of 100 base pairs (of DNA), two people will generally differ by one base pair. Since there are 3 billion base pairs in human DNA, one individual's DNA will differ from another's by 3 million base pairs. To examine an individual's DNA fingerprint, the DNA sample is cut with a restriction endonuclease, and the fragments are separated by gel electrophoresis. The fragments are then transferred to a nylon membrane, where they are incubated in a solution containing a radioactive DNA probe that is complementary to specific polymorphic sequences.

What **methods of analysis** are used for **DNA profiling**?

Both RFLP (restriction fragment length polymorphism) and PCR (polymerase chain reaction) techniques can be used for DNA profiling. The success of either method depends on identifying where the DNA of two individuals varies the most and how this variation can be used to discriminate between two different DNA samples. Since RFLP uses markers in regions that are highly variable, it is unlikely that the DNA of two unrelated individuals will be identical. However, this method requires at least 20 nanograms of purified, intact DNA. PCR-based DNA fingerprinting is a rapid, less expensive method and requires only a very small amount of DNA (as little as fifty white blood cells).

When was **DNA fingerprinting** developed?

Sir Alec Jeffreys (1950–), a geneticist, developed DNA fingerprinting in the early 1980s, when he was studying inherited genetic variations between people. He was one of the first scientists to describe small DNA changes, referred to as single nucleotide polymorphisms (SNPs). From SNPs, he began to look at tandem repeat DNA sequences, in which a short sequence of DNA was consecutively repeated many times.

What types of **samples** can be used for **DNA fingerprinting**?

Any body fluid or tissue that contains DNA can be used for DNA fingerprinting, including hair follicles, skin, ear wax, bone, urine, feces, semen, or blood. In criminal cases, DNA evidence may also be gathered from dandruff; from saliva on cigarette butts, chewing gum, or envelopes; and from skin cells on eyeglasses.

What is the largest **forensic DNA investigation** in U.S. history?

The identification of the remains of the victims from the September 11, 2001, terrorist attacks in New York City has comprised the largest and most difficult DNA identification to date. After 1.6 million tons of debris were removed from the site of the attacks on the World Trade Center, only 239 intact bodies (out of 2,795) were found, along with about 20,000 pieces of human remains. In order to match DNA profiles to the bodies, personal items such as razor blades, combs, and toothbrushes were collected from the victim's homes. When possible, cheek swabs were taken from the victim's family members for comparison with remains. The identification process was still ongoing as of mid-2004.

What was the first use of DNA fingerprinting in a criminal investigation?

The first case of DNA fingerprinting occurred in 1986 in England. Between 1983 and 1986 two schoolgirls were murdered in a village near Leicestershire, England. The only clue left at the respective crime scenes was semen, and a suspect had already been arrested. However, police asked a scientist to help them positively identify whether the DNA profiles from both crime scenes matched that of the suspect. Using VNTR analysis, the scientist showed that although the DNA profiles from both crime scenes matched, they did not match that of the suspect. The suspect was released, and police then obtained a blood sample from every adult male in the area of the village. A local bakery worker, who had paid another coworker to provide his DNA sample, was eventually arrested and forced to provide a blood sample. His DNA sample matched the samples from the crime scene, and the worker confessed to the murders.

What is the accuracy of **DNA fingerprinting**?

The accuracy of DNA fingerprinting depends on the number of VNTR or STR loci that are used. At present the FBI uses thirteen STR loci in its profile, with the expected frequency of this profile to be less than one in 10 billion. As the number of loci analyzed increases, the probability of a random match becomes smaller.

What factors might influence the interpretation of **DNA fingerprinting**?

Ruling out procedural sources of error (such as contamination from another sample or improper sample preparation), DNA obtained from siblings will share both alleles at a particular locus, and so the DNA profile may not be conclusive. DNA from identical twins will have identical DNA profiles. The frequency of identical twins in a population is one in every 250 births. In all cases the probability calculations that are used in DNA profiling are based on a large population; therefore, if a population is small or reflects interbreeding, probabilities must be adjusted.

What are applications of **DNA fingerprinting**?

DNA fingerprinting is used to determine paternity; in forensic crime analysis; in population genetics to analyze variation within populations or ethnic groups; in conservation biology to study the genetic variability of endangered species; to test for the presence of specific pathogens in food sources; to detect genetically modified organisms either within plants or food products; in evolutionary biology to compare DNA extracts from fossils to modern day counterparts; and in the identification of victims of a disaster.

Can an **innocent person** be convicted based on **DNA analysis**?

Current methods of DNA analysis are very sensitive, as only a few cells are needed for DNA extraction. However, it is possible for an innocent person's DNA to be found at any crime scene, either from accidental deposition or by direct deposition by a third party. Also, a partial DNA profile from a crime scene could match that of an innocent person whose DNA is already in a DNA data bank. In addition, close relatives of suspected criminals could also be partial matches to a DNA profile.

APPLICATIONS

What is **gene therapy**?

Gene therapy involves replacement of an "abnormal" disease-causing gene with a "normal" gene. The normal gene is delivered to target cells using a vector, which is usually a virus that has been genetically engineered to carry human DNA. The virus genome is altered to remove disease-causing genes and insert therapeutic genes. Target cells are infected with the virus. The virus then integrates its genetically altered material containing the human gene into the target cell, where it should produce a functional protein product.

What **setbacks** have occurred in the use of **gene therapy**?

During the 1990s there were more than 4,000 patients involved in gene therapy trials. Unfortunately, most of these trials were not successful, and a further setback occurred in 1999 with the death of eighteen-year-old Jesse Gelsinger. Gelsinger suffered from ornithine transcarbamylase deficiency, a rare metabolic disease in which toxic levels of ammonia can be produced by the liver. He died of complications due to a severe immune response to the viral carrier used to deliver the therapeutic genes.

What viruses are commonly used as vectors in human gene therapy?

Type of virus	RNA or DNA?	Example	Target cells
Retroviruses	RNA	Human immunodeficiency virus (HIV)	any cell
Adenoviruses	Double-strand DNA	Viruses responsible for common respiratory, intestinal, and eye infections	lung
Adeno-associated	Single-strand DNA	Can insert genetic material at specific site on chromosome 19	varies
Herpes simplex	Double-strand DNA	Infects neurons; causes cold sores	neurons

How can genes be used to detect **single gene disorders**?

Genetic testing can be used to determine those at risk for a particular inherited condition. There are more than 200 single gene disorders that can be diagnosed in prenatal individuals using recombinant DNA techniques. Also, since some genetic disorders appear later in life, children and adults can be tested for genetic disorders before becoming symptomatic. If the locus of the disease-causing gene is known, gene markers can be used to determine which family members are at risk. An example of an adult onset genetic disorder is polycystic kidney disease, which occurs between the ages of thirty-five and fifty. These cysts produced by the disease will eventually destroy the kidneys. Prior knowledge of the condition allows both patient and doctor to closely monitor any changes in the kidneys.

How can genes be used to detect **multigene disorders**?

DNA chips and microarray technology can be used as a diagnostic tool when searching for multiple mutations that can be part of a multigene disease, such as some types of cancers. DNA chip technology is now being used to screen for mutations in the p53 gene. Approximately 60 percent of all cancers are linked to mutations of the p53 gene. Eventually, these methods will be used to collectively generate an individual genetic profile of all mutations associated with known genetic disorders.

How can **biotechnology** be used to manufacture **vaccines**?

Vaccine development is risky using conventional methods because vaccines must be manufactured inside living organisms, and the diseases themselves are

431

What genetic diseases are a routine focus of newborn screening techniques in the United States?

All states in the United States require newborn screening for the metabolic genetic disease called phenylketonuria. Many states have more inclusive newborn screening that may include sickle cell disease and homocystinuria.

extremely dangerous and infectious. Using genetic engineering, specific pathogen proteins that trigger antibody production are isolated and inserted into a bacterial or fungal vector. The organisms are then cultured to produce large quantities of the protein.

What is **xenotransplantation**?

Xenotransplantation is the transplantation of tissue or organs from one species to another. The development of this technique has led to the breeding of animals specifically for use as human organ donors. Because humans would reject a nonhuman organ as foreign, transgenic animals (e.g., pigs) are genetically altered with human DNA with the hope of suppressing eventual rejection.

What are the dangers involved with **transgenic animal transplants**?

One of the major risks of xenotransplantation is the risk of transplanting animal viruses along with the transfer cells or organs. Since the patient is already immuno-suppressed, the patient could die from the viral infection, or the virus could be spread to the general population.

What is **germline therapy**?

Germline therapy involves gene alteration within the gametes (eggs or sperm) of an organism. Alterations in gamete genes transfer the gene to all cells in the next generation. As a result, all members of future generations will be affected.

What is **pharmacogenomics**?

Pharmacogenomics is the use of DNA technology to develop new drugs and optimize current drug treatment to individual patients. For example, the interaction of a drug with a specific protein can be studied and then compared to a cell in which a genetic mutation has inactivated that protein. Its potential is to tailor drug therapy to an individual's genome, a tailoring that could reduce adverse drug reactions and increase the efficacy of drug treatment.

> ## What are examples of xenotransplantation commonly used in humans?
>
> **X**enotransplantation in humans is not new; it began in 1971. Inert heart valves from pigs are used in human heart valve replacement operations. Tissue for human bones and skin is being grown and developed from pigs. Testing is underway using fetal pig cell treatments for Parkinson's disease, stroke, epilepsy, Huntington's disease, and spinal cord injuries.

How has biotechnology helped **thalidomide** become popular again?

The drug thalidomide was developed during the 1950s. It was found to be effective in combating the symptoms of morning sickness in pregnant women. Unfortunately, if taken during the first trimester of pregnancy, thalidomide causes serious limb deformities because it interferes with the normal growth process of the fetus. Recently, it was discovered that the drug also has the ability to interfere with the replication of HIV in human cells. Thalidomide also may prevent the weight loss commonly associated with AIDS. Thalidomide actually has two mirror image forms (chirality), only one of which causes birth defects. Researchers hope that by using biotechnology they will be able to separate the forms and return thalidomide to their arsenal of disease-fighting chemicals once again.

What is **biopharming**?

Biopharming is a relatively new technology in which transgenic plants are used to "grow" pharmaceuticals. In this technique, scientists bioengineer medically important proteins into a corn plant, thus spurring the corn plant to produce large amounts of a particular medicinal protein. This is less expensive than using a microbial fermentation chamber.

What is a **gene patent**?

A gene patent is a patent issued for a specific DNA sequence of a specific organism. It is common for companies to obtain patents for genes important to medicine and agriculture.

What is **bioremediation**?

Bioremediation is the use of organisms to remove toxic materials from the environment. Bacteria, protists, and fungi are good at degrading complex molecules into waste products that are generally safe and recyclable. Sewage treatment plants perform bioremediation in a limited way. An example of bioremediation is the massive cleanup in Alaska following the Exxon *Valdez* oil spill in 1989. The superficial layer of

> ### What is an example of a drug developed through pharmacogenomics?
>
> Gleevec (approved by the FDA in 2001) is a drug developed for a rare type of genetically caused chronic myeloid leukemia. In this type of leukemia, pieces of two different chromosomes break off and reattach on the opposite chromosome, causing a chromosome translocation. This abnormality causes a gene for a blood cell enzyme to continually manufacture the enzyme, resulting in high levels of white blood cells in the bone marrow and blood. Gleevec was specifically engineered to inhibit the enzyme created by the translocation mutation and to thus block the rapid growth of white blood cells.

oil was removed by suction and filtration, but the oil-soaked beach was cleaned by bacteria that could use oil as an energy source.

Can **DNA** be extracted from a **mummy**?

Yes, DNA can be (and has been) extracted from a mummy. However, the problem with extraction of ancient DNA lies within contamination from modern DNA. In order to minimize contamination, researchers usually try to get DNA from inside teeth or bone. Ancient DNA is being used to study the genealogy of the pharaohs of Egypt.

What are **GMO**?

GMO are genetically modified organisms that have been created using new techniques of recombinant DNA technology. However, the term is misleading because almost all domesticated animals and crop plants have been genetically modified over thousands of years by human selection and cross-breeding. GMO are viewed with concern because of public debate over the safety of the products and the fear that GMO foods represent a type of "biological" pollution. The battle has been particularly intense in Europe, where GMO have been banned since 1998.

How are **GMO** regulated?

In the United States, GMO are regulated by the Food and Drug Administration (FDA), Environmental Protection Agency (EPA), and United States Department of Agriculture (USDA).

What are some examples of **genetic engineering** in plants?

Genetically engineered plants include transgenic crop plants that are resistant to herbicides used in weed control. These transgenic crops carry genes for resistance to her-

Is golden rice in use today?

As of 2004 objection to genetically modified crops has prevented worldwide marketing of golden rice. Also, it will take an estimated five to ten years before golden rice can be bred into the local varieties found in rice-growing countries.

bicides such that all plants in a field are killed with the exception of the modified plant. Transgenic soybeans, corn, cotton, canola, papaya, rice, and tomatos are used by many farmers in the United States. Plants resistant to predatory insects have also been genetically engineered.

How widespread is the use of **GMO crops**?

As of 2002 more than 120 million acres (40 million hectares) of fertile farmland were planted with GMO crops. The acreage is confined to four countries: United States (containing 68 percent of the total acreage), Argentina (22 percent), Canada (6 percent), and China (3 percent).

What is a **Flavr Savr tomato**?

The Flavr Savr tomato was produced in response to consumer complaints that tomatoes were either too rotten to eat when they arrived at the store or too green. Growers had found that they could treat green tomatoes in the warehouse with ethylene, a gas that causes the tomato skin to turn red. However, the tomato itself stayed hard. In the late 1980s researchers at Calgene (a small biotech company) discovered that the enzyme polygalactouronase (PG) controlled rotting in tomatoes. The scientists reversed the DNA sequence of PG; the effect was that tomatoes turned red on the vine and yet the skin of the tomatoes remained tough enough to withstand the mechanical pickers. However, before the Flavr Savr tomato was introduced to the market, Calgene disclosed to the public how the tomato was bioengineered. This caused a public protest that led to a worldwide movement against genetically modified organisms (GMO).

What is **golden rice**?

In 2000 scientists used three genes, two from a daffodil and one from a bacterium, to create a rice variety that contained beta carotene. The importance of this is that in the human body, beta carotene is converted to vitamin A, a crucial vitamin missing from the diet of millions of poor people around the world. Lack of vitamin A causes the death of an estimated one million of Asia's poorest children due to weakened immune

systems; vitamin A deficiency is also linked to blindness. The rice was called "golden
rice" because it turned yellow like a daffodil.

What is **Starlink corn**?

Starlink is a bioengineered corn variety that was genetically modified to include a
gene from the bacterium *Bacillus thuringiensis* (Bt), which produces a protein (called
an endotoxin) that kills some types of insects. Bt endotoxin has been registered as a
biopesticide in the United States since 1961, and the Bt endotoxin has been used by
organic farmers for biological pest control. The endotoxins only become activated in
the guts of susceptible insects. Because of the significant losses to corn crops caused
by the European corn borer, scientists targeted the corn plant itself as a candidate for
insertion of the Bt gene.

What is the controversy surrounding **Bt corn** and the **monarch butterfly**?

Bt corn was specifically engineered to control the European corn borer, which in
2003 caused an estimated $1 billion worth of damage to U.S. farmers. In 1999 a study
was released, based on controlled laboratory feeding experiments, that showed that
corn pollen from Bt-altered plants would kill monarch butterflies. In the study, three-
day-old monarch butterfly larvae were fed milkweed leaves dusted with Bt corn
pollen. The larvae ate less, grew slower, and had a higher mortality rate than those fed
milkweed with no corn pollen or milkweed coated with non-Bt corn pollen. However,
the laboratory study did not provide information on the number of Bt pollen grains
that were consumed by the monarch larvae in order to observe the lethal effects. Also,
no information was provided on the effects on older, larger larvae, which would be
expected to have a higher tolerance to Bt toxicity. Headlines such as "Attack of the
Killer Corn" and "Nature at Risk" triggered regulatory action on the part of the European Union to ban the importation and use of Bt corn varieties in Europe. In
response, other researchers described the use of Bt as a biocontrol agent since 1938
because of its selective toxicity to certain species within a given insect order. After
extensive studies dealing with the likelihood that Bt corn pollen would be found on
milkweed plants near cornfields that are close to the habitats of monarch butterflies,
certain safeguards were set to decrease the risk of Bt corn pollen to monarchs: 1)

The best way that dinosaurs could be recreated would probably be by altering bird DNA, since dinosaurs and birds are genetically related.

farmers in monarch-rich areas should choose to grow Bt corn with lower toxicity levels; 2) plant a border of non-Bt corn around a Bt cornfield to decrease the problem of pollen drift to milkweed plants; and 3) plant milkweeds at sites away from cornfields to increase the probability of female monarchs encountering milkweed plants that are uncontaminated with corn pollen.

What is **Frankenfood**?

Frankenfood is a term invented by environmental and health activist groups to denote any food that has been genetically modified (GM) or that contains genetically modified organisms (GMO). Opposition to GM food is based on concerns that the gene pool of "natural" plants could be altered permanently if exposed to pollen from genetically altered plants. There is also fear that people and animals that consume GM food might have allergic reactions to altered protein or could develop health problems later.

Could the events depicted in **Jurassic Park** actually happen?

In this novel (later adapted into a film) about dinosaur reincarnation, scientists bring dinosaurs back to life by using dinosaur DNA (obtained from insects embedded in fossilized amber). There are several flaws in this scenario. One is that prehistoric mosquitoes were most likely to have digested the dinosaur blood, making the DNA unusable. Also, one would need a complete genome (not one that has been augmented with amphibian DNA) to genetically reconstruct a dinosaur. The best way that

437

dinosaurs could be recreated would probably be by altering bird DNA, since dinosaurs and birds are genetically related.

What is **biopreservation**?

Biopreservation refers to the preservation and enhanced safety of food using biological materials. An example of this is nisin, a bacterial protein that can act as a broad-spectrum antibiotic. Nisin cannot be synthesized chemically, so the nisin-producing bacteria *Lactobacillus* must be used to generate the protein.

What is a **biopesticide**?

A biopesticide is a chemical derived from an organism that interferes with the metabolism of another species. An example is the Bt toxin *Bacillus thuringiensis*, which interferes with the absorption of food in insects but does not harm mammals.

What are **biosensors**?

A biosensor is a unique combination of biological substances (e.g., microbe, cell, enzyme, antibody) linked to a detector. It can be used to measure very low concentrations of a particular substance. An example of a biosensor currently on the market is the insulin pump, which maintains correct blood glucose concentrations for diabetics.

What are some examples of **genetic engineering** in **animals and microbes**?

One of the earliest applications of biotechnology was the genetic engineering of a growth hormone (bovine GH) produced naturally in the bovine pituitary. Bovine GH can increase milk production in lactating cows. Using biotechnology, scientists bioengineered the gene that controls bovine GH production into *E. coli*, grew the bacteria in fermentation chambers, and thus produced large quantities of bovine GH. The bioengineered bovine GH, when injected into lactating cows, resulted in an increase of up to 20 percent in national milk production. Using bovine GH, farmers are able to stabilize milk production in their herds, avoiding fluctuations in production levels. A similar regimen was adapted using the pig equivalent of growth hormone (porcine GH) Injected in pigs, porcine GH reduced back fat and increased muscle (meat) gain.

The first transgenic animal available as a food source on a large scale was the salmon, which reached U.S. food markets in 2001, following rigid evaluations of consumer and environmental safety. These salmon have the capability of growing from egg to market size (6 to 10 lbs) in eighteen months, as compared to conventional fish breeding, which takes up to thirty-six months to bring a fish to market size. The use of transgenic salmon can help reduce overfishing of wild salmon stocks.

What is **bioterrorism**?

Bioterrorism is the use of biological substances or toxins with the goal of causing harm to humans. Biotechnology can be used to manufacture biological weapons such as large amounts of anthrax spores. However, biotechnology can also be used positively to identify bioweapons. A new faster method of PCR, called continuous flow PCR, uses a biochip and requires only nanoliter amounts of DNA to detect a bioweapon.

What is **proteomics**?

Proteomics is the study of proteins encoded by a genome. This field extends the Human Genome Project and is a far more complex study than finding where genes are located on chromosomes. Proteins are dynamic molecules that can change according to the needs of a cell, and complete understanding of cell metabolism requires that scientists understand all of the proteins involved as well as their genes.

What is a **nucleotide analog**?

Nucleotide analogs are compounds that look like the nucleotides in DNA; they are used as antiviral compounds because the nucleic acids assembled with these analogs fall apart. Therefore, the viral genome cannot be copied and the infection cycle is broken. AZT is an example of a drug that interferes with HIV's ability to replicate its genome by substituting azidothymidine for thymidine, thus terminating viral DNA reproduction.

What is **tissue engineering**?

Tissue engineering is used to create semisynthetic tissues that are used to replace or support the function of defective or injured body parts. It is a broad field, encompassing cell biology, biomaterial engineering, microscopic engineering, robotics, and bioreactors, where tissues are grown and nurtured. Tissue engineering can improve on current medical therapies by designing replacements that mimic natural tissue function. Commercially produced skin is already in use for treating patients with burns and diabetic ulcers.

What is the **Innocence Project**?

The Innocence Project is a public law clinic at the Benjamin Cardozo Law School in New York that uses biotechnology, specifically DNA evidence, to reopen cases of people who have been wrongly convicted of crimes.

What is *Gattaca*?

The 1997 film *Gattaca* was one of the first movies to examine the idea that one's station in life is dependent on the "perfection" of one's genes. The movie explores genetic discrimination and previews the bioethics of using genetic information as a means of identification.

What is a **biomemetic**?

A biomemetic is a chemical reagent that can perform the function of a biological molecule. An example is a small molecular mimic of erythropoietin, the protein that causes release of red blood cells.

439

Dolly the sheep was the first mammal to be successfully cloned.

What is **cloning**?

A clone is a group of cells derived from the original cell by fission (one cell dividing into two cells) or by mitosis (cell nucleus division with each chromosome splitting into two). Cloning perpetuates an existing organism's genetic make-up. Gardeners have been making clones of plants for centuries by taking cuttings of plants to make genetically identical copies. For plants that refuse to grow from cuttings, or for the animal world, modern scientific techniques have greatly extended the range of cloning. The technique for plants starts with taking a cutting of a plant that best satisfies the criteria for reproductive success, beauty, or some other standard. Since all of the plant's cells contain the genetic information from which the entire plant can be reconstructed, the cutting can be taken from any part of the plant. Placed in a culture medium having nutritious chemicals and a growth hormone, the cells in the cutting divide, doubling in size every six weeks until the mass of cells produces small white globular points called embryoids. These embryoids develop roots, or shoots, and begin to look like tiny plants. Transplanted into compost, these plants grow into exact copies of the parent plant. The whole process takes eighteen months. This process, called tissue culture, has been used to make clones of oil palm, asparagus, pineapples, strawberries, Brussels sprouts, cauliflower, bananas, carnations, ferns, and others. Besides making highly productive copies of the best plant available, this method controls viral diseases that are passed through normal seed generations.

Whatever happened to Dolly?

Dolly died on February 14, 2003. She was suffering from progressive lung disease, and scientists at the Roslin Institute (near Edinburgh, Scotland) made the decision to euthanize her. Her body was preserved and is now on display at the National Museum of Scotland.

Can **human beings** be **cloned**?

In theory, yes. There are, however, many technical obstacles to human cloning, as well as moral, ethical, philosophical, religious, and economic issues to be resolved before a human being could be cloned. At the present time most scientists would agree that cloning a human being is unsafe under current conditions.

How could a **human be cloned**?

Nuclear transplantation or somatic cell nuclear transfer is used to move the cell nucleus and its genetic material from one cell to another. Somatic cell nuclear transfer may be used to make tissue that is genetically compatible with that of the recipient and could be used in the treatment of specific disease. Or, if the material is moved to an egg cell lacking its own nucleus, the transfer could result in the formation of a clone embryo.

What was the **first animal** to be successfully **cloned**?

In 1970 the British molecular biologist John B. Gurdon (1933–) cloned a frog. He transplanted the nucleus of an intestinal cell from a tadpole into a frog's egg that had had its nucleus removed. The egg developed into an adult frog that had the tadpole's genome in all of its cells and was therefore a clone of the tadpole.

What was the **first mammal** to be successfully **cloned**?

The first mammal cloned from adult cells was Dolly, a ewe born in July 1996. Dolly was born in a research facility in Scotland. Ian Wilmut (1944–) led the team of biologists that removed a nucleus from a mammary cell of an adult ewe and transplanted it into an enucleated egg extracted from a second ewe. Electrical pulses were administered to fuse the nucleus with its new host. When the egg began to divide and develop into an embryo, it was transplanted into a surrogate mother ewe. Dolly was the genetic twin of the ewe that donated the mammary cell nucleus. On April 13, 1998, Dolly gave birth to Bonnie—the product of a normal mating between Dolly and a Welsh mountain ram. This event demonstrates that Dolly was a healthy, fertile sheep, able to produce healthy offspring.

What is **CODIS**?

CODIS is the FBI Laboratory's Combined DNA Index System, which allows federal, state, and local police agencies to compare DNA profiles electronically. CODIS uses two indexes: 1) the Forensic Index, which contains DNA profiles from crime scenes; and 2) the Offender Index, which contains DNA profiles of individuals convicted of sex offenses and other violent crimes.

What is the **ENCODE project**?

The ENCODE (ENCyclopedia Of DNA Elements) project refers to a long-term project sponsored by the National Human Genome Research Institute to identify and locate all genes (including protein-coding and non-protein-coding ones) and other functional elements within the human genome.

What is **nanotechnology**?

The term "nanotechnology" was coined in 1974 by Norio Taniguchi (1912–1999) at the University of Tokyo. It includes a number of technologies that deal with the miniaturization of existing technology down to the scale of a nanometer (one-billionth of a meter) in size, about the size of molecules and atoms. Potential effects of nanotechnology include microcomputers capable of storing trillion of bytes of information in the size of a sugar cube; portable fluids containing nanobots that are programmed to destroy cancer cells; and airborne nanobots that are programmed to rebuild the thinning ozone layer.

Can **pearls** be **genetically engineered**?

Pearls can now be genetically engineered because scientists have isolated and characterized the protein nacrein, which is the primary component of pearls. When nacrein is combined with calcium ions, it forms an organic matrix similar to that found in an oyster's inner shell.

Can **genetic engineering** be used to save **endangered species**?

As endangered species disappear from natural habitats and are only found in zoos, researchers are looking for ways to conserve these species. Using cryopreservation,

How big is a nanometer?

A nanometer is about the width of six bonded carbon atoms, 1/40,000th the width of a human hair. DNA, our genetic material, is in the 2.5 nanometer range, while red blood cells are approximately 2.5 micrometers (2,500 nanometers).

the Zoological Society of San Diego has created a "frozen zoo" that stores viable cell lines from more than 3,200 individual mammals, birds, and reptiles, representing 355 species and subspecies. Researchers maintain that there should still be continued efforts to preserve species in their natural habitats, but by preserving and studying animal DNA, scientists can learn genetic aspects crucial to the species' survival.

EVOLUTION

INTRODUCTION AND HISTORICAL BACKGROUND

What is **evolution**?

Although it was originally defined in the nineteenth century as "descent with modification," evolution is currently described as the change in frequency of genetic traits (also known as the allelic frequency) within populations over time.

What were **early ideas** on **evolution**?

While some Greek philosophers had theories about the gradual evolution of life, Plato (427–347 B.C.E.) and Aristotle (384–322 B.C.E.) were not among them. In the 1700s, "natural theology" (the explanation of life as the manifestation of the creator's plan) held sway in Europe. This idea was the motive force behind the work of Carl Linnaeus (1707–1778), who was the first to classify all known living things by kingdom. Also popular prior to the work of Charles Darwin (1809–1882) were the theories of "special creation" (creationism), "blending inheritance" (that offspring were always the mixture of the traits of their two parents), and "acquired characteristics."

What is the **"ladder of nature"**?

Aristotle (384–322 B.C.E.) tried to use logic to build a diagram of living things, but he realized that living things could not be so easily apportioned. Instead, he constructed a "ladder of nature" or "scale of perfection," which began with humans at one end and proceeded through animals and plants to minerals at the other end. Aristotle ranked these groups on a scale based on the four classical elements: fire, water, earth, and air.

What is **Lamarckian evolution**?

The French biologist Jean-Baptiste de Lamarck (1744–1829) is credited as the first person to propose a theory that attempts to explain how and why evolutionary change occurs in living organisms. The mechanism Lamarck proposed is known as "the inheritance of acquired characteristics," meaning that what individuals experience during their lifetime will be passed along to their offspring as genetic traits. This is sometimes referred to as the theory of "use and disuse." A classic example of this would be the giraffe's neck. Lamarckian evolution would predict that as giraffes stretch their necks to reach higher branches on trees, their necks grow longer. As a result, this increase in neck length will be transmitted to egg and sperm such that the offspring of giraffes whose necks have grown will also have long necks. While Lamarck's idea was analytically based on available data (giraffes have long necks and give birth to offspring with long necks as well), he did not know that, in general, environmental factors do not change genetic sequences in such a direct fashion.

Who coined the term "**biology**"?

Jean-Baptiste de Lamarck (1744–1829) is credited with coining the term "biology" (from the Greek terms *bios*, meaning "life," and *logy*, meaning "study of") in 1802 to describe the science of life. He was also the first to publish a version of an evolutionary tree, which describes the ancestral relationships among species.

Who **disproved Lamarck's** theory?

In the 1880s the German biologist August Weismann (1834–1914) formulated the germ-plasm theory of inheritance. Weismann reasoned that reproductive cells (germ cells) were separate from the functional body cells (soma or somatic cells). Therefore, changes to the soma would not affect the germ-plasm and would not be passed on to the offspring. In order to prove that the disuse or loss of somatic structures would not affect the subsequent offspring, Weismann removed the tails of mice and then allowed them to breed. After twenty generations of this experimental protocol, he found that mice still grew tails of the same length as those who had never been manipulated. This not only disproved Lamarck's theory of use and disuse, it also increased understanding of the new field of genetics.

Who was **Comte de Buffon**?

Georges Louis Leclerc, Comte de Buffon (1707–1788) was an early proponent of natural history, which is the study of plants and animals in their natural settings. He was also known for his work as a mathematician. Buffon was interested in the modes by which evolutionary change could occur. A prolific writer (his work *Natural History* comprised thirty-five volumes), he pondered the meaning of the term "species" and whether such groupings were immutable (unchanging) over time. In addition, he served as mentor to Jean-Baptiste de Lamarck (1744–1829).

Who was **Erasmus Darwin**?

Erasmus Darwin (1731–1802) was the grandfather of Charles Darwin (1809–1882). Erasmus was a physician, inventor, and natural scientist who published a book on his ideas (*Zoonomia*) between 1794 and 1796. *Zoonomia* contained poetic couplets describing Erasmus Darwin's ideas about science and, in particular, the evolution of life. Erasmus Darwin's hypothesis was that all the animals on the planet had their origin in a "vital spark" that set in motion life as we know it. In his posthumously published book *Temple of Nature* (1803), Erasmus Darwin further speculated on a theory of evolution that included basic ideas on the unity of organic life and the importance of both sexual selection and the struggle for existence in the evolutionary process. Years later, in writing a biography about his grandfather, Charles Darwin acknowledged that his grandfather's ideas had influenced his thinking.

The theory of natural selection proposed by Charles Darwin revolutionized all aspects of natural science.

Who was **Charles Darwin**?

The theory of natural selection proposed by Charles Darwin (1809–1882) revolutionized all aspects of natural science. Darwin was born into a family of physicians and planned to follow his father and grandfather in that profession. Unable to stand the sight of blood, he studied divinity at Cambridge and received a degree from the university in 1830.

What were the *Beagle* voyages?

The HMS *Beagle* was a naval survey ship that left England in December 1831 to chart the coastal waters of Patagonia, Peru, and Chile. On a voyage that would last five years, Darwin's job as unpaid companion to the captain on board the *Beagle* allowed him to satisfy his interests in natural history. On its way to Asia, the ship spent time in the Galapagos Islands off the coast of Ecuador; Darwin's observations there caused him to generate his theory of natural selection.

What is the significance of **Darwin's study of finches**?

In his studies on the Galapagos Islands, Darwin observed patterns in animals and plants that suggested to him that species changed over time to produce new species. Darwin collected several species of finches. The species were all similar, but each had developed beaks and bills specialized to catch food in a different way. Some species had heavy bills for cracking open tough seeds. Others had slender bills for catching insects. One species used twigs to probe for insects in tree cavities. All the species resembled one species of South American finch. In fact, all the plants and animals of the Galapagos Islands were similar to those of the nearby coast of South America. Darwin felt that the simplest explanation for this similarity was that a few species of plants and animals from South America must have migrated to the Galapagos Islands. These few plants and animals then changed as they adapted to the environment on their new home, eventually giving rise to many new species. Evolutionary theory proposes that species change over time in response to environmental challenges.

How did **geology** influence **Darwin**?

While traveling aboard the HMS *Beagle*, Charles Darwin read the *Principles of Geology* by Charles Lyell (1797–1875). Catastrophism was the popular theory of the time about the forces driving geological change. Lyell's theory suggested that geologic change was not solely the result of random catastrophes. Rather, he proposed that geologic formations were most often the result of everyday occurrences like storms, waves, volcanic eruptions, and earthquakes that could be observed within an individual lifetime. This idea, that the same geologic processes at work today were also present during our evolutionary past, is known as Uniformitarianism. This conclusion also led Lyell and, before him, James Hutton (1726–1797), to suggest that Earth must be much older than the previously accepted age of 6,000 years, because these uniform processes would have required many millions of years to generate the structures he observed. Reading Lyell's work gave Darwin a new perspective as he traveled through South America and sought a mechanism by which he could explain his thoughts on evolution.

Who proposed that **Earth** was only **6,000 years old**?

In the seventeenth century Archbishop James Ussher (1581–1656) of the Irish Protestant Church calculated from the Old Testament genealogies that Earth was created on

Sunday, October 26, 4004 B.C.E. He was also able to calculate exact dates for when Adam and Eve were forced from Paradise (Monday, November 10, 4004 B.C.E.) and when Noah's Ark touched down on Mt. Ararat (Wednesday, May 5, 1491 B.C.E.).

Who was **Alfred Russel Wallace**?

Alfred Russel Wallace (1823–1913) was a naturalist whose work was presented with Charles Darwin's (1809–1882) at the Linnaean Society of London in 1858. After extensive travels in the Amazon basin, Wallace independently came to the same conclusions as Darwin on the significance of natural selection in driving the diversification of species. Wallace also worked as a natural history specimen collector in Indonesia. Wallace, like Darwin, also read the work of Thomas Malthus (1766–1834). During an attack of malaria in Indonesia, Wallace made the connection between the Malthusian concept of the struggle for existence and a mechanism for change within populations. From this, Wallace wrote the essay that was eventually presented with Darwin's work in 1858.

Why was **Wallace** not as **well known** as **Darwin**?

While Darwin was well connected to the scientific establishment of the time, Wallace entered the scene somewhat later and so was less well known. Although Darwin would become far more famous than Wallace in subsequent decades, Wallace became quite well known during his own time as a naturalist, writer, and lecturer. He was also honored with numerous awards for his work.

What was Alfred Russel Wallace's contribution to **phrenology**?

Alfred Russel Wallace became interested in a number of subjects outside of natural science, including the suffragette movement and spiritualism. He is also known for his approval of phrenology, which uses the shape of an individual's skull to make predictions about his or her mental functions. Wallace pointed out that while other scientists talked about the connection between the mind and the body, phrenologists had actually attempted to link physical structures (the skull) with mental attributes (thought processes). Although phrenology has been largely discounted as a valid scientific endeavor, in Wallace's time it was just one of a number of areas of investigation into the natural world.

Who was "**Darwin's bulldog**"?

Thomas Huxley (1825–1895) was a staunch supporter of Darwin's work; in fact, Huxley wrote a favorable review of Darwin's *On the Origin of Species* that appeared soon after its publication. When the firestorm of controversy began after the appearance of Darwin's book, Huxley was ready and able to defend Darwin, whose chronic public reticence about his theories was at that time exacerbated by illness. Huxley's defense of Darwin was so vigorous during a debate with Bishop Samuel Wilberforce

(1805–1873) at the British Association for the Advancement of Science in 1860 that he earned the title "Darwin's bulldog."

How did **Malthus** influence **Darwin**?

Both Darwin and Wallace read the work of Thomas Malthus (1766–1834), who in 1798 had published *Essay on the Principle of Population*. In it Malthus argued that the reproductive rate of humans grows geometrically, far outpacing the available resources. This means that individuals must compete for a share of the resources in order to survive in what has become known as the "struggle for existence." Darwin and Wallace incorporated this idea as part of natural selection—that is, adaptations that made for a more successful competitor would be passed on to subsequent generations, leading to greater and greater efficiency.

What is the "**struggle for existence**"?

As noted by Thomas Malthus (1766–1834), more individuals are born than can possibly survive due to limited resources. Therefore, there is a struggle for existence. We can think of this as the competition between individuals (of the same species or between species) for food or hiding places or other necessary resources. The origin of this phrase can be traced back to the Greek philosophers but is perhaps best described by Alfred, Lord Tennyson's (1809–1892) phrase, "Nature red in tooth and claw."

What is the **significance** of *On the Origin of Species*?

Charles Darwin (1809–1882) first proposed a theory of evolution based on natural selection in his treatise *On the Origin of Species*. The publication of *On the Origin of Species* ushered in a new era in our thinking about the nature of man. The intellectual revolution it caused and the impact it had on man's concept of himself and the world were greater than those caused by the works of Isaac Newton (1642–1727) and other individuals. The effect was immediate—the first edition sold out on the day of publication (November 24, 1859). *Origin* has been referred to as "the book that shook the world." Every modern discussion of man's future, the population explosion, the struggle for existence, the purpose of man and the universe, and man's place in nature rests on Darwin.

The work was a product of his analyses and interpretations of his findings from his voyages on the HMS *Beagle*. In Darwin's day the prevailing explanation for organic diversity was the story of creation in the book of Genesis in the Bible. *Origin* was the first publication to present scientifically sound, well-organized evidence for the theory of evolution. Darwin's theory was based on natural selection in which the best, or fittest, individuals survive more often than those who are less fit. If there is a difference in the genetic endowment among these individuals that correlates with fitness,

Did Charles Darwin have any nicknames?

Darwin had several nicknames. As a young naturalist on board the HMS *Beagle*, he was called "Philos" because of his intellectual pursuits and "Flycatcher" when his shipmates tired of him filling the ship with his collections. Later in his life, when he became a leader in the scientific community, journalists referred to him as the "Sage of Down" or the "Saint of Science," but his friend Thomas Henry Huxley privately called him the "Czar of Down" and the "Pope of Science." His own favorite nickname was "Stultis the Fool," and he often signed letters to scientific friends with "Stultis." This name referred to his habit of trying experiments most people would prejudge to be fruitless or fool's experiments.

the species will change over time and will eventually resemble more closely (as a group) the fittest individuals. It is a two-step process: the first consists of the production of variation, and the second, of the sorting of this variability by natural selection in which the favorable variations tend to be preserved.

Who coined the phrase "**survival of the fittest**"?

Although frequently associated with Darwinism, this phrase was coined by Herbert Spencer (1820–1903), an English sociologist. It is the process by which organisms that are less well adapted to their environment tend to perish and better-adapted organisms tend to survive.

What **books** did **Darwin** publish?

- *Journal of Researches into the Geology and Natural History of the Various Countries visited by HMS "Beagle" under the Command of Capt. FitzRoy, R.N., from 1832 to 1836* (1839)

- *Geological Observations on Coral Reefs, Volcanic Islands, and on South America: Being the Geology of the Voyage of the "Beagle," under the Command of Capt. FitzRoy, during the Years 1832–36* (1846)

- *A Monograph on the Sub-Class Cirripedia* (1851–1854)

- *A Monograph on the Fossil Lepadidae, or, Pedunculated Cirripedes of Great Britain* (1851)

- *A Monograph on the Fossil Balanidae and Verrucidae of Great Britain* (1854)

- *On the Origin of Species by Means of Natural Selection; Or, the Preservation of Favoured Races in the Struggle for Life* (1859)

- *On the Various Contrivances by which British and Foreign Orchids Are Fertilised by Insects, and on the Good Effects of Intercrossing* (1861)
- *The Movements and Habits of Climbing Plants* (1865)
- *The Variation of Animals and Plants under Domestication* (1868)
- *The Descent of Man, and Selection in Relation to Sex* (1871)
- *The Expression of the Emotions in Man and Animals* (1872)
- *Insectivorous Plants* (1876)
- *The Effects of Cross and Self Fertilisation in the Vegetable Kingdom* (1876)
- *The Different Forms of Flowers on Plants of the Same Species* (1877)
- *The Power of Movement in Plants* (1880)
- *The Formation of Vegetable Mould, through the Action of Worms, with Observations on their Habits* (1881)
- *The Movements and Habits of Climbing Plants* (1882)

DARWIN–WALLACE THEORY

What is the **Darwin-Wallace** theory?

The Darwin-Wallace theory can be summarized as the following: Species as a whole demonstrate descent with modification from common ancestors, and natural selection is the sum of the environmental forces that drive those modifications. The modifications or adaptations make the individuals in the population better suited to survival in their environment, more "fit" as it were.

The four postulates presented by Darwin in *On the Origin of Species by Natural Selection* are as follows: 1) Individuals within species are variable. 2) Some of these variations are passed on to offspring. 3) In every generation more offspring are produced than can survive. 4) The survival and reproduction of individuals are not random; the individuals who survive and go on to reproduce the most are those with the most favorable variation. They are naturally selected. It follows logically from these that the characteristics of the population will change with each subsequent generation until the population becomes distinctly different from the original; this process is known as evolution.

What is **Darwinian fitness**?

Darwinian fitness is measured as the average representation of one allele or genotype in the next generation of a population as compared to other alleles or genotypes. In other words, those alleles or genotypes that become more common within the population are more fit.

Why is **evolution a theory**?

A scientific theory is an explanation of observed phenomena that is supported by the available scientific data. The term "theory" is used as an indication that the explanation will be modified as new data becomes available. For example, the Darwin-Wallace theory was proposed prior to the discovery of the molecular nature of genetics but has since been expanded to encompass that information as well.

Which **scientific** disciplines provide **evidence for evolution**?

Although information from any area of natural science is relevant to the study of evolution, there are several in particular that directly support the work of Darwin and Wallace. Paleobiology, geology, and organic chemistry provide insight on how living organisms have evolved. Ecology, genetics, and molecular biology also demonstrate how living species are currently changing in response to their environments and therefore undergoing evolution.

What is **gradualism**?

The Darwin-Wallace theory of evolution is based on gradualism—the idea that speciation occurs by the gradual accumulation of new traits. This would allow one species to gradually evolve into a different-looking one over many, many generations, which is the scale of evolutionary time.

What is an **adaptation**?

This term refers to how well an organism adapts to its environment. Adapted individuals survive and reproduce better than individuals without those adaptations. An example of an adaptation would be the long ears and limbs of rabbits living in desert-like conditions. These adaptations allow the rabbits to radiate heat more efficiently over a large surface area, thus making it easier to survive in a harsh climate.

What is **scientific creationism**?

Scientific creationism is an attempt to promote the teaching of creation theory in schools. By designating creationist theory "scientific," proponents hope to gain equal time with evolutionary theory in school curricula. Creationist theory proposes that species observed today are the result of "intelligent design" or special creation rather than the result of the effects of natural selection.

What is **artificial selection**?

Artificial selection is the selective breeding of organisms for a desired trait, such as breeding a rose plant to produce larger flowers or a chicken to lay more eggs. Darwin cited artificial selection as evidence that species are not immutable—that is, unable to be changed by selection.

What is **homology**?

Homology is the similarity in traits between two species that is indication of their common ancestry. For example, the general characteristics of cheetahs, lions, tigers, and house cats are whiskers, retractable claws, tooth structure, and so forth. These similarities indicate that each of these traits was inherited from a feline ancestor.

What is **analogy**?

To an evolutionist, an analogous structure is one that looks similar or has the same purpose but is definitely not the result of common inheritance. For example, bats and birds both use wings to fly, and the wings have the same general shape (thin but broad in width). However, the structures were not inherited from the same ancestors. Bats were four-legged mammals before their front limbs became modified for flight, while birds are not descended from mammals at all.

How do scientists **differentiate** between **homology and analogy**?

Scientists can determine whether a trait is homologous or analogous by comparing it in species thought to be of common origin and contrasting it to traits of unrelated species in similar habitats.

What is meant by "**common origin**"?

The term "common origin" refers to the shared ancestors of the Darwin-Wallace theory. It explains the commonality we see among different species of birds, for example. Blue jays, starlings, sparrows, and other birds share traits like beaks, wings, and general body structure that are of common origin.

Why was the **Darwin-Wallace** theory so **ridiculed**?

The work of Darwin and Wallace generated controversy on at least two fronts. First, their theory directly countered Christian teaching on the immutability of species and their special creation by God. Second, by presenting arguments for common descent, Darwin and Wallace showed that humans were related to other animals—apes in particular. This insulted those who felt that humans were unique and not part of the animal kingdom. However, it should be noted that as the work was disseminated among scientists, it gained general acceptance in the late decades of the nineteenth century.

POST-DARWINIAN EVOLUTION

What is the **modern synthesis**?

In 1942 Julian Huxley (1887–1975), the grandson of Thomas Huxley (a.k.a. Darwin's bulldog), published *Evolution: The Modern Synthesis*. This work, which used discoveries in the areas of population genetics and Mendelian inheritance to re-introduce Darwinian evolution, did much to reassert natural selection as the mechanism of evolution.

What is a **species**?

There are several ways of defining a species, and scientists will use different definitions depending on whether they are referring to a fossil (extinct) species or a living (extant) one. For example, an extant species can be defined as all the individuals of all the populations capable of interbreeding. A group of populations that are evolutionarily distinct from all other populations may also be defined as a species, even if they are incapable of interbreeding due to extinction.

What is a **subspecies**?

A subspecies is another way of describing a distinct population or variety. This term is used to describe the generation of hybrids that can occur when two different populations meet and interbreed.

What is a **strain**?

A strain or variety is a subcategory of a species. For example, Gregor Mendel's (1822–1884) work with garden peas involved various strains; one strain had purple flowers while another had white.

What are different ways of **defining** the term "**species**"?

A species may be defined in a number of ways. The biological species concept defines any two individuals that can breed and produce fertile offspring as belonging to the

Mules—the offspring of a horse and a donkey—are not a separate species because they cannot reproduce.

same species. This would mean that lions and tigers would be considered the same species, as they can produce hybrid offspring—there is at least one case of a tigon (the offspring of a male tiger and a female lion) producing offspring after an unplanned mating with a tiger. Other concepts include the phylogenetic species concept, which bases species determination on the shared evolutionary history of the populations in question.

What is **speciation**?

Speciation is the process by which new species are formed. This occurs when populations become separated from the rest of the species. At this point the isolated group will respond independently to natural selection until the population becomes reproductively isolated. The group is then considered a new species.

How can **speciation occur**?

If a population becomes reproductively isolated, then individuals within the population will no longer exchange genetic material with the rest of the species. At that point environmental factors (i.e., natural selection) will work on the genetic variation within that population until it has become a new species.

How can a population become **reproductively isolated**?

Reproductive isolation means that individuals of one population are unable to exchange gene sequences (eggs and sperm) with individuals from another. This means that natural selection will work on the isolated population independently from the rest of the species, therefore increasing the likelihood that the isolates will become a separate species. Methods by which this can occur include geographic isolation, habitat isolation, and temporal isolation. In other words, two populations can become physically separated by a barrier like an ocean or mountain range; they can use different parts of the same habitat (birds that visit only the tops of trees as opposed to the lower branches); or they may be active at different times—nocturnal and diurnal insects, for example.

How did **Hugo de Vries** demonstrate **sympatric speciation**?

Hugo de Vries (1848–1935) discovered a way in which a population could become a separate species while still sharing the same environment with other members of the species. The process, known as sympatric speciation, occurs almost exclusively in plants rather than in animals and involves a series of rare genetic accidents that can occur during the formation of gametes (eggs and sperm). As a result, gametes are formed as polyploids—that is, they have extra copies of each chromosome and thus are unable to match their chromosomes to others of the same species. Since these polyploids are forced to mate only with other polyploids in the population, they are reproductively isolated and considered a new species.

What is the **evolutionary time scale**?

Trends in the fossil record require at least one million years to resolve themselves, and so paleontologists tend to work on a scale of 10 to 20 million years. Evolutionary biologists working on living species tend to describe selection as it occurs over a decade or less.

What separates **humans** from **primates**?

Morphologic and molecular studies suggest that our closest living relative species is the chimpanzee, although some of the evidence is conflicting. We do know that analysis of protein structure in both chimps and humans indicates that approximately 98 percent of our gene sequences are functionally identical, meaning that if the gene sequences differ it is not enough to radically change the proteins produced from them. It is also estimated that our last common ancestor with the chimp would have lived at least five million years ago.

What is **adaptive radiation**?

As populations move into new environments and adapt to those local conditions, there is an increase in diversity. This splitting creates a divergence from the original population. When diagrammed on paper, the new populations appear to be radiating outward from the original like the spokes of a wheel.

An iguana in the Galapagos Islands. Because island populations tend to be reproductively isolated from the mainland, they are more likely to demonstrate adaptations that are particular to that island's ecology.

What is **cladogenesis**?

Cladogenesis is the formation of a group of species that share a common ancestor. Cladogenesis can occur as a result of adaptive radiation, which is the divergence or splitting of one species into several.

What is **anagenesis**?

When a species gradually changes over time to the extent that it becomes a "new" species but does not give rise to additional species (no divergence), this is described as anagenesis.

Why are **islands** a good place to **study evolution**?

Island populations tend to be reproductively isolated from the mainland. Therefore, they are more likely to demonstrate adaptations that are particular to that island's ecology. This provides researchers an opportunity to study not only how natural selection works on populations but also what forces may be at work on the mainland populations as well.

What are the different **types** of **natural selection**?

There are several ways in which natural selection can cause a population to change. Natural selection can cause a trait to change in one direction only; for example when

What did the introduction of rabbits do to Australia?

Domestic rabbits were first introduced into Australia by a wealthy landowner who missed the rabbits of his native England. The rabbits (twenty-four) were released in 1859 on an estate in southern Victoria. In 1866, only seven years after the introduction, 14,253 rabbits were shot while hunting on the property. By 1869 it was estimated that there were 2,033,000 on the property, and from there they spread out across the continent. Their populations were so dense that they would literally eat all vegetation. And with no predators, they literally carpeted the ground. In desperation, the government built a 2,000-mi (3,219-km) long fence in 1907 to try to stop rabbits from entering southwest Australia. Eventually, the rabbits got through the holes in the fence and kept increasing in overwhelming numbers. In the 1950s a rabbit virus (the Myxoma virus) was introduced into the population and caused some decrease in numbers. Since then, other viruses have been used in an attempt to control the rabbits.

individuals within a population grow taller with each generation, this is known as directional selection. Diversifying selection can cause the loss of individuals in the mid-range of a trait. For example, if there is a range in color of a certain prey species from very dark to very light, those individuals in the mid-color range may not be able to hide from predators. The mid-color prey will then be selectively removed from the population, leaving the population with only two forms, the very dark and the very light. In stabilizing selection, those at either end of the range are removed more often, creating selection pressure for the mid-range. Selection can also work on traits important to sexual reproduction; this is known as sexual selection.

What is **sexual selection**?

When individuals of the same sex differ in their mating success and that difference is correlated to the presence or absence of a particular trait, that trait is said to be the result of sexual selection. Traits that provide no benefit to survival but that increase the likelihood of the male acquiring a mate are examples of sexual selection.

What does **species diversity** have to do with **evolution**?

Species diversity is direct evidence that evolution has occurred. When species can be identified in which individuals share a number of significant traits while also having some unique adaptations, it is logical to assume that the common traits are the result of common origin while the unique ones demonstrate adaptive radiation. An example of the significance of species diversity would be Darwin's finches of the Galapagos Islands. While the species share a common body structure that is inherited from a common ancestor, each species also demonstrates variations in beak size and structure that are indicative of their adaptations to local environments and the type of food available.

What is the **controversy** surrounding the **midwife toad**?

During the 1920s an Austrian biologist, Paul Kammerer (1880–1926), rose to prominence with claims that he had been able to demonstrate Larmarckian evolution, that is the inheritance of acquired characteristics. Kammerer had bred midwife toads (a land-breeding species) in water-filled aquaria for generations and reported that the toads had acquired the same structures that water-breeding toads use to hold onto each other. When Kammerer's toads were examined, they were found to have been injected with dye to mimic the pads found on the water-breeding toad species. Although Kammerer insisted that he was innocent, his reputation was ruined, and he committed suicide shortly thereafter.

Why did **sexual reproduction** evolve?

The appearance and maintenance of energetically expensive traits like sexual reproduction will occur only if there is a net benefit to the fitness of individuals with those traits. There are costs to sexual reproduction; for example, finding a mate can demand considerable time and energy. More importantly, individuals engaging in sexual reproduction only pass on half of their genes to each offspring, so their fitness is half that of asexually reproducing individuals. Scientists have examined the costs and benefits of sexual reproduction and determined that it is most likely to have evolved as a way to maintain genetic diversity. Experiments have shown that populations in erratic environments or those who are reproductively isolated from the rest of their species are at an advantage when they reproduce sexually rather than asexually. By being able to mix and match alleles, individuals within these populations can maintain genetic diversity and phenotypic variation, expanding their toolkit as a hedge against an unpredictable future.

Do all vertebrates reproduce sexually?

Among the vertebrates, there are some fishes, amphibians, and lizards that reproduce parthenogenically. Parthenogenesis is the formation of a viable embryo from an unfertilized egg. One group of species in particular, the whiptailed lizards, have no males. Females take turns playing the "male" role in mating behavior, which stimulates hormonal release within the "female," causing her eggs to begin development. During the next mating cycle the females will switch roles so that eventually each female reproduces parthenogenically.

What is **balanced polymorphism**?

When a trait exists in several forms within a population, it is said to be polymorphic. Polymorphisms that maintain a stable distribution within the population over generations are known as balanced polymorphisms. Balanced polymorphism can be maintained if heterozygotes (mixtures of two types) have a fitness advantage. When this occurs, both types of alleles are maintained in the population. A classic example of this is sickle cell anemia. Individuals who are heterozygous (Hh) are resistant to malaria, dominant homozygotes (HH) are susceptible to malaria, and recessive homozygotes (hh) have sickle cell anemia. Because those who have both types of alleles and who live in malaria-prone regions are the most likely to survive long enough to produce children, both types are maintained in the population at a relatively stable rate.

What is **industrial melanism**?

Industrial melanism is the change in the coloration of species that occurs as a result of industrial pollution. Increased air pollution as a result of the Industrial Revolution in Great Britain during the eighteenth and nineteenth centuries led to an accumulation of soot on many structures including tree trunks. As a result, organisms whose coloration allowed them to use the trees to hide from predators lost that advantage and were eaten more often by predators. A classic example of this was the peppered moth (*Biston betularia*), whose coloration is polymorphic. Prior to the Industrial Revolution, collection records indicate that the darker or melanistic form was almost unknown, but by 1895 it constituted about 98 percent of the moths collected. The two forms eventually reached a state of balanced polymorphism. Because the change in morphology could be directly linked to the change in industry, this process is described as industrial melanism.

What is **Müllerian mimicry**?

Fritz Müller (1821–1897), a German-born zoologist, described a phenomenon in 1878 in which a group of species with the same adaptations against predation was also of

Lions are the only cat with a mane. Charles Darwin was the first to suggest that the mane might be a result of sexual selection.

similar appearance. This phenomenon is now called Müllerian mimicry. Müllerian mimics include wasps and bees, all of which have similar yellow-and-black-striped patterns that serve as a warning to potential predators.

What is **Batesian mimicry**?

In 1861 Henry Walter Bates (1825–1892), a British naturalist, proposed that a nontoxic species can evolve (especially in color and color pattern) to look or act like a toxic or unpalatable species in order to avoid being eaten by a predator. The classic example is the viceroy butterfly, which resembles the unpalatable monarch butterfly. This is called Batesian mimicry. Bates is also well known as a colleague of Alfred Russel Wallace. In fact, it was Bates who introduced Wallace to botany and field collecting of animals and plants.

What is **convergent evolution**?

Convergent evolution occurs when diverse species develop similar adaptations in response to the same environmental pressure. For example, dolphins and sharks are descended from different ancestors, but as a result of sharing an aquatic environment, they have similar adaptations in body shape.

What is **divergent evolution**?

When two species move away from the traits that they share with a common ancestor as they adapt to their own environments, the result is called divergent evolution. As an example, imagine the diversity among bird species. Ducks, hummingbirds, ostriches, and penguins are all descended from an ancestral bird species, yet they have all diverged as they adapted to their particular environments.

What is **microevolution**?

Microevolution is the change in allelic frequencies that occurs at the level of the population or species. When individuals with certain traits are more successful at reproduction, the ensuing generation will have more copies of that trait. Should the trend continue, eventually the traits will become so common in the population that the population profile will change. This is microevolution.

What is **macroevolution**?

Macroevolution is large-scale change that can generate entire new groups of related species, also known as a clade. One example would be the movement of plants onto land; all terrestrial plants are descended from that event, which occurred during the Devonian period 400 million years ago.

What is the **founder effect**?

When a small group of individuals forms the genetic basis for a new population, only the variations within that founding group will be found in the resultant population. This is known as the founder effect. An example of the founder effect is found among the finches of the Galapagos Islands.

What is the **bottleneck effect**?

The bottleneck effect is a term used to describe a population that has undergone some kind of temporary restriction that has severely reduced its genetic diversity. A bottle-necking event could be an epidemic or natural disaster like fire or flood. It is hypothesized that the lack of genetic diversity among African cheetahs is due to some bottle-necking event in the species' past.

What is **parsimony**?

Parsimony is a method for choosing the simplest explanation among a variety of possible explanations for phenomena when decisive evidence is unavailable. In evolutionary terms this means grouping organisms or their traits in hierarchies that minimize the occurrence of special events. For example, all mammals are descendants of a fur-bearing, milk-producing ancestor, even though some mammals (e.g., platypus, echidna) produce eggs instead of live young. The most parsimonious explanation is that live bearing is a secondary adaptation that occurred after mammals separated from the other groups of animals.

463

What is **biogeography**?

The geographic distribution of species is known as biogeography. By observing similar organisms and their adaptations to different habitats, researchers can ask and then answer important questions about evolutionary change. In fact, the diversity of habitats visited by Darwin on his voyages with the HMS *Beagle* helped him to solidify his ideas on evolution.

EVOLUTIONARY HISTORY
OF LIFE ON EARTH

What is the meaning of the phrase "**ontogeny recapitulates phylogeny**"?

Ontogeny is the course of development of an organism from fertilized egg to adult; phylogeny is the evolutionary history of a group of organisms. The phrase "ontogeny recapitulates phylogeny" originated with Ernst Haeckel (1834–1919). It means that as an embryo of an advanced organism grows, it will pass through stages that look very much like the adult phase of less-advanced organisms. For example, at one point each human embryo has gills and resembles a tadpole. Although further research demonstrated that early stage embryos are not representative of our evolutionary ancestors, Haeckel's general concept that the developmental process reveals some clues about evolutionary history is certainly true. Animals with recent common ancestors tend to share more similarity during development than those that do not. A dog embryo and a pig embryo will look more alike through most stages of development than a dog embryo and a salamander embryo, for example.

Who was **Ernst Haeckel**?

Ernst Haeckel (1834–1919) was a physician who became a fervent evolutionist after reading *On the Origin of Species*, although he differed with Charles Darwin over natural selection as the primary mode of evolution. Haeckel is best known for his attempts to tie the stages of development to the stages of evolution ("ontogeny recapitulates phylogeny"); he thought that each stage of the developmental process was a depiction of an evolutionary ancestor. Haeckel is also credited with coining the terms "phylum," "phylogeny," and "ecology."

What is the **Oparin-Haldane** hypothesis?

In the 1920s, while working independently, Alexandr Oparin (1894–1980) and John Haldane (1892–1964) both proposed scenarios for the "prebiotic" conditions on Earth (the conditions that would have allowed organic life to evolve). Although they differed on details, both models described an early Earth with an atmosphere containing ammonia and water vapor. Both also surmised that the assemblage of organic molecules began in the atmosphere and then moved into the seas. The steps of the Oparin-Haldane model are described below.

1) Organic molecules including amino acids and nucleotides are synthesized abiotically (without living cells).

2) Organic building blocks in the prebiotic soup are assembled into polymers of proteins and nucleic acids.

3) Biological polymers are assembled into a self-replicating organism that fed on the existing organic molecules.

How did **cells evolve**?

The central criteria for living cells are a membrane capable of separating the inside of the cell from its surroundings, genetic material capable of being reproduced, and the ability to acquire and use energy (metabolism). The Miller-Urey experiment demonstrated that organic molecules, including genetic material, could arise abiotically. Phospholipids, the molecules that comprise all cellular membranes, can spontaneously form spheres when exposed to water. Although no one knows exactly how living cells evolved, these data demonstrate how a part of the scenario may have unfolded.

What is **phylogeny**?

Phylogeny is the evolutionary history of a group of species. This history is often displayed as a phylogenetic tree, in which individual species or groups of species are listed at the end of the branches. The nexuses where branches join indicate common ancestors.

What is **systematics**?

Systematics is the area of biology devoted to the classification of organisms. Originally introduced by Carl Linnaeus (1707–1778), who based his classification system on physical traits, systematics now includes the similarities of DNA, RNA, and proteins across species as criteria for classification.

What is **punctuated equilibrium**?

Punctuated equilibrium is a model of macroevolution first detailed in 1972 by Niles Eldredge (1942–) and Stephen Jay Gould (1941–2002). It can be considered either a rival or supplementary model to the more gradual-moving model of evolution posited by neo-Darwinism. The punctuated equilibrium model essentially asserts that most of geological history shows periods of little evolutionary change, followed by short (geologically speaking, a few million years) periods of rapid evolutionary change. Gould and Eldredge's work has been buttressed by the discovery of the Hox genes that control embryonic development. Hox genes are found in all vertebrates and many other species as well; they control the placement of body parts in the developing embryo. Relatively minor mutations in these gene sequences could result in major body changes for species in a short period of time, thereby giving rise to new forms of organisms and therefore new species.

When was the great period of **expansion** for **species**?

The Cambrian explosion was the relatively brief period (approximately 40 million years long) when all of the major animal groups recognized today first appeared. This event, which occurred about 500 million years ago, is probably the result of changes in genome organization. These changes led to variations in the types of proteins produced, ultimately leading to a change in the structures built from those proteins.

How is the **Cambrian** explosion related to "**evo-devo**"?

Mutations in the suite of genes that control developmental processes, known as the Hox genes, would have caused the radical changes in the body structure of animals that occurred during the Cambrian explosion. These genes have been identified in every major animal group (both vertebrates and invertebrates), so it seems likely that they date back to the pre-Cambrian period. Thus, the mechanisms that control the development of the animal embryo ("devo") also probably played a role in the evolution of animals as a group, thus the term "evo-devo."

What is the value of **fossils** to the study of **evolution**?

Fossils are the preserved remains of once-living organisms. The value of fossils comes not only from the information they give us about the structures of those animals. The placement of common fossils in the geologic layers also gives researchers a method for dating other, lesser known, samples.

How do **fossils form**?

Fossils form rarely, since an organism is usually consumed totally or scattered by scavengers after death. If the structures remain intact, fossils can be preserved in amber (hardened tree sap), Siberian permafrost, dry caves, or rock. Rock fossils are the most common. In order to form a rock fossil, three things must happen: 1) the organism must be buried in sediment; 2) the hard structures must mineralize; and 3) the sediment surrounding the fossil must harden and become rock. Many rock fossils either erode before they can be discovered, or they remain in places inaccessible to scientists.

How are **fossils dated**?

There are two methods for dating fossils. The first is known as relative dating. By determining the age of the surrounding rock, scientists can give an approximate age to the fossils therein. Rocks are dated by their distance from the surface, with older rocks generally deeper from the surface. Using the data from other fossils found in the same rock stratum can also be used to give an approximate date to a new sample. The second method is known as absolute dating. Absolute dating relies on the known rates of radioactive decay within rocks. By measuring the ratio between the radioactive forms of an element like uranium-238 to its nonradioactive, "decayed" form, scientists can determine when the rock formed, because that would be when the radioactive isotope was acquired. Amino acids also convert gradually from one form (the left-handed) to another (right-handed) after an organism dies and so may be used as a method of dating some fossils.

Why can **fossils** be **misleading**?

The fossil record is biased toward organisms who were very common, who may have had hard shells or bony structures, and whose species lasted a long time. Therefore, it does not give us a full picture of what species may have been active in the evolutionary past. Fossils are also unlikely to record soft structure changes such as an increase in muscle mass or the development of new organ systems.

Why are **mass extinctions** a part of evolution?

A mass extinction is named for a time period in which at least 60 percent of the living species present become extinct over a period of one million years. Mass extinctions are considered biological catastrophes because of the relative speed and range of their effects. The loss of so many species allows surviving populations to exploit their adaptations in new ways; they can adapt to new parts of the environment without facing competition from other species.

What is the **Endosymbiosis theory**?

Certain organelles (e.g., chloroplasts and mitochondria) within eukaryotic cells share a number of similarities with bacteria. Because of this, scientists surmise that early versions of eukaryotic cells had symbiotic relationships with certain bacteria; the eukaryote provided protection and resources, while the prokaryote specialized in converting energy (either sunlight or chemical) into forms that could be used by the eukaryotic cell (sugar or ATP)—thus the term "endosymbiote," meaning "shared internal life." Similarities between chloroplasts, mitochondria, and free-living bacteria include:

- Genetic material: all have DNA in chromosomes

- Protein synthesis: all are capable of making proteins

- Energy transduction: all are capable of acquiring and using energy to perform reactions.

What is the **tree of life**?

Darwin envisioned a tree of life with branch tips representing currently living species and the base as the common ancestor of all. As one moved from the tips toward the trunk, one would pass the common ancestors for each species group. This idea is still under debate today as researchers try to determine how and how often life has evolved on Earth.

What is **Darwin's warm pond**?

Darwin's idea of a suitable milieu for the origin of life on Earth was a little pond gently warmed by sunlight. This can be contrasted to the Oparin-Haldane hypothesis of an electrically active methane and carbon dioxide atmosphere for the origin of organic evolution.

Which **biological events** occurred during the **geologic time divisions**?

Period	Epoch	Beginning date in est. millions of years	Plants and microorganisms	Animals
Cenozoic Era (Age of Mammals)				
Quaternary	Holocene (Recent)	10,000 years ago	Decline of woody plants and rise of herbaceous plants	Age of *Homo sapiens*; humans dominate

Period	Epoch	Beginning date in est. millions of years	Plants and microorganisms	Animals
	Pleistocene	1.9	Extinction of many species (from four ice ages)	Extinction of many large mammals (from four ice ages)
Tertiary	Pliocene	6	Development of grasslands; decline of forests; flowering plants	Large carnivores; many grazing mammals; first known human-like primates
	Miocene	25		Many modern mammals evolve
	Oligocene	38	Spread of forests; flowering plants, rise of monocotyledons	Apes evolve; all present mammal families evolve; saber-toothed cats
	Eocene	55	Gymnosperms and angiosperms dominant	Beginning of age of mammals; modern birds
	Paleocene	65		Evolution of primate mammals

Mesozoic (Age of Reptiles)

Cretaceous		135	Rise of angiosperms; gymnosperms decline	Dinosaurs reach peak and then become extinct; toothed birds become extinct; first modern birds; primitive mammals
Jurassic		200	Ferns and gymnosperms common	Large, specialized dinosaurs; insectivorous marsupials
Triassic		250	Gymnosperms and ferns dominate	First dinosaurs; egg-laying mammals

Paleozoic Era (Age of Ancient Life)

Permian		285	Conifers evolve	Modern insects like reptiles; extinction of many Paleozoic invertebrates

469

Period	Epoch	Beginning date in est. millions of years	Plants and microorganisms	Animals
Carboniferous (divided into Mississippian and Pennsylvanian periods by some in the U.S.)		350	Forests of ferns and gymnosperms; swamps; club mosses and horsetails	Ancient sharks abundant; many echinoderms, mollusks, and insect forms; first reptiles; spread of ancient amphibians
Devonian		410	Terrestrial plants established; first forests; gymnosperms appear	Age of fishes; amphibians; wingless insects and millipedes appear
Silurian		425	Vascular plants appear; algae dominant	Fishes evolve; marine arachnids dominant; first insects; crustaceans
Ordovician		500	Marine algae dominant; terrestrial plants first appear	Invertebrates dominant; first fishes appear
Cambrian		570	Algae dominant	Age of marine invertebrates
Precambrian Era				
Archeozoic and Proterozoic Eras		3800.0	Bacterial cells; then primitive algae and fungi; marine protozoans	Marine invertebrates at end of period
Azoic		4600.0	Origin of Earth	

Who originated the idea called **panspermia**?

Panspermia is the idea that microorganisms, spores, or bacteria attached to tiny particles of matter have traveled through space, eventually landing on a suitable planet and initiating the rise of life there. The word itself means "all-seeding." The British scientist Lord Kelvin (1824–1907) suggested in the nineteenth century that life may have arrived here from outer space, perhaps carried by meteorites. In 1903 the Swedish chemist Svante Arrhenius (1859–1927) put forward the more complex panspermia idea that life on Earth was "seeded" by means of extraterrestrial spores, bacteria, and microorganisms coming here on tiny bits of cosmic matter.

How did **humans evolve**?

Evolution of the *Homo* lineage of modern humans (*Homo sapiens*) has been proposed to originate from a hunter of nearly 5 ft (1.5 m) tall, *Homo habilis*, who is widely presumed to have evolved from an australopithecine ancestor. Near the beginning of the Pleistocene (2 million years ago), *Homo habilis* is thought to have transformed into *Homo erectus* (Java Man), who used fire and possessed culture. Middle Pleistocene populations of *Homo erectus* are said to show steady evolution toward the anatomy of *Homo sapiens* (Neanderthals, Cro-Magnons, and modern humans), 120,000 to 40,000 years ago. Premodern *Homo sapiens* built huts and made clothing.

Are **humans** still **evolving**?

In order for humans to continue evolving in a measurable way, we would predict that those individuals with certain variations (alleles) of traits must reproduce more efficiently than those without those alleles, such that eventually the alleles in question become more common in the population. For example, if nearsighted individuals were less likely to produce healthy offspring (for example, perhaps they have trouble finding enough food), then we would expect that eventually there would be very few nearsighted individuals in the population (assuming that nearsightedness is genetic). Because of our uniquely human skill of inventing things like eyeglasses and corrective surgery, which counteract genetic programming without actually changing the genes, there is little selective pressure against being nearsighted, so it is unlikely to be lost from the human population gene pool. The same could be said for many other traits. So while our gene pool could certainly change with time, it is unclear whether it will change in a clearly measurable direction such that we physically evolve.

APPLICATIONS

What is **social Darwinism**?

Social Darwinism is one of a number of perversions of Darwin-Wallace theory. These movements attempt to use evolutionary mechanisms as excuses for social change. Followers of social Darwinism believe that the "survival of the fittest" applies to socioeconomic environments as well as evolutionary ones. By this reasoning, the weak and the poor are "unfit" and should be allowed to die without societal interven-

What are the genetic differences between two brothers and eight cousins?

On average each of the gene sequences in our cells has a 50 percent chance of being found in our siblings. Therefore, two siblings could contain copies of all of the sequences of a third. First cousins would share about one eighth. So, as JBS Haldane (1892–1964) wrote, "Would I lay down my life to save my brother? No, but I would to save two brothers or eight cousins."

tion. This idea has nothing to do with Charles Darwin and Alfred Russel Wallace but was promoted by Herbert Spencer (1820–1903) and is related to the works of Thomas Malthus, whose work did indeed inspire Darwin. Although social Darwinism has faded as a movement, it did help to spur the eugenics movement of Nazi Germany as well as a number of laws and policies in the United States in the twentieth century.

What is a **race**?

The term "race" was originally used to describe subspecies. However, as genetic analysis of humans has shown, there is greater genetic variation within geographic sub-populations (races) than among the human population as a whole. In other words, there is so much genetic overlap between groups formerly designated as races that the term is meaningless and biologically indefensible.

What is **kin selection**?

Kin selection is a version of natural selection that measures success not just by individual reproductive effort (how many fertile offspring produced) but that also includes the number of copies of those genes that are located in close relatives. A parent and child, for example, share about 50 percent of their gene sequences on average.

What is **molecular evolution**?

The study of molecular evolution is the study of how proteins and nucleic acids change with time. Early studies on evolution were based on biogeography and the fossil record. As scientists developed techniques for studying DNA and proteins, evolutionary researchers soon followed their lead. Molecular evolution involves two broad areas: how the forces at work on individuals (e.g., natural selection, sexual selection) affect gene sequences and proteins, and how entire genomes evolve.

What is **Darwinian medicine**?

Darwinian medicine is the application of Darwinian principles (descent with modification via natural selection) to the disease process. The purpose of the study of medicine from a Darwinian point of view is not to determine who is "fittest" or healthiest, but instead to determine the evolutionary underpinnings of why, as such highly adapted organisms, we as a species are still prone to diseases like atherosclerosis (hardening of the blood vessels), nearsightedness, or cancer.

Is **antibiotic resistance** an **evolutionary trend**?

Antibiotic resistance is the loss of susceptibility in bacteria to drugs like penicillin and erythromycin. The reason that it is interesting from an evolutionary perspective is that it demonstrates evolution happening in real time—that is, within a period easily observable by humans. The variation in response to antibiotics within a population of bacteria is similar to that described by Darwin and Wallace. When some bacterial cells survive after an incomplete course of antibiotic treatment, they form the basis for a new drug-resistant strain. What is troubling, however, is that different types of bacteria can actually share the genes that make them drug resistant, so that this ability is becoming more prevalent among different strains of disease-causing organisms.

What is **coevolution**?

Coevolution is a rare form of evolution. By definition, it requires that two species adapt to evolutionary changes occurring in each other. An example of this reciprocal adaptation would be the development of mutualistic relationships between plants and the insects that prey upon them. As the plant develops defenses (like the oils of species in the mustard family), the insects develop counter weapons (cabbage butterflies have metabolic adaptations that can safely break down these toxic compounds).

What is the **Darwin station** in the **Galapagos Islands**?

The Charles Darwin Research Station in the Galapagos Islands is a biological field station that serves as a center for research on the flora and fauna of the islands. Work on the station, which is located on the island of Santa Cruz (Indefatigable on the British atlas), began in 1960, two years after the formation of the Darwin Foundation. The Darwin Foundation, UNESCO, and the government of Ecuador were the primary contributors to the building of the station, although many other organizations and individuals helped as well.

What is **cultural evolution**?

Instead of changing at the molecular level, in the sequence of our genes, the history of *Homo sapiens* has been marked by what is rare in other animals: cultural evolution. Culture is passed from one generation to the next, and by changing our culture, we have been able to adapt to our environment without requiring structural changes. One example of cultural evolution is the emphasis placed on the use of soap and water to prevent disease. Adopting this cultural tradition has had a definite positive impact on the survival of human populations, and yet there have been no changes that would be visible in the fossil record of our species.

Is it possible to **observe evolutionary** change in other species?

The change in allele frequencies within populations, which is driven by natural selection, is occurring around us constantly. However, whether we can actually observe those changes is a different matter. For example, elephants may be adapting to changing conditions in their environment, but since they tend to be extremely long-lived, it is unlikely that we will be able to follow enough generations to observe a trend. On the other hand, it is possible to notice such changes in populations of individuals with very short life spans, particularly when they are experimentally manipulated. Bacteria and guppies are just two examples of species that have been observed evolving in response to changing environmental conditions.

What is a **molecular clock**?

The molecular clock is a hypothesis that assumes that random mutation, in terms of the substitution of one nucleotide for another, occurs at a linear rate. By comparing the number of nucleotide substitutions between two species, scientists could then estimate the amount of time that has elapsed since they shared that sequence with their most recent common ancestor.

What is the **Darwin award**?

The most popular award by that name is a humorous award given posthumously to those who improve the human gene pool by removing themselves from it.

What is the **Wallace award**?

In 2004 the International Biogeography Society established the Alfred Russel Wallace Award for lifetime contributions to biogeography.

How can **behavior evolve**?

If behavior has a genetic basis, and certain behaviors provide greater reproductive advantages than others, then those genes will become more common in the population. Therefore, while the behavior could be said to evolve, it is actually the genetic source of the behavior that is evolving.

When was the **Scopes (monkey) trial**?

John T. Scopes (1900–1970), a high-school biology teacher, was brought to trial by the State of Tennessee in 1925 for teaching the theory of evolution. He challenged a law passed by the Tennessee legislature that made it unlawful to teach in any public school any theory that denied the divine creation of man. He was convicted and sentenced, but the decision was reversed later and the law repealed in 1967.

In the early twenty-first century, pressure against school boards still affects the teaching of evolution. Recent drives by anti-evolutionists either have tried to ban the teaching of evolution or have demanded "equal time" for "special creation" as described in the biblical book of Genesis. This has raised many questions about the separation of church and state, the teaching of controversial subjects in public schools, and the ability of scientists to communicate with the public. The gradual improvement of the fossil record, the result of comparative anatomy, and many other developments in biological science has contributed toward making evolutionary thinking more palatable.

What is the **Red Queen hypothesis**?

This hypothesis, also called the law of constant extinction, is named after the Red Queen in Lewis Carroll's *Through the Looking Glass*, who said "now here, you see, it takes all the running you can do to keep in the same place." The idea is that an evolutionary advance by one species represents a deterioration of the environment for all remaining species. This places pressure on those species to advance just to keep up.

ENVIRONMENT

ENVIRONMENTAL CYCLES AND RELATED CONCEPTS

What is the **Gaia hypothesis**?

The Gaia hypothesis (named for Gaia, the Earth goddess of ancient Greece) is a controversial idea that the world is a single living organism that is capable of self-maintenance and regulation. It was first proposed by James Lovelock (1919–) and Lynn Margulis (1938–) in 1974. Many scientists regard it as a useful analogy but a difficult theory to test scientifically.

Why do we have **seasons**?

Seasons are due to two factors: 1) Earth's inclination on its axis; and 2) the distance of Earth from the Sun, making days longer for the hemisphere that is tilted away from the Sun during any given period.

What is a **climate**, and how is it **characterized**?

Climate refers to the long-term weather conditions of a region, based on long-term averaging of temperature. Climates often undergo cyclic changes over decades, centuries, and millennia, but it is difficult to predict future climate changes. A climate diagram summarizes seasonal variation in temperature, precipitation, length of wet/dry seasons, and portion of the year spent in specific temperature ranges. Weather and climate are important because they are the determining factors of biomes and ecosystems.

What is a **microclimate**?

When you notice that the temperature forecast in your local media is consistently warmer or colder than that which occurs in your neighborhood, you have identified a

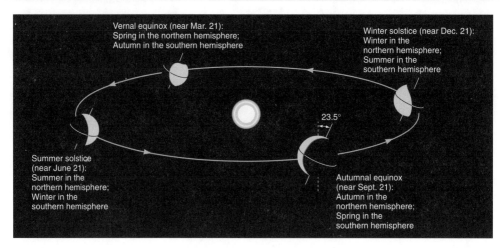

The seasons.

microclimate. Light, temperature, and moisture may all vary from one area to another within a biome because of changes in altitude, vegetation, or other factors.

What is **El Niño**?

Along the west coast of South America, near the end of each calendar year, a warm current of nutrient-poor tropical water moves southward, replacing the cold, nutrient-rich surface water. Because this condition frequently occurs around Christmas, local residents call it El Niño (Spanish for "child"), referring to the Christ child. In most years the warming lasts for only a few weeks. However, when El Niño conditions last for many months, the economic results can be catastrophic. It is this extended episode of extremely warm water that scientists now refer to as El Niño. During a severe El Niño, large numbers of fishes and marine plants may die. Decomposition of the dead material depletes the water's oxygen supply, which leads to the bacterial production of huge amounts of smelly hydrogen sulfide. A greatly reduced fish (especially anchovy) harvest affects the world's fishmeal supply, leading to higher prices for poultry and other animals that normally are fed fishmeal. Anchovies and sardines are also major food sources for marine mammals such as sea lions and seals. When the food source is in short supply, these animals travel further from their homes in search of food. Not only do many sea lions and seals starve, but also a large proportion of the infant animals die. During the winter of 1997–1998 El Niño effects resulted in the second warmest and seventh wettest winter since 1895. Severe weather events included flooding in the U.S. Southeast, ice storms in the Northeast, flooding in California, and tornadoes in Florida. The 1997–1998 event indirectly caused 2,100 deaths and $33 billion in damage globally.

What is a **limiting factor** in an **ecosystem**?

A limiting factor is any environmental factor that restricts the ecological niche of an organism. Limiting factors are based on the law of supply and demand. Those factors (resources) whose supply is less than demand can influence the distribution of species within a community. Examples of limiting factors include soil, minerals, temperature extremes, and water availability.

Where does **rain** come from?

Solar energy drives winds that evaporate water from the surface of the oceans. The water vapor cools as it rises and then falls to the ground as rain, snow, or some other form of precipitation. Rain is part of the hydrologic cycle, which describes dynamic changes in the aquatic environment.

What is the **hydrologic cycle**?

The hydrologic cycle takes place in the hydrosphere, which is the region containing all the water in the atmosphere and the earth's surface. It involves five phases: condensation, infiltration, runoff, evaporation, and precipitation.

What is a **biogeochemical cycle**?

The elements that organisms need most (carbon, nitrogen, phosphorus, and sulfur) cycle through the physical environment, the organism, and then back to the environment. Each element has a distinctive cycle that depends on the physical and chemical properties of the element. Examples of biogeochemical cycles include the carbon and nitrogen cycles, both of which have a prominent gaseous phase. Examples of biogeochemical cycles with a prominent geologic phase include phosphorus and sulfur, where a large portion of the element may be stored in ocean sediments. Examples of cycles with a prominent atmospheric phase include carbon and nitrogen.

What is the **carbon cycle**?

To survive, every organism must have access to carbon atoms. Carbon makes up about 49 percent of the dry weight of organisms. The carbon cycle includes movement of carbon from the gaseous phase (carbon dioxide [CO_2] in the atmosphere) to solid phase (carbon-containing compounds in living organisms) and then back to the atmosphere via decomposers. The atmosphere is the largest reservoir of carbon, containing 32 percent CO_2. Biological processes on land shuttle carbon between atmospheric and terrestrial compartments, with photosynthesis removing CO_2 from the atmosphere and cell respiration returning CO_2 to the atmosphere.

How do **plants obtain nitrogen**?

The primary way that plants obtain nitrogen compounds is via the nitrogen cycle, which is a series of reactions involving several different types of bacteria, including nitrogen-fixing bacteria and denitrifying bacteria. During nitrogen fixation, symbiotic bacteria, which live in association with the roots of legumes, are able through a series of enzymatic reactions to make nitrogen available for plants. Nitrogen is crucial to all organisms because it is an integral element of proteins and nucleic acids. Although Earth's atmosphere is 79 percent nitrogen, molecular nitrogen is very stable and does not easily combine with other elements. Plants must use nitrogen in its fixed form, such as ammonia, urea, or the nitrate ion.

BIOMES AND RELATED CONCEPTS

What is **biogeography**?

Biogeography is the study of the distribution, both current and past, of individual species in specific environments. One of the first biogeographers was Carolus Linnaeus (1707–1778), a Swedish botanist who studied the distribution of plants. Biogeography specifically addresses the questions of evolution, extinction, and dispersal of organisms in specific ecosystems.

What are the **general characteristics** of **biomes**?

A biome is a one of the world's prominent ecosystems, characterized by both vegetation and organisms particularly adapted to that environment.

Biome summary

Biome	Temperature	Precipitation	Vegetation	Animals
Arctic tundra	-40°C to 18°C (-40°F to 64°F)	Dry season, wet season	Shrubs, grasses, lichens, mosses	Birds, insects, mammals
Deciduous forest	Warm summers, cold winters	Low, distributed throughout year	Trees, shrubs herbs, lichens, mosses	Mammals, birds, insects, reptiles
Desert	Hottest; great daily range	Driest <10 in (25 cm) of rain per year	Trees, shrubs, succulents, forbs	Birds, small mammals, reptiles
Taiga	Cold winters, cool summers	Moderate	Evergreens, tamarack	Birds, mammals

Biome	Temperature	Precipitation	Vegetation	Animals
Tropical Rainforest	Hot	Wet season, short dry season	Trees, vines, epiphytes, fungi	Small mammals, birds, insects
Tropical Savannah	Hot	Wet season, dry season	Tall grasses shrubs, trees	Large mammals, birds, reptiles
Temperate grassland	Warm summers, cold winters	Seasonal drought, occasional fires	Tall grasses	Large mammals, birds, reptiles

What is a **cline**?

A cline refers to geographic variation due to a gradient of climatic features. For example, a north-south gradient may include a range of temperature and a range of plant sizes, with taller plants at the southern end and shorter plants at the northern end.

How do **trees grow** out of **rocks**?

If you didn't mow your lawn for a whole summer, eventually your lawn would become a grassy meadow. The change in community structure observed over time is known as ecological succession. When the succession begins in an area that is previously unoccupied or unchanged by other species, the process is known as primary succession. While the tree cannot actually grow out of bare rock, it can begin to grow from small amounts of soil and debris that collect in pockets of the rock. Over time the tree may grow well enough to send roots into the rock itself, causing the rock to split and making it appear that the tree has sprung from within the rock instead of from its surface.

What is a **wetland**?

A wetland is an area that is covered by water for at least part of the year and has characteristic soils and water-tolerant plants. Examples of wetlands include swamps, marshes, bogs, and estuaries.

Type of Wetland	Typical Features
Swamp	Tree species such as willow, cypress, and mangrove
Marsh	Grasses such as cattails, reeds, and wild rice
Bogs	Floating vegetation including mosses and cranberries
Estuaries	Specially adapted flora and fauna such as crustaceans, grasses, and certain types of fishes

The structure of a eutrophic lake.

What is an **estuary**?

Estuaries are places where freshwater streams and rivers meet the sea. The salinity of such areas is less than that of the open ocean but greater than that of a typical river, so organisms living in or near estuaries have special adaptations. Estuaries are rich sources of invertebrates such as clams, shrimps, and crabs, as well as fishes such as striped bass, mullet, and menhaden. Unfortunately, estuaries are also popular locations for human habitation and businesses. Contamination from shipping, household pollutants, and power plants (which are carried to the sea by rivers and streams) threaten the ecological health of many estuaries.

What is **eutrophication**?

Eutrophication is a process in which the supply of plant nutrients in a lake or pond is increased. In time, the result of natural eutrophication may be dry land where water once flowed, caused by plant overgrowth. Natural fertilizers, washed from the soil, result in an accelerated growth of plants, producing overcrowding. As the plants die off, the dead and decaying vegetation depletes the lake oxygen supply, causing fish to die. The accumulated dead plant and animal material eventually changes a deep lake to a shallow one, then to a swamp, and finally it becomes dry land. While the process of eutrophication is a natural one, it has been accelerated enormously by human activities. Fertilizers from farms, sewage, industrial wastes, and some detergents all contribute to the problem.

482

What is **limnology**?

Limnology is the study of freshwater ecosystems—especially lakes, ponds, and streams. These ecosystems are more fragile than marine environments since they are subject to greater extremes in temperature. The study of limnology includes the chemistry, physics, and biology of these bodies of water. F. A. Forel (1848–1931), a Swiss professor, has been called the father of limnology.

What does it mean when a **lake** is **brown** or **blue**?

When a lake is brown, it usually indicates that eutrophication is occurring. This process refers to premature "aging" of a lake, when nutrients are added to the water, usually due to run-off, which may be either agricultural or industrial in origin. Due to this rich supply of nutrients, blue-green algae begin to take over the green algae in the lake, and food webs within the lake are disturbed, leading to an eventual loss of fish. When a lake is blue, this usually means that the lake has been damaged by acid precipitation. The gradual drop in pH caused by exposure to acid rain causes disruption of the food webs, eventually killing most organisms. The end result is clear water, which reflects the low productivity of the lake.

What is a **red tide**?

A red tide occurs when there is a population explosion among toxic red dinoflagellates (members of the genera *Gymnodidium* and *Gonyaulax*, both protistans that have an unusual cellular plate or armor). The population explosion, referred to a as a "bloom," may tint the water orange, red, or brown and can be toxic to shellfish, birds, and humans who eat red tide–contaminated food.

What is a **community**?

Suppose you wanted to study not just the sparrows living in your backyard but also the insects they feed to their young and the plants that those insects eat. The term used by ecologists to describe a group of populations of different species living in the same place at the same time is "community." This is a concise way of describing the organisms likely to be affected by any change to the local environment.

What is a **niche**?

A niche is similar to the job description of a population within a given species. However, instead of describing how many hours individuals will work, it describes features such as whether the species is active at night or during the day, what it eats, where it lives, and other aspects of day-to-day life. From one environment (or community) to another, the niche may vary depending on how much competition the species faces.

Can **two populations occupy** the **same niche**?

According to G. F. Gause (1910–1988), an ecologist, two species that are in direct competition for the same resource cannot coexist if that resource is limited in some way. The work of Gause and others predicts that under such conditions one population will drive the other to extinction in that local area.

What is the difference between a **food chain** and a **food web**?

A food chain refers to transfer of energy from producers through herbivores through carnivores in a community. An example of a food chain would be little fishes eating plankton and those little fishes in turn being eaten by bigger fishes. The term food web is broader, as it includes interconnected food chains within a specific ecosystem (perhaps the bigger fishes eat the plankton as well!). A food web describes a nutritional portrait of the ecosystem. In 1927 Charles Elton (1868–1945) was one of the first scientists to diagram a food web, in his case a description of feeding relationships on Bear Island in the Arctic.

What is a **trophic level**?

A trophic level represents a step in the dynamics of energy flow through an ecosystem. The first trophic level is made up of the producers, those within the ecosystem that harvest energy from an outside source like the Sun (or deep-sea thermal vents) and stabilize or "fix" it so that it remains in the system. The second level would comprise those who consume the producers, also known as the primary consumers. The next level would contain the secondary consumers (those who consume the primary consumers), and so on. Because of the limited amount of energy available to each level, these trophic pyramids rarely rise above a third or fourth level of structure. R. Lindeman (1915–1942) was one of the first ecologists to refer to the "trophic dynamics" of ecosystems, doing so in 1942.

What is a **keystone species**?

A keystone species is a species that is crucial or essential to the ecosystem's community structure. Originally, a keystone species was always thought to be the top predator, such as the gray wolf. Scientists have found that wolf population sizes influence populations of both their prey and other species in the environment. However, a more recent viewpoint recognizes that less conspicuous species are also very important, as all species are interconnected in a biological community. Other examples of keystone species include the sea star, *Pisaster*, found along the coast of Washington state, and the black-tailed prairie dog of the prairie ecosystem. The sea star feeds on mussels and prevents the mussels from crowding out other species. The prairie dog is a critical source of food for larger predators, its burrowing loosens the soil, and its burrows act as home for other creatures.

Do **animals** ever **run out of energy**?

Because the amount of energy available to an animal at any particular trophic level is limited by the distance between that level and primary producers, animals can obtain only a finite amount of energy from their habitats. Therefore, over evolutionary time, species make concessions in terms of their lifestyle attributes in order to best utilize their limited resources. Examples of items within the energy budget are activities like reproduction, growth, and maintenance. While the energy expended on growth and reproduction can remain in the system as biomass, the energy used for maintenance is lost as heat.

What is the **most efficient animal**?

Efficiency may be judged by a variety of measures, and thus it is difficult to pick the "most efficient" animal. For example, small animals tend to be much less efficient than larger ones at conserving the heat required to run their metabolic reactions. Conversely, many small animals are much more efficient at reproduction (compare the reproductive output of a roach to that of an elephant!). Therefore, determining the comparative efficiency of different animals depends upon what attributes are selected for measurement.

What are **producers** and **consumers**?

"Producers" and "consumers" are terms used to describe the different roles played by species within ecosystems. Producers are those who "fix" energy—that is, they take energy from one source and convert it into a form (their biomass) that makes it accessible to others within the system (the consumers). Consumer levels are numbered according to their reliance on producers as a main source of energy. So primary consumers are those that rely heavily on producers, while secondary and tertiary (and even quaternary) consumers exploit other consumers as their preferred energy sources.

What is an **ecological pyramid**?

If the organisms in a food chain are arranged according to trophic levels, they form a pyramid, with a broad base representing the primary producers and usually only a few individuals in the highest part of the pyramid. Also known as a "pyramid of numbers," an ecological pyramid is a way of describing the distribution of energy, biomass, or individuals among the different levels of ecosystem structure.

Why do **ecosystems need decomposers**?

While energy flows through ecosystems in only one direction, entering at the producer level and exiting as heat and the transfer of energy (as biomass) to consumers, chemical compounds can be reused over and over again. In a well-functioning ecosystem, some organisms make their living (their niche) by breaking down structures and recycling the compounds. These organisms are known as decomposers. Without these organisms, the chemicals used to build a tree would remain locked in the tree biomass for eternity instead of being returned to the soil after the tree's death. From this soil will spring new growth, beginning the cycle once again.

What is a **climax community**?

Terrestrial communities of organisms move through a series of stages from bare earth or rock to forests of mature trees. This last stage is described as the "climax" because it is thought that, if left undisturbed, communities can remain in this stage in perpetuity. However, more recent studies suggest that climax may be only one part of a continuous cycle of successional stages in these communities.

Why are **big and fierce animals rare**?

Because the transfer of energy from one individual to another is inefficient, only about 10 percent of the energy used to build a tasty worm actually is available to the hungry robin who eats it. As we move through the levels of the food web, each predator (or predator grouping) tends to become bigger and more aggressive (fiercer) than the last. However, the amount of energy available to each level continues to decline, so there is very little room at the top of the web for a predator large enough and fierce enough to consume all the others. In fact, it is estimated that only 1/1000th of the energy brought into the system by photosynthesis actually makes it to the hawk or owl at the top of such a system. Less energy available means that fewer individuals can be supported, so big, fierce animals tend to be rare in their ecosystems.

Do **animals** ever **help each other**?

Populations of organisms within an environment may engage in a variety of relationships with each other. For example, in a relationship known as mutualism, each species provides a benefit to the other. Mutualism can occur between two animal species like the relationship between large coral reef fishes and the smaller species like the wrasses that swim into their mouths and eat the parasites that may have taken up residence there. Mutualistic relationships also allow bean plants to grow better because they exchange nutrients with *Rhizobium* fungi living on their roots.

What is a **population**?

A population is a group composed of all members of the same species that live in a specific geographical area at a particular time. An example of a population might include all the gray squirrels that live in a certain urban park. The areas occupied by a population could include the small area (measured in square millimeters) occupied by bacteria in a rotting apple to the vast areas of ocean (square kilometers) that include the territory of migrating sperm whales. Population ecology is the branch of ecology that studies the structure and changes within a population. Studies of specific populations will indicate the dynamics of the population, in terms of active, ongoing growth; declining growth; or stability.

Who was the **first person** to **study populations mathematically**?

Thomas Robert Malthus (1766–1834) in 1798 attempted to inform people that the human population, like any other population, had the potential to increase exponentially. Malthusian ideas were not well received, as he predicted the rate of population growth would exceed the ability of the land to produce food. His work was later used by Charles Darwin to explain his theory of natural selection.

What is a **survivorship curve**?

A survivorship curve can indicate how long individuals survive in a population. There are three distinct types of curves. In a Type I curve, the young have a high survival rate and typically live a long life. An example of this curve can be seen in the Daal sheep that live in Mt. Denali National Park in Alaska. Humans are also an example of a Type I curve. In a Type II curve, individuals have a relatively constant death rate throughout their life span. An example of this curve can be found in populations of American robins. A Type III curve includes those species that have a large number of young—most of which die at a high rate, at an early age—but have a lower death rate at later ages. An example of this survivorship curve can be found in lobsters and crabs.

What is a **life history table**?

A life history table, also referred to as a life table, is a table that shows both survival and death rates in a specific population or organisms. The life table is patterned after actuarial tables used by insurance companies.

Can one **predict** how **populations will grow**?

Mathematical models can predict growth when populations are growing at their maximum rates. There are two distinct patterns: logistic or exponential growth. In logistic population growth, the population grows in cycles, responding to limiting factors in the environment. An example would be a population of insects that is limited by the amount of available food. Exponential growth is growth at a constant rate of increase per unit of time and is used to model continuous population growth in an unlimited environment. An example of exponential growth would be the doubling rate of bacterial growth on the turkey left unrefrigerated after Thanksgiving dinner.

Why do **lemmings rush to the sea**?

Lemmings are a type of rodent whose "plague" behavior has been studied for decades. Every four years or so, lemming populations in Canada explode to the point where fur trappers describe the tundra as being alive with the little brown mice. In fact, the population density varies from less than twenty mice per hectare (2.5 acres) in some years to as many as 200 in that same space the next year. What is puzzling to scientists is that while it is possible to observe this population boom first hand, the impetus for the growth and rapid decline (the population drops again within a few months) has not yet been determined. Although it has been documented that these population explosions cause an overconsumption of the available food, there is no evidence that such overcrowding can cause the lemmings to become suicidal (or even incredibly thirsty) and rush en masse into the sea. Instead, the lemming migrations begin slowly with small groups traveling at night, and gradually build so that animals travel in larger groups in the day time. They tend to avoid water but will swim if necessary: lemmings can cross a 656-ft (200-m) body of water on a calm night, but most will drown on a windy night, which may be the source of the myth.

Why are there **so many flies in the summer**?

Another way to ask this question would be: What is an opportunistic species? Species have evolved two general strategies for success in life. Either they specialize in adaptations that allow them to survive the boom and bust cycles within the environment, or they concentrate their efforts in taking advantage of short-term opportunities to succeed. Species utilizing these two strategies have various names including *opportunist* and *equilibrium* species. Opportunists move into newly opened habitats (for example, as dandelions overwhelm spring lawns) and reproduce rapidly with abandon. Other common examples of opportunistic species include insect species like mosquitoes and flies that save their reproductive efforts to exploit advantageous changes in their habitat like a hard rain or fresh roadkill. When conditions are just right, these species go full tilt from mere existence to population explosion, which explains why fly populations boom so suddenly in the summer.

EXTINCT AND ENDANGERED PLANTS AND ANIMALS

How is **biodiversity measured**?

Biodiversity or biological diversity refers to the breadth of species represented within ecosystems or even on Earth as a whole. Biodiversity may be defined at three levels: genetic diversity, species diversity, and ecosystem diversity. Genetic diversity refers to the variety of genes found within a population or *between* populations of the same species. Species diversity may also be described as species richness. In other words, how many different species are there within a habitat? Finally, measuring ecosystem diversity is an attempt to keep track of the loss of different types of habitats. This in turn gives scientists a sense of what types of species are going extinct at any given time.

What is the difference between an **endangered species** and a **threatened species**?

An "endangered" species is one that is in danger of extinction throughout all or a significant portion of its range. A "threatened" species is one that is likely to become endangered in the foreseeable future.

Did **dinosaurs** and **humans** ever **coexist**?

No. Dinosaurs first appeared in the Triassic period (about 220 million years ago) and disappeared at the end of the Cretaceous period (about 65 million years ago). Modern humans (*Homo sapiens*) appeared only about 25,000 years ago. Movies that show humans and dinosaurs existing together are only Hollywood fantasies.

When did the **last passenger pigeon die**?

At one time, 200 years ago, the passenger pigeon (*Ectopistes migratorius*) was the world's most abundant bird. Although the species was found only in eastern North America, it had a population of between 3 and 5 billion birds (25 percent of the North American land bird population). Overhunting caused a chain of events that reduced their numbers below minimum threshold for viability. In the 1890s several states passed laws to protect the pigeon, but it was too late. The last known wild bird was shot in 1900. The last passenger pigeon, named Martha, died on September 1, 1914, in the Cincinnati Zoo.

How did the **dodo** become **extinct**?

The dodo became extinct around 1800. Thousands were slaughtered for meat, but pigs and monkeys, which destroyed dodo eggs, were probably most responsible for the dodo's extinction. Dodos were native to the Mascarene Islands in the Central Indian Ocean. They became extinct on Mauritius soon after 1680 and on Reunion about 1750. They remained on Rodriguez until 1800.

What is the **status** of the **African elephant**?

From 1979 to 1989 Africa lost half of its elephants from poaching and illegal ivory trade, with the population decreasing from an estimated 1.3 million to 600,000. This led to the transfer of the African elephant from threatened to endangered status in October 1989 by CITES (the Convention on International Trade in Endangered Species). An ivory ban took effect on January 18, 1990. Botswana, Namibia, and Zimbabwe have agreed to restrict the sale of ivory to a single, government-controlled center in each country. All countries have further pledged to allow independent monitoring of the sale, packing, and shipping process to ensure compliance with all conditions. Finally, all three countries have promised that all net revenues from the sale of ivory will be directed back into elephant conservation for use in monitoring, research, law enforcement, other management expenses, or community-based conservation programs within elephant range.

How can you **avoid buying items** made from **endangered species**?

The 1975 CITES (Convention on International Trade in Endangered Species of Wild Fauna and Flora) prohibits trade in endangered species. TRAFFIC, the wildlife trade monitoring network, suggests that travelers buy wisely; although it may be legal to buy products in certain popular tourist locations, it is often illegal or may require a permit to bring these objects home. Tourists should check the TRAFFIC guide for specific countries and regulations

How many **species** of **plants** and **animals** are **threatened** or **endangered** in the **United States**?

The total number of U.S. species listed as endangered by the U.S. Fish and Wildlife Service is 990 (391 animals, 599 plants). The total number of U.S. species listed as threatened is 275 (128 animals, 147 plants).

Summary of Listed Species: Species and Recovery Plans as of 6/22/2004

Group	Endangered U.S. species	Endangered Foreign species	Threatened U.S. species	Threatened Foreign species	Total Species	U.S. Species with Recovery Plans
Mammals	69	251	9	17	346	55
Birds	77	175	14	6	272	78
Reptiles	14	64	22	15	115	33
Amphibians	12	8	9	1	30	14
Fishes	71	11	43	0	125	95
Clams	62	2	8	0	72	69
Snails	21	1	11	0	33	23
Insects	35	4	9	0	48	31
Arachnids	12	0	0	0	12	5
Crustaceans	18	0	3	0	21	13
Animal SubTotal	391	516	128	39	1074	416
Flowering Plants	571	1	144	0	716	577
Conifers and Cycads	2	0	1	2	5	2

Group	Endangered U.S. species	Endangered Foreign species	Threatened U.S. species	Threatened Foreign species	Total Species	U.S. Species with Recovery Plans
Ferns and Allies	24	0	2	0	26	26
Lichens	2	0	0	0	2	2
Plant SubTotal	599	1	147	2	749	602
Grand Total	990	517	275	41	1823	1023

Which **species** in the United States have become **extinct** since the **Endangered Species Act** was passed in 1973?

Seven domestic species have become extinct:

First listed	Date delisted (declared extinct)	Species name
3/11/1967	9/2/1983	Cisco, longjaw (*Coregonus alpenae*)
4/30/1980	12/4/1987	Gambusia, Amistad (*Gambusia amistadensis*)
6/14/1976	1/9/1984	Pearlymussel, Sampson's (*Epioblasma sampsoni*)
3/11/1967	9/2/1983	Pike, blue (*Stizostedion vitreum glaucum*)
10/13/1970	1/15/1982	Pupfish, Tecopa (*Cyprinodon nevadensis calidae*)
3/11/1967	12/12/1990	Sparrow, dusky seaside (*Ammodramus maritimus nigrescens*)
6/4/1973	10/12/1983	Sparrow, Santa Barbara song (*Melospiza melodia graminea*)

Have any **species recovered** and been **removed** from the list?

Sixteen species have been removed from the Endangered Species List because they have recovered:

First listed	Date delisted	Species name
3/11/1967	6/4/1987	Alligator, American (*Alligator mississippiensis*)
6/2/1970	9/12/1985	Dove, Palau ground (*Gallicolumba canifrons*)
6/2/1970	8/25/1999	Falcon, American peregrine (*Falco peregrinus anatum*)
9/17/1980	8/27/2002	Cinquefoil, Robbins' (*Potentilla robbinsiana*)
3/11/1967	7/24/2003	Deer, Columbian white-tailed (*Odocoileus virginianus leucurus*)

491

The American alligator was removed from the endangered species list in 1987.

First listed	Date delisted	Species name
6/2/1970	10/5/1994	Falcon, Arctic peregrine (*Falco peregrinus tundrius*)
6/2/1970	9/12/1985	Flycatcher, Palau fantail (*Rhipidura lepida*)
3/11/1967	3/20/2001	Goose, Aleutian Canada (*Branta canadensis leucopareia*)
12/30/1974	3/9/1995	Kangaroo, eastern gray (*Macropus giganteus*)
12/30/1974	3/9/1995	Kangaroo, red (*Macropus rufus*)
12/30/1974	3/9/1995	Kangaroo, western gray (*Macropus fuliginosus*)
4/26/1978	9/14/1989	Milk-vetch, Rydberg (*Astragalus perianus*)
6/2/1970	9/12/1985	Owl, Palau (*Pyroglaux podargina*)
6/2/1970	2/4/1985	Pelican, brown (U.S. Atlantic coast, FL, AL) (*Pelecanus occidentalis*)
6/2/1970	6/16/1994	Whale, gray (except where listed) (*Eschrichtius robustus*)
7/19/1990	10/7/2003	Wooly-star, Hoover's (*Eriastrum hooveri*)

The status of five species has been changed due to taxonomic revision. New information has been discovered for four other species.

CONSERVATION, RECYCLING, AND OTHER APPLICATIONS

Who is considered the **founder** of **modern conservation?**

American naturalist John Muir (1838–1914) is the father of conservation and the founder of the Sierra Club. He fought for the preservation of the Sierra Nevada in California and the creation of Yosemite National Park. He directed most of the Sierra Club's conservation efforts and was a lobbyist for the Antiquities Act, which prohibited the removal or destruction of structures of historic significance from federal lands. Another prominent influence was George Perkins Marsh (1801–1882), a Vermont lawyer and scholar. His book *Man and Nature* emphasized the mistakes of past civilizations that resulted in destruction of natural resources. As the conservation movement swept through the country in the last three decades of the nineteenth century, a number of prominent citizens joined the efforts to conserve natural resources and to preserve wilderness areas. Writer John Burroughs (1837–1921), forester Gifford Pinchot (1865–1946), botanist Charles Sprague Sargent (1841–1927), and editor Robert Underwood Johnson (1857–1937) were early advocates of conservation.

Who started **Earth Day?**

The first Earth Day, April 22, 1970, was coordinated by Denis Hayes at the request of Gaylord Nelson (1916–), U.S. Senator from Wisconsin. Nelson is sometimes called the father of Earth Day. His main objective was to organize a nationwide public demonstration so large it would get the attention of politicians and force the environmental issue into the political dialogue of the nation. Important official actions that began soon after the celebration of the first Earth Day were the establishment of the Environmental Protection Agency (EPA); the creation of the President's Council on Environmental Quality; and the passage of the Clean Air Act, establishing national air quality standards. In 1995 Gaylord Nelson received the Presidential Medal of Freedom for his contributions to the environmental protection movement. Earth Day continues to be celebrated each spring.

How is **Henry David Thoreau** associated with the **environment?**

Henry David Thoreau (1817–1862) was a writer and naturalist from New England. His most familiar work, *Walden*, describes the time he spent in a cabin near Walden Pond in Massachusetts. He is also known for being one of the first to write and lecture on the topic of forest succession. His work, along with that of John Muir (1838–1914) and others, has served to inspire those others to understand the natural world and provide for its conservation.

What was the **United States' first national park**?

On March 1, 1872, an act of Congress signed by Ulysses S. Grant established Yellowstone National Park as the first national park. The action inspired a worldwide national park movement.

What are the **five largest national parks** in the **United States**?

Park	Location	Acreage
Wrangell–St. Elias	Alaska	7,662,670
Gates of the Arctic	Alaska	7,266,102
Denali	Alaska	4,724,787
Katmai	Alaska	3,611,608
Death Valley	California	3,291,779

How many **acres** of **wetlands** have been **lost** in the **United States**?

Since access to water is important to industrial development, many cities are located in areas including wetlands. In the urbanization process, wetlands have been drained, filled, or used as dumps. Each wetland area serves as a habitat to many different plants and animals, with special regard to spawning and nursery habitats. The Wetlands Restoration Act (HR1474, enacted November 29, 1990) refers to *wetland mitigation banking* and provides that any person who discharges dredged or fill material into the waters of the United States must have a permit from the Army Corps of Engineers. This act is an attempt to preserve the complex communities that are found within wetlands. Wetlands are the lands between aquatic and terrestrial areas, such as bogs, marshes, swamps, and coastal waters. Although wetlands were at one time considered wastelands, scientists now recognize the importance of wetlands to improve water quality, stabilize water levels, prevent flooding, regulate erosion, and sustain a variety of organisms. The United States has lost approximately 100 million acres of wetland areas between colonial times and the 1970s. The 1993 Wetlands Plan established a goal of reversing the trend of 100,000 acres of wetland loss to 100,000 acres of wetland recovery.

What is the **importance** of the **rain forest**?

Half of all medicines prescribed worldwide are originally derived from wild products, and the United States National Cancer Institute has identified more than 2,000 tropical rain forest plants with the potential to fight cancer. Rubber, timber, gums, resins and waxes, pesticides, lubricants, nuts and fruits, flavorings and dyestuffs, steroids, latexes, essential and edible oils, and bamboo are among the products that would be drastically affected by the depletion of the tropical forests. In addition, rain forests greatly influence patterns of rain deposition in tropical areas; smaller rain forests mean less rain. Large groups of plants, like those found in rain forests, also help control levels of carbon dioxide in the atmosphere.

How **rapidly** is **deforestation occurring**?

Agriculture, excessive logging, and fires are the major causes of deforestation.

CHANGE IN FOREST AREA, 1990–2000

Area	Total forest, 1990 ('000 ha)	Total forest, 2000 ('000 ha)	Annual change ('000 ha)	Annual rate of change (percent)
Africa	702,502	649,866	-5,262	-0.8
Asia	551,448	547,793	-364	-0.1
Europe	1,030,475	1,039,251	881	0.1
North and Central America	555,002	549,304	-570	-0.1
Oceania	201,271	197,623	-365	-0.2
South America	922,731	885,618	-3,711	-0.4
Total world	3,963,429	3,869,455	-9,391	-0.2

When was the symbol of **Smokey Bear** first used to **encourage forest fire prevention**?

The origin of Smokey Bear can be traced to World War II, when the U.S. Forest Service, concerned about maintaining a steady lumber supply for the war effort, wished to educate the public about the dangers of forest fires. They sought volunteer advertising support from the War Advertising Council, and on August 9, 1944, Albert Staehle, a noted illustrator of animals, created Smokey Bear. Since 1944 Smokey Bear has been a national symbol of forest fire prevention not only in America, but also in Canada and Mexico, where he is known as Simon in both countries. This public service advertising (PSA) campaign is the longest running PSA campaign in U.S. history. In 1947 a Los Angeles advertising agency coined the slogan "Only you can prevent forest fires." On April 23, 2001, after more than fifty years, the famous ad slogan was revised to "Only you can prevent wildfires," in response to the wildfire outbreaks during 2000. The campaign gained a living mascot in 1950 when a firefighting crew rescued a male bear cub from a forest fire in the Capital Mountains of New Mexico. Sent to the National Zoo in Washington, D.C., to become Smokey Bear, the animal was a living symbol of forest fire protection until his death in 1976. His remains are buried at the Smokey Bear State Historical Park in Capitan, New Mexico.

In what way can **forest fires** be **good for the environment?**

Wildfires are critical to maintaining the integrity of forest and grassland ecosystems. Forest and grass fires, usually started by lightning, act as an ecologically renewing force by creating necessary conditions for plant germination and continued healthy growth to occur. The primary goal of fire management is to simulate the revitalizing aspects of natural fire cycles. Fire management also attempts to prevent large catastrophic wildfires from occurring by removing accumulated debris from forests. Seen throughout the American West every summer, these extremely intense fires are caused primarily by decades of fire suppression, which has allowed heavy fuels—accumulated debris—to build up. Ironically, by attempting to prevent natural fires, humans have only increased their prevalence.

The American peregrine falcon is a success story, as its overall population recovered enough for it to be removed from the Endangered Species list in 1994. However, local populations of the species remain threatened.

When was the **EPA** created and what does it do?

In 1970 President Richard M. Nixon (1913–1994) signed an executive order that created the Environmental Protection Agency (EPA) as an independent agency of the U.S. government. The creation of a federal agency by executive order rather than by an act of the legislative branch is somewhat uncommon. The EPA was established in response to public concern about unhealthy air, polluted rivers and groundwater, unsafe drinking water, endangered species, and hazardous waste disposal. Responsibilities of the EPA include environmental research, monitoring, and enforcement of legislation regulating environmental activities. The EPA also manages the cleanup of toxic chemical sites as part of a program known as Superfund.

What is the **Pollutant Standard Index**?

The U.S. Environmental Protection Agency and the South Coast Air Quality Management District of El Monte, California, devised the Pollutant Standard Index to monitor concentrations of pollutants in the air and inform the public concerning related health effects. The scale, which measures the amount of pollution in parts per million, has been in use nationwide since 1978.

PS Index	Health effects	Cautionary status
0	Good	
50	Moderate	
100	Unhealthful	
200	Very unhealthful	Alert: elderly or ill should stay indoors and reduce physical activity.
300	Hazardous	Warning: general population should stay indoors and reduce physical activity.
400	Extremely hazardous	Emergency: all people should remain indoors with windows shut and no physical exertion.
500	Toxic	Significant harm; same as above.

What is the **Toxic Release Inventory**?

Toxic Release Inventory (TRI) is a government-mandated, publicly available compilation of information on the release of over 650 individual toxic chemicals and toxic chemical categories by manufacturing facilities in the United States. The law requires manufacturers to state the amounts of chemicals they release directly to air, land, or water, or state that they transfer to off-site facilities that treat or dispose of wastes. The U.S. Environmental Protection Agency compiles these reports into an annual inventory and makes the information available in a computerized database. In 2000, 23,484 facilities released 7.1 billion lb (3.2 billion kg) of toxic chemicals into the environment. Over 260 million lb (118 million kg) of this total were released into surface water; 1.9 billion lb (.86 million kg) were emitted into the air; over 4.13 billion lb (1.87 billion kg) were released to land; and over 278 million lb (126 million kg) were injected into underground wells. The total amount of toxic chemicals released in 2000 was 6.7 percent lower than the amount released in 1999.

Why were such **dangerous chemicals** as **DDT, PCBs,** and **CFCs released** into the **environment?**

Dichlorodiphenyl-trichloro-ethene (DDT), polychlorinated biphenyls (PCBs), and chlorofluorocarbons (CFCs) were once widely used. Although DDT was synthesized as early as 1874 by Othmar Zeidler, it was the Swiss chemist Paul Müller (1899–1965) who recognized its insecticidal properties in 1939. He was awarded the 1948 Nobel Prize in Physiology or Medicine for his development of DDT. Unlike the arsenic-based compounds then in use, DDT was effective in killing insects and seemed not to harm plants and animals. In the following twenty years it proved to be effective in controlling disease-carrying insects (mosquitoes that carry malaria and yellow fever, and lice that carry typhus), and in killing many plant crop destroyers. Publication of Rachel Carson's book *Silent Spring* in 1962 alerted scientists to the detrimental effects of DDT. Increasingly DDT-resistant insect species and the accumulative hazardous effects of DDT on plant and animal life cycles led to its disuse in many countries during the 1970s. In fact, DDT and PCBs have been added to the list of chemicals known as estrogenic compounds—that is, synthetic substances in the environment that cause the mammalian body to respond as if to estrogen.

A group of chemicals with the same general chemical structure and physical properties as DDT are known as the polychlorinated biphenyls or PCBs. Because of their physical properties (nonflammability, chemical stability, high boiling point, and electrical insulating properties), PCBs can be used in a variety of applications. Formerly, many products contained these compounds. From electrical circuitry to the dyes and pigments used in paint, to carbonless copy paper, all were manufactured with PCBs. Before production was ceased in 1977, the United States produced about 1.5 billion lb (6.8 billion kg) of PCBs.

Chlorofluorocarbons (CfCs) are commonly used as aerosol sprays, refrigerants, solvents, and foam blowing agents. They are in and of themselves nontoxic and nonflammable molecules containing chlorine, fluorine, and carbon. However, they are thought to have a deleterious effect on ozone concentrations in the atmosphere.

Which **industries release** the **most toxic chemicals?**

As of 2001 the metal mining industry released the most toxic chemicals, although overall it's output was less than that for the year 2000. Of the total 6.15 billion lb (2.8 billion kg) of material released on and off site by industry, food production, commonly thought of as a mostly environmentally benign industry, was responsible for 2 percent (125 million lb or 57 million kg), or about eight times as much as the coal mining industry.

What are the **components** of **smog?**

Smog, the most widespread pollutant in the United States, causes a photochemical reaction that produces ground-level ozone, an odorless and tasteless gas. Ozone, in the presence of light, can initiate a chain of chemical reactions. However, while ozone is desirable in the stratospheric layer of the atmosphere (defending Earth's surface from radiation), it can be hazardous to health when found near Earth's surface in the troposphere. The hydrocarbons, hydrocarbon derivations, and nitric oxides emitted from such sources as automobiles are the raw materials for these photochemical reactions. In the presence of oxygen and sunlight, the nitric oxides combine with organic compounds such as the hydrocarbons from unburned gasoline to produce a whitish haze, sometimes tinged with a yellow-brown color. In the process, a large number of new hydrocarbons and oxyhydrocarbons are produced. These secondary hydrocarbon products may compose as much as 95 percent of the total organics in a severe smog episode.

Industry	Total on site and off site releases (lb)	Percent of total
Metal mining	2,782,554,257	45.2
Electric utilities	1,062,247,281	17.2
Hazardous waste/solvent Recovery	219,872,011	3.5
Coal mining	16,117,710	0.3
Petroleum terminals/bulk storage	21,340,651	0.3
Chemical wholesale distributors	1,468,270	0.02

How are **hazardous waste materials classified?**

There are four types of hazardous waste materials: corrosive, ignitable, reactive, and toxic.

- *Corrosive* materials can wear away or destroy a substance. Most acids are corrosive and can destroy metal, burn skin, and give off vapors that burn the eyes.

- *Ignitable* materials can burst into flames easily. These materials pose a fire hazard and can irritate the skin, eyes, and lungs. Gasoline, paint, and furniture polish are ignitable.

- *Reactive* materials can explode or create poisonous gas when combined with other chemicals. Combining chlorine bleach and ammonia, for example, creates a poisonous gas.

- *Toxic* materials or substances can poison humans and other life. They can cause illness or death if swallowed or absorbed through the skin. Pesticides and household cleaning agents are toxic.

What is **biomagnification?**

Some compounds are not recycled by decomposers, nor are they released into the atmosphere like energy. Instead, they remain in the ecosystem in virtually unchanged form as they are passed from one organism to another by predation. If a larger fish con-

Rachel Carson is often credited with launching the modern environmental awareness movement with her 1962 publication, *Silent Spring*.

sumes five smaller ones every day for several years, some of the compounds in the flesh of those little fishes will be transferred to the larger fish as it builds and repairs its own structures. Over time, the larger fish will accumulate many units of such compounds. An example of these compounds is the pesticide DDT. The toxic effects of DDT may not be apparent in the small concentrations found in the little fishes, but the accumulation over time in the larger fish will allow the effects to be magnified. This may become even more apparent as the chemicals move up the trophic pyramid to the top predators like the birds (or humans) that eat those larger fishes. To describe this phenomenon, ecologists use the terms *biomagnification* and *bioaccumulation* in recognition of the disproportional effect these toxins have on the upper levels of the ecological pyramid.

Who was **Rachel Carson**?

Rachel Carson (1907–1964) was one of the first to describe to the general public the consequences of chemical contamination in the environment. In her book *Silent Spring*, published in 1962, Carson exposed the dangers of hydrocarbons, particularly DDT, to the reproduction of species that prey upon the insects for whom the pesticide was intended.

What is meant by **ecoterrorism**?

Ecoterrorism is the term used to describe actions taken by individuals or organizations to prevent what may loosely be termed as "environmental change." This change may be seen as the clear-cutting of forests for wood products or land for housing or the use

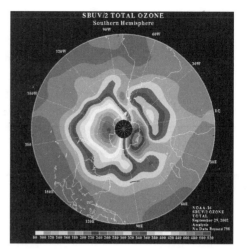

of genetically modified plants or animals for human consumption. Ecoterrorists are those who are willing to take violent and potentially harmful action in order to prevent these types of changes. Beginning in the 1980s industrial sabotage such as tree-spiking (the process of inserting metal spikes into trees so that they cannot be cut down by chainsaws) was used to prevent logging. However, tree-spiking can seriously injure the loggers who are cutting down such trees. Ecoterrorists have also used arson as a tactic. In 1998 ecoterrorists burned down a major portion of a new ski resort in Colorado, which resulted in $12 million in damages.

A satellite image of the Antarctic ozone hole from 2002. This image showed that the hole had split into two distinct sections.

What is a **"green"** building?

It is estimated that just building the typical wood-framed house can use enough wood to denude an entire acre of forest while generating up to 7 tons of waste material. Green building involves an integrated approach to design and construction that minimizes resource waste while emphasizing energy conservation and efficiency. Some additional approaches to green building involve solar water-heating systems and more efficient heating and cooling methods.

How does **ozone benefit life** on Earth?

Ozone, a form of oxygen with three atoms instead of the normal two, is highly toxic; less than one part per million of this blue-tinged gas is poisonous to humans. In Earth's upper atmosphere (stratosphere), ozone is a major factor in making life on Earth possible. About 90 percent of the planet's ozone is in the ozone layer. The ozone belt shields and filters Earth from excessive ultraviolet (UV) radiation generated by the Sun. Scientists predict that a diminished or depleted ozone layer could lead to increased health problems for humans, such as skin cancer, cataracts, and weakened immune systems. Increased UV can also lead to the reduced crop yield and disruption of aquatic ecosystems including the marine food chain. While beneficial in the stratosphere, near ground level ozone is a pollutant that helps form photochemical smog and acid rain.

What is the **greenhouse effect**?

The greenhouse effect is a warming near Earth's surface that results when Earth's atmosphere traps the Sun's heat. The atmosphere acts much like the glass walls and roof of a greenhouse. This effect was first described by John Tyndall (1820–1893) in 1861. It was given the greenhouse analogy much later, in 1896 by the Swedish chemist Svante Arrhenius (1859–1927). The greenhouse effect is what makes Earth habitable. Without the presence of water vapor, carbon dioxide, and other gases in the atmosphere, too much heat would escape, and Earth would be too cold to sustain life. Carbon dioxide, methane, nitrous oxide, and other greenhouse gases absorb the infrared radiation rising from Earth and hold this heat in the atmosphere instead of reflecting it back into space.

Various independent historical measurements conclude that the global average near surface temperature has increased by about 33°F (0.5°C) over the past 100 years. In the twentieth century the increased build-up of carbon dioxide, caused by the burning of fossil fuels, has been linked to this increase. There is some controversy concerning whether the increase noted in Earth's average temperature is due to the increased amount of carbon dioxide and other gases or is due to other causes. In addition to the burning of fossil fuels, volcanic activity, destruction of the rain forests, use of aerosols, and increased agricultural activity may also be contributing factors.

What is **zero population growth**?

Zero population growth, or ZPG, is the estimation of the birth rate necessary to maintain the size of the human population at its current level. As of now, the rate is estimated as 2.1, which means that each set of existing parents would need to have (on average) slightly more than two children during their lifetime. The extra 0.1 allows for infant mortality.

What was the **distribution** of **radioactive fallout** after the **1986 Chernobyl accident**?

On April 25–26, 1986, the world's worst nuclear power accident occurred at Chernobyl in the former USSR (now Ukraine). While scientists were testing one of the four reactors at the Chernobyl nuclear power plant, located 80 mi (129 km) north of Kiev,

An atmosphere with natural levels of greenhouse gases (left) compared with an atmosphere of increased greenhouse effect (right).

an unusual chain reaction occurred in the reactor. This subsequently led to explosions and a fireball that blew the heavy steel and concrete lid of the reactor. Radioactive fallout, containing the isotope cesium 137, and nuclear contamination covered an enormous area, including Byelorussia, Latvia, Lithuania, the central portion of what was then the Soviet Union, the Scandinavian countries, the Ukraine, Poland, Austria, Czechoslovakia, Germany, Switzerland, northern Italy, eastern France, Romania, Bulgaria, Greece, Yugoslavia, the Netherlands, and the United Kingdom. The fallout, extremely uneven because of the shifting wind patterns, extended 1,200 to 1,300 mi (1,930 to 2,090 km) from the point of the accident. The accident led to the release of roughly 5 percent of the reactor fuel, or 7 tons of fuel containing 50 to 100 million curies. Estimates of the effects of this fallout range from 28,000 to 100,000 deaths from cancer and genetic defects within the subsequent fifty years. In particular, livestock in high rainfall areas received lethal dosages of radiation.

What is **acid rain**?

The term *acid rain* was coined by the British chemist Robert Angus Smith (1817–1884), who in 1872 published *Air and Rain: The Beginnings of a Chemical Climatology*. Since then acid rain has unfortunately become an increasingly used term for rain, snow, sleet, or other precipitation that has been polluted by acids such as sulfuric and nitric acids. When gasoline, coal, or oil are burned, their waste products of sulfur dioxide and nitrogen dioxide combine in complex chemical reactions with water vapor in clouds to form acids. The United States alone discharges 40 million metric tons of sulfur and nitrogen oxides into the atmosphere. This, combined with natural emissions of sulfur and nitrogen compounds, has resulted in severe ecological damage. Hundreds of lakes in North America (especially in northeastern

Canada and the United States) and in Scandinavia are so acidic that they cannot support fish life. Crops, forests, and building materials such as marble, limestone, sandstone, and bronze have been affected as well, but the extent is not as well documented as it is with fish life. However, in Europe, where many trees are stunted or have been killed, a new word—*Waldsterben* ("forest death")—has been coined to describe this phenomenon.

In 1990 amendments to the U.S. Clean Air Act contained provisions to control emissions that cause acid rain. The amendments called for the reductions of sulfur dioxide emissions from 19 million tons to 9.1 million tons annually and the reduction of industrial nitrogen oxide emissions from 6 to 4 million tons annually.

Year	Sulfur dioxide emissions (million tons)	Nitrogen oxide emissions (million tons)
1990	15.73	6.66
1995	11.87	6.09
1996	12.51	5.91
1997	12.96	6.04
1998	13.13	5.97
1999	12.45	5.49
2000	11.28	5.11
2001	N/A	4.7
2002	11.28	N/A

Was the Exxon *Valdez* spill the **largest oil spill** of the **twentieth century**?

Although the Exxon *Valdez* was widely publicized as a major spill of 255,500 barrels (35,000 tons) in 1989, it was not the largest of the century. The first major commercial oil spill occurred on March 18, 1967, when the tanker *Torrey Canyon* grounded on the Seven Stones Shoal off the coast of Cornwall, England, spilling 830,000 barrels (119,000 tons) of Kuwaiti oil into the sea. This was the first major tanker accident. However, during World War II German U-boat attacks on tankers between January and June 1942 spilled 4.3 million barrels (590,000 tons) of oil off the East Coast of the United States. Even this spill is dwarfed by the deliberate dumping of oil from Sea Island into the Persian Gulf during the first Gulf War in 1991. It is estimated that the Sea Island spill equaled almost 10.9 million barrels (1.5 million tons) of oil. A major spill also occurred in Russia in October 1994 in the Komi region of the Arctic. The size of the spill was reported to be as much as 2 million barrels (286,000 tons). In addition to the large disasters, day-to-day pollution occurs from drilling platforms where waste generated from platform life, including human waste, and oils, chemicals, mud, and rock from drilling are discharged into the water.

Date	Cause	Thousands tons spilled
1942	German U-boat attacks on tankers off the East Coast of U.S. during World War II	590
1967	Tanker *Torrey Canyon* grounds off Land's End in the English Channel	119
1970	Tanker *Othello* collides with another ship in Tralhavet Bay, Sweden	60–100
1972	Tanker *Sea Star* collides with another ship in Gulf Gulf of Oman	115
1976	*Urquiola* grounds at La Coruña, Spain	100
1978	Tanker Amoco *Cadiz* grounds off northwest France	223
1979	Itox I oil well blows in southern Gulf of Mexico	600
1979	Tankers *Atlantic Express* and *Aegean Captain* collide off Trinidad and Tobago	300
1983	Blowout in Norwuz oil field in the Persian Gulf	600
1983	Fire aboard *Castillo de Beliver* off Cape Town, South Africa	250
1989	Exxon *Valdez* runs aground in Prince William Sound, Alaska	35
1991	Iraq begins deliberately dumping oil into Persian Gulf from Sea Island, Kuwait	1,450
1994	A structure to prevent pipeline leaks fails, spilling oil in the Komi Republic in northern Russia	almost 102,000
1999	Tanker *New Carissa* spills some of its oil in Coos Bay, Oregon	70,000 gallons (238 tons)

What are **Operation Ranch Hand** and **Agent Orange**?

Operation Ranch Hand was the tactical military project for the aerial spraying of herbicides in South Vietnam during the Vietnam Conflict (1961–1975). In these operations, Agent Orange, the collective name for the herbicides 2,4-D and 2,4,5-T, was used for the defoliation. The name derives from the color-coded drums in which the herbicides were stored. In all, U.S. troops sprayed approximately 19 million gallons (72 million liters) of herbicides over 4 million acres (1.6 million hectare). Concerns about the health effects of Agent Orange were initially voiced in 1970, and since then the issue has been complicated by scientific and political debate. In 1993 a sixteen-member panel of experts reviewed the existing scientific evidence and found strong evidence of a statistical association between herbicides and soft-tissue sarcoma, non-Hodgkin's lymphoma, Hodgkin's disease, and chloracne. On the other hand, the panel concluded that no connection appeared to exist between exposure to Agent Orange and skin cancer, bladder cancer, brain tumors, or stomach cancer.

What is **indoor air pollution**, and how is it caused?

Indoor air pollution, also known as "tight building syndrome," results from conditions in modern, high energy-efficiency buildings, which have reduced outside air exchange, or have inadequate ventilation along with chemical contamination and microbial contamination. Indoor air pollution can produce various symptoms, such as headaches, nausea, and eye, nose, and throat irritation. In addition, houses are affected by indoor air pollution emanating from consumer and building products and from tobacco smoke. Below are some pollutants found in houses:

Pollutant	Sources	Effects
Asbestos	Old or damaged insulation, fireproofing, or acoustical tile	Many years later, chest and abdominal cancers and lung diseases
Biological pollutants	Bacteria, mold and mildew, viruses, animal dander and cat saliva, mites, cockroaches, and pollen	Eye, nose, and throat irritation; shortness of breath; dizziness; lethargy; fever; digestive problems; asthma; influenza and other infectious diseases
Carbon monoxide	Unvented kerosene and gas heaters; leaking chimneys and furnaces; wood stoves and fireplaces; gas stoves; automobile exhaust from attached garages; tobacco smoke	At low levels, fatigue; at higher levels, impaired vision and coordination; headaches; dizziness; confusion; nausea. Fatal at very high concentrations
Formaldehyde	Plywood, wall paneling, particle board, and fiber-board; foam insulation; fire and tobacco smoke; textiles and glues	Eye, nose, and throat irritations; wheezing and coughing; fatigue; skin rash; severe allergic reactions; may cause cancer
Lead	Automobile exhaust; sanding or burning of lead paint; soldering	Impaired mental and physical development in children; decreased coordination and mental abilities; kidneys, nervous system and red blood cell damage

Pollutant	Sources	Effects
Mercury	Some latex paints	Vapors can cause kidney damage; long-term exposure can cause brain damage
Nitrogen dioxide	Kerosene heaters and unvented gas stoves and heaters; tobacco	Eye, nose, and throat irritation; may impair lung function and increase respiratory infections in smoke young children
Organic gases	Paints, paint strippers, solvents, and wood preservatives; aerosol sprays; cleansers and disinfectants; moth repellents; air fresheners; stored fuels; hobby supplies dry-cleaned clothing	Eye, nose and throat irritation; headaches; loss of coordination; nausea; damage to liver, kidney, and nervous system; some organics cause cancer in animals and are suspected of causing cancer in humans
Pesticides	Products used to kill household pests and products used on lawns or gardens that drift or are tracked inside the house	Irritation to eye, nose, and throat; damage to nervous systems and kidneys; cancer
Radon	Earth and rock beneath the home; well water, building materials	No immediate symptoms; estimated to cause about 10 percent of lung cancer deaths; smokers at higher risk

What is the **Superfund Act**?

In 1980 the United States Congress passed the Comprehensive Environmental Response, Compensation, and Liability Act, commonly known as the Superfund program. This law (along with amendments in 1986 and 1990) established a $16.3-billion Superfund financed jointly by federal and state governments and by special taxes on chemical and petrochemical industries (which provide 86 percent of the funding). The purpose of the Superfund is to identify and clean up abandoned hazardous-waste dump sites and leaking underground tanks that threaten human health and the environment. To keep taxpayers from footing most of the bill, cleanups are based on the "polluter-pays principle." The EPA is charged with locating dangerous dump sites, finding the potentially liable culprits, ordering them to pay for the entire cleanup, and suing them if they don't. When the EPA can find no responsible party, it draws money out of the Superfund for cleanup.

What is the **NIMBY** syndrome?

NIMBY is the acronym for "Not In My Back Yard." It refers to major community resistance to construction of new incinerators, landfills, prisons, roads, and so forth. NIMFY is "Not In My Front Yard."

How is **nuclear waste stored** and **regulated**?

Nuclear wastes consist either of fission products formed from atom splitting of uranium, cesium, strontium, or krypton, or from transuranic elements formed when uranium atoms absorb free neutrons. Wastes from transuranic elements are less radioactive than fission products; however, these elements remain radioactive far longer than fission products. Transuranic wastes include irradiated fuel (spent fuel) in the form of 12-ft (4-m) long rods, high-level radioactive waste in the form of liquid or sludge, and low-level waste (nontransuranic or legally high-level) in the form of reactor hardware, piping, toxic resins, water from fuel pools, and other items that have become contaminated with radioactivity.

Currently, most spent nuclear fuel in the United States is safely stored in specially designed pools at individual reactor sites around the country. If pool capacity is reached, licensees may move toward use of above-ground dry storage casks. The three low-level radioactive waste disposal sites are Barnwell, South Carolina; Hanford, Washington; and Envirocare, Utah. Each site accepts low-level radioactive waste from specific regions of the country, but only Envirocare uses above-ground storage.

Most high-level nuclear waste has been stored in double-walled stainless-steel tanks surrounded by 3 ft (1 m) of concrete. The current best storage method, developed by the French in 1978, is to incorporate the waste into a special molten glass mixture, then enclose it in a steel container and bury it in a special pit. The Nuclear Waste Policy Act of 1982, as amended in 1987, specified that high-level radioactive waste would be disposed of underground in a deep geologic repository. Yucca Mountain, Nevada, was chosen as the single site to be developed for disposal of high-level radioactive waste. On July 23, 2002, President George W. Bush signed House Joint Resolution 87, allowing the Department of Energy to establish a repository in Yucca Mountain to safely store nuclear waste. However, some scientists still expressed concerns about the estimates of how long it would take for rainwater and snow to infiltrate the mountain and corrode the containers.

How much **garbage** does the **average American generate**?

According to the Environmental Protection Agency, nearly 232 million tons of municipal waste was generated in 2000. This is equivalent to 4.6 lb (2.1 kg) per person per day, or approximately 1,700 lb (770 kg) per person per year. The total amount of waste is distributed as follows:

Waste Product	Percent of total
Paper and paperboard	38.1
Glass	5.5
Metals	7.8
Plastics	10.5

When was metal recycling started?

The first metal recycling in the United States occurred in 1776 when patriots in New York City toppled a statue of King George III which was melted down and made into 42,088 bullets.

Waste Product	Percent of Total
Rubber and leather	2.7
Textiles	3.9
Wood	5.3
Food wastes	10.9
Yard wastes	12.1
Other wastes	3.2

How **critical** is the problem of **landfills** in the United States?

Landfilling has been an essential component of waste management for several decades. In 1960, 62 percent of all garbage was sent to landfills, and by 1980 the figure had risen to 81 percent. By 1990, 84 percent of the 269 million tons of municipal solid waste that was generated was sent to landfills. An increased awareness of the benefits of recycling has brought a decline in the actual number of landfills from 4,482 in 1995 to 2,142 in 2000 as well as a decrease in the amount of municipal solid waste that is sent to landfills. Figures for 2000 indicate that only 60 percent of the municipal solid waste generated was sent to landfills. The total amount of recycled waste increased from 8 percent to 33 percent between 1990 and 2000.

How much **newspaper** must be **recycled** to **save one tree**?

One 35 to 40 ft (10.6 to 12 m) tree produces a stack of newspapers 4 ft (1.2 m) thick; this much newspaper must be recycled to save a tree.

How can **plastics** be made **biodegradable**?

Plastic neither rusts nor rots. This is an advantage in its usage, but when it comes to disposal of plastic, the advantage turns into a liability. Degradable plastic has starch in it so that it can be attacked by starch-eating bacteria to eventually disintegrate the plastic into bits. Chemically degradable plastic can be broken up with a chemical solution that dissolves it. Used in surgery, biodegradable plastic stitches slowly dissolve in the body fluids. Photodegradable plastic contains chemicals that disintegrate over a period of one to three years when exposed to light. Twenty-five percent of the plastic yokes used to package beverages are made from a plastic called Ecolyte, which is photodegradable.

509

What do the **numbers** inside the **recycling symbol** on **plastic containers** mean?

The Society of the Plastics Industry developed a voluntary coding system for plastic containers to assist recyclers in sorting plastic containers. The symbol is designed to be imprinted on the bottom of the plastic containers. The numerical code appears inside a three-sided triangular arrow. A guide to what the numbers mean is listed below. The most commonly recycled plastics are polyethylene terephthalate (PET) and high-density polyethylene (HDPE).

Code	Material	Examples
1	Polyethylene terephthalate (PET/PETE)	2-liter soft drink bottle
2	High-density polyethylene (HDPE)	Milk and water jugs
3	Vinyl	(PVC) Plastic pipes, shampoo bottles
4	Low-density polyethylene (LDPE)	Produce bags, food storage containers
5	Polypropylene (PP)	Squeeze bottles, drinking straws
6	Polystyrene (PS)	Food packaging

What **products are made from** recycling plastic?

Resin	Common uses	Products made from recycled resin
HDPE	Beverage bottles, milk jugs, milk and soft drink crates, pipes and pails pipe, cable, film	Motor oil bottles, detergent bottles
LDPE	Film bags such as trash bags, coatings, and	New trash bags, pallets, carpets, fiberfill, nonfood containers plastic bottles
PET	Soft drink, detergent, and juice bottles	Bottles and containers
PP	Auto battery cases, screw on caps and lids, some yogurt and margarine tubs, plastic film	Auto parts, batteries, carpets
PS	Housewares, electronics, fast food carry-out packaging, plastic utensils	Insulation board, office equipment, reusable cafeteria trays
PVC	Sporting goods, luggage, pipes, auto parts; packaging for shampoo bottles, blister products, and films	Drainage pipes, fencing, house siding

A worker stacks recycled cans at a recycling plant in Brazil.

A new clothing fiber called Fortrel EcoSpun is made from recycled plastic soda bottles. The fiber is knit or woven into garments such as fleece for outerwear or long underwear. The processor estimates that every pound of Fortrel EcoSpun fiber results in ten plastic bottles being kept out of landfills.

When offered a choice between **plastic** or **paper bags** for your **groceries**, which should you choose?

The answer is neither. Both are environmentally harmful, and the question of which is the more damaging has no clear-cut answer. On one hand, plastic bags degrade slowly in landfills and can harm wildlife if swallowed, and producing them pollutes the environment. On the other hand, producing the brown paper bags used in most supermarkets uses trees and pollutes the air and water. Overall, white or clear polyethylene bags require less energy for manufacture and cause less damage to the environment than do paper bags not made from recycled paper. Instead of having to choose between paper and plastic bags, you can bring your own reusable canvas or string containers to the store, and you can save and reuse any paper or plastic bags you get.

What is the **Kyoto Protocol**?

The Kyoto Protocol was an international summit held in Kyoto, Japan, in December 1997. Its goal was for governments around the world to reach an agreement regarding emissions of carbon dioxide and other greenhouse gases. The Kyoto Protocol called for the industrialized nations to reduce national emissions over the period 2008–2012 to 5 percent below the 1990 levels. The protocol covers these greenhouse gases: carbon dioxide, methane, and nitrous oxide. Other chemicals such as hydrofluorocarbons, perfluorocarbons, and sulfur hexa fluoride were to be added in subsequent years.

What is **biopiracy?**

Biopiracy refers to the development of pharmaceutical products without compensation to the native communities that protected and nurtured the organisms on which these products are based.

What is a **bioinvader**?

A bioinvader is an exotic organism usually introduced into an ecosystem accidentally. These bioinvaders are nonnative plants and often overwhelm the native species. Examples of bioinvaders include the kudzu vine. Kudzu was first introduced in the 1930s by the United States Soil Conservation Service for a good purpose- to control erosion. Kudzu now grows uncontrolled in the southeastern United States, pulling down powerlines and killing trees. Other bioinvader species include zebra mussels (Great Lakes), purple loosestrife (northern United States and Canada), and the Asian long-horned beetle (first reported in New York but now spreading into the Midwest).

LABORATORY TOOLS AND TECHNIQUES

SCIENTIFIC METHOD

What is the **scientific method**?

The scientific method is the basis of scientific investigation. A scientist will pose a question and formulate a hypothesis as a potential explanation or answer to the question. The hypothesis will be tested through a series of experiments. The results of the experiments will either prove or disprove the hypothesis. Hypotheses that are consistent with available data are conditionally accepted.

What are the **steps** of the **scientific method**?

Research scientists follow these steps:

1) State a hypothesis.

2) Design an experiment to "prove" the hypothesis.

3) Assemble the materials and set up the experiment.

4) Do the experiment and collect data.

5) Analyze the data using quantitative methods.

6) Draw conclusions.

7) Write up and publish the results.

What is a **variable**?

A variable is something that is changed or altered in an experiment. For example, to determine the effect of light on plant growth, growing one plant in a sunny window and one in a dark closet will provide evidence as to the effect of light on plant growth. 513

How does an **independent variable** differ from a **dependent variable**?

An independent variable is manipulated and controlled by the researcher. A dependent variable is the variable that the researcher watches and/or measures. It is called a dependent variable because it depends upon and is affected by the independent variable.

What is a **control group**?

A control group is the experimental group tested without changing the variable. For example, to determine the effect of temperature on seed germination, one group of seeds may be heated to a certain temperature. The researcher will then compare the percent of seeds in this group that germinate and the time it takes them to germinate to another group of seeds (the control group) that have not been heated. All other variables, such as light and water, will remain the same for each group.

What is a **double-blind study**?

In a double-blind study, neither the subjects of the experiment nor the persons administering the experiment know the critical aspects of the experiment. This method is used to guard against both experimenter bias and placebo effects.

How does **deductive reasoning** differ from **inductive reasoning**?

Deductive reasoning, often used in mathematics and philosophy, uses general principles to examine specific cases. Inductive reasoning is the method of discovering general principles by close examination of specific cases. Inductive reasoning first became important to science in the 1600s, when Francis Bacon (1561–1626), Sir Isaac Newton (1642–1727), and their contemporaries began to use the results of specific experiments to infer general scientific principles.

How does an **in vivo** study differ from an **in vitro** study?

An in vivo study uses living biological organisms and specimens. In contrast, an in vitro biological study is carried out in isolation from a living organism, such as in a petri dish or test tube.

BASIC LABORATORY CHEMISTRY

Why are **dilution techniques** important to biologists?

Dilution techniques provide simple and accurate procedures to: 1) change the concentration of a solution; 2) indirectly "weigh" a solute whose weight is significantly below the usual limits of analytical balances; and 3) determine the quantity of bacteria in a culture.

What food can be used to determine whether a solution is acidic or basic?

Red cabbage may be used to determine whether a solution is acidic or basic. Red cabbage contains a pigment called flavin (an anthocyanin). This water-soluble pigment is also found in apple skin, plums, poppies, cornflowers, and grapes. Very acidic solutions will turn anthocyanin a red color. Neutral solutions result in a purplish color. Basic solutions appear in greenish-yellow. Therefore, it is possible to determine the pH of a solution based on the color it turns the anthocyanin pigments in red cabbage juice.

To prepare a solution of red cabbage juice indicator, chop some red cabbage into small pieces and cover them with boiling water. Allow the mixture to sit for approximately ten minutes. The indicator may now be used to test various solutions as to their acidity.

How does one **prepare** a **1:10 dilution**?

A 1:10 dilution means one part in a total of ten. There are three different ways to prepare a 1:10 dilution: 1) the weight-to-weight (w:w) method; 2) the weight-to-volume (w:v) method; and 3) the volume to volume (v:v) method. In the weight-to-weight method, 1.0 g of solute is dissolved in 9.0 g of solvent, yielding a total of ten parts by weight, one of which is solute.

In the weight-to-volume method, enough solvent is added to 1.0 g of solute to make a total volume of 10 ml. In this method, one part (by weight) is dispersed in ten total parts (by volume). Since most biological solutions are very dilute, the accuracy of most research is not affected if a previously weighed solute is dissolved in the desired volume of solvent.

The volume to volume method is preferred when the solute is a liquid. One milliliter of solute, such as ethanol, added to 9.0 ml of water yields a ten-part solution, one part of which is the solute.

How is it possible to **measure** the **pH** of a solution?

An easy way to measure the pH of a solution is with pH paper. This paper is treated with a chemical indicator that changes colors depending on the concentration of H^+ (hydrogen ions) in the solution. Examples of pH indicators are:

Indicator	Range (pH scale)	Color change
Methyl violet	0.2–3.0	Yellow to blue-violet
Bromphenol blue	3.0–4.6	Yellow to blue

Indicator	Range (pH scale)	Color change
Methyl red	4.4–6.2	Red to yellow
Litmus	4.5–8.3	Red to blue
Bromcresol purple	5.2–6.8	Yellow to purple
Phenol red	6.8–8.0	Yellow to red
Thymol blue	8.0–9.6	Yellow to blue
Phenolphthalein	8.3–10.0	Colorless to red

What are some of the **tests** scientists use to **identify** major types of **organic compounds** in living organisms?

Scientists use different tests to detect the presence of carbohydrates, lipids, proteins, and nucleic acids. Commonly used tests include the Benedict's test for reducing sugars, the iodine test for starch, the Biuret test for proteins, the Sudan IV test and the grease-spot test for lipids, and the Dische diphenylamine test for nucleic acids.

What is the **Benedict's test** for reducing sugars?

A commonly used test to detect carbohydrates is the Benedict's test for reducing sugars such as glucose and fructose. Benedict's reagent, containing sodium bicarbonate, sodium citrate, and copper sulfate, is added to a solution and heated. The Benedict's test identifies reducing sugars based on their ability to reduce the cupric ions to cuprous oxide at high pH values (basic solutions). Cuprous oxide is green to reddish orange. A green solution indicates a small amount of reducing sugars, while a reddish-orange solution indicates an abundance of reducing sugars. If the solution contains sucrose, a nonreducing sugar, there is no change in color in the solution, and it remains blue.

How does the **iodine test** detect starch?

Starch is a coiled polymer of glucose. Iodine reacts with the coiled molecules and turns bluish-black when added to a solution. A solution that remains a yellowish-brown color is a negative test for starch, whereas one that turns bluish-black is a positive test for starch.

How does the **Biuret test** indicate the presence of **protein**?

The bond between the amino group and the carboxyl acid group on adjacent amino acids in a protein is a peptide bond. When the Biuret reagent (1 percent solution of copper sulfate) is added to a solution containing peptide bonds, the solution turns a violet color. The violet color is a positive test for the presence of protein. The more intense the color, the greater the number of peptide bonds that react.

What **tests** can be used to determine the presence of **lipids**?

A very simple test to determine the presence of lipids is the grease-spot test. Place a drop of solution on an unglazed, clean sheet of brown paper. Once the liquid evaporates, a spot containing lipids will remain visible. The Sudan IV test also tests for lipids. It is based on the ability of lipids to selectively absorb pigments in fat-soluble dyes.

MEASUREMENT

What is the **SI system** of measurement?

French scientists as far back as the seventeenth and eighteenth centuries questioned the hodgepodge of the many illogical and imprecise standards used for measurement, and they began a crusade to make a comprehensive, logical, precise, and universal measurement system called Système Internationale d'Unités, or SI for short. It uses the metric system as its base. Since all the units are in multiples of 10, calculations are simplified. Today, all countries except the United States, Burma, and Liberia use this system. However, some elements within American society do use SI—scientists, exporting/importing industries, and federal agencies (as of November 30, 1992).

The SI or metric system has seven fundamental standards: the meter (for length), the kilogram (for mass), the second (for time), the ampere (for electric current), the kelvin (for temperature), the candela (for luminous intensity), and the mole (for amount of substance). In addition, two supplementary units—the radian (plane angle) and steradian (solid angle)—and a large number of derived units compose the current system, which is still evolving. Some derived units, which use special names, are the hertz, newton, pascal, joule, watt, coulomb, volt, farad, ohm, siemens, weber, tesla, henry, lumen, lux, becquerel, gray, and sievert. The SI's unit of volume or capacity is the cubic decimeter, but many still use liter in its place. Very large or very small dimensions are expressed through a series of prefixes, which increase or decrease in multiples of ten. For example, a decimeter is 1/10 of a meter, a centimeter is 1/100 of a meter, and a millimeter is 1/1000 of a meter. A dekameter is 10 meters, a hectometer is 100 meters, and a kilometer is 1,000 meters. The use of these prefixes enables the system to express these units in an orderly way and avoid inventing new names and new relationships.

How are U.S. customary measures **converted** to metric measures and vice versa?

Listed below is the conversion process for common units of measure:

To convert from	To	Multiply by
centimeters	inches	0.394
centimeters	feet	0.0328
centimeters, cubic	inches, cubic	0.06

To convert from	To	Multiply by
centimeters, square	inches, square	0.155
feet	meters	0.305
feet, square	meters, square	0.093
grams	ounces (avoirdupois)	0.035
inches	centimeters	2.54
inches	millimeters	25.4
inches, cubic	centimeters, cubic	16.387
inches, cubic	liters	0.016387
inches, cubic	meters, cubic	0.0000164
inches, square	centimeters, square	6.4516
inches, square	meters, square	0.0006452
kilograms	ounces, troy	32.15075
kilograms	pounds (avoirdupois)	2.205
liters	fluid ounces	33.815
liters	pints	2.113
liters	quarts	1.057
meters	feet	3.281
ounces (avoirdupois)	grams	28.35
ounces (avoirdupois)	kilograms	0.0283495
ounces, fluid	liters	0.03
pints, liquid	liters	0.473
pounds (avoirdupois)	grams	453.592
pounds (avoirdupois)	kilograms	0.454
quarts	liters	0.946

What is **scientific notation**?

Scientific notation allows scientists to manipulate very large or very small numbers. It is based on the fact that all numbers can be expressed as the product of two numbers, one of which is the power of the number ten (written as the small superscript next to the number 10 and called the exponent). Positive exponents indicate how many times the number must be multiplied by 10, while negative exponents indicate how many time a number must be divided by 10.

Numbers greater than one (the exponent to which the power of 10 is raised is equal to the number of zeros to the right)	**Numbers less than one** (the exponent to which the of power of 10 is raised is equal to the number of zeros to the left of 1 plus 1)
$1,000,000,000 = 1 \times 10^9$	$0.000000001 = 1 \times 10^{-9}$

Numbers greater than one (the exponent to which the power of 10 is raised is equal to the number of zeros to the right)	Numbers less than one (the exponent to which the of power of 10 is raised is equal to the number of zeros to the left of 1 plus 1)
$100,000,000 = 1 \times 10^8$	$0.00000001 = 1 \times 10^{-8}$
$10,000,000 = 1 \times 10^7$	$0.0000001 = 1 \times 10^{-7}$
$1,000,000 = 1 \times 10^6$	$0.000001 = 1 \times 10^{-6}$
$100,000 = 1 \times 10^5$	$0.00001 = 1 \times 10^{-5}$
$10,000 = 1 \times 10^4$	$0.0001 = 1 \times 10^{-4}$
$1,000 = 1 \times 10^3$	$0.001 = 1 \times 10^{-3}$
$100 = 1 \times 10^2$	$0.01 = 1 \times 10^{-2}$
$10 = 1 \times 10^1$	$0.1 = 1 \times 10^{-1}$
$1 = 1 \times 10^0$	$1 = 1 \times 10^0$

Who invented the thermometer?

The Greeks of Alexandria knew that air expanded as it was heated. Hero of Alexandria (first century C.E.) and Philo of Byzantium (*fl. ca.* 250 B.C.E.) made simple "thermo-scopes," but they were not real thermometers. In 1592 Galileo (1564–1642) made a kind of thermometer that also functioned as a barometer, and in 1612 his friend Santorio Santorio (1561–1636) adapted the air thermometer (a device in which a colored liquid was driven down by the expansion of air) to measure the body's temperature change during illness and recovery. Still, it was not until 1713 that Daniel Fahrenheit (1686–1736) began developing a thermometer with a fixed scale. He worked out his scale from two "fixed" points: the melting point of ice and the heat of the healthy human body. He realized that the melting point of ice was a constant temperature, whereas the freezing point of water varied. Fahrenheit put his thermometer into a mixture of ice, water, and salt (which he marked off at 0°) and, using this as a starting point, marked off melting ice at 32° and blood heat at 96°. In 1835 it was discovered that normal blood measured 98.6°F. Sometimes Fahrenheit used spirit of wine as the liquid in the thermometer tube, but more often he used specially purified mercury. Later, the boiling point of water (212°F) became the upper fixed point of the thermometer.

How did the original Celsius temperature scale differ from the one in use now?

In 1742 the Swedish astronomer Anders Celsius (1701–1744) set the freezing point of water at 100°C and the boiling point of water at 0°C. It was Carolus Linnaeus (1707–1778) who reversed the scale, but a later textbook attributed the modified scale to Celsius, and the name has remained.

How are centigrade temperatures converted into Fahrenheit temperatures?

The formula for converting centigrade temperatures into Fahrenheit is °F = (°C X 9/5) + 32. The formula for converting Fahrenheit temperatures into centigrade is °C = (°F - 32) X 5/9.

Some comparisons between the two scales are:

Temperature	Fahrenheit	Centigrade
Absolute zero	-459.67	-273.15
Point of equality	-40	-40
Zero Fahrenheit	0	-17.8
Freezing point of water	32	0
Normal human blood temperature	98.4	36.9
100°F	100	37.8
Boiling point of water (at standard pressure)	212	100

What is the **Kelvin temperature** scale?

Temperature is the level of heat in a gas, liquid, or solid. The freezing and boiling points of water are used as standard reference levels in both the metric (centigrade) and the English system (Fahrenheit). In the metric system, the difference between freezing and boiling is divided into 100 equal intervals called degree centigrade (°C). In the English system, the intervals are divided into 180 units, with one unit called degree Fahrenheit (°F). But temperature can be measured from absolute zero (no heat, no motion); this principle defines thermodynamic temperature and establishes a method to measure it upward. This scale of temperature is called the Kelvin temperature scale, after its inventor, William Thomson, Lord Kelvin (1824–1907), who devised it in 1848. The Kelvin (symbol K) has the same magnitude as the degree centigrade (the difference between freezing and boiling water is 100 degrees), but the two temperatures differ by 273.15 degrees (absolute zero, which is -273.15 °C on the centigrade scale).

Characteristic	K	°C	°F
Absolute zero	0	-273.15	-459.67
Freezing point of water	273.15	0	32
Normal human body temperature	310.15	37	98.6
Boiling point of water	373.15	100	212

MICROSCOPY

What **elements** are **common** to all types of **microscopes**?

Three elements are needed to form an image: a source of illumination, a specimen to be examined, and a system of lenses that focuses the illumination on the specimen and forms the image.

What distinguishes the different types of **microscopes**?

Microscopes have played a central role in the development of cell biology, allowing scientists to observe cells and cell structures that are not visible to the human eye. The two basic types of microscopes are light microscopes and electron microscopes. The major differences between light and electron microscopes are the source of illumination and the construction of the lenses. Light microscopes utilize visible light as the source of illumination and a series of glass lenses. Electron microscopes utilize a beam of electrons emitted by a heated tungsten filament as the source of illumination. The lens system consists of a series of electromagnets.

Recent advances using optical techniques have led to the development of specialized light microscopes, including fluorescence microscopy, phase-contrast microscopy, and differential interference contrast microscopy. In fluorescence microscopy, a fluorescent dye is introduced to specific molecules. Both phase-contrast microscopy and differential interference contrast microscopy utilize techniques that enhance and amplify slight changes in the phase of transmitted light as it passes through a structure that has a different refractive index than the surrounding medium.

How do **dissecting microscopes** differ from compound microscopes?

Compared to compound microscopes, dissecting microscopes—also called stereoscopic microscopes—provide a much larger working distance between the lens and stage in order to dissect and manipulate specimens. The light source on a dissecting microscope is above the specimen since the specimen is often too thick to allow light to be transmitted from a light source below the specimen. Dissecting microscopes are always binocular, which provides a three-dimensional image.

What is the difference between **magnification** and **resolution**?

Magnification—making smaller objects seem larger—is the measure of how much an object is enlarged. Resolution is the minimum distance that two points can be separated and still be seen as two distinct points.

What is the **resolving power** of various **lens systems**?

Lens system	Resolving power
Human eye	0.1 mm = 100 μm
Light microscope	0.4 μm
Oil immersion light microscope	0.2 μm
Ultraviolet microscope	0.1 μm
Scanning electron microscope	10 nm = 0.010 μm
Transmission electron microscope	0.2 nm = 0.0002 μm

Who invented the **compound microscope**?

The principle of the compound microscope, in which two or more lenses are arranged to form an enlarged image of an object, occurred independently, at about the same time, to more than one person. Certainly many opticians were active in the construction of telescopes at the end of the sixteenth century, especially in Holland, so it is likely that the idea of the microscope may have occurred to several of them independently. In all probability the date may be placed within the period 1590–1609, and the credit should go to three spectacle makers in Holland. Hans Janssen, his son Zacharias (1580–1638), and Hans Lippershey (1570–1619) have all been cited at various times as deserving chief credit. An Englishman, Robert Hooke (1635–1703), was the first to make the best use of a compound microscope, and his book *Micrographia*, published in 1665, contains some of the most beautiful drawings of microscopic observations ever made.

Who invented the **electron microscope**?

The theoretical and practical limits to the use of the optical microscope were set by the wavelength of light. When the oscilloscope was developed, it was realized that cathode-ray beams could be used to resolve much finer detail because their wavelength was so much shorter than that of light. In 1928 Ernst Ruska (1906–1988) and Max Knoll (1897–1969), using magnetic fields to "focus" electrons in a cathode-ray beam, produced a crude instrument that gave a magnification of 17, and by 1932 they had developed an electron microscope having a magnification of 400. By 1937 James Hillier (1915–) had advanced this magnification to 7,000. The 1939 instrument Vladimir Zworykin (1889–1982) developed gave 50 times more detail than any optical microscope ever could, with a magnification up to 2 million. The electron microscope revolutionized biological research: for the first time scientists could see the molecules of cell structures, proteins, and viruses.

How does a **transmission electron microscope** differ from a **scanning electron microscope**?

The electrons used to visualize the specimens in transmission electron microscopes are transmitted by the material. The scanning electron microscope beams the electrons onto the surface of the specimen from a fine probe that passes back and forth rapidly. Electrons reflected back from the surface of the specimen, along with other electrons emitted by the specimen itself, are amplified and transmitted to a television screen for viewing.

What is **scanning tunneling microscopy**?

Scanning tunneling microscopy (STM), also called a scanning probe microscopy, was developed in the 1980s to explore the surface structure of specimens at the atomic

What tool was used to take the first photograph of a DNA molecule?

The first photograph of a DNA molecule was made using a scanning tunneling microscope. It showed the two strands of DNA magnified a million times! The DNA molecule is so slender that it would take 50,000 of them to equal the diameter of a human hair.

level. This technique uses electronic methods to move a metallic tip (a conducting material such as platinum-iridium), composed ideally of a single atom, across the surface of a specimen. As the tip is moved across the surface of the specimen, electrical voltage is applied to the surface. If the tip is close enough to the surface and the surface is electrically conductive, electrons will begin to leak or "tunnel" across the gap between the probe and the sample. The tip of the probe is automatically moved up and down to maintain a constant rate of electron tunneling across the gap as the probe scans the sample. The movement is presented on a video screen. Successive scans then build up an image of the surface at atomic resolution.

What is **microscopic autoradiography**?

Microscopic autoradiography is a technique used to localize radioactive molecules within cells. It utilizes photographic emulsion to determine where a specific radioactive compound is located within a cell at the time the cell is fixed and sectioned for microscopy.

Which **radioisotope** is **most commonly used** for biological specimens?

The most widely used radioisotope in autoradiography is tritium (^{3}H). Tritium allows a resolution of about 1 micrometer with the light microscope and close to 0.1 micrometer with the electron microscope. Since hydrogen is common in biological molecules, a wide-range of ^{3}H-labeled compounds are potentially available for use in autoradiography. ^{3}H-amino acids are used for locating newly synthesized proteins, ^{3}H-thymidine is used to monitor DNA synthesis, ribonucleiodies such as ^{3}H-uridine or ^{3}H-cytidine are used to localize newly made RNA molecules, and ^{3}H-glucose is used to study the synthesis of polysaccharides.

What are the **steps** of **microscopic autoradiography**?

In microscopic autoradiography, the desired radioactively labeled compound is first added to cells or organisms, and the material is then incubated. After a period of time to allow the radioactive compound to become incorporated into newly forming intracellular molecules and structures, the incubation is stopped, and the biological specimen is rinsed to wash away the excess radioactive compound. The specimen is prepared in a conventional manner (sectioned and placed on a microscopic slide). The slide is then covered with a thin layer of photographic emulsion. Next, the prepared slide is placed in a sealed box for an appropriate period of time (several days to several weeks) to allow the radioactivity in the cell to expose the emulsion above it. Upon removal from the sealed box, the emulsion is developed, and the specimen is examined under the microscope.

MICROTECHNIQUE

Who is considered the **first histologist**?

Marcello Malpighi (1628–1694) is considered the first histologist for his pioneering work in the science of microscopic anatomy. He used the newly developed tool of the time, the microscope, to examine living things including both plants and animals. His observations included that blood passed through the capillaries and extensive work on insect larvae.

What are the **common types** of **slide preparations** for investigation with a microscope?

Commonly prepared slide preparations are wholemounts, smears, squashes, and sections. Smears, squashes, and sections are techniques used to make specimens thinner or smaller. Wholemounts are often used to examine an entire organism or specific organ structure in some detail. Smears are mostly prepared for bacteriological and blood specimens. Squashes are prepared to study chromosomes. Sections are prepared to examine tissues and cells.

When was the **microtome invented**?

Although there were cutting engines as early as the eighteenth century, the first microtome was invented by Wilhlem His (1831–1904) in 1870. Its development allowed scientists to prepare very thin, uniformly sized slices of tissue for examination under a microscope instead of the imprecise ones previously prepared using a hand-held razor or cutting engines.

What are the **three main types** of **microtomes**?

The three main types of microtomes are the rocking, the rotary, and the sliding microtome. Each type of microtome has a special sharp steel knife to cut the specimen, which is embedded in a wax block. In a rocking microtome the knife is in a fixed horizontal position. The wax block is attached to the end of an arm pivoted near the knife

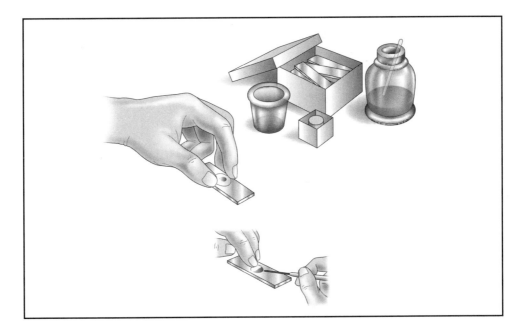

Wholemounting is a common slide preparation technique. Wholemounts are often used to examine an entire organism or specific organ structure in some detail.

and is moved or rocked in an arc past the knife edge. On a rotary microtome the specimen moves up and down in a vertical plane. A large hand wheel in which one rotation produces a complete cutting cycle advances the specimen. A common type of sliding microtome has the specimen mounted on a moving carriage while the knife is fixed.

How does an **ultramicrotome** differ from a standard microtome?

The ultramicrotome, developed after 1950 to prepare specimens for examination under the electron microscope, enables technicians to cut very thin sections (50–150 nanometers). Instead of embedding specimens in wax, small (0.5–1.0 cubic mm) biological specimens are embedded in very hard synthetic resins such as epoxy. In order to cut these very hard materials, special knives—sharper than steel knives—are used. The knives are either diamond knives or pieces of plate glass broken in a controlled way to produce a fine edge.

What is the **optimum thickness** for specimens to be examined under a microscope?

Biological and medical techniques require specimens of 1 to 50 micrometers, with the usual thickness being 4 to 5 micrometers for examination under a light microscope. Since an electron microscope has greater resolution, it requires thinner sections of 20 to 100 nanometers for biological specimens.

525

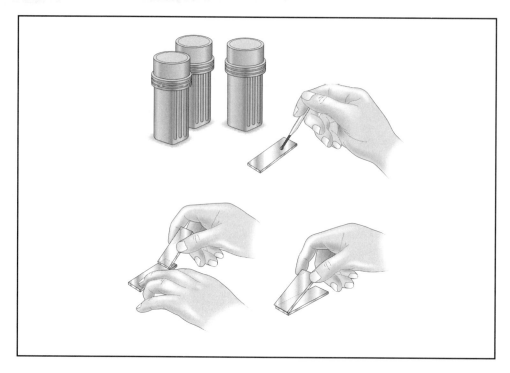

Smear preparation is another standard technique used to prepare slides.

What are the **steps** to **prepare a specimen** for examination?

The three basic steps to prepare a specimen are fixation (preservation), staining, and mounting. Preservation prevents destruction or decay of the specimen as well as inhibiting microbiological growth. Different stains and dyes attach to different parts of a cell, such as the nucleus. Specimens are mounted in a medium that is also a preservative and often covered with a cover slip.

Why is it necessary to "**fix**" a biological specimen?

"Fixing" a biological specimen retains a reasonably good semblance of the object as it appeared when it was alive. It allows the scientists to observe details of the external and internal anatomy of the specimen.

What are **simple stains**?

Simple stains highlight an entire microorganism so that cellular shapes and basic structures are visible. Simple stains commonly used include methylene blue, carbolfuchsin, crystal violet, and safranin. A stain is applied to a fixed smear for a certain amount of time and then washed off, and the slide is dried and examined.

What is the purpose of a **mordant**?

A mordant is a chemical added to the solution used to stain a specimen in order to intensify the stain. Two major functions of a mordant are to increase the affinity of a stain for a biological specimen and to coat a structure to make it thicker and easier to see upon observation under the microscope.

What are the steps of the **Gram stain**?

Crystal violet, a primary stain (one that imparts its color to all cells), is applied to a heat-fixed smear. After a short time the crystal violet is washed off, and the smear is covered with iodine, a mordant. When the iodine is washed off, the bacteria appear dark violet or purple. The slide is then washed in an alcohol or alcohol-acetone solution. This decolorizing agent removes the purple from the cell of some species but not from others. The alcohol is rinsed off, and the slide is then stained with safranin, a basic red dye. The smear is then washed again, blotted dry, and examined under a microscope. Gram-positive bacteria retain the purple dye, while those that lose the purple color are classified as gram-negative.

What was Camillo **Golgi's** (1843–1926) contribution to **histology**?

In 1873 Camillo Golgi (1843–1926) devised a way to stain tissue samples with inorganic dye using silver salts. When he applied this technique and stained nerve tissue, he was able to see details previously not visible.

CENTRIFUGATION

What is **centrifugation**?

Centrifugation is the separation of immiscible liquids or solids from liquids by applying centrifugal force. Since the centrifugal force can be very great, it speeds the process of separating these liquids instead of relying on gravity.

How has **centrifugation** been used in **biological applications**?

Biologists primarily use centrifugation to isolate and determine the biological properties and functions of subcellular organelles and large molecules. They study the effects of centrifugal forces on cells, developing embryos, and protozoa. These techniques have allowed scientists to determine certain properties about cells, including surface tension, relative viscosity of the cytoplasm, and the spatial and functional interrelationship of cell organelles when redistributed in intact cells.

Who first developed the **modern technique** for the **isolation** of **cell parts**?

Robert R. Bensley (1867–1956) and Normand Louis Hoerr (1902–1958) disrupted the liver cells in a guinea pig and isolated mitochondria in 1934. Between 1938 and 1946 Albert Claude (1899–1983) continued the work of Bensley and Hoerr and isolated two fractions—a heavier fraction consisting of mitochondria and another fraction of lighter submicroscopic granules, which he called microsomes. Further developments led to the development of centrifugal techniques of cell fractionation commonly used now. The development of this procedure was one of the earliest examples of differential centrifugation. It initiated the era of modern experimental cell biology.

What is the standard procedure for **differential centrifugation**?

Differential centrifugation is a technique commonly used by biochemists. Tissue, such as liver, is homogenized at 32°F (0°C) in a sucrose solution that contains both a buffer to stabilize the pH and a salt. The homogenate is then placed in a centrifuge and spun at a constant centrifugal force at a constant temperature. After a predetermined amount of time, a sediment forms at the bottom of the centrifuge tube and is covered by an overlying solution called the supernatant. The overlying solution is then placed in another centrifuge tube. The sediment after the first centrifugation is called the nuclear fractions and consists mainly of nuclei, which are the largest and densest organelles present in cells. The supernatant is centrifuged at a higher centrifugal force for a longer period of time. Another group of particles, usually the mitochondria, form the sediment. The process is repeated several times, each time at a higher centrifugal force for longer periods of time, until the sediment only contains enzymes and other substances not associated with any cell organelles.

CHROMATOGRAPHY

What are the **uses** of **chromatography**?

Chromatography is used to separate and identify the chemicals in a mixture. It is useful to: 1) separate and identify the chemicals in a mixture; 2) check the purity of a chemical product; 3) identify impurities in a product; and 4) purify a chemical product (on a laboratory or industrial scale).

How is **chromatography** used to **identify individual compounds**?

Chromatography is another technique used to separate mixtures into their individual components. All methods of chromatography share common characteristics. The process is based on the principle that different chemical compounds will stick to a solid surface, or dissolve in a film of liquid, to different degrees. Chromatography

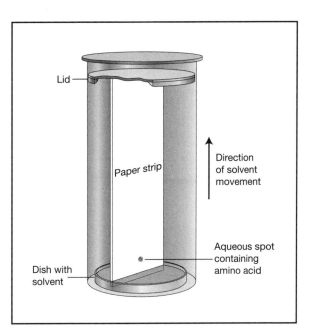

Ascending paper chromatography is a common chromatographic technique.

involves a sample (or sample extract) being dissolved in a mobile phase (which may be a gas, a liquid, or a supercritical fluid). The mobile phase is then forced through an immobile, immiscible stationary phase. The phases are chosen such that components of the sample have differing solubilities in each phase. The least soluble component is separated first, and as the separation process continues, the components are separated by increasing solubility.

What are the most **common chromatographic techniques**?

The most common chromatographic techniques are paper chromatography, gas-liquid chromatography (also called gas chromatography), thin layer chromatography, and high pressure (or high performance) liquid chromatography (HPLC).

What are some **biological applications** of chromatography?

Chromatography has many applications in biology. It is used to separate and identify amino acids, carbohydrates, fatty acids, and other natural substances. Environmental testing laboratories use chromatography to identify trace quantities of contaminants such as PCBs in waste oil and pesticides such as DDT in groundwater. It is also used to test drinking water and test air quality. Pharmaceutical companies use chromatography to prepare quantities of extremely pure materials. The food industry uses chromatography to detect contaminants such as aflatoxin.

Who **coined the term** "chromatography"?

The Russian-Italian biochemist Mikhail Semyonovich Tsvet (or Tswett) (1872–1919) coined the term "chromatography" and published the first paper on the method in 1903. The term comes from the Greek words *chroma*, meaning "color," and *graphein*, meaning "writing or drawing."

What are the stationary and mobile phases of **paper chromatography**?

The stationary phase of paper chromatography is filter paper, and the mobile phase is an organic solvent or mixture. A small drop of the liquid mixture is placed at one end of the paper. The end of the paper is then immersed in a solvent. As the solvent moves up the paper, any molecules in the mixture that are soluble in the solvent will move with the solvent. Based on their solubility to the mobile phase and attraction to the stationary phase, some molecules will move faster than others. Each different molecule in the mixture will move at a different speed and will be at a different position on the finished chromatogram.

How are the **results** of **paper chromatography deciphered**?

Substances that have color are easily seen on a chromatogram. A completed chromatogram is sprayed with ninhydrin and may be heated to detect amino acids. Amino acids turn pink or purple when sprayed with ninhydrin. Molecules of reducing sugars become gray-black when treated with analine phthalate. Many organic substances become brown spots on a yellow background when exposed to iodine vapors. The distance of the various spots from the origin is measured and used to quantify the movement of the sample. Under standardized conditions a substance moves at a characteristic rate in a particular solvent system. A researcher can calculate the distance the solvent travels with the distance the substance travels. The result is known as the R_f value. Consulting tables of R_f values for various substances in various solvents permits preliminary identification of the substance.

What is **gas chromatography**?

Gas chromatography, specifically gas-liquid chromatography, involves a sample being vaporized and injected onto the head of the chromatographic column. The sample is

What are the most common adsorbents used in column chromatography?

Silica gel and alumina are the most common adsorbents.

transported through the column by the flow of inert, gaseous mobile phase. The column itself contains a liquid stationary phase that is adsorbed onto the surface of an inert solid.

What **gases** are **commonly used** for the mobile phase of gas chromatography?

The carrier gas must be chemically inert. Commonly used gases include nitrogen, helium, argon, and carbon dioxide.

What are some **applications** of **gas chromatography**?

Gas-liquid chromatography is the most widely used chromatographic technique for environmental analyses. Analysis of organic compounds is possible for a variety of matrices such as water, soil, soil gas, and ambient air. It is often used to test hazardous waste sites for determining personal protective equipment (PPE) level and emergency response testing.

How does **column chromatography** differ from other chromatographic methods?

Column chromatography is generally used as a purification technique: it isolates desired compounds from a mixture. Column chromatography separates molecules according to their size and shape. The stationary phase, a solid adsorbent, is placed in a vertical glass column and the mobile phase, a liquid, is added to the top and flows down through the column by either gravity or external pressure. The mixture to be analyzed by column chromatography is applied to the top of the column. The liquid solvent (the eluent) is passed through the column by gravity or by the application of air pressure. An equilibrium is established between the solute adsorbed on the adsorbent and the eluting solvent flowing down through the column. Because the different components in the mixture have different interactions with the stationary and mobile phases, they will be carried along with the mobile phase to varying degrees, and a separation will be achieved. The individual components, or elutants, are collected as the solvent drips from the bottom of the column.

How does **thin layer chromatography** differ from other chromatographic techniques?

Thin layer chromatography does not use a column. The stationary phase is thinly coated on a substrate, most frequently a glass plate. A binding agent such as calcium sulfate is

531

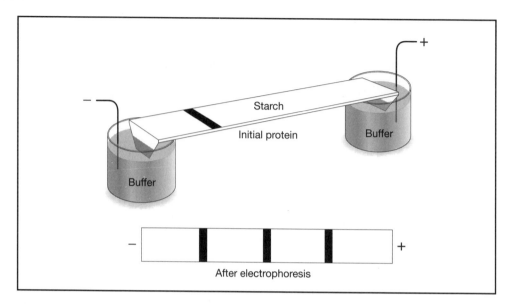

Electrophoresis has become an important tool in DNA identification ("genetic fingerprinting").

usually added to make the adsorbent adhere to the substrate. A sample is placed on the plate and the mobile phase migrates through the stationary phase by capillary action, similar to paper chromatography. After separation, the spots may be scraped from the glass for detailed analysis, or they can be subjected to further chromatographic study.

What is the **advantage** of **thin layer chromatography**?

Thin layer chromatography is a much quicker technique than paper chromatography. To separate a mixture using paper chromatography may take twenty-four hours, but less than one hour using thin layer chromatography.

What is **high pressure liquid chromatography**?

High pressure liquid chromatography (HPLC) was developed in the mid-1970s. It differs from other chromatographic techniques in that the liquid is pumped through a short packed column under pressure, instead of relying only on gravity.

ELECTROPHORESIS

What is **electrophoresis**?

Electrophoresis is a technique used to separate biological molecules, such as nucleic acids, carbohydrates, and amino acids, based on their movement due to the influence of a direct electric current in a buffered solution. Positively charged molecules move toward the negative electrode, while negatively charged molecules move toward the positive electrode.

What are the advantages of **gel electrophoresis**?

In gel electrophoresis the molecules are forced across a span of gel, usually agarose. Gels are easy to modify and have an excellent power of separation. The frictional force of the gel material acts as a "molecular sieve," separating the molecules based on their size.

Who **introduced electrophoresis** as an analytical technique?

Electrophoresis was introduced in 1937 by Arne Tiselius (1902–1971), a Swedish biochemist. His work pioneered the use of electrophoretic methods to separate chemically similar proteins of blood serum.

Why is **electrophoresis** important for **biological research**?

Gel electrophoresis has become an important tool to identify DNA molecules. Genetic differences between different species of plants and animals have been identified using gel electrophoresis. Protein molecules that differ by a single amino acid can be identified. This process is useful for determining the "genetic fingerprint" of an individual.

SPECTROSCOPY

What is **spectroscopy**?

Spectroscopy includes a range of techniques to study the composition, structure, and bonding of elements and compounds. The different methods of spectroscopy use different wavelengths of the electromagnetic spectrum to study atoms, molecules, ions, and the bonding between them.

Type of spectroscopy	Wavelength used
Nuclear magnetic resonance spectroscopy	Radio waves
Infrared spectroscopy	Infrared radiation
Atomic absorption spectroscopy, atomic emission spectroscopy, and ultraviolet spectroscopy	Visible and UV radiation
X-ray spectroscopy	X-rays

What technique may be used to **determine** the **presence** and **concentration** of **various chemicals** in a solution?

Spectrophotometry is based on the principle that every different atom, molecule, or chemical bond absorbs a unique pattern of wavelengths of light. Using a spectrophotometer, scientists are able to measure the amount of light that is absorbed or transmitted by molecules in solution. This value of this measurement allows scientists to identify the chemical.

What is the light source for a **spectrophotometer**?

Most commonly, a spectrophotometer uses white light as its light source. White light is a combination of all *visible* wavelengths. Ultraviolet and infrared wavelengths may also be used.

What is the **electromagnetic spectrum**?

The electromagnetic spectrum is the range of wavelengths. It ranges from gamma rays with very short wavelengths (10^{-5} nanometers) and high energy through radio waves with longer wavelengths (10^3 meters) and less energy. Visible light, seen as color, occurs between 380 and 750 nanometers; ultraviolet light has a shorter wavelength, while infrared light has a greater wavelength.

What are **X-rays**?

X-rays are electromagnetic radiation with short wavelengths (10^{-3} nanometers) and a great amount of energy. They were discovered in 1898 by William Conrad Roentgen (1845–1923). X-rays are frequently used in medicine because they are able to pass through opaque, dense structures such as bone and form an image on a photographic plate.

X–RAY DIFFRACTION

What is **X-ray crystallography**?

X-ray crystallography, also called X-ray diffraction, is used to determine crystal structures by interpreting the diffraction patterns formed when X-rays are scattered by the electrons of atoms in crystalline solids. X-rays are sent through a crystal to reveal the pattern in which the molecules and atoms contained within the crystal are arranged.

What important scientific discoveries were made using **x-ray diffraction**?

In 1951 the protein α-helix was discovered by Linus Pauling (1901–1994). X-ray diffraction was used to reveal the double helix structure of DNA in 1953 by Maurice Wilkins (1916–), Francis Crick (1916–2004), Rosalind Franklin (1920–1958), and James Watson (1928–). Dorothy Mary Crowfoot Hodgkin (1901–1994) used the technique to determine the structure of vitamin B_{12} in 1956.

NUCLEAR MAGNETIC RESONANCE AND ULTRASOUND

What is **nuclear magnetic resonance**?

Nuclear magnetic resonance (NMR) is a process in which the nuclei of certain atoms absorb energy from an external magnetic field. Scientists use NMR spectroscopy to identify unknown compounds, check for impurities, and study the shapes of molecules. They use the knowledge that different atoms will absorb electromagnetic energy at slightly different frequencies.

What is **nuclear magnetic resonance imaging**?

Magnetic resonance imaging (MRI), sometimes called nuclear magnetic resonance imaging (NMR), is a noninvasive, nonionizing diagnostic technique. It is useful in detecting small tumors, blocked blood vessels, or damaged vertebral disks. Because it does not involve the use of radiation, it can often be used where X-rays are dangerous. Large magnets beam energy through the body, causing hydrogen atoms in the body to resonate. This produces energy in the form of tiny electrical signals. A computer detects these signals, which vary in different parts of the body and according to whether an organ is healthy or not. The variation enables a picture to be produced on a screen and interpreted by a medical specialist.

What distinguishes MRI from computerized X-ray scanners is that most X-ray studies cannot distinguish between a living body and a cadaver, while MRI "sees" the difference between life and death in great detail. More specifically, it can discriminate between healthy and diseased tissues with more sensitivity than conventional radiographic instruments like X-rays or CAT scans. CAT (computerized axial tomography) scanners have been around since 1973 and are actually glorified X-ray machines. They offer three-dimensional viewing but are limited because the object imaged must remain still.

Who **proposed** using **magnetic resonance imaging** for **diagnostic purposes**?

The concept of using MRI to detect tumors in patients was proposed by Raymond Damadian (1936–) in a 1972 patent application. The fundamental MRI imaging concept used in all present-day MRI instruments was proposed by Paul Lauterbur (1929–) in an article in *Nature* in 1973. Lauterbur and Peter Mansfield (1933–) were awarded the Nobel Prize in Physiology or Medicine in 2003 for their discoveries concerning magnetic resonance imaging. The main advantages of MRI are that it not only gives superior images of soft tissues (like organs), but it can also measure dynamic physiological changes in a noninvasive manner (without penetrating the body in any way). A disadvantage of MRI is that it cannot be used for every patient. For example, patients with implants, pacemakers, or cerebral aneurysm clips made of metal cannot be examined using MRI because the machine's magnet could potentially move these objects within the body, causing damage.

What is **ultrasound**?

Ultrasound is another type of 3-D computerized imaging. Using brief pulses of ultra-high frequency acoustic waves (lasting 0.01 second), it can produce a sonar map of the imaged object. The technique is similar to the echolocation used by bats, whales, and dolphins.

METHODS IN MICROBIAL STUDIES

How do **aseptic procedures** prevent contamination?

The goal of aseptic procedures is to keep the organisms used in an experiment separate from the millions of other organisms in the environment. These procedures focus on the way to transfer organisms from test tube to test tube, from test tube to flask or petri dish, and from petri dish to petri dish or flask.

What are the **basic steps** to transfer bacteria using **aseptic procedures**?

First, plug the test tube or flask with cotton or plastic cap. Sterilize the inoculating loop or needle (tool used to transfer the bacteria) by heating the wire in the flame of a Bunsen burner or alcohol lamp until red hot. Remove the cotton plugs from the test tubes or flasks and briefly flame the mouths of the test tubes or flasks. Pick up the organisms to be transferred by dipping the inoculating loop or needle into the liquid culture or touching the growth lightly. Transfer the culture to a new test tube by dipping the loop gently into the broth or drawing it across the agar. Pass the mouths of the tubes through the flame again and then reinsert the cotton plugs to prevent the entrance of other microbes. Heat the inoculating loop or needle again to destroy any remaining organisms. The procedure to transfer a sample to a petri dish is similar except the culture is streaked along the agar in a pattern to isolate the bacteria on the petri dish.

What are the two **most common media** to **grow bacteria**?

Beef extract and peptones (hydrolyzed protein) are the basic ingredients of nutrient broth. These materials supply a variety of carbon sources, nitrogen compounds in the form of amino acids, and a mixture of cofactors such as vitamins. The addition of agar (a complex carbohydrate extracted from seaweed) results in a solid medium. Agar is an ideal solidifying agent for microbiological media because of its melting properties and because it has no nutritive value for the vast majority of bacteria. Solid agar melts at 194–212°F (90–100°C); liquid agar solidifies at about 103°F (42°C).

What is **lypholization**?

Lypholization is a freeze-drying technique for preservation and storage of bacteria and other microorganisms. Bacteria can be stored for extended periods of time (three to five years) as frozen cell suspensions or as freeze-dried (lypholized) cultures. This technique is achieved by placing bacteria in a nutrient broth containing 15 to 25 percent glycerol and freezing at temperatures of -94°F (-70°C) or lower. The glycerol reduces ice crystal formation that would cause subsequent cell damage and disrupt biological structures.

METHODS IN ECOLOGICAL STUDIES

What **methods** are used to **estimate wildlife populations**?

Since it is usually impossible (and often impractical) to count all individuals in a population, researchers use a variety of sampling techniques to estimate population densities. One method is to count the individuals in a certain area. The larger the number and size of sample plots, the more accurate the estimates. Population densities may also be estimated based on indirect indicators such as animal droppings or tracks, nests, or burrows.

How does the **mark-recapture method** estimate a wildlife population?

Researchers set traps to capture a population sampling in a given area. The captured animals are "marked" (or tagged) and then released. After a certain amount of time the traps are set again. The second time, both marked and unmarked animals will be captured. The proportion of marked to unmarked animals gives an estimate of the size of the entire population. The equation used to estimate the number of individuals in the population is:

$$N = \frac{\text{Marked individuals X Total catch second time}}{\text{Recaptured marked individuals}}$$

where N is the number of individuals in the population. A disadvantage of this method is that it assumes that a marked individual has the same chance of being trapped as an unmarked individual. Individuals that were trapped once may be wary of traps, or they may seek traps since they have learned that traps provide food.

How do **conservationists predict** whether a **species** will become **extinct**?

Conservation biologists use population viability analysis (PVA), a relatively new method, to predict the viability of a species in a particular habitat. Computer modeling generates PVAs using life history data, genetic variability, and a population's response to environmental conditions—especially disturbances—to predict viability of a species.

What is a **MVP**?

The MVP (minimum viable population size) is the smallest number of individuals needed to perpetuate a population, subpopulation, or species. PVAs (population viability analyses) are especially helpful in predicting the MVP.

Further Reading

General

Books

Barnes-Svarney, Patricia L., and Thomas E. Svarney. *The Oryx Guide to Natural History: The Earth and All Its Inhabitants*. Phoenix, AZ: Oryx Press, 1999.

Bazler, Judith. *Biology Resources in the Electronic Age*. Westport, CT: Greenwood Press, 2003.

Biology. Richard Robinson, editor-in-chief. New York: Macmillan Reference USA, 2002.

Biology Data Book. 2nd ed. Bethesda, MD: Federation of American Societies for Experimental Biology, 1972–1974.

Bruno, Leonard C. *U.X.L. Complete Life Science Resource*. Detroit: U.X.L., 2001.

Bryson, Bill. *A Short History of Nearly Everything*. New York: Broadway Books, 2003.

Campbell, Neil A., and Jane B. Reece. *Biology*. 6th ed. San Francisco: Benjamin Cummings, 2002.

Chapman, Carolyn. *Basic Chemistry for Biology*. 2nd ed. New York: WCB/McGraw-Hill, 1999.

The Facts on File Dictionary of Biology. New York: Facts on File, 1999.

The Gale Encyclopedia of Science. 2nd ed. Detroit: Gale Group, 2001.

Guttman, Burton S. *Biology*. New York: WCB/McGraw-Hill, 1999.

Indge, Bill. *Dictionary of Biology*. Chicago: Fitzroy Dearborn Publishers, 1999.

Innovations in Biology. Santa Barbara, CA: ABC-CLIO, 1999.

Krogh, David. *Biology: A Guide to the Natural World*. 2nd ed. Upper Saddle River, NJ: Prentice Hall, 2002.

Layman, Dale Pierre. *Biology Demystified*. New York: McGraw-Hill, 2003.

Margulis, Lynn, and Karlene V. Schwartz. *Five Kingdoms: An Illustrated Guide to the Phyla of Life on Earth*. 3rd ed. New York: W.H. Freeman, 1998.

McGraw-Hill Encyclopedia of Science and Technology. 9th ed. New York: McGraw-Hill, 2002.

Mertz, Leslie A. *Recent Advances and Issues in Biology*. Phoenix, AZ: Oryx Press, 2000.

Nature Encyclopedia. New York: Oxford University Press, 2001.

The New Book of Popular Science. Danbury, CT: Grolier, 2003.

Raven, Peter, George B. Johnson, Jonathon Losos, and Susan Singer. *Biology*. 7th ed. New York: McGraw-Hill, 2005.

Science and Technology Encyclopedia. Chicago: University of Chicago Press, 2000.

Solomon, Eldra P., Linda R. Berg, and Diana W. Martin. *Biology*. 7th ed. Belmont, CA: Brooks/Cole, 2004.

Tudge, Colin. *The Variety of Life: A Survey and a Celebration of All the Creatures That Have Ever Lived*. New York: Oxford University Press, 2000.

Van Nostrand's Scientific Encyclopedia. 9th ed. New York: John Wiley and Sons, 2002.

World of Biology. Detroit: Gale Group, 1999.

Web Sites

American Association for the Advancement of Science
http://aaas.org

Ask a Scientist
http://www.newton.dep.anl.gov/archive.htm

Biology Online
http://www.biology-online.org

The Biology Project
http://www.biology.arizona.edu

National Academy of Sciences
http://www.nas.edu

National Institutes of Health
http://www.nih.gov

National Science Foundation
http://www.nsf.gov

Online Conversion of Measurements
http://www.onlineconversion.com

Physical Anthropology Tutorials
http://anthro.palomar.edu/tutorials/physical.htm

Science News
http://www.sciencenews.org

Smithsonian Institution
http://www.si.edu

Who Named It?
http://www.whonamedit.com

Journals and Periodicals

American Biology Teacher

American Health

American Scientist

Audubon Magazine

Biocycle

Bioscience

Buzzworm: The Environmental Journal

Discover

E: The Environmental Magazine

Environment

FDA Consumer

Horticulture

International Wildlife

National Geographic

National Geographic World

National Wildlife

Natural History

Nature

New Scientist

Popular Science

Recycling Today

Science

Science News

Science Teacher

Scientific American

Smithsonian

Animals

Books

Animal. David Burnie and Don E. Wilson, editors-in-chief. Washington, DC: Smithsonian Institution; New York: DK, 2001.

The Concise Oxford Dictionary of Zoology. New York: Oxford University Press, 1991.

Drickhamer, Lee C., Stephen H. Vessey, and Elizabeth M. Jakob. *Animal Behavior: Mechanisms, Ecology, Evolution*. 5th ed. New York: McGraw-Hill, 2002.

Eisner, Thomas. *For Love of Insects*. Cambridge, MA: Belknap Press of Harvard University Press, 2003.

541

The Encyclopedia of Mammals. New York: Facts on File, 2001.

Firefly Encyclopedia of Insects and Spiders. Buffalo, NY: Firefly Books, 2002.

Firefly Encyclopedia of Reptiles and Amphibians. Buffalo, NY: Firefly Books, 2002.

Friend, Tim. *Animal Talk: Breaking the Code of Animal Language*. New York: Free Press, 2004.

Grzimek's Animal Life Encyclopedia. 2nd ed. Detroit: Gale, 2003.

Hickman, Cleveland P., Jr., Larry S. Roberts, Allan Larson, and Helen Ianson. *Integrated Principles of Zoology*. 12th ed. New York: McGraw-Hill Higher Education, 2003.

International Wildlife Encyclopedia. 3rd ed. New York: Marshall Cavendish, 2002.

Jessop, Nancy M. *Zoology, the Animal Kingdom: A Complete Course in 1000 Questions and Answers*. New York: McGraw-Hill, 1995.

Lorenz, Konrad. *King Solomon's Ring*. London, New York: Routledge, 2002.

Magill's Encyclopedia of Science: Animal Life. Carl W. Hoagstrom, editor. Pasadena, CA: Salem Press, 2002.

Miller, Stephen A., and John P. Harley. *The Animal Kingdom*. Dubuque, IA: Wm. C. Brown Publisher, 1996.

National Audubon Society. *The Sibley Guide to Bird Life and Behavior*. New York: Alfred A. Knopf, 2001.

Pough, F. H., Christine M. Janis, and John B. Heiser. *Vertebrate Life*. 7th ed. New York: Prentice-Hall, 2004.

Ruppert, Edward E., Richard S. Fox, and Robert D. Barnes. *Invertebrate Zoology*. 7th ed. Hinsdale, IL: Dryden Press, 2003.

Schmidt-Nielsen, Knut. *Animal Physiology: Adaptation and Environment*. 5th ed. Cambridge, MA: Cambridge University Press, 1997.

Svarney, Thomas, and Patricia Barnes-Svarney. *The Handy Dinosaur Answer Book*. Detroit: Visible Ink Press, 2000.

Wilson, Edward O. *Sociobiology: The New Synthesis*. Cambridge, MA: Belknap Press of Harvard University Press, 2000.

Wootton, Anthony. *Insects of the World*. New York: Facts on File, 2002.

Wynnem, Clive D. L. *Do Animals Think?* Princeton, NJ: Princeton University Press, 2004.

Web Sites

Animal Diversity Web
http://animaldiversity.ummz.umich.edu

Dinosaurs: Facts and Fiction
http://pubs.usgs.gov/gip/dinosaurs

Duke University Primate Center
http://www.duke.edu/web/primate

The Gorilla Foundation
http://www.koko.org

Iowa State Entomology Index of Internet Resources
http://www.ent.iastate.edu/list

Primate Info Net
http://www.primate.wisc.edu/pin

Smithsonian National Museum of Natural History, Department of Entomology
http://entomology.si.edu

The Tree of Life Web Project
http://tolweb.org/tree

Bacteria, Viruses, and Protists

Books

Alcamo, I. Edward. *Fundamentals of Microbiology*. 6th ed. Boston: Jones and Bartlett Publishers, Inc., 2000.

Encyclopedia of Microbiology. 2nd ed. Joshua Lederberg, editor-in-chief. San Diego: Academic Press, 2000.

Encyclopedia of Virology. 2nd ed. Allan Granoff and Robert G. Webster, editors. San Diego: Academic Press, 1999.

Lim, Daniel. *Microbiology*. 3rd ed. Dubuque, IA: Kendall/Hunt Publishing Co., 2002.

Madigan, Michael M., John M. Martinko, and Jack Parker. *Brock Biology of Microorganisms*. 10th ed. New York: Prentice Hall, 2002.

Postgate, John. *Microbes and Man*. 4th ed. Cambridge, MA: Cambridge University Press, 2000.

Sankaran, Neeraja. *Microbes and People*. Phoenix, AZ: Oryx Press, 2000.

Tortora, Gerard J., Berdell R. Funke, and Christine L. Case. *Microbiology: An Introduction*. 7th ed. San Francisco: Benjamin Cummings, 2002.

World of Microbiology and Immunology. K. Lee Lerner and Brenda Wilmoth Lerner, editors. Detroit: Gale, 2003.

Web Sites

All the Virology on the Web
http://www.tulane.edu/-dmsander/garryfavweb.html

American Society for Microbiology
www.microbe.org

Centers for Disease Control
www.cdc.gov

Food and Drug Administration
www.fda.gov

543

MicrobeWorld
www.microbeworld.org

Biography

Books

Biographical Dictionary of Scientists. 3rd ed. New York: Oxford University Press, 2000.

The Biographical Dictionary of Women in Science: Pioneering Lives from Ancient Time to the Mid-Twentieth Century. Marilyn Ogilivie and Joy Harvey, editors. New York: Routledge, 2000.

Biographical Encyclopedia of Scientists. Richard Olson, editor. New York: Marshall Cavendish, 1998.

The Cambridge Dictionary of Scientists. Cambridge, U.K.: Cambridge University Press, 2002.

Concise Dictionary of Scientific Biography. 2nd ed. New York: Scribner's, 2000.

Dictionary of Scientific Biography. New York: Charles Scribner's Sons, 1973.

The Grolier Library of Science Biographies. Danbury CT: Grolier Educational, 1997.

Kurian, George Thomas. *The Nobel Scientists: A Biographical Encyclopedia*. Amherst, NY: Prometheus Books, 2002.

Life Sciences before the Twentieth Century: Biographical Portraits. Everett Mendelsohn, editor. New York: Charles Scribner's Sons, 2002.

Life Sciences in the Twentieth Century: Biographical Portraits. Everett Mendelsohn, editor. New York: Charles Scribner's Sons, 2001.

The Nobel Prize Winners: Physiology or Medicine. Pasadena, CA: Salem Press, 1991.

Notable Scientists: From 1900 to the Present. Brigham Narins, editor. Farmington Hills, MI: Gale Group, 2001.

Oakes, Elizabeth H. *Encyclopedia of World Scientists*. New York: Facts on File, 2001.

Web Sites

Biography.com
http://www.biography.com

Eric Weisstein's World of Biography
http://scienceworld.wolfram.com/biography

Nobel e-Museum
http://www.nobel.se

Biotechnology

Books

Acquaah, George. *Understanding Biotechnology; An Integrated and Cyber-Based Approach*. New York: Pearson Prentice Hall, 2004.

Borém, Aluízio, Fabrício R. Santos, and David E. Bowen. *Understanding Biology*. Upper Saddle River, NJ: Prentice Hall, 2003.

Bowring, Finn. *Science, Seeds, and Cyborgs: Biotechnology and the Appropriation of Life*. New York: Verso, 2003.

Fumento, Michael. *BioEvolution: How Biotechnology Is Changing Our World*. San Francisco: Encounter Books, 2003.

Genetically Modified Foods: Debating Biotechnology. Amherst, NY: Prometheus Books, 2002.

Human Cloning and Human Dignity: The Report of the President's Council on Bioethics. New York: Public Affairs, 2002.

Kreuzer, Helen, and Adrianne Massey. *Recombinant DNA and Biotechnology: A Guide for Teachers*. 2nd ed. Washington, DC: ASM Press, 2001.

Lambrecht, Bill. *Dinner at the New Gene Café: How Genetic Engineering Is Changing What We Eat, How We Live, and the Global Politics of Food*. New York: Thomas Dunne Books, 2001.

Le Vine, Harry. *Genetic Engineering: A Reference Handbook*. Santa Barbara, CA: ABC-CLIO, 1999.

Maienschein, Jane. *Whose View of Life? Embryos, Cloning, and Stem Cells*. Cambridge, MA: Harvard University Press, 2003.

Pence, Gregory E. *Designer Food: Mutant Harvest or Breadbasket of the World?* Lanham, MD: Rowman and Littlefield, 2002.

President's Council on Bioethics (U.S.). *Beyond Therapy: Biotechnology and the Pursuit of Happiness*. New York: Dana Press, 2003.

Steinberg, Mark, and Sharon D. Cosloy. *The Facts on File Dictionary of Biotechnology and Genetic Engineering*. New York: Checkmark Books, 2001.

Wilmut, Ian. *The Second Creation: Dolly and the Age of Biological Control*. New York: Farrar, Straus and Giroux, 2000.

Winston, Mark L. *Travels in the Genetically Modified Zone*. Cambridge, MA: Harvard University Press, 2002.

Web Sites

Agricultural Biotechnology
http://agnic.umd.edu/

BioLink: Center for Biotechnology Information
http://www.bio-link.org

BioTech
http://biotech.icmb.utexas.edu

National Center for Biotechnology Information
http://www.ncbi.nlm.nih.gov

545

Cells

Books

Becker, Wayne M. *The World of the Cell*. 5th ed. San Francisco: Benjamin/Cummings Pub. Co., 2003.

Cooper, Geoffrey M. *The Cell: A Molecular Approach*. 3rd ed. Sunderland, MA: Sinauer Associates, Inc., 2003.

The Facts on File Dictionary of Cell and Molecular Biology. Robert Hine, editor. New York: Facts on File, 2003.

Harold, Franklin M. *The Way of the Cell: Molecules, Organisms, and the Order of Life*. New York: Oxford University Press, 2001.

Harris, Henry. *The Birth of the Cell*. New Haven, CT: Yale University Press, 1999.

Hughes, Arthur Frederick William. *A History of Cytology*. Ames, IA: Iowa State University Press, 1989.

Web Sites

Cells Alive!
http://www.cellsalive.com

Environment

Books

The Atlas of Endangered Species. John A. Burton, editor. New York: Macmillan Library Reference, 1999.

The Atlas of U.S. and Canadian Environmental History. Char Miller, editor. New York: Routledge, 2003.

Beacham's Guide to the Endangered Species of North America. Detroit: Gale Group, 2001.

Callahan, Joan R. *Recent Advances and Issues in Environmental Science*. Phoenix, AZ: Oryx Press, 2000.

Carson, Rachel. *Silent Spring*. 40th anniversary ed. Boston: Houghton Mifflin, 2002.

Cunningham, William P., and Barbara Woodworth Saigo. *Environmental Science: A Global Concern*. 6th ed. New York: McGraw-Hill, 2001.

Ecology Basics. Pasadena, CA: Salem Press, 2004.

Encyclopedia of Environmental Biology. William A. Nierenberg, editor-in-chief. San Diego: Academic Press, 1995.

Encyclopedia of Environmental Issues. Pasadena, CA: Salem Press, 2000.

Encyclopedia of the Biosphere. Detroit: Gale Group, 2000.

Encyclopedia of World Environmental History. Shepard Krech III, J.R. McNeill, and Carolyn Merchant, editors. New York: Routledge, 2004.

Endangered Animals. Danbury, CT: Grolier Educational, 2002.

Endangered Animals: A Reference Guide to Conflicting Issues. Richard P. Reading and Brian Miller, editors. Westport, CT: Greenwood Press, 2000.

Enger, Eldon D., and Bradley F. Smith. *Environmental Science: A Study of Interrelationships*. 7th ed. New York: McGraw-Hill, 2000.

Environmental Literature: An Encyclopedia of Works, Authors, and Themes. Santa Barbara, CA: ABC-CLIO, 1999.

Great Events from History II: Ecology and the Environment Series. Frank N. Magill, editor. Pasadena, CA: Salem Press, 1995.

Greive, Bradley Trevor. *Priceless: The Vanishing Beauty of a Fragile Planet*. Kansas City, MO: Andrews McMeel Pub., 2003.

Groombridge, Brian. *World Atlas of Biodiversity: Earth's Living Resources in the Twenty-First Century*. Berkeley, CA: University of California Press, 2002.

Hosansky, David. *The Environment A to Z*. Washington, DC: CQ Press, 2001.

Jukofsky, Diane. *Encyclopedia of Rainforests*. Westport, CT: Oryx Press, 2002.

Merchant, Carolyn. *The Columbia Guide to American Environmental History*. New York: Columbia University Press, 2002.

Molles, Manuel C., Jr. *Ecology: Concepts and Applications*. New York: WCB/McGraw-Hill, 1999.

Mongillo, John, and Linda Zierdt-Warshaw. *Encyclopedia of Environmental Science*. Phoenix, AZ: Oryx Press, 2000.

Muir, John. *Nature Writings*. New York: Library of America, 1997.

Weigel, Marlene. *U.X.L. Encyclopedia of Biomes*. Detroit: U.X.L., 2000.

Web Sites

EnviroLink Network
http://www.envirolink.org

Environmental Protection Agency
http://www.epa.gov

International Union for Conservation of Nature and Natural Resources
http://www.redlist.org

National Audubon Society
http://www.audubon.org

Rainforest Action Network
ttp://www.ran.org

Sierra Club
http://www.sierraclub.org

U.S. Fish and Wildlife Service
http://www.fws.gov

WWF: The Conservation Organization
http://www.panda.org

Evolution

Books

Darwin, Charles. *On the Origin of Species*. New York: The Modern Library, 1993.

———. *Voyage of the Beagle*. Mineola, NY: Dover Publications, 2002.

Encyclopedia of Evolution. Mark Pagel, editor-in-chief. New York: Oxford University Press, 2002.

Gould, Stephen Jay. *The Structure of Evolutionary Theory*. Belknap Press of Harvard University Press, 2002.

Jones, Steve. *Darwin's Ghost: The Origin of Species Updated*. New York: Random House, 2000.

Keynes, Randal. *Darwin, His Daughter, and Human Evolution*. New York: Riverhead Books, 2002.

Larson, Edward J. *Evolution: The History of a Scientific Theory*. New York: Modern Library, 2004.

Mills, Cynthia. *The Theory of Evolution: What It Is, Where It Came From, and Why It Works*. Hoboken, NJ: Wiley, 2004.

Zimmer, Carl. *Evolution: The Triumph of an Idea*. New York: HarperCollins, 2001.

Web Sites

Evolution
http://www.ucmp.berkeley.edu/history/evolution.html

Understanding Evolution
http://evolution.berkeley.edu

Fungi

Books

The Great Encyclopedia of Mushrooms. Cologne: Könemann, 1999.

Hall, Ian R., et al. *Edible and Poisonous Mushrooms of the World*. Portland, OR: Timber Press, 2003.

Hudler, George W. *Magical Mushrooms, Mischievous Molds*. Princeton, NJ: Princeton University Press, 1998.

Laessøe, Thomas. *Mushrooms*: New York: DK Pub., 1998.

Money, Nicholas P. *Mr. Bloomfield's Orchard: The Mysterious World of Mushrooms, Molds, and Mycologists*. New York: Oxford University Press, 2002.

Moore, David. *Slayers, Saviors, Servants, and Sex: An Exposé of Kingdom Fungi*. New York: Springer, 2001.

Purvis, Walter. *Lichens*. Washington, DC: Smithsonian Institution Press, 2000.

Schaechter, Elio. *In the Company of Mushrooms: A Biologist's Tale*. Cambridge, MA: Harvard University Press, 1997.

Web Sites

The American Chestnut Foundation
http://www.chestnut.acf.org

Fun Facts about Fungi
http://www.herbarium.usu.edu/fungi/funfacts.factindx.htm

Mycological Society of America
http://msafungi.org

The WWW Virtual Library: Mycology
http://mycology.cornell.edu/

University of Michigan Fungus Collection (MICH)
www.herbarium.lsa.umich.edu/

Genetics

Books

Cummings, Michael R. *Human Heredity: Principles and Issues*. 6th ed. Belmont, CA: Thomson Brooks/Cole, 2003.

Encyclopedia of Genetics. Eric C. R. Reeve, editor. Chicago: Fitzroy Dearborn, 2001.

Encyclopedia of Genetics. Sydney Brenner and Jeffrey H. Miller, editors-in-chief. San Diego, CA: Academic Press, 2002.

Genetics. Richard Robinson, editor-in-chief. New York: Macmillan Reference USA, 2003.

Genetics: From Genes to Genomes. 2nd ed. Leland Hartwell, et al., editors. New York: McGraw-Hill, 2004.

Hartl, Daniel, and Elizabeth W. Jones. *Genetics: Analysis of Genes and Genomes*. 5th ed. Boston: Jones and Bartlett, Publishers, 2001.

The Human Genome. Carina Dennis and Richard Gallagher, editors. New York: Nature/Palgrave, 2001.

King, Robert C. and William D. Stansfield. *A Dictionary of Genetics*. 6th ed. New York: Oxford University Press, 2002.

Moore, David S. *The Dependent Gene: The Fallacy of "Nature vs. Nurture."* New York: Times Books, 2002.

Ridley, Matt. *Genome: The Autobiography of a Species in 23 Chapters*. New York: HarperCollins, 1999.

Russell, Peter J. *iGenetics*. San Francisco: Benjamin Cummings, 2002.

Shreeve, James. *The Genome War: How Craig Venter Tried to Capture the Code of Life and Save the World*. New York: Alfred A. Knopf, 2004.

Stock, Gregory. *Redesigning Humans: Our Inevitable Genetic Future*. Boston: Houghton Mifflin, 2002.

Tudge, Colin. *The Impact of the Gene: From Mendel's Peas to Designer Babies*. New York: Hill and Wang, 2001.

Watson, James D. *DNA: The Secret of Life*. New York: Alfred A. Knopf, 2003.

Witherly, Jeffre L., Galen P. Perry, and Darryl L. Leja. *An A to Z of DNA Science: What Scientists Mean when They Talk about Genes and Genomes*. Cold Spring Harbor, NY: Cold Spring Harbor Laboratory Press, 2001.

World of Genetics. Detroit, MI: Gale Group/Thomson Learning, 2002.

Zhang, Yong-he, and Meng Zhang. *A Dictionary of Gene Technology Terms*. New York: Parthenon, 2001.

Web Sites

Access Excellence
http://www.accessexcellence.org

Human Genome Project
http://www.ornl.gov/hgnis/

The Institute for Genomic Research
http://www.tigr.org

MendelWeb
http://www. MendelWeb.org

Virtual Library on Genetics
http://www.ornl.gov/sci/techresources/Human Genome/genetics.shtml

Plants

Books

Bagust, Harold. *The Firefly Dictionary of Plant Names: Common and Botanical*. Buffalo, NY: Firefly Books, 2003.

Bailey, Jill. *Plants and Plant Life*. Danbury, CT: Grolier Educational, 2001.

Elpel, Thomas J. *Botany in a Day: Thomas J. Elpel's Herbal Field Guide to Plant Families*. 4th ed. Pony, MT: HOPS Press, 2000.

The Facts on File Dictionary of Botany. Jill Bailey, editor. New York: Facts on File, 2003.

Hallé, Francis. *In Praise of Plants*. Portland, OR: Timber Press, 2002.

Magill's Encyclopedia of Science: Plant Life, Bryan D. Ness, editor. Pasadena, CA: Salem Press, 2003.

Moore, Randy, W. Dennis Clark, and Darrell S. Vodopich. *Botany*. 2nd ed. New York: McGraw-Hill, 1997.

Plant Biology. Thomas L. Rost, Michael G. Barbour, C. Ralph Stocking, and Terence M. Murphy, editors. Belmont, CA: Wadsworth Publishing Company, 1998.

Plant Sciences. Richard Robinson, editor-in-chief. New York: Macmillan Reference USA, 2001.

Raven, Peter. *Biology of Plants*. 7th ed. New York: W.H. Freeman and Co., 2004.

Taiz, Lincoln, and Eduardo Zeiger. *Plant Physiology*. Sunderland, MA: Sinauer Associates, Pub., 2002.

Uno, Gordon, Richard Storey, and Randy Moore. *Principles of Botany*. New York: McGraw-Hill, 2001.

Web Sites

American Fern Society
http://www.amerfernsoc.org

Botanical Society of America
http://www.botany.org

International Carnivorous Plant Society
http://www.carnivorousplants.org

Internet Directory for Botany
http://www.botany.net/IDB

National Biological Information Infrastructure
http://www.nbii.gov/disciplines/botany

PlantFacts
http://plantfacts.osu.edu

Plants Database
http://www.plants.usda.gov

The Royal Horticultural Society
http://www.rhs.org.uk/index.asp

Index

Note: (ill.) indicates photographs and illustrations; (chart) indicates tables and charts.

A

Acharius, Erik, 142
Acidophilus milk, 123
Acid rain, 503-4, 504 (chart)
Acquired immunodeficiency syndrome. *See* AIDS.
Adaptation, 453
Adaptive radiation, 457
Addicott, Frederick T., 223
African elephant. *See* Elephants.
Africanized honeybees. *See* Bees.
Agassiz, Louis, 250
Agent Orange, 505
Aging, 377
Agnathans, 261
AIDS, 122, 433, 439, 474
Akhenaten, 409-10
Alcoholic beverages, 28
Algae, 123-24, 166, 470
Alkaptonuria, 13, 393-94
Alleopathy, 232
Allergies, 307
Alligators and crocodiles, 314, 319, 491, 492 (ill.)
Alpini, Prospero, 165
Alternation of generations, 164-65
Altman, Sidney, 370-71
Altruism, animal, 337-38
Alvarez, Luis, 266
Alvarez, Walter, 266
Alzheimer's disease, 72, 301

Amber, 163, 164, 255
Amino acids, 11
Amoebas, 111
Amoebic dysentery, 111
Amphibians, 263, 264, 314, 470, 490. *See also* Frogs.
Anabolic steroids. *See* Steroids.
Anagenesis, 458
Analogy in evolution, 454
Aneuploidy, 372
Angiogenesis, 79
Angiosperms, 175-83, 469, 490-91
Animal behavior. *See* Behavior, animal.
Anise, 187
Antarctic ozone hole, 501 (ill.), 502
Antelope, four-horned, 272
Anthrax, 91, 100-1, 101 (ill.), 439
Anthropomorphism, 325-26
Antibacterial products, 119-20
Antibiotics, 114-17, 114 (chart), 473
Antisense technology, 421
Antoinette, Marie, 430
Ants, 317, 349-50
Apical dominance, 204
Archaebacteria, 88
Archaeopteryx, 268
Aristotle, 85, 233, 323, 445

Aromatherapy, 195-96, 196 (chart)
Arrhenius, Svante, 471, 502
Arthropods, 249-59, 250-51 (chart). *See also* Insects; Spiders.
Artificial selection, 453-54
Aseptic procedures, 536
Asexual reproduction, 95, 307-8
Aspirin, 28
Atom, 1 (chart), 1-2
ATP (adenosine triphosphate), 14, 62
Avery, Oswald T., 360
Avogadro, Amedeo, 4

B

Baboons, 337
Bacillus thuringiensis, 91, 436-37, 438
Bacon, Francis, 514
Bacteria, 87-101, 88-89 (chart). *See also* Antibiotics; Prokaryotic cells.
    antibiotic-resistant, 115-16
    disease and, 98, 99, 121 (chart)
    in humans, 76-77
    production of foods and, 122 (chart)
    reproduction of, 95-96
    rickettsiae and chlamydiae, 99-100 (chart), 100

553

role in fighting pollution, 122

temperature and, 96

dormancy of spores, 97-98

Bacterial population growth curve, 93-94

Bacteriophage, 106, 107

Bakker, Robert T., 268

Balbiani, Edouard, 40-41

Bald eagles, 268 (ill.), 268-69

Balloons, 506

Bambi, 325-26

Bamboo, 179

Banyan trees, 178, 213, 213 (ill.)

Bark. *See* Plants.

Barnacles, 250

Barr, Murray, 393

Barr body, 393

Bartram, John, 160

Bartram, William, 160

Baseball bats, 185

Bates, Henry Walter, 462

Batesian mimicry, 462

Bateson, William, 387, 393

Bats, 321, 349, 353

Bayliss, William, 295

Beadle, George, 378, 394-95

Beagle voyages, 448, 450-51, 464

Bears, 353

Beer, 144-45, 415

Bees, 192 (ill.), 216-17, 217 (ill.), 218, 256-58, 344 (ill.), 345, 348, 404

Behavior, animal, 323-57. *See also* Ethology.

Behavioral ecology, 325

Beneden, Edouard van, 40

Benedict's test, 516

Bensley, Robert R., 528

Benthem, George, 162

Berg, Paul, 414

Bergey, David H., 88

Bernard, Claude, 277

Bessey, Charles, 163

Beta-blockers, 81

Binary fission, 95

Biochemistry, 1

Biodiversity, 488

Biogeochemical cycle, 479

Biogeography, 464, 480

Bioindicators, 167

Bioinvader, 512

Biological clocks and rhythms, 351 (chart), 351-52

Bioluminescence, 98, 425

Biomagnification, 499-500

Biomemetic, 439

Biomes, 480, 480-81 (chart)

Biopesticide, 438

Biopharming, 433

Biopiracy, 512

Bioprospecting, 422

Bioreactor, 422, 422 (chart)

Bioremediation, 433-34

Biosensors, 438

Biotechnology, 413-15, 414 (chart), 423, 431-33

Bioterrorism, 438-39

Birds, 267-69, 304, 314-15, 319-20, 320 (chart), 324-26, 333, 336, 341, 341-42 (chart), 348, 356-57, 459-60, 462-63, 469, 480-81 (chart), 490. *See also* Finches; Geese; Hummingbirds; Parrots; Swallows.

Bison, 273

Biuret test, 516

Blackman, F. F., 64

Black widow spider, 252-53

Blood, animal, 290-92

Bloodroot wildflower, 187-88

Blood types, 185, 290-91 (chart)

Blue lobsters. *See* Lobsters.

Body temperature, animal, 305 (chart)

Bonds, chemical, 3-4, 4 (chart), 8

Bordet, Jules, 92

Bosenberg, Henry F., 196

Botany subdisciplines, 159 (chart)

Botox, 99, 100 (ill.)

Bottleneck effect, 463

Botulism, 92, 99

Bovine spongiform encephalopathy. *See* Mad cow disease.

Boyer, Herb, 414

Bragg, William Henry, 534

Bragg, William Lawrence, 534

Brain comparisons, 298 (ill.)

Bread molds. *See* Fungi.

Brenner, Sydney, 244

Bridges, Calvin, 404

Brown, Robert, 32, 38

Brown algae. *See* Algae.

Brown fat, 281

Brown recluse spider, 252

Bryophytes, 165-68

Bt foods. *See Bacillus thuringiensis*.

Bubonic plague, 91, 92

Buffon, Comte de, 446-47

Burbank, Luther, 196

Burroughs, John, 493

Bush, George W., 508

Butcher's blocks, 185

Butterflies, 256, 256 (chart), 257, 304, 333, 436-37

C

Caesalpinus, 162

Caffeine, 82-83, 83 (chart), 229 (chart)

Calories, 22, 23, 28. *See also* Metabolism.

Calvin, Melvin, 14

Calvin cycle, 14, 63

Cambrian explosion, 466

Camels, 273

Camouflage behavior, 348

Cancer, 377, 414-15, 434, 494, 505

Candolle, Augustin Pyrame de, 162

Cannon, Walter Bradford, 277

Carbohydrates, 10, 11, 23, 27, 28, 130

Carbon, 5, 6

Carbon cycle, 479

Carbon dioxide, 15, 501-2

Carbon monoxide, 81

Carnivorous plants, 180

Carroll, Lewis, 475

Carson, Rachel, 498, 500, 500 (ill.)

Catastrophism, 448

Catnip, 231

Cato the Censor, 162

Cats, domestic, 343, 354-55, 398

Cattails, 176

Cave moss, 166-67

Cech, Thomas, 370-71

Cell culture, 422-23

Cell division, 32, 58, 58-59 (chart). *See also* Meiosis; Mitosis.

Cells. *See also* Cloning; Eukaryotic cells; Prokaryotic cells; Stem cells.
bladder, 76
brain, 71-72
cancer, 77, 78-79, 79 (chart), 80
cell junctions, 50
cell communication, 66-67, 66-67 (chart)
cell death, 62
cell theory, 31, 32
cell wall, 47-50
cytoplasm, 37, 41-47
cytosol, 37
effects of alcohol on, 81-82
egg, human, 76
electrical conduction by, 65-66
energy storage, 13, 68
fat, 82
in animals, 51-52, 52 (ill.), 60
injured, 73
life cycle, 57-58
life span, 52-53
movement of, 56-57
muscle, 69-71, 70 (ill.)
nucleus, 1-2, 32, 33, 36, 37, 38 (ill.), 38-41, 51, 372
plasma membrane, 47-50, 50 (chart), 61
receptor, 302
red, 68, 69, 75
reproduction of, 58

shape of, 73
size of, 55
skin, 76
stomach, 75-76
structure of, 31-53
totipotent, 58

Cellular respiration, 15

Celsius, Anders, 519

Celsius scale, 519, 520 (chart)

Centipedes, 250

Central vacuole, 51

Centrifugation, 527-28

Centrioles, 51, 130

Cereal, 177

CFCs, 498

Chain, Ernst Boris, 151

Chang, Annie, 414

Chestnut blight, 147

Chickenpox, 122

Chimeras, 422

Cilia. *See* Eukaryotic cells.

Chagas's disease, 111

Chargaff, Edwin, 361, 362, 364

Charlemagne, 149

Chatton, Edouard, 33

Cheese
Brie, 149-50
Camembert, 128, 149-50
Roquefort, 128, 149 (ill.)

Chemical bonds. *See* Bonds, chemical.

Chemical elements. *See* Elements, chemical.

Chernobyl nuclear disaster, 143, 144, 502-3

Chicken egg, 32

Chimpanzees, 326, 330, 347

Chinese desert cat, 272

Chlorophyll, 65, 128-29, 197, 225

Chloroplasts, 45 (ill.), 45-47, 51, 63, 64

Chocolate, 29, 83, 183

Cholera, 91, 400

Cholesterol, 10, 21, 25, 26, 65, 81

Chordates, 260-75. *See also* Vertebrates.

Chromatography, 528-32, 529 (ill.)

Chromoplasts, 47

Chromosomes, 32, 33, 36, 39, 40-41, 370-77, 391, 395-96, 398, 414, 420-21. *See also* Genes; Genetics; Mutations, genetic.

Cinnamon, 186, 191

Circulation, animal, 289-92

Cladogenesis, 458

Clams, 247-48, 249

Classification
of angiosperms, 176, 176 (chart)
of animals, 234, 284-85
of arthropods, 249-50 (chart)
of bacteria, 85-86, 85-86 (chart), 87-88, 95, 97
of bacteriophages, 107
of fungi, 125, 127
of plants, 161, 162-63
of protists, 108-10, 108-9 (chart), 109-10, 109-10 (chart)

Claude, Albert, 43, 528

Climate, 477. *See also* Microclimate.

Climax community, 486

Cline, 481

Cloning, 77, 424, 440-41

Clover, 188

Club moss, 168, 170-71

Coal formation, 168-69, 169 (ill.)

Cobras, 344

Cockroaches, 259

CODIS, 442

Coelenterates. *See* Sponges and coelenterates.

Coevolution, 473

Coffee, 187, 187 (ill.)

Cohen, Stanley, 414

Collins, R. C., 160

Columbus, Christopher, 186

Communication, animal, 330, 333, 334-35, 339-40, 346, 352-53, 356-57

Community, 483

Compound microscope, 521, 522

Conditioned reflexes, 330-31
Connective tissue, 280-81
Conservation, 493-95
Control group, 514
Convergent evolution, 462
Cookson, Isabel, 168
Copepods, 259
Coral bleaching, 242-43
Coral reefs, 242, 243
Cork, 208-9
Correns, Carl, 392
Cotton, 226, 227 (ill.)
Cows, 285-86, 438
Creutzfeldt-Jakob disease, 108
Crick, Francis, 360-61, 361 (ill.), 363, 366, 370, 465, 535
Crocodiles. See Alligators and crocodiles.
Cryopreservation, 442-43
Courtship, animal, 345
Cultural evolution, 473-74
Cyanide, 80-81
Cypress trees, 179 (chart)
Cystic fibrosis, 400
Cytology, 32
Cytoskeleton, 47

D

Damadian, Raymond, 535
Danielli, James F., 49
Darwin, Charles, 162, 222, 245, 323, 332, 387-88, 410, 445, 447 (ill.), 447-55, 468, 472, 473, 487
Darwin, Erasmus, 447
Darwin, Francis, 222
Darwinian medicine, 472-73
Darwin Station, 473
Darwin-Wallace theory, 452-55, 471-72
da Vinci, Leonardo. See Leonardo da Vinci.
Davson, Hugh, 49
DDT, 498, 500, 529
Decomposers, 485-86
Deductive reasoning, 514
Deforestation, 495, 495 (chart)
Dementia, 301-2

Dengue fever, 121
Dethier, Vincent, 346
Deuteromycetes. See Fungi.
deVries, Hugo, 392, 455
d'Hérelle, Felix, 107
Diabetes, 67, 78, 297
Diatoms, 111-12
Dietary fiber, 25
Digestion, animal, 284-86
Dill, 186-87, 193
Dilution techniques, 514-15
Dinosaurs, 265-66, 268, 437 (ill.), 437-38, 469, 489
Dioscorides, 193
Diphtheria, 91
Divergent evolution, 462-63
DNA
    applications, 372
    base pairs, 375
    B-DNA, 366
    cDNA, 368
    chromosomes and, 40
    description, 13, 31, 39, 360, 362-63
    discovery of, 39, 359, 360-62
    effect of smoking on DNA in lungs, 368
    enzyme function and, 19
    extraction from mummy, 434
    law of complementary base pairing, 363-64
    mutated, 77, 365-66
    photographing, 523
    recombinant DNA, 414, 417
    RNA and, 369 (ill.), 369-70, 380, 380 (chart)
    structure of, 25, 39, 363-65, 364 (ill.), 367, 465, 535
    transduction, 420
    transformation, 419
    viruses and, 104
DNA acetylation, 367
DNA amplification, 419
DNA fingerprinting, 403, 427-30, 532, 533
DNA methylation, 368
DNA profiling, 428

DNA replication, 365 (ill.), 365-66, 375
DNA sequence analysis, 424-25, 427
DNA supercoiling, 364-65
Dodo birds, 489
Dogs, domestic, 330-31, 355, 356, 396, 397 (chart)
Dolly, 414, 440 (ill.), 441
Dolphins, 271
Dominance, animal, 338-39
Doppler, Christian, 386
Dormancy, 202, 203
Double-blind study, 514
Dover, England, white cliffs, 123, 124 (ill.)
Down syndrome, 371-72
Dutch elm disease, 147-48, 148 (ill.)
Dutrochet, Henri, 32
Duve, Christian de, 42-43
Dyes, 188, 188 (chart)

E

Earth Day, 493
Earthworms. See Worms.
East Coast fever, 111
Eberth, Karl, 92
Ebola hemorrhagic fever, 121
Echinoderms, 259-60
Ecological pyramid, 485
Ecoterrorism, 500-1
Ectotherms, 305, 327-28
Eel, electric, 304-5
Ehrlich, Paul, 91
Eldredge, Niles, 466
Electron microscope, 521, 522
Electrophoresis, 532 (ill.), 532-33
Elements, chemical, 2, 2-3 (chart)
Elephants, 274, 274 (chart), 339, 346, 347, 489-90
ELISA test, 425
El Niño phenomenon, 478
Elton, Charles, 484
Emerging infectious diseases, 120, 121 (chart)

Emotion, animal, 332, 347
ENCODE project, 442
Endangered species, 247-48, 267, 429, 442-43, 488-92, 490-91 (chart), 491-92 (chart), 538
Endocrine system, animal, 294-97
Endocytosis, 55-56
Endomembrane system, 41-42
Endoplasmic reticulum, 41-42
Endosymbiosis theory, 468
Endotherms, 305, 327-28
Energy
    expenditures in animals, 327-28, 328 (chart), 485
    sources in living systems, 10, 24-25
    stored in body, 27 (chart)
Engler, Adolf, 162
Environmental Protection Agency, 496-97
Enzymes
    deficiencies, 17, 18 (chart)
    effect of aspirin on, 28
    function, 19
    in HIV treatment, 18
    role of mathematics to predict behavior, 19
Ephedra, 195
Ephrussi, Boris, 378
Epigenetics, 401-2
Epilepsy, 301
Epithelial cells, 58, 59, 278, 278 (ill.), 279 (ill.), 279-80
Ereky, Karl, 414
Erysipelas, 91
Escherich, Theodor, 92
Essential oils. See Aromatherapy.
Estivation, 353
Estuary, 482
Ethology, 323-24
Eugenics, 410-11, 472
Eukaryotic cells, 33-34 (chart), 33-36, 35 (chart), 36 (chart), 40, 42, 44, 47, 50-51, 53, 57-58, 382, 468
European starling, 269

Eutrophication, 482, 482 (ill.), 483
Evolution, 445-75
Excretory system, animal, 292-94
Exocytosis, 56
External fertilization, 308-9
Exxon Valdez, 122, 123 (ill.), 433-34, 504, 505 (chart)

F

Fahrenheit, Daniel, 519
Fahrenheit scale, 519, 520 (chart)
Fats (molecules)
    as energy sources, 24-25
    good vs. bad, 10-11
    saturated v. unsaturated, 20, 21
    trans fats, 20
    used in body, 26 (chart)
Fehleisen, Friedrich, 91
Fermentation, 15-16, 16 (ill.), 28, 92, 122, 123
Fernald, Merritt Lyndon, 160
Ferns, 168-71, 469, 470, 491
Fertilizer, 482
Fever, 26-27
Fiddlehead ferns, 170
Finches, 448, 459, 463
Fingerprints, 305-6
Fireflies, 255
Fir trees, 172 (chart), 179 (chart)
Fisher, R. A., 388
Fishes, 261-62, 293-94, 304, 319, 342, 344, 352, 470, 481 (chart), 490. See also Sharks.
Fish oil, 23, 24 (ill.)
Fixed action pattern, 329
Flagella. See Eukaryotic cells.
Flashlight fish, 98
Flatworms. See Worms.
Flavr Savr tomatoes, 435
Fleas, 318-19
Fleming, Alexander, 151
Flemming, Walther, 32-33, 41

Flies, 488. See also Fruit flies.
Flight, animal, 319-21
Florey, Howard, 151
Flow cytometry, 426
Flowering plants. See Angiosperms.
Flowers, 215-18, 258
Fluorescence, 68
Fluorescent in situ hybridization (FISH), 426
Folkman, Jonah, 79
Food chain, 484
Food web, 484
Forel, F. A., 483
Forensic botany, 185-86
Forest fires, 495, 496
Fossey, Dian, 326-27
Fossils and fossilization, 163-64, 176, 466-67
Founder effect, 463
Fragile X syndrome, 373
Frankenfood, 437
Frankincense, 185
Franklin, Benjamin, 269
Franklin, Rosalind, 360-62, 535
Fries, Elias, 125
Froehlich syndrome, 409-10
Frogs, 313 (ill.), 441
Fruit flies, 313, 394
Fungi. See also Lichens; Yeasts.
    antifungal drugs, 150-52
    ants and, 129-30
    appearance on Earth, 470
    Armillaria mushrooms, 154 (ill.)
    Aztec religious ceremonies and, 142
    biocontrol and, 157
    bread molds, 133, 133 (ill.), 153
    characteristics of, 127 (chart)
    cheese and, 149-50
    dikaryotic phase of life, 133, 134 (ill.)
    dimorphic, 132
    diseases in humans, 150 (chart)

diseases in trees and
plants, 147-49, 155-57
dispersal of spores, 132
edible fungi, 135-42, 135-
36 (chart), 137 (chart)
ergot, 146, 147 (ill.)
fairy rings, 139-40
food digestion, 128-29
fossilized, 141
fruiting bodies, 131
imperfect, 128, 130-31,
153
largest organism on
Earth, 129
life cycle of, 126 (ill.), 133
location of, 128
medicinal uses of, 152,
152-53 (chart), 154
morels, 138 (ill.)
mycorrhiza, 133-35
poisonous, 137-39
pollution and, 129
portobellos, 139
recycling and, 135
reproduction of, 131, 132
sclerotia, 132, 146
shape of, 131
stinkhorn fungus, 139
structure of, 130-32
toxins produced by, 138
truffles, 140-41, 141 (ill.)
use in cosmetics industry,
155, 156 (ill.)
World War I and, 154

G

Gaertner, August, 92
Gaffky, Georg, 92
Gaia hypothesis, 477
Galapagos Islands, 354, 448,
458 (ill.), 459, 463, 473
Galileo, 519
Galton, Francis, 410-11
Gamow, George, 370
Garbage, 508, 508-9 (chart),
509
Garlic, 214 (ill.)
Garrod, Archibald, 13, 393-
94
Gattaca, 439

Gattefosse, Rene, 195-96
Gause, G. F., 484
Geese, 326 (ill.), 340 (ill.),
341
Gelsinger, Jesse, 430
Gene chip, 424
Gene deficiencies, 404-5
Gene gun, 421
Gene library, 420
Gene patent, 433
Gene probe, 421
Genes. *See also* Chromo-
somes; Genetics.
alcoholism and, 407
alleles, 397, 397 (chart),
452-53
chromosomes and, 371-
77, 395-96
description of, 392
environment and, 398
homeobox genes, 375-76
homosexuality and, 407-8
in bacterial cells, 94
jumping genes, 395
linked genes, 401
location in genome, 420
obesity and, 407
one gene–one enzyme
hypothesis, 378-79
role in cancer, 79
skin color and, 411
structure of, 374 (ill.),
374-75
telomeres, 376-77
Gene therapy, 415, 430-31,
431 (chart)
Genetically modified organisms (GMO), 434, 435, 437
Genetic determinism, 391
Genetic diseases, 409, 409
(chart), 431, 432
Genetic engineering, 415-17,
434-37, 442-43
Genetic fingerprinting. *See*
DNA fingerprinting.
Genetic imprinting, 402
Genetics. *See also* Chromosomes; Genes; Human
Genome Project.
applications, 404-11
autosomes, 391

behavior and, 335
blending theory, 385
crosses, 389-90 (chart),
390, 391
heterozygote advantage,
399-400
homologous chromo-
somes, 391, 454
homunculus, 385
in Nazi Germany, 411
in the Soviet Union, 410
law of independent
assortment, 387
law of segregation, 387
Mendelian genetics, 385-
92, 455
New genetics, 392-402
pedigree, 389 (ill.), 390-
91
sex chromosomes, 392
sex determination, 400
(chart)
traits, 386, 396-97, 401,
406, 445, 446, 454
Gene vectors, 417, 430, 431
Genome, cellular, 93, 403,
403-4 (chart). *See also*
Human Genome Project.
Gengou, Octave, 92
Geologic time divisions, 468-
70 (chart)
Germination, 202, 203, 211
Germline therapy, 432
Gessner, Conrad, 233
Gestation periods, 309
Giant tube worms. *See*
Worms.
Gilbert, Walter, 371, 413, 414
Ginkgo biloba, 171, 194
Giraffes, 291, 317, 446
Gleevec, 434
GloFish, 414, 426
GMO. *See* Genetically modi-
fied organisms.
Goiter, 297
Golden rice, 435-36
Golgi, Camillo, 42, 527
Golgi apparatus, 42
Gonorrhea, 91, 121
Goodall, Jane, 326, 332
Gorillas, 327, 333, 334

Gorter, Evert, 49
Gould, Stephen Jay, 466
Gradualism, 453
Gram, Hans-Christian, 87
Gram stain, 87, 527
Grapes, seedless, 181, 224
Gray, Asa, 160
Gray fox, 272
Gray wolf, 484
Great white shark, 262
Green algae. *See* Algae.
Green building, 501
Greenhouse effect, 501-2, 503 (ill.)
Grendel, F., 49
Grew, Nehemiah, 45
Griffith, Frederick, 359-60
Grooming, animal, 354
Growth factor, 58-59, 58-59 (chart)
Guillain-Barrè syndrome, 299-300
Gurdon, John B., 441
Gymnosperms, 171-75, 195, 469, 470
Gypsy moths, 255-56

H

Haberlandt, Gottlieb, 224
Habituation, 333
Haeckel, Ernst, 32, 85, 108, 464
Hagfishes. *See* Lampreys and hagfishes.
Haldane, John, 464-65, 472
Hantavirus pulmonary syndrome, 121
Hardy, Godfrey, 394
Hardy-Weinberg theorem, 392-93
Harrison, Francis C., 88
Hauptmann, Bruno, 186
Hawley bog, 167 (ill.)
Hayes, Denis, 493
Hazardous waste, 499
Hemlock, 173-74, 195, 229 (chart)
Hemolyticuremic syndrome, 121
Hemophilia, 388 (ill.)

Hemp, 226
Hermaphroditism, 308
Hero of Alexandria, 519
Herpesvirus, 104 (ill.)
Hertwig, Oscar, 32
Hepatitis C, 121, 122
Hesse, Fannie E., 124
Hesse, Walther, 124
Hibernation, 353
High throughput screening, 427
Hill, Robert, 64
HIV. *See* AIDS.
Hodgkin, Dorothy Mary Crowfoot, 535
Hoerr, Normand Louis, 528
Hoffman, P. Erich, 92
Hofmeister, Wilhelm, 165
Homeobox genes. *See* Genes.
Homeostasis, 277
Homo sapiens, 469, 471, 473-74, 489
Honeybees. *See* Bees.
Hooke, Robert, 32, 33
Hooker, Joseph Dalton, 162
Hoppe-Seyler, Felix, 359
Hormones, 21, 67, 295-96 (chart), 295-97, 438. *See also* Plants.
Hornworts. *See* Bryophytes.
Horses, 355-56
Horsetails, 170
Hox genes, 466
Human Genome Project, 402-4, 414, 427, 439
Human immunodeficiency virus. *See* AIDS.
Hummingbirds, 320-21
Humpback whales, 271
Huntington's disease, 433
Hutton, James, 448
Huxley, Andrew, 69
Huxley, Hugh, 69
Huxley, Julian, 455
Huxley, Thomas, 449-50, 451, 455
Hydras, 234, 243
Hydrogen, 2, 6-7
Hydrologic cycle, 479
Hydrolysis, 9
Hyman, Flo, 408

I

Ice, 6. *See also* Water.
Immune system, animal, 306-7
Imprinting, 326 (ill.), 335
Indian elephants. *See* Elephants.
Indoor air pollution, 506, 506-7 (chart)
Inductive reasoning, 514
Industrial melanism, 461
Influenza, 102, 104-5
Ingenhousz, Jan, 64
Innocence Project, 439
Insects, 253-58, 303, 469, 470, 480-81 (chart), 490
Institute for Genomic Research, 413
Insulin, 67
Intelligence, animal, 330, 332, 410
Interferons, 106
Internal fertilization, 308-9
Interphase, 57-58
Invertebrates, 247, 259, 298, 303, 330, 470
In vitro studies, 514
In vivo studies, 514
Iodine test, 516
Irish potato famine, 112-13, 186
Isomer, 4, 5 (ill.)
Isotope, 4

J

Jeffreys, Alec, 428
Jellyfishes, 68, 235, 240, 241, 425
Jimson weed, 188-89
Johannsen, Wilhelm, 392, 393
Johnson, Robert Underwood, 493
Jones, Meredith, 246
Joshua trees, 178
Jost, A., 417
Jussieu, Antoine Laurent de, 162

K

Kammerer, Paul, 460
Karyotypes, 396
Kelvin, Lord. *See* Thomson, William.
Kelvin scale, 520, 520 (chart)
Keystone species, 484
Kidney dialysis, 294
Killer bees. *See* Bees.
Killer whales. *See* Orcas.
King, Mary Claire, 414-15
King George III of England, 408
King Louis XV of France, 430
Kin selection, 472
Kitasato, Shibasaburo, 92
Klebs, Edwin, 91
Knockout mouse, 425-26
Koch, Robert, 90-91, 93, 98, 124
Koehler, Arthuer, 186
Koko, 334
Kölliker, Rudolf Albert von, 44
Kossel, Albrecht, 360, 369
Krebs cycle, 14, 15, 16, 62
Krebs, Hans, 14
Kudzu, 182, 183 (ill.), 512
Kuhne, William, 17
Kurosawa, Eiichi, 222
Kwashiorkor, 24
Kyoto Protocol, 511-12

L

Lacks, Henrietta, 80
Lactose intolerance, 405, 405-6 (chart)
Lamarck, Jean-Baptiste de, 410, 446, 447
Lampreys and hagfishes, 261
Langmuir, Irving, 48-49
Language, animal. *See* Communication, animal.
Lauterbur, Paul, 536
Laveran, Charles-Louis Alphonse, 91
Lavoisier, Antoine, 65
Leakey, Louis B., 326

Leeuwenhoek, Antoni van, 45, 89-90
Legionnaires' disease, 121
Lemmings, 487-88
Leonardo da Vinci, 229
Levene, Phoebus, 368-69
Lichens, 142-44, 155, 156 (ill.), 166, 491
Life history table, 487
Life spans, animal, 310-12 (chart), 313
Limiting factors, 478-79
Limnology, 483
Lincoln, Abraham, 407
Lindbergh, Anne Morrow, 186
Lindbergh, Charles, 186
Lindeman, R., 484
Linnaeus, Carolus, 85, 162, 224, 445, 466, 480
Lipids, 11, 23, 27, 28, 48, 49, 61-62. *See also* Fats (molecules).
Lister, Joseph, 91, 119
Liverworts. *See* Bryophytes.
Lobsters, 250
Locomotion, animal, 317-21, 329 (chart)
Locoweed, 232
Locusts, 255
Löffler, Friedrich, 91
Lorenz, Konrad, 324, 325, 326 (ill.), 335
Love, chemistry of, 28-29
Lovelock, James, 477
Luffa sponges, 190
Lyell, Charles, 448
Lyme disease, 121, 122
Lyon, Mary, 393
Lyon hypothesis, 393
Lypholization, 537
Lysenko, Trofim, 410
Lysosomes, 42-43, 51, 107-8, 528

M

MacLeod, Colin M., 360
Macroevolution, 463
Macromolecules, 9, 42
Mad cow disease, 108

Malaria, 91, 111, 230, 461, 498
Malayan sun bear, 273
Malnutrition, 25 (ill.)
Malpighi, Marcello, 524
Malthus, Thomas, 449, 450, 472, 487
Mammals, 269-70 (chart), 271-75, 310-12 (chart), 315-16, 321, 469, 480-81, 490
Mammals, marine, 288, 288-89 (chart)
Mammoths, 274, 318, 318 (chart)
Manatees, 271-72
Mansfield, Peter, 536
Maple syrup, 228-29
Marasmus, 24
Marfan syndrome, 408, 409-10
Margulis, Lynn, 477
Marijuana, 29
Mark-recapture method, 537-38
Marsh, George Perkins, 493
Marsupials, 315, 469
Martin, Arthur Porter John, 530
Mass extinctions, 467
Mastodons, 274
Mating, animal, 349, 356-57
Maxam, Allan, 414
McCarty, Maclyn, 360
Medawar, Peter, 105
Medicinal uses of plants. *See* Plants.
Meiosis, 60-61, 390
Mendel, Gregor, 385-86, 387-88, 455
Meristems, 201
Metabolism. *See also* Krebs cycle.
    anabolic reactions, 14
    bacterial classification and, 97
    catabolic reactions, 14
    diet drugs and, 82
    life span and, 28
    metabolic disorders and genetics, 393-94
    metabolic pathways, 14, 17

Metamorphosis, 254

Metchnikoff, Elie, 91

Metric conversions, 517-18 (chart). *See also* SI system.

Mexican jumping beans, 319

Microarray, 423

Microclimate, 477-78

Microevolution, 463

Microorganisms, control by chemical agents, 117-18 (chart)

Microscopes and microscopy, 32, 33, 520-24, 521 (chart)

Microscopic autoradiography, 523-24

Microtechnique, 524-27

Microtome, 524-25

Microtubules, 52

Midwife toads, 460

Migration, animal, 336, 351

Miescher, Johann Friedrich, 39, 359

Miller, Carlos O., 223

Miller, Stanley, 465

Millipedes, 250

Mimosa plants, 179

Minerals, 19-20, 19-20 (chart)

Mitochondria, 43 (ill.), 44-45, 51, 61

Mitosis, 57-58, 59-60, 424 (ill.), 441

Mole (in chemistry), 4

Molecular clock, 474

Molecular evolution, 472

Molecules, 3, 8-13

Mollusks, 247-49. *See also* Clams; Octopods; Snails and slugs; Squids.

Monarch butterflies, 257

Monkey ball trees, 178

Monogamy. *See* Mating, animal.

Mordant, 527

Morgan, Thomas Hunt, 376, 392, 394, 404, 405 (ill.)

Morphine, 229, 232

Mosses. *See* Bryophytes.

Moths, 256, 256 (chart). *See also* Gypsy moths.

Movement, animal. *See* Locomotion, animal.

Muir, John, 493

Müller, Fritz, 461-62

Müller, Paul, 498

Müllerian mimicry, 461-62

Mullis, Kerry, 414, 418

Multiple sclerosis, 299, 301

Muscles, 69, 70 (ill.)

Muscle tissue, 281-82

Mushrooms. *See* Fungi.

Mussels. *See* Clams.

Mutations, genetic, 365-66, 366 (chart), 398 (chart), 398-99

Mutualism, 133-34, 269, 486. *See also* Symbiosis.

Mycology. *See* Fungi.

Mycoplasmas, 100

Mycorrhiza. *See* Fungi.

Myelin, 298-300

Myrrh, 185

N

Names, animal, 236-37 (chart)

Names, juvenile animal, 237-39 (chart)

Nanotechnology, 442

Natural selection, 325, 447, 448, 449, 450-51, 452-53, 455, 456-57, 458-59, 472, 474

Navel oranges, 181

Negative feedback, 13

Neisser, Albert L. S., 91

Neljubov, Philip F., 223

Nelson, Gaylord, 493

Nematocysts, 241-42

Nerve tissue, 282-83

Nervous system, animal, 297-306

Neuberg, Carl, 154

Neurons, 71-72, 73, 282 (ill.), 282-83

Newton, Isaac, 450, 514

Niche, 483, 484

Nicholson, Garth L., 49

Nitrogen cycle, 479-80

Nitrogenous waste, 292, 292 (chart), 293

Nixon, Richard M., 496

Nuclear magnetic resonance, 535-36

Nuclear waste, 508

Nucleic acids, 13. *See also* DNA; RNA.

Nucleolus, 40. *See also* Cells.

Nuttall, George Henry Falkiner, 92

O

Octopods, 247, 248 (ill.), 330, 350

Ogata, Masaki, 91

Ogston, Sir Alexander, 91

Oil spills, 504-5, 505 (chart). *See also* Exxon *Valdez*.

"Ontogeny recapitulates phylogeny," 464

Oogenesis, 60-61

Oparin, Alexandr, 464-65

Oparin-Haldane hypothesis, 464-65, 468

Operant conditioning, 331-32, 333

Operation Ranch Hand, 505

Opium War, 186

Opossums, 343-44

Orange trees, 181

Orcas, 271

Organic chemistry, 1, 516

Organs, animal, 283-84, 283-84 (chart)

Ostrom, John H., 268

Overton, Charles, 48

Owen, Richard, 265

Oxpeckers, 269

Oxygen, 15, 62, 64-65

Ozone, 207, 501, 501 (ill.), 502

P

Paganini, Niccolò, 408

Palade, George, 43

Panspermia, 470-71

Paper manufacturing, 184, 226-27

Papyrus, 226-27
Parasitism, 340, 347-48
Parental care, 350
Parkinson, James, 301
Parkinson's disease, 301, 433
Parrots, 356, 357
Parsimony, 463
Parvovirus, 103
Passenger pigeons, 489
Pasteur, Louis, 90-91, 92, 113
Pasteurization, 90, 113
Pauling, Linus, 25, 27 (ill.), 360, 535
Pauling, Peter, 360-61
Pavlov, Ivan, 330-31
PCBs, 498, 529
PCR. See Polymerase chain reaction.
Pearls, 248-49, 442
Peat moss, 167 (ill.), 167-68
Pedigree. See Genetics.
Penguins, 315
Penicillin, 115, 128, 151-52, 153, 473
Pepys, Samuel, 334
Peregrine falcons, American, 496 (ill.)
Perfumes, plants used in, 190, 190-91 (chart)
Peroxisomes, 43
Pertusis. See Whooping cough.
Petri, Julius Richard, 93
Petri dish, 93
Petrified wood, 164
Pfungst, Oskar, 355
Phagocytosis, 56, 91
Pharmacogenomics, 432, 434
Phenylketonuria, 432
Philo of Byzantium, 519
pH level, 6-7, 7 (chart), 8, 94-95, 220, 515, 515-16 (chart)
Photoperiodism, 224
Photosynthesis, 14, 45, 62-64, 65, 68
Phrenology, 449
Phylogeny, 464, 465
Physiology, 277

Pinchot, Gifford, 493
Pine trees, 172 (chart), 173, 175, 179 (chart), 212 (ill.)
Pinocytosis, 55
Plants. See also Flowers; Soils; Trees.
    bark of, 209 (ill.), 209-10, 210 (chart)
    breeding of, 196
    bulbs, 214 (ill.), 215
    cells of, 46 (ill.), 47, 48, 51, 60, 197-202
    cloning of, 440
    fibers of, 201, 225-26 (chart), 226
    growth of, 201-2
    hormones in, 222-24
    land, origins of, 163
    leaves of, 204-6
    medicinal uses of, 192-95, 194-95 (chart), 206
    nutrients of, 219, 219-20 (chart)
    parasitic, 225
    periderm, 208
    pith, 211
    poisonous, 181-82, 190, 195
    polyploidy in, 373-74, 373-74 (chart)
    reproduction of, 164-65
    roots of, 197-98, 211-18, 212 (ill.), 213 (ill.)
    secondary metabolites in, 229, 229-31 (chart)
    seeds of, 202-3
    shoots of, 197-98, 203-11
    stems of, 203-4
    stomata, 206-7, 207 (chart), 208
    tissues of, 199-201, 199-201 (chart)
    trichomes, 208
Plasmids, 96-97
Plasmodesmata, 51
Plastics, 509-11, 510 (chart)
Plastids, 46
Plato, 445
Platypus, duck-billed, 271, 315
Play behavior, animal, 331

Pleiotropy, 399
Pliny the Elder, 162, 323
Poisonous plants. See Plants.
Poisonous snakes. See Snakes.
Polar molecule, 8
Polio, 103
Pollination, 216-17, 218
Pollutant Standard Index, 497, 497 (chart)
Polyacrylamide gel electrophoresis (PAGE), 427
Polygyny. See Mating, animal.
Polymerase chain reaction (PCR), 414, 418, 419 (ill.), 423, 428, 439
Polymorphism, 461
Polyploidy, 372, 373-74, 373-74 (chart), 400-1, 457
Population, 486-87, 537, 538
Porcupines, 273-74
Porpoises, 271
Portuguese man-of-war, 241, 242
Potter, Beatrix, 126
Prairie dogs, 484
Prantl, Karl, 162
Predation, animal, 347, 348, 349
Priestley, Joseph, 64
Prince Albert, 388, 390
Prions, 107-8
Prokaryotic cells, 33-36, 34 (ill.), 34-36 (chart), 40, 44, 48, 53, 86 (ill.), 87-101, 114, 468
Protease inhibitors, 18
Proteins
    aging's effect on, 27
    diseases due to inadequate intake, 24
    function, 11, 11 (chart)
    importance in diet, 23
    in eggs, 26
    in hair, 26
    storage location in human body, 27
    structure of, 12 (ill.)
Protists, 108-13, 328-29, 433-34, 470

Proteomics, 439
Protozoans, see Protists.
Prusiner, Stanley, 107
Punctuated equilibrium, 466
Punnett, Reginald, 387, 394

Q

Queen Victoria, 388, 390

R

Rabbits, 459
Rabies, 102
Rachmaninoff, Sergey, 408
Racial designations, 472
Rain, 479
Rain forests, 494-95
Ramón y Cajal, Santiago, 42
Randall, John, 361-62
Ray, John, 162
Recombinant DNA. *See* DNA.
Recycling, 509-12, 510
  (chart), 511 (ill.)
Red algae. *See* Algae.
Red cabbage, 515
Redi, Francesco, 90
Redox reaction, 15
Red Queen hypothesis, 475
Red tide, 483
Redwood trees, 172-73, 174
  (ill.), 174-75, 179 (chart)
Reflexes, 300
Remak, Robert, 32
Reproduction, 307-16, 348,
  460-61
Reptiles, 263-67, 350, 354,
  470, 480-81 (chart)
Respiration, animal, 286-89,
  287-88 (chart)
Restriction enzymes, 417,
  417-18 (chart)
Restriction fragment length
  polymorphism (428)
Rhinoceros, black, 269
Ribosomes, 43, 44, 51, 379-80
Ribozymes, 370-71
RNA, 13, 104, 359, 360, 368-
  71, 369 (ill.), 377-83, 380
  (chart), 381 (chart)

Roentgen, William Conrad,
  534
Romanes, George, 332
Roots. *See* Plants.
Ross, Betsy, 226
Roundworms. *See* Worms.
Russow, Edmund, 40

S

Saffron, 191, 192, 192 (ill.)
Saint-Hilaire, Isidore Geof-
  froy, 324
Salem witch trials, 146, 409
Salmon, 24 (ill.), 294, 337,
  438
Salmonellosis, 92, 121
Salten, Felix, 325-26
Sand cat, 272
Sanger, Frederick, 413, 414,
  425
Santorio, Santorio, 519
Sargent, Charles Sprague, 493
SARS, 103, 121
Scanning tunneling micro-
  scope, 522-23
Schaller, George, 326
Schaudinn, Fritz R., 92
Schieffelin, Eugene, 269
Schleiden, Matthias, 32
Schlesinger, Martin, 107
Schwann, Theodor, 32
Sciatic nerve, human, 300
Scientific method, 513-14
Scientific notation, 518, 518-
  19 (chart)
Scopes, John T., 475
Scopes monkey trial, 475
Seasons, 477, 478 (ill.)
Sea stars. *See* Starfishes.
Sea urchins, 259-60, 310
Segmented worms. *See*
  Worms.
Sequoia trees. *See* Redwood
  trees.
Sertürner, Friedrich Wil-
  helm A., 232
Severe acute respiratory syn-
  drome. *See* SARS.
Sexual selection, 459, 462
  (ill.)

Sharks, 261-62, 317, 470
Shiga, Kiyoshi, 91, 92
Shigellosis, 92, 122
Shoots. *See* Plants.
Sickle cell anemia, 399, 400,
  432
Silicon, 6
Simple stains, 526
Singer, Seymour J., 49
SI system, 517
Skeletal system, animal,
  316-17
Skinner, B. F., 331-32
Skoog, Folke, 223
Skunk cabbage, 189
Sleeping sickness, 111
Slime molds, 112
Smallpox, 102, 102 (ill.)
Smear preparation, 524, 526
  (ill.)
Smells, 302-3
Smith, Michael, 418
Smog, 498-99, 499 (chart)
Smokey Bear, 495
Snails and slugs, 247, 490
Snakes, 264-65 (chart), 344.
  See also Cobras.
Social Darwinism, 471-72
Sociobiology, 326, 327
Socrates, 174, 229 (chart)
Soils, 218-21
Sound, 303-4
Spallanzani, Lazzaro, 90
Species and speciation, 455-
  57
Spectroscopy, 533 (chart),
  533-34
Spencer, Herbert, 451, 472
Sperm, 76
Spermatogenesis, 60
Sperry, Roger, 300-1
S phase, 59
Spices, 191, 191 (chart). *See
  also* Cinnamon; Saffron.
Spiders, 251-53, 252 (ill.),
  258-59, 309-10, 470, 490
Spinach, 28
Sponges and coelenterates,
  239-43, 334. *See also* Jelly-
  fishes; Portuguese man-of-
  war.

Spontaneous generation, 90
Spruce trees, 172 (chart), 175
Squids, 247
Stanley, Wendell, 104
Starfishes, 259, 484
Starling, Ernest, 295
Starlink corn, 436
Statin drugs, 81. *See also*
   Cholesterol.
Stem cells, 57, 57 (ill.), 78
   (chart)
Steroids, 21 (ill.), 21-22, 297
Steward, Frederick Campion,
   58, 225
Sticklebacks, 324, 325 (ill.)
Stone plants, 183
Succulents, 182-83
Sudan IV test, 516, 517
Sumiki, Yasuke
Superfund Act, 507
Survivorship curve, 487
Swallows, 336
Symbiosis, 133-34
Symmetry, body, 234
Synge, Richard Laurence
   Millington, 530
Syphilis, 91, 92, 122
Systematics, 466

T

Taniguchi, Norio, 442
Tapeworms. *See* Worms.
Taq polymerase, 419
Tatum, Edward, 378, 394-95
Taxol, 172, 231
Taxonomy. *See* Classification.
Telephone poles, 184
Telomeres, 41
Tennyson, Alfred, 450
Terminator gene technology,
   423
Termites, 258, 350
Territorialism, animal, 342-43
Tetanus, 92
Thalidomide, 433
Theophrastus, 160
Thermocycler, 419
Thermodynamics, 8
Thermometer, 519
Thomson, William, 471, 520

Thoreau, Henry David, 493
Thought, animal, 329-30
Tight building syndrome.
   See Indoor air pollution.
Tinbergen, Niko, 323, 324,
   325
Tiselius, Arne, 533
Tissue, animal, 278-83
Tissue engineering, 439
TNT (trinitrotoluene), 28
Tobacco mosaic, 102
Tobacco seeds, 231 (ill.)
Tomatoes, 191, 435
Tournefort, J. P. de, 162
Toxic Release Inventory, 497
Toxic shock syndrome, 121
Toxins, 74, 74 (chart)
Traits. *See* Genetics.
Transcription, 377-78, 378
   (ill.)
Translation, 378
Transposon, 427
Trees
   diseases, 256
   growth from rocks, 481
   hardwoods vs. softwoods,
      227
   leaves of, 228, 228
      (chart), 230 (ill.)
   rings of, 227, 228 (ill.),
      229
Trembley, Abraham, 234
Trophic level, 484, 485
Tropism, 221
Trouvelot, Leopold, 255-56
Tsvet, Mikhail Semyonovich,
   530
Tuberculosis, 91, 122
Turgor movements, 221-22
Turner syndrome, 371-72
Turtles, 266-67, 317
Twort, Frederick W., 107
Tyndall, John, 502
Typhoid fever, 92

U

Ultrasound, 536
Uniformitarianism, 448
Urey, Harold, 465
Ussher, James, 448-49

V

Vaccines, 431-32
Vancomycin, 115-16
Van Ermengem, Emile, 92
Van Helmont, Jan Baptista,
   64
Van Niel, C. B., 64
Van Overbeek, Johannes, 223
Variable, 513-14
Varicella. *See* Chickenpox.
Vascular plants, 197, 198,
   199
Vavilov, Nikolai, 410
Vectors. *See* Genes.
Venter, Craig, 413
Venus fly trap, 180, 180 (ill.)
Vertebrates, 235, 235-36
   (chart), 260-75, 298, 300,
   330, 461
Victoria amazonica, 180
Virchow, Rudolph, 32
Viruses
   and bacteria, 101-2
      (chart)
   DNA, 104, 105 (chart)
   origin of, 103-4
   reproduction of, 103
   RNA, 104, 105 (chart)
   size of, 102, 102-3 (chart)
   structure of, 101
   treatment with medica-
      tions, 106, 106 (chart)
Vision, 303, 304
Vitamins, 19-20, 19-20
   (chart), 74, 77
von Frisch, Karl, 324, 325,
   345, 352
von Tschermak, Erich, 392
von Uexkuell, Jacob, 325
Vuillemin, Paul, 114

W

Wagner, Rudolph, 40
Waldeyen-Hartz, Wilhelm
   von, 32
Wallace, Alfred Russel, 449,
   452-55, 462, 472, 473, 474
Wareing, Philip F.
Water, 5-6, 7

Watermelons, seedless, 181

Watson, James, 360-61, 361 (ill.), 363, 366, 370 (ill.), 465, 535

Weinberg, William, 394

Weismann, August, 446

Weizmann, Chaim, 113

Welch, William Henry, 92

Went, Frits W., 222

Wetlands, 481, 481 (chart), 494

Whales, 262, 271, 340

Wheat, 177 (ill.), 189, 190

Wheeler, William Morton, 324

Whitaker, R. H., 85

White fat, 281

Wholemounting, 524, 525 (ill.)

Whooping cough, 92, 122

Wilberforce, Samuel, 449-50

Wilkins, Maurice, 360-62, 361 (ill.), 362 (ill.), 535

Williams, George C., 325

Wilmut, Ian, 414, 441

Wilson, Edmund Beecher, 404

Wilson, Edward O., 327, 330, 337-38

Woese, Carl, 85

World War I, 113-14, 154

Wordsworth, William, 246

Worms
earthworms, 245
flatworms, 243
giant tube worms, 245-46
leeches used in medicine, 246
roundworms, 244-45, 245 (chart)
segmented worms, 245
tapeworms, 243, 244, 244 (chart)

Wormwood, 188

X

Xenotransplantation, 432, 433

X-ray diffraction, 534-35

X-rays, 534

Y

Yabuta, Teijiro, 222

Yeasts, 144-45

Yellowstone National Park, 494

Yersin, Alexandre, 92

Z

Zebra mussels, 248

Zeidler, Othmar, 498

Zero population growth, 502

Zoology, 233, 234